RESEARCH IN THE BIOMEDICAL SCIENCES

Transparent and Reproducible

RESEARCH IN THE BIOMEDICAL SCIENCES

Transparent and Reproducible

Edited by

MICHAEL WILLIAMS
Adjunct Professor, Feinberg School of Medicine
Northwestern University
Chicago, IL, USA

MICHAEL J. CURTIS
Reader in Pharmacology, King's College London
Rayne Institute, St. Thomas' Hospital
London, UK

KEVIN MULLANE
Director, Corporate Liaison and Ventures
J. David Gladstone Institutes
San Francisco, CA, USA

Academic Press is an imprint of Elsevier
125 London Wall, London EC2Y 5AS, United Kingdom
525 B Street, Suite 1800, San Diego, CA 92101-4495, United States
50 Hampshire Street, 5th Floor, Cambridge, MA 02139, United States
The Boulevard, Langford Lane, Kidlington, Oxford OX5 1GB, United Kingdom

Library of Congress Cataloging-in-Publication Data
A catalog record for this book is available from the Library of Congress

British Library Cataloguing-in-Publication Data
A catalogue record for this book is available from the British Library

ISBN: 978-0-12-804725-5

For information on all Academic Press publications visit our website at https://www.elsevier.com/books-and-journals

Publisher: Mica Haley
Senior Content Strategist: Kristine Jones
Editorial Project Manager: Kathy Padilla
Production Project Manager: Julia Haynes and Priya Kumaraguruparan
Designer: Matthew Limbert

Typeset by Thomson Digital

Contents

5. Addressing Reproducibility: Peer Review, Impact Factors, Checklists, Guidelines, and Reproducibility Initiatives

MICHAEL WILLIAMS, KEVIN MULLANE AND MICHAEL J. CURTIS

6. Biomedical Research in the 21st Century: Multiple Challenges in Resolving Reproducibility Issues

KEVIN MULLANE, MICHAEL J. CURTIS AND MICHAEL WILLIAMS

List of Contributors

Michael J. Curtis King's College; The Rayne Institute, St. Thomas' Hospital, London, United Kingdom

Michael J. Marino Symptomatics, Merck & Co., Inc. West Point, PA, United States

Kevin Mullane J. David Gladstone Institutes, San Francisco, CA, United States

Michael Williams Feinberg School of Medicine, Northwestern University, Chicago, IL, United States

Preface

The systematic inability to reproduce research findings has become a topic of considerable concern in the biomedical research community in the second decade of the 21st century. This is the result of key commentaries in the scientific literature by Ioannidis (2005), Prinz et al. (2011), and Begley and Ellis (2012) that formed the basis of a widely-cited article on the reproducibility crisis published in the mainstream weekly, the Economist (2013). Reproducibility issues have been variously ascribed to fraud, investigator bias, statistical misuse, substandard peer review, perverse career incentives, and an absence of appropriate researcher training and mentoring.

Frequently described in terms of a crisis, an inability to reproduce findings is a long-standing issue in biomedical research that dates back to the 17th century (Bishop, 2015) when the first scientific journals were published under the auspices of the Royal Society of London and the Académie Royale des Sciences in Paris. Since then, researchers in academia and industry have sought to validate research findings by their independent reproduction, a process known as scientific self-correction. Given the exploratory nature of scientific research, many findings are often not amenable to reproduction, a simple reflection of the complexity of Mother Nature and "business as usual" in biomedical research (McClain 2013).

While proactively addressing known issues related to reproducibility can be a logical and necessary first step in identifying facile solutions, it does not alter an additional facet of reproducibility, the impact of as yet to be discovered unknowns in research (Mullane and Williams, 2015). As an example, the utility of the animal models used in translational research can be markedly impacted by the environment, for example, temperature, humidity, bedding, the microbiome of the animal, and investigator pheromones, and in transgenic animals, passenger mutations, developmental adaption, and physiological compensation (Drucker, 2016). The existence of such phenomena was unknown 15–20 years ago and can increase variability and outcomes, confounding data output and its interpretation.

In the decade or so since the seminal Ioannidis paper on research findings being false, the most commonly cited causes for an inability to reproduce findings in biomedical research include: (1) the misuse of statistics; (2) poor experimental design, execution, analysis, and reporting; (3) naïve inference/extrapolation from animal models of human disease; and (4) experimenter bias and/or fraud in the original study. The latter is a cultural issue that has been ascribed to what have been termed "perverse career incentives" that "...pressure ... authors to rush into print, cut corners, exaggerate ... and overstate the significance of their work" (Alberts et al., 2014).

In terms of solutions, statisticians will inevitably argue that all reproducibility problems can be solved by the appropriate design, powering, and statistical analysis of experiments; other researchers will argue that blinding and randomization of experiments together with antibody validation and the authentication of cell lines and animals, especially transgenics, are imperatives in resolving reproducibility issues; while those involved in more strategic, "big picture" aspects of reproducibility, including the exponentially expanding field of "research-on-research" (Ioannidis, 2016), would propose that improved peer review (including open data and postpublication peer review/commentary), a more transparent and inclusive replacement for the misappropriated, misused, and abused journal Impact Factor (JIF), and formal reproducibility initiatives (RIs) to replace the informal process of scientific self-correction, are critical in addressing detrimental research practices and improving reproducibility metrics.

As discussed in this work, many of these solutions are iterations on existing processes that appear to have failed in addressing reproducibility issues either due to their inappropriate use or a need to improve their relevance in the age of the Internet. In some instances, the solutions have led to minor improvements; overall however, they have tended to confound and compound existing problems while creating new ones, some of which only make the problem worse.

With this background, the editors, with a cumulative century or more of experience as researchers, authors, peer reviewers, and editors in academia and the pharmaceutical industry, have prepared this book to address the full spectrum of experimental activities, a lack of attention to which can contribute immeasurably to irreproducibility in biomedical experimentation. With the exception of Chapter 3 by Michael Marino, the editors have also authored the work which, in addition to defining the reproducibility problem and discussing progress on its possible solutions, provides chapters that address best practices in experimentation, data analysis and reporting, and a consideration of the cultural aspects of 21st century biomedical research as these pertain to drug discovery and human health. From these, the authors argue that practical solutions to improving reproducibility lie in needed improvements in the training and mentoring of scientists, the proactive, responsible and timely investigation of fraud, and bias and the mandatory enforcement of consequences by granting bodies, universities, and research institutions that are appropriate to the ethical lapses and dishonesty involved.

These issues have also been addressed in the recently updated guidelines from the National Academies of Science, Engineering, and Medicine (NASEM, 2017) that specifically emphasize the often-overlooked obligations on the part of research institutions to promote a culture of integrity, train students in the fundamentals of good experimental practices, and continually implement improvements to the research environment beyond merely complying with federal regulations. These guidelines highlight the need for simple yet effective solutions that address the current concerns with reproducibility that are critical to set the stage for the responsible conduct of research, a role that many of these bodies tend to ignore as the "businessification" of science (Lazebnik, 2015) continues unabated. As a consequence, this results in effectively condoning detrimental research practices that waste billions of dollars in funding, investigative opportunities, and researcher time (Freedman et al., 2015; Glasziou et al., 2014). Thus while granting agencies, universities and research institutions are entrusted with the oversight and administration of public funds to conduct research, it has been left to individual researchers like Adam Marcus, Ivan Oransky, Arturo Casadavell, Ferric Fang and their colleagues, to constructively address the issue of reproducibility, find solutions, and provide leadership, the former in the form of *retractionwatch*.com and the latter, most recently, in the American Academy of Microbiology's "A Framework for Improving the Quality of Research in the Biological Sciences" (Casadevall et al., 2016).

The present book resulted from a meeting held in March 2015 at the Experimental Biology Meeting in Boston, MA. Convened at the request of Martin Michel (Johannes Gutenberg-Universität Mainz) by Rich Dodenoff of the American Society for Pharmacology and

Experimental Therapeutics (ASPET), this meeting was attended by editorial representatives (including MW and MJC) from the peer-reviewed journals, *Biochemical Pharmacology*, the *British Journal of Pharmacology*, the *European Journal of Pharmacology*, the *Journal of Pharmacology and Experimental Therapeutics*, and *Naunyn-Schmiedeberg's Archives of Pharmacology* as well as journals in the field of physiology and representatives from the scientific publishers, ASPET, Elsevier, Springer Nature, and Wiley-Blackwell.

The purpose of the meeting was to identify common themes and initiatives that could form the basis of a unified approach to comprehensively address issues of reproducibility in the biomedical sciences, specifically in pharmacology, that could be presented to the International Union of Pharmacology (IUPHAR) for support and further action. Inevitably, however, the discussion became preoccupied with a lively discussion on appropriate statistical usage with the predictable outcome being n^2 differing viewpoints originating from the n participants present at the meeting.

At the conclusion of the meeting, Dr. Jaap Van Harten (Elsevier, Amsterdam) extended a formal invitation to MW to produce this book. Some 24 months later, with the insights and support of KM and MJC, it is finally available with however, the inevitable caveats that the topic of reproducibility is in a constant state of flux with often daily opinions in the scientific literature and mainstream media that are myriad and conflicted. These often result in initiatives that propagate increasingly bureaucratic solutions that frequently are lacking in logic and value which in time, stagnate or disappear.

The editors would like to thank Len Freedman (Global Biological Standards Institute), Mike Jarvis (AbbVie), Harvey Motulsky (GraphPad), Jacques Piette (Université de Liège) and the brothers Triggle, David (Professor Emeritus, SUNY) and Chris (Weill Cornell Medical College in Qatar) for helpful discussions. They would also like to acknowledge the many publications, books, and blogs that have helped inform the book. Authors of these include: Kent Anderson, Glenn Begley, Arturo Casadevall, David Colquhoun, Ferric Fang, David Goodstein, David Gorski, Richard Harris, the late David Horrabin, John Ioannidis, Terry Kenakin, Yuri Lazebnik, Gary Pisano, Drummond Rennie, and last, but far from least, Derek Lowe, the prolific, and highly networked author of the influential drug discovery/pharma industry blog, *In the Pipeline* (http://blogs.sciencemag.org/pipeline/) which has been cited extensively throughout the book.

Finally, the editors would like to thank Michael Marino for contributing Chapter 3, the indefatigable Kristine Jones (Elsevier) for her support, enthusiasm, and infinite patience in guiding this book to completion despite her new responsibilities at Elsevier, as well as Kathy Padilla (Elsevier, Boston), Julia Haynes (Elsevier, Oxford) and Priya Kumaraguruparan (Elsevier, Chennai) for their excellent support and equally infinite patience during the production process. Thank you all.

June, 2017

Michael Williams
Lake Forest, IL

Michael J. Curtis
London

Kevin Mullane
San Francisco, CA

References

Alberts, B., Kirschner, M.W., Tilghman, S., Varmus, H., 2014. Rescuing US biomedical research from its systemic flaws. Proc. Natl. Acad. USA 111, 5773–5777.

Begley, C.G., Ellis, L.M., 2012. Raise standards for preclinical cancer research. Nature 483, 531–533.

Bishop, D., 2015. Publishing replication failures: some lessons from history. Bishop Blog. Available from: http://deevybee.blogspot.co.uk/2015/07/publishing-replication-failures-some.html.

Casadevall, A., Ellis, L.M., Davies, E.W., McFall-Ngai, M., Fang, F.C., 2016. A framework for improving the quality of research in the biological sciences. mBio 7, e01256-16.

Drucker, D.J., 2016. Never Waste a Good Crisis: Confronting Reproducibility in Translational Research. Cell Metab. 24, 348–360.

Economist, 2013. Unreliable research. Trouble at the lab. The Economist, Available from: http://www.economist.com/news/briefing/21588057-scientists-think-science-self-correcting-alarming-degree-it-not-trouble.

Freedman, L.P., Cockburn, I.M., Simcoe, T.S., 2015. The economics of reproducibility in preclinical research. PLoS Biol. 13, e1002165.

Glasziou, P., Altman, D.G., Bossuyt, P., Boutron, I., Clarke, M., Julious, S., et al., 2014. Research: increasing value, reducing waste 5. Reducing waste from incomplete or unusable reports of biomedical research. Lancet 383, 267–276.

Ioannidis, J.P.A., 2005. Why most published research findings are false. PLoS Med. 2, e12.

Ioannidis, J.P.A., 2016. The mass production of redundant, misleading, and conflicted systematic reviews and meta-analyses. Millbank Quart. 94, 485–514.

Lazebnik, Y., 2015. Are scientists a workforce? Or, how Dr Frankenstein made biomedical research sick. EMBO Rep. 16, 1592–1600.

McClain, S., 2013. Not breaking news: many scientific studies are ultimately proved wrong! Guardian. Available from: http://www.theguardian.com/science/occams-corner/2013/sep/17/scientific-studies-wrong.

Mullane, K., Williams, M., 2015. Unknown unknowns in biomedical research: does an inability to deal with ambiguity contribute to issues of irreproducibility? Biochem. Pharmacol. 97, 133–136.

NASEM (National Academies of Science Engineering and Medicine), 2017. Fostering Integrity in Research. The National Academies Press, Washington, DC, Available from: http://www.nap.edu/21896.

Prinz, F., Schlange, T., Asadullah, K., 2011. Believe it or not: how much can we rely on published data on potential drug targets? Nat. Rev. Drug Discov. 10, 712–713.

CHAPTER

1

Reproducibility in Biomedical Research

Kevin Mullane, Michael J. Curtis, Michael Williams

OUTLINE

Research in the Biomedical Sciences. http://dx.doi.org/10.1016/B978-0-12-804725-5.00001-X

1.1 INTRODUCTION

The remarkable progress and benefits scientific research has brought to human well-being are built on a foundation created by previous researchers. The robustness and consequent value of this edifice is predicated on the notion that researchers adhere to sound practices—the *responsible conduct of research* (NASEM, 2017)—and communicate openly their complete and accurate findings and the methods used to derive them. It is a cumulative process, dependent on both successes and failures, which builds a reliable body to advance scientific knowledge. The process includes the initial discovery, its reproduction by others, and then extension of the initial finding, and where appropriate, correction. Discoveries initially considered as significant or even important might ultimately be tempered or revised as new and improved techniques, tools, and analytical methods are applied that lead to modifications to the initial interpretations. This is a normal part of the research process, and has given rise to the concept that research is validated through the process of scientific self-correction, where "self" refers to the research community at large. That is, independent reproduction of scientific findings provides validity, whereas poorly conceived or badly executed studies will not be reproduced and will be rejected.

The present monograph focuses on the current, highly visible debate on what is viewed as a systematic lack of reproducibility in biomedical research (Economist, 2013a), what it is, its scope, and perceived cause(s), why it matters, and the approaches that are being used to deal with it more effectively. While frequently described as a crisis, Gorski (2016) has taken issue with this terminology noting that "Reproducibility in science is ... a chronic problem.... not a crisis," a viewpoint shared by McClain (2013) and many others. Indeed, concerns have been expressed ever since the advent of the first scientific journals in the 17th century, when the eminent scientist philosopher Robert Boyle noted "you will find...many of the Experiments publish'd by Authors, or related to you by the persons you converse with, false or unsuccessful" (Bishop, 2015).

The importance of the reproducibility problem is reflected in the adages, "the demarcation between science and nonscience" (Braude, 1979), "Science isn't science if it isn't reproducible" (Roth and Cox, 2015), "reproducibility, a bedrock principle in the conduct and validation of experimental science" (Casadevall and Fang, 2010), the "coin of the scientific realm" (Loscalzo, 2012), and "If a scientific finding cannot be independently verified, then it cannot be regarded as an empirical fact. And if a literature contains illusory evidence rather than real findings, the efficiency of the scientific process can be compromised" (Bollen et al., 2015).

The self-correction process of independently reproducing research findings to validate them is an honor system, "built upon a foundation of trust and verification, trust among scientists, who rely on each other's data and conclusions, and trust between scientists and the public, which funds much of the science" (Alberts et al., 2015; Kraus, 2014). The continued effectiveness of self-correction has been questioned as the biomedical research enterprise grows, leading some to argue for a more systematic and transparent approach (Estes, 2012; Morrison, 2014; Pattinson, 2012), since many negative studies are not published, and irreproducible results often remain in the scientific canon (Begley, 2017a).

Given that society has come to depend on the application of findings from basic biomedical research to understand the causes of human diseases, to identify new drug targets, and develop therapeutics to treat disease—not necessarily in that order—if experimental findings cannot be repeated, research becomes a veritable house of cards with a diminution in both its value and its impact on the primary goal of improving human health. If that goal is compromised, and the bond of trust between biomedical research and the public is eroded, it can have deleterious consequences for the continuation of research support (Alberts et al., 2014; Begley and Ioannidis, 2015).

This by no means implies that irreproducible findings should automatically be viewed as fraudulent. As noted by Flier (2017), "Although scientists should and most often do seek to publish reliable results, to expect a standard of certainty before publication, and/or to excessively stigmatize or penalize claims later found honestly to be in error, would diminish progress by replacing a spirit of scientific excitement and daring with professional fear of error.... [requiring researchers to be] ....appropriately tolerant of tentative conclusions and honest errors, while continuously seeking to reduce the latter."

A failure to reproduce an experimental finding can result from a variety of factors, including:

- honest errors;
- the adoption of "detrimental" or "questionable" research practices, colloquially referred to as "sloppy science" that are often enabled by "perverse incentives";
- the introduction of biases, conscious, or unconscious; and
- unknown variables—frequently described as "noise" (Kass et al., 2016)—that by definition cannot be controlled, but include limitations in the tools and techniques for addressing a research question, the skills of the researcher, and multiple undefined factors that can emerge and influence scientific outcomes.

Many of these factors are touched on in this chapter and developed in more detail in subsequent chapters of this monograph.

1.2 DEFINING REPRODUCIBILITY

An unambiguous understanding of precisely what constitutes a problem is essential in its solution. In biomedical research, the term *reproducibility* is used interchangeably with that of *replicability*, often in the same sentence despite the fact that there are subtle but important differences in their meaning (Baker, 2016a). In fact, there are so many definitions for the activities related to reproducibility that the reader can end up being totally confused as to which term means what in different research disciplines and when/how they should be used (Baker, 2016a).

It is also important to note that much of the discussion on improving reproducibility is informed by experiences in the clinical and social/psychological sciences that may not appear to have immediate relevance to other areas of biomedical research. With respect to clinical studies, due to the inherent responsibility and accountability in testing new drug candidates in human subjects, the research infrastructure is, by design, infinitely more complex, transparent, and multidisciplinary as well as better funded than preclinical research. The randomized controlled trial (RCT) format of clinical trials represents the gold standard in biomedical research (Bothwell et al., 2016) and has been proposed as a model for animal experimentation (Henderson et al., 2013; Muhlhausler et al., 2013) while being used as the basis to develop the *ARRIVE* and related animal research guidelines (Chapter 5.10). However, the resources required to implement RCT-like standards in preclinical research will be a major challenge; one obvious source may be the estimated $28 billion that is wasted annually in the United States on irreproducible research due to poor design, execution, and reporting (Freedman et al., 2015). In the field of social psychology, reproducibility initiatives were, until recently, rare (Earp and Trafimow, 2016; Schmidt, 2009), the result of both a perceived "lack of prestige" in "copying" the work of another researcher, or a lack of respect for the competence/eminence/prestige of the scientist who had conducted the original study.

Data that are *reproducible* represent those that are *similar* to one another and reflect "the extent to which consistent results are obtained when produced repeatedly" (Casadevall and Fang, 2010) with Fang further noting that these results need to be "robust enough to survive various sorts of analysis" (Baker, 2016a). Data that are *replicable* are *exact* copies of one another. However, in biomedical research only reproducible data has value since replication, a term used in the field of biomarkers and clinical chemistry is, in practical terms, unfeasible as few, if any, experiments can actually be replicated due to biological variability attributed to known and unknown differences in experimentation (Nosek and Errington, 2017). Adding to this confusion in terminologies, in the biomarker field an additional term, *repeatability,* is used that "refers to the outcome of measurements that are performed under *the same conditions*, while.. [using]... reproducibility to... [describe]... the outcome of measurements performed *under different conditions*" (Chau et al., 2008).

In their seminal editorial on reproducibility, Casadevall and Fang (2010) noted, based on definitions used in the field of information technology (Drummond, 2009), that "reproducibility refers to a phenomenon that can be predicted to recur even when experimental conditions may vary to some degree. On the other hand, replicability describes the ability to obtain an identical result when an experiment is performed under precisely identical conditions." Another term that is often used instead of reproducibility, is that of *generalizability* "the persistence of an effect in settings different from and outside of an experimental framework" (Goodman et al., 2016). Thus the terms "reproducible" and "generalizable" also speak to the *robustness* of a finding—that the same qualitative outcome is observed under different conditions to give confidence that the effect is real, and not an artifact consequent to some particular feature of a test system.

The reader may wonder about the relevance of these apparently semantic distinctions for addressing the issue of irreproducibility and why it matters. The answer lies in the expectations from research findings. If the expectation is that reconstructing a published experiment will yield quantitatively *exactly* the same results, then the experimenter will likely be disappointed. Indeed, Nosek and Errington (2017) have noted "There is no such thing as exact replication because there are always differences between the original study and the replication.... As a consequence, repeating the methodology does not mean an exact replication, but rather the repetition of what is presumed to matter for obtaining the original result."

If the broad research finding can be repeated independently to reach the same conclusions, then it bodes well that the effect is real and could be extended into additional research, and possibly clinical, settings. In this regard, Begley and Ioannidis (2015) also concluded that while "There is no clear consensus as to what constitutes a reproducible study..... it seems completely reasonable that the one or two big ideas or major conclusions that emerge from a scientific report should be validated and withstand close interrogation."

Clearly an area exists between adequately adopting key elements of the original study to ensure reproducibility and determining the validity and robustness of the findings by exploring alternative settings. Thus the intention is to reproduce in a closely matched setting and then extend the observations using alternative approaches/techniques that are still appropriate to meet the goal of a study (Flier, 2017).

Attempts to reproduce a finding often tend to overlook key elements of the original study and have proven to be a topic of debate regarding the Open Science Collaboration (OSC) report, "Estimating the reproducibility of psychological science" (OSC, 2015) (Chapter 5.13.2.3). One major issue has been the use of a different subject population in the reproducibility studies from that used in the original study (Gilbert et al., 2016). For instance, the replication/reproducibility attempt for an original study that asked Israelis to imagine the consequences of military service was replicated by asking Americans to imagine the consequences of a honeymoon, which, perhaps unsurprisingly, failed (Mathur, 2016). Reproducible *findings* represent the goal, while the relevance of a *study* is, by definition, limited until the findings can be independently and robustly reproduced in other different, populations. Replication, once again can be considered to have limited intrinsic value in biomedical research.

1.2.1 A New Lexicon for Reproducibility

As a consequence of the extensive semantic confusion in the use of the terms *reproducibility* and *replicability*, with *repeatability* being used to cover all eventualities, Goodman et al. (2016) have proposed a "new lexicon" to better describe issues in reproducibility using qualifiers for methods, results, and inferential reproducibility.

1.2.1.1 Methods Reproducibility

Methods reproducibility was defined as the reproduction of experimental procedures including design, execution and data analysis in reports of an original study. For example, it is often assumed that the purchase of a product (e.g., a cell-line or antibody) from a catalog implies that the vendor has performed some quality control on the product, but they are often just a repository and distribution center for multiple products from different sources and make no commitments to the accuracy or integrity of the materials provided. It is up to the investigator to ensure the validity of any reagent used (Chapter 2.5.2). Moreover, cell-lines drift, become contaminated, undergo genomic and epigenomic alterations, etc., all of which can modify experimental results without careful attention to their fidelity. This level of detail is frequently missing from publications (Vasilevsky et al., 2013) often for the simple reason that it has not been done (Freedman and Gibson, 2015; Freedman and Inglese, 2014; Nardone, 2008; Neimark, 2014, 2015).

1.2.1.2 *Results Reproducibility*

Results reproducibility was defined as the situation where an experiment is conducted to confirm a previously reported finding, the procedures used for which are as closely matched to those used in the original experiment as is possible. However, when a study is not reproduced, frequently the original study is found to be underpowered and/or provides a low signal-to-noise ratio and is incapable of discerning a true effect, such that the criteria for defining results that are "the same" need to be so loose that they lose meaning. Instead, "the paradigm for accumulating evidence…[is] more appropriate than any binary criteria for successful or unsuccessful replication," a point often overlooked by proponents of formalized reproducibility initiatives (Chapter 5.12). If the effects reported in a study are weak and group sizes are inadequate then there is a high probability that a finding will be false, regardless of whether it reproduces data from a previous finding.

Another important aspect is the overreliance on, and limited understanding of, what precisely a p value denotes. While this issue is developed more fully in Chapter 3, it should be noted that the probability of repeating a study that is just significant at the 5% level (i.e., $p = 0.05$) is only 50% due to random variation of the p value, but does not necessarily undermine the credibility of the first experiment (Button et al., 2013; Goodman et al., 2016; Horton, 2015).

1.2.1.3 *Inferential Reproducibility*

Inferential reproducibility was described by Goodman et al. (2016) as the "most important" of the three types of reproducibility since it "refers to the drawing of qualitatively similar conclusions for either an independent replication … (i.e., reproduction)… of a study or a reanalysis of the original study," the latter assuming that the information from the original study is available which is infrequently the case (Engber, 2016). The authors coined the term inferential reproducibility "because scientists might draw the same conclusions from different sets of studies and data or could draw different conclusions from the same original data… even if they agree on the analytical results." The issue of drawing different conclusions from the same data serves to illustrate that this new terminology may take time to be accepted into mainstream biomedical research especially as it invokes the use of Bayesian paradigms, an area of statistics where many researchers are either uncomfortable or dismissive (Colquhoun, 2014).

1.2.1.4 *Bayesian Paradigms in Reproducibility*

Goodman et al. (2016) concluded their proposal for a new lexicon in scientific reproducibility on the topic of "operationalizing truth" noting that "replication" per se is not the objective of reproducing an experiment but rather "whether scientific claims based on scientific results are true since if a finding can be reliably repeated, it is likely to be true." Unfortunately, terms such as "true" and "false" can provoke an emotive connotation that impugns the integrity of the researcher while it is only addressing the reproducibility of an outcome.

As the traditional frequentist approach to statistics (Chapter 3.6) cannot assign any probability to the truth of a finding, Goodman et al., advocate the use of Bayesian statistics where "The probability that a claim is true after an experiment is a function of the strength of the new experimental evidence combined with how likely it was to be true before the experiment," an important concept that is often overlooked in the reproducibility debate. Thus, rather than relying on a single reproduction of a research finding (or not), the current *modus operandi* of the various formal *Reproducibility Initiatives* (Baker and Dolgin, 2017) (Chapter 5.12) reflect the hypercompetitive aspects of 21st century biomedical research (Alberts et al., 2014), where the deference given to the first published, peer reviewed-finding (its "priority") leads to its reproduction being viewed as a test of the original published finding rather than a part of the continued testing of the original hypothesis via the process of self-correction (Alberts et al., 2015). In this context, Goodman et al. note that "the aim of repeated experimentation is to increase the amount of evidence, measured on a continuous scale, either for or against the original claim ...[resulting in]...strong *cumulative* evidence with each independent data set adding to a final outcome that can confirm or refute a hypothesis." Such a strategy was espoused in 1620 by Francis Bacon, considered the grandfather of the empirical scientific method. Based on performing experiments, making observations, and analyzing the outcomes, Bacon observed that "Now my method, though hard to practice, is easy to explain, and it is this. I propose to establish progressive stages of certainty" (http://www.constitution.org/bacon/nov_org.htm).

1.3 DISCIPLINE SPECIFIC TERMINOLOGY IN THE BIOMEDICAL SCIENCES?

As previously discussed, there is considerable confusion in the definitions used when discussing reproducibility (Baker, 2016a), much of it discipline specific. In 2015, a National Science Foundation (NSF) workshop on the topic of enhancing the reliability of research in the social and behavioral sciences (Bollen et al., 2015) defined *reproducibility* as "the ability of a researcher to duplicate the results of a prior study using the same materials and experimental design and analysis as were used by the original investigator"; an additional condition of this definition was the reuse of the same raw data set that was used to generate the original finding. *Replicability* was defined as "the ability of a researcher to duplicate the results of a prior study if the same procedures are followed *but new data are collected*" (author emphasis added). Of additional note, both definitions include the term "duplicate" adding to the confusion since the latter is defined as "exactly like something else" such that these definitions of reproducibility and replicability appear to differ on only a single point—whether new data is generated. This appears to be a situation unique to the social, behavioral, and economic sciences—the subject of the NSF workshop—and may be reflective of the behavioral sciences having lower methodological consensus and higher noise (Fanelli and Ioannidis, 2013). Interestingly, the NSF recommendations were also "opposed" (or ignored) in the OSC paper on reproducibility in the psychological sciences where *new studies* that attempted to repeat original studies were described in terms of the NSF definition of *reproducibility* rather than *replicability* (OSC, 2015).

Finally, in response to issues with reproducibility failures in the initial reports from the *Reproducibility Project: Cancer Biology* initiative (Baker and Dolgin, 2017), a successor to the

OSC study on reproducibility in the psychological sciences (OSC, 2015), Nosek and Errington (2017) qualified their replication (e.g., reproducibility) process with the terms direct and conceptual. *Direct replication* was defined as "attempting to reproduce a previously observed result with a procedure that provides no a priori reason to expect a different outcome [using] protocols from the original study [that] are followed with different samples of the same or similar materials: as such, a direct replication reflects the current beliefs about what is needed to produce a finding"; the definition of *Conceptual replication* concerned "a different methodology (such as a different experimental technique or a different model of a disease) to test the same hypothesis: as such, by employing multiple methodologies conceptual replications can provide evidence that enables researchers to converge on an explanation for a finding that is not dependent on any one methodology."

In summary, reproducibility and replicability, as well as repeatability, are terms that mean the same thing to many researchers and are used interchangeably in the literature. Since there is a difference between the exact copying of a study, and the confirmation of a finding, there have been many attempts, mostly unsuccessful, to imbue a clear distinction between the two and the implications this has in real world experimentation (Nosek and Errington, 2017). Since reproducibility is what matters in assessing the broader truth of a research finding (Goodman et al., 2016), this is the descriptor that will be used by the authors to describe the independent confirmation of a research finding—its self-correction; however, the terms replication and replicability will inevitably appear in citations from the literature, with the majority of this usage being the result of common usage rather than scientific accuracy.

1.4 EXPERIMENTAL FACTORS IN ADDITION TO STATISTICS THAT AFFECT REPRODUCIBILITY

The ongoing debate on the misuse of statistical analysis as a key—or more often *the* key—factor in the reproducibility debate (Colquhoun, 2014; Kass et al., 2016; Marino, 2014; Motulsky, 2014) has tended to overshadow other causal factors some of which are binary and have a greater impact on research outcomes than an inadequate appreciation of statistical methodology. This includes (Chapter 2) flawed experimental design (Curtis et al., 2015; Glass, 2014; Holder and Marino, 2017; Ruxton and Colegrave, 2011) that includes the lack of a credible hypothesis, inadequate powering (Marino, 2014; Motulsky, 2014); the absence of appropriate controls, both positive and negative (Begley, 2013; Begley and Ioannidis, 2015); the use of unvalidated antibodies (Baker, 2015; Bradbury and Pluckthun, 2015) and inappropriate/nonselective concentrations/doses of reference tool compounds (Arrowsmith et al., 2015); an absence of cell line authentication (Almeida et al., 2016; Geraghty et al., 2014; Nardone, 2008; Neimark, 2014, 2015), lack of blinding and randomization (Hirst et al., 2014; Ruxton and Colegrave, 2011; Suresh, 2011), and an overinterpretation/overgeneralization of the results obtained from animal studies (Section 1.11.1). These are what are referred to as "detrimental research practices" or simply as "sloppy science."

As noted not all irreproducible studies are due to fraud, bias, or detrimental research practices and that resolution of these issues will minimize all questions regarding reproducibility.

1.4.1 Unknown Unknowns Affecting Reproducibility

Despite best efforts and taking extraordinary measures to limit potential sources of variability, attempts to replicate studies can still lead to statistically significant differences in outcomes, for reasons that have yet to be identified, and are referred to as unknown unknowns.

1.4.1.1 The Crabbe Mouse Study

In an examination of the behavioral phenotypes of several genetically inbred mouse strains and one null mutant at three different laboratories in Albany, New York, Edmonton, Alberta and Portland, Oregon, Crabbe et al. (1999) took steps to control for; animal strains (by using the same supplier); test apparatus; testing protocols; handling and cage environment (including materials, bedding; incandescent light exposure; number of littermates; cage changing frequency) etc., with behaviors being assessed simultaneously in the same sequence at the same time of day at all three sites. Performance of the mice was assessed in six validated behaviors—locomotor activity, anxiety in the elevated plus maze (EPM), rotarod performance, swim test, cocaine-induced locomotor activation and ethanol preference, and powered to a 90% level. The expectation, given the extensive planning involved, was that the data generated by each laboratory would be similar, but this was not what was observed.

In the cocaine-induced locomotor activation test, mice treated with cocaine in Portland moved 600 cm more than control animals did. In Albany, cocaine-treated mice moved 701 cm, not markedly different from the results obtained in Portland. However, those tested in Edmonton moved 5,000 cm (nearly an order of magnitude difference that did not require any estimate of standard deviation or statistical analysis to provoke consternation). Other digressions were observed in the EPM, leading the authors to conclude that the results were idiosyncratic to a particular laboratory (Crabbe et al., 1999).

The findings of this study resulted in many behavioral scientists concluding that the standardization of experimental conditions was an exercise in futility (Van der Staay and Steckler, 2002). However, in the 17 years since these studies were conducted, previously unknown variables that may have contributed to the experimental differences have come to light. These include the influence of pheromones produced by researchers on animal behavior (Sorge et al., 2014), the animal microbiome (Bahrndorff et al., 2016; Cryan and O'Mahoney, 2011; Dinan et al., 2015; Ezenwa et al., 2012) which differs in its composition, and effects on host phenotype based on environmental factors (Rogers et al., 2014), background passenger mutations in genetically modified mice (Vanden Berghe et al., 2015) as well as more mundane causes including circadian rhythms (Drucker, 2016). These factors, all of which have been reported to alter animal phenotypes, especially behavior, may potentially explain the discrepancies in the results seen in the Crabbe et al. study, while there may still be other unknowns yet to be identified.

1.4.1.2 Reproducibility Confounds in RNA Interference

While it can be argued that in vivo experiments in general, and behavioral studies in particular, can be subjected to marked variations in outcomes that muddy interpretations due to the large number of uncontrolled and uncontrollable variables, in vitro and even molecular biology studies are not immune to similar vagaries. Two identically designed RNA interference-based whole genome screens to identify host factors that support yellow fever virus propagation in human cells using high-content cell-based imaging conducted 5 months apart by the same investigators revealed different hit lists with only approximately 40% overlap (Barrows et al., 2010). An additional confounder, the method of analysis also significantly impacted measures of intra- and interassay reproducibility despite the four analytical tools utilized being accepted and used routinely in the field. Reasons for the low reproducibility of hits identified from the two studies remain speculative, while the analytical inconsistencies are an example of a known unknown, where the issues surrounding the analytical methods are known but the information necessary to identify the "correct answer" is not.

One problematic issue of such RNA interference assays is "false discovery rates"; false negatives that miss valid hits, and, in particular, false positives that erroneously assign activities that are not real. A second-generation RNA interference library computationally optimized to decrease the incidence of false discoveries was compared to results previously obtained by the same research team in 2005 using the same cell line, reporters, and experimental design, but that had used a first-generation library to study the *Drosophila* JAK/STAT pathway (Fisher et al., 2012). Since analytical tools had also improved in the intervening period, the results from 2005 were also subjected to reanalysis. While from the original study, 134 hits were identified, there were 42 in the follow-on study, but only 12 targets were common to both screens. While this can partially be explained by improved library design to reduce off-target effects, the second-generation library still showed 31% false positive hit rates upon rescreening, and other unknown factors likely contributed to the variability. The study highlights the difficulties in reproducing such studies, particularly as older libraries become superseded by newer ones and are no longer available. As the authors (Fisher et al., 2012) concluded, "even the most sophisticated screening approaches are only a tool to identify genes that potentially interact with a chosen assay system," and these require further validation and confirmation.

1.4.1.3 *Caenorhabditis* Lifespan

A consortium, the *Caenorhabditis Intervention Testing Program* (Lucanic et al., 2017) was established to assess the impact of genetic backgrounds and compounds on lifespan in different *Caenorhabditis* strains and species rigorously adopted common procedures across three research sites that minimized intersite variability. However, variation in reproducibility at any one site was around 15%, which proved to be of a similar or even greater magnitude to the impact of different genetic backgrounds on the longevity of 12% among species and 8% among strains within species, that might obfuscate their influence or that of chemicals intended to prolong lifespan. The investigators attribute this intralaboratory variation "to unidentified and apparently subtle differences in the assay environment, which vary similarly within each laboratory."

Few investigators are willing to conduct confirmatory studies using new tools or methods that might repudiate their earlier work or conclusions, and which could result in findings that cannot be explained, are difficult to publish, and question the direction and fidelity of their research efforts. Consequently, the full impact of unanticipated, unidentified, and unknown variables on the reproducibility problem cannot be quantified. However, the few examples described here show that problems of reproducibility are not due only to bias or poor experimental techniques or design, but there is a wide range of mitigating factors. While these must be distinguished from issues of "sloppy science," they are important regarding the robustness of scientific discoveries and their translation to the clinic. Only the most robust findings that transcend the minutiae of animal husbandry or incubation conditions or the like stand a chance of going on to benefit patients.

1.4.2 Known Unknowns: Tacit Expertize

As in other domains of human performance, whether it is knitting a scarf, *haute* cuisine *a la* Bocuse or Robuchon, crafting a musical instrument (think Stradivarius), tuning a piano or performing complex surgery, some biomedical researchers are better than others in the routine execution of experiments. Such individuals reliably derive outcomes that others cannot, a phenomenon described as "superior technical skills" (Stroebe et al., 2012). This applies to animal surgery in the areas of pain, stroke, tumor xenografts, etc. and in some laboratories, this has led to a designated researcher being tasked with conducting a particular procedure. While this may suggest that there are irregularities in experimental procedures, in such instances, independent oversight has repeatedly determined that this is not the case. Rather, the researcher as a result of their experience, continued practice, and intrinsic mental and physical coordination abilities has developed skill sets that others lack and are not easily taught—a gift of nature as it were. Thus, in the field of biomedical research, some laboratories are well known for their unique ability to produce certain types of data and equally widely known for not manipulating their results. As a result, their techniques and the data they produce become the benchmark for all others working in their field even though in a strictly rational sense, their findings can be described as irreproducible.

1.4.3 Diminishing Effects: Regression to the Mean

As studies progress and additional information is provided it is not unusual for previously reported or observed activities (e.g., of a therapeutic agent) to have smaller effects. This phenomenon, first noted by Francis Galton in 1886, describes the tendency for data-points that are outliers to move toward a population mean over the course of repeated measurements—an event termed "regression to the mean" (Sen, 2011). Serikawa (2015) has made the analogy to baseball players who, during the season, can go through "hot" or "cold" streaks in batting averages or on-base percentages, while the full season is the better indicator of true performance.

In biomedical research, this trend is particularly evident in clinical trials of new drug candidates since the number of participants in clinical studies escalates as the therapeutic progresses, so the increased number of data-points reveal the true effect of the drug, and outcomes might differ markedly from the earlier, smaller Phase II trials, and clearly have a major impact on their reproducibility.

Several explanations have been proposed to rationalize the diminishing effects observed in randomized clinical trials. The first, as indicated, simply relates to sample size, and this is the key factor in explaining the decline effect or "regression to the mean." Ioannidis (2006) calculated that a median sample size of 80 patients (the average Phase IIa study), and prestudy odds of 1:10, using $\alpha = 0.05$ with 20% power, the outcome has only a 28% likelihood of being true. Consequently, the chances of repeating the outcome in a second study, or a larger Phase IIb or Phase III trial, are slim. Moreover, there is also a "clinical trial effect" where patients are monitored continuously and adherence to treatment is fostered, that can differ from "real world" situations in a more heterogeneous group of patients (Menezes et al., 2011) (Section 1.11.2.2.1).

However, diminishing effects in clinical trials, and hence poor reproducibility, can also be attributed to other influences. One such factor is *patient selection*, based on entry criteria for a specific study and random (biological) variation. Many chronic disorders, such as arthritis, asthma, hypertension, depression and back pain, are not static, but fluctuate in terms of their severity. Clinical trials recruit patients based on specific entry criteria, and often patients seek out new treatments when their disease has flared up and they are not getting adequate relief from their standard medications. However, during the course of the study, the natural variation might cause their symptoms to recover from that peak, which cannot be differentiated from the effectiveness of the treatment.

This is exemplified in Fig. 1.1 (McGorry et al., 2000) as discussed elsewhere (http://www.dcscience.net/2015/12/11/placebo-effects-are-weak-regression-to-the-mean-is-the-main-reason-ineffective-treatments-appear-to-work/). These researchers followed the natural history of patients with low back pain over 5 months, recording their daily pain perception on a 10-point scale. Despite having free access to pain medication as needed, the patients show marked fluctuations in their symptoms, and this background variability or "biological noise" makes it extremely difficult to discern the effect of a new treatment, or obtain reproducible data.

Some, but not all, clinical trials have a "run in" period to stabilize existing treatment and ensure patients still meet the entry criteria. However, for practical considerations the "run in" period is often short and not based on any detailed knowledge or consideration of the dynamics of the individual's disease, while there is also pressure to maintain recruitment even if participants now fall a little outside the criteria.

Clearly in the example provided in Fig. 1.1, outcomes and conclusions are going to vary depending on whether or not patients were recruited or drug effects measured at a peak, trough, or midpoint. Indeed, a metaanalysis of 118 clinical trials related to treatments for nonspecific lower back pain reported a similar modest improvement in pain scores over a few weeks, regardless of treatment type or even if any treatment was involved (Artus et al., 2010). A corollary is that posthoc analyses of clinical trial data often interpret this biological fluctuation as indicative of a patient "subgroup" that derives significant benefit, when it has nothing to do with the treatment (http://www.dcscience.net/2015/12/11/placebo-effects-are-weak-regression-to-the-mean-is-the-main-reason-ineffective-treatments-appear-to-work/).

Biological noise is supposedly offset by inclusion of a placebo-control group as a statistical comparator. However, small sample sizes can also skew the placebo data, while patient selection according to specific entry criteria can bias the dataset and contribute to misleading conclusions. This problem is exemplified by Sen (2011) modeling hypertension based on diastolic blood pressure readings taken at baseline and again at a later time point, in the absence of any treatment. If the data were cut according to a baseline reading for hypertension greater than 95 mm Hg, then all that could be measured were subjects who began the study hypertensive and either remained hypertensive or became normotensive. In this group, a drop in blood pressure was observed. But this analysis missed out the patients who started out normotensive and then became hypertensive. Including this group showed there was no change in mean blood pressure. So, the entry criterion of patients needing to be hypertensive to participate in the study introduced a bias that erroneously indicated that a fall in blood pressure occurred.

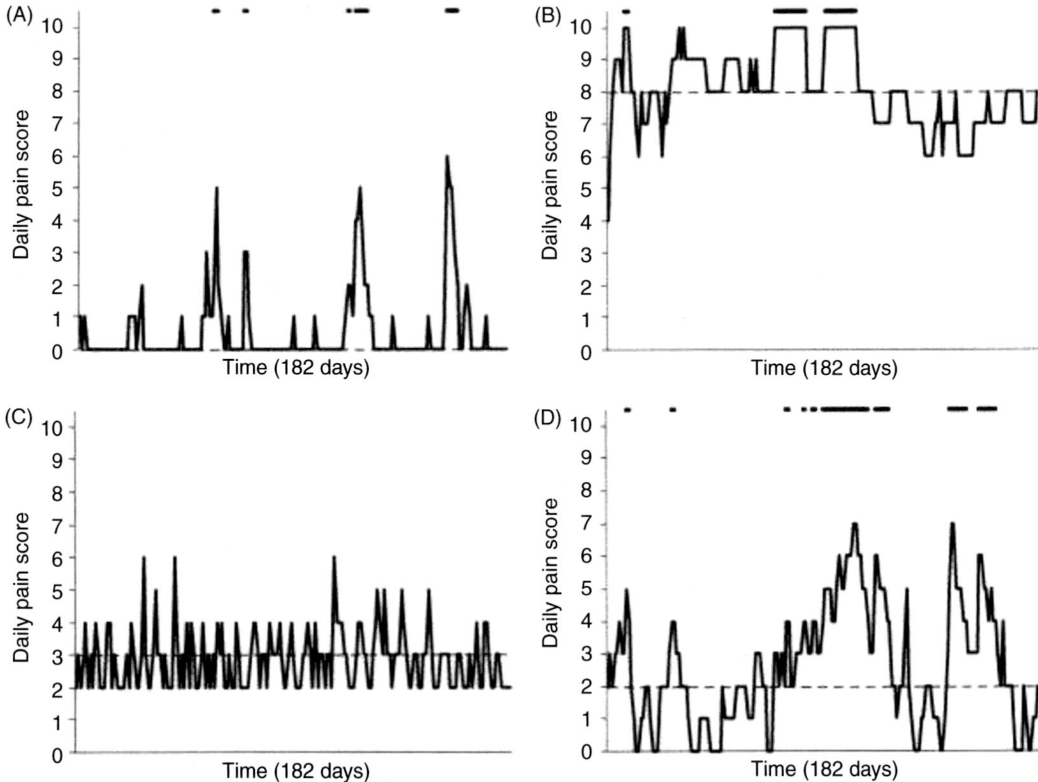

FIGURE 1.1 **Daily recordings of pain intensity, measured on a scale from 0 (no pain) to 10 (as bad as could be), in 4 of the 94 subjects with low back pain (LBP) tracked for 6 months by McGorry et al. (2000).** (A–D) The *dashed bars* at the top of each graph depict the occurrence and duration of episodic "flare-ups," defined as 2–9 consecutive days where the pain score was at least two units above the median score for the 6 months. Despite free access to pain medications as needed, the high degree of variability between subjects and over time is readily apparent, with the authors concluding that LBP is not static and that both its intensity and episodic nature profoundly affect the patient's ability to function. *Source: From McGorry, R.W.,Webster, B.S., Snook, S.H., Hsiang, S.M., 2000. The relation between pain intensity, disability, and the episodic nature of chronic and recurrent low back pain. Spine 25, 834–841, used with permission of Woulters Kluwer.*

Changes observed in patients taking a placebo are often referred to as "the placebo effect," thought to confound many clinical trials and have often been utilized by early stage biotechnology CEOs to justify why their treatment did not work as they predicted. However, a comprehensive review of placebo versus no treatment from 202 clinical trials covering 60 conditions (Hróbjartsson and Gøtzsche, 2010) found that placebo interventions could influence patient-reported outcomes, especially in the areas of pain and nausea, but in general had no important effects. The large placebo effects reported in clinical studies were largely due to issues of experimental design and bias, limited sample sizes, and large outlier effects.

Other explanations for a decline effect over time include changing treatment patterns, since new therapies are provided on a background of the standard of care; while a publication bias of reporting only positive studies might lead to the perception of a diminishing effect as corresponding negative data necessary for regression to the mean determinations are not available. Funnel plots of the treatment effect charted against precision (e.g., standard error) for individual studies have been proposed to help determine publication bias based on whether they are symmetrical (suggestive of inclusion of negative or neutral studies) or asymmetrical (omission of such studies); although caution must be exercised in their application and interpretation (Lau et al., 2006).

1.4.4 Overinterpretation of Effects on Secondary Endpoints in Studies

Another cause of poor reproducibility between clinical trials relates to overinterpretation of the effects on secondary endpoints. Moyé and Deswal (2002) provide excellent examples where drugs to treat congestive heart failure, including vesnarinone, losartan, amlodipine, and carvedilol showed a benefit in one clinical trial that disappeared or reversed in a second study specifically designed to confirm the first observation. In most of these cases, mortality was used in the first trial as a secondary endpoint, while the study was principally designed and powered around the primary endpoint. The authors pointed out that "Type 1 errors are unacceptably high when subsidiary analysis results are proclaimed as positive when there was no prospective type 1 error statement," and that "the cumulative type 1 error increases with the number of statistical evaluations"—that is, looking at different endpoints; and they reiterate the need to distinguish between exploratory and confirmatory analyses. It is likely that the same confound occurs in preclinical in vivo studies where multiple endpoints are used while the study is designed and powered around a single primary endpoint.

1.5 THE IMPACT OF THE INTERNET ON THE EVOLUTION OF RESEARCH PRACTICES—DATABASES AND THIRD PARTY ANALYSES

The issue of reproducibility is somewhat different when dealing with projects involving "big science," like the NIH Big Data to Knowledge (BD2K) Initiative (https://datascience.nih.gov/bd2k). Large-scale datasets from genomic, epigenomic, transcriptomic, and proteomic analyses (to name but a few) are routinely deposited in publicly accessible databases, as mandated by granting authorities and journals (Chapter 5.3.2.6). Two genomic databases, the *Gene Expression Omnibus* (GEO) at the US National Center for Biotechnology Information, and *ArrayExpress* at the European Bioinformatics Institute, together house the data and links for over 30,000 published studies comprising well over 1,000,000 samples, from more than 13,000 laboratories—and with double digit increases each year (Rung and Brazma, 2013; https://academic.oup.com/nar/article/41/D1/D991/1067995/NCBI-GEO-archive-for-functional-genomics-data-sets). In 2011, the *ArrayExpress* database alone was accessed by approximately 1,000 different users each day (Rung and Brazma, 2013) while another major database, the *Kyoto Encyclopedia of Genes and Genomes* (KEGG), was visited 400,000–500,000 times a month in 2014—a figure that had doubled over the preceding 3 years (http://www.kegg.jp/kegg/docs/plea.html). Similarly, the *Protein Data Bank* and *Uniprot* had nearly 750,000 visitors accessing 20 million pages in just 1 month, back in 2008 (Howe et al., 2008). Another way of measuring the impact of these databases is to look at the number of third party publications that emanate from the archived material. A *PubMed* search for articles referencing any of the 2,711 datasets deposited in GEO in 2007, led to the estimate that there were more than 1,150 such articles by the end of 2010 (Piwowar et al., 2011). In just 1 year, 2011, 90 publications were identified that cited and used data from any of the five *ArrayExpress* studies published in database issues of the journal *Nucleic Acid Research* (Rung and Brazma, 2013), and it was concluded that the actual number was probably much higher since many studies do not cite the original publication, only the data accession number. The number of third party publications is increasing and it is suggested that one quarter of studies now use publicly available datasets rather than conducting new experiments when addressing biological questions (Rung and Brazma, 2013). Although some pejoratively call such scientists "research parasites" (Longo and Drazen, 2016), and only approximately 3% of metaanalyses are considered well conducted and clinically useful (Ioannidis, 2016a), this is a burgeoning field that has become a practical reality, that brings with it several unique issues surrounding issues in reproducibility.

Interrogating such enormous datasets allows questions to be asked beyond that of any original study with its more limited samples that were designed for a specific, limited query. One common application is to increase analytical power by pooling data from multiple genetic analyses to unmask disease-associated genes and mutations that might not be identified in any single study (Panagiotou et al., 2013; Price et al., 2015; Torrente et al., 2016). Selecting a cancer gene expression dataset of 28,000 samples in *GEO* and *ArrayExpress* obtained using the same Affymetrix chip led to the identification of 1,285 potential cancer related genes in solid tissue samples, many of which were novel, while gene enrichment analysis identified some cancer pathways that are tissue-specific (Torrente et al., 2016). Gene variants with small effect sizes identified by genome-wide association studies (GWAS) include *HMGCo-A reductase* and *PCSK9* (Proprotein convertase subtilisin/kexin type 9) for LDL cholesterol levels (Global Lipids Genetics Consortium, 2013), and *PPAR*γ (Peroxisome proliferator-activated receptor gamma), and *KCNJ11* (Potassium Voltage-Gated Channel Subfamily J Member 11) for Type 2 diabetes (Morris et al., 2012), representing important drug targets that would not be identified by single studies but required the increased power of enlarged datasets.

Entry into the era of predictive, personalized medicine requires information not only on the individual, but how they fit into specific groups. It has been estimated that within the next 10–15 years, the genes of 1 billion people around the world will have been sequenced, and, in many cases coupled with electronically captured medical records (Price et al., 2015). Moreover, smartphone-app enabled health research has just begun (Chapter 6.4.2.6), for example, with curated data shared by Sage Bionetworks from more than 9,000 participants in a study of Parkinson's disease (Wilbanks and Friend, 2016). This is a tool that has the potential to add enormous amounts of patient data into the mix. If issues of data integrity, accessibility, integration, and interrogation can be overcome, this will enable determination of how a particular mutation can influence disease risk and progression, define treatment options and predict outcomes. To realize such a goal requires overcoming several impediments that weaken all aspects of "Big Data" and publicly available databases—namely issues surrounding inputted data quality and integrity; annotation, curation, and accessibility of information within the databases; cross-database integration for a holistic and amalgamated view of a patient; and defined analytical tools that provide accurate and relevant interpretation of the datasets. As articulated by Howe et al. (2008) "Biology today needs more robust, expressive, computable, quantitative, accurate and precise ways to handle data."

1.5.1 Data Input

In the context of archiving experimental data for further use it is particularly important to ensure that it is accurate and complete, since any biases can infest the dataset and be perpetuated in any secondary analyses. As discussed in detail later (Section 1.10.6.5), the data from microarray analyses of gene expression in cancer cell-lines and sensitivity to treatments (Hsu et al., 2007; Potti et al., 2006), was uploaded to GEO, where independent reanalysis of the datasets revealed significant flaws (Baggerly and Coombes, 2009). While this particular example proved to be due to overt fraud, it nonetheless represents an important application of the databases to check for reproducibility concerns. Attempts to reproduce 18 microarray-based studies published in *Nature Genetics* in 2005–2006, with data deposited in *GEO* and *ArrayExpress*, were only successful in 2 cases, while 6 others were "partially" reproduced and 10 not at all, leading to the conclusion that reproduction of such studies is problematic and that stricter rules around the quality and completeness of the data deposited in such databases were required (Ioannidis et al., 2009). Steps have been taken to improve and enforce guidelines for data submission, but there is still need for further improvement (Rung and Brazma, 2013). Moreover, it is apparent that the same control datasets are sometimes used for multiple experiments, so reappear in the database, and can give a distorted perspective. The *Cancer Proteomics Consortium* spent their first 10 years developing standards and ensuring reproducibility before beginning to publish studies, giving an indication of the efforts required.

1.5.2 Data Curation

Many of the databases are not merely repositories, but process, analyze, and annotate the information so that it can be recalled and interrogated effectively. Often the submission of results to the database is performed by the bioinformatician who analyzed the data, rather than the experimentalist who conducted the study, the former of whom might lack detailed knowledge regarding the experiments and protocols. Relying on authors to annotate their own data failed (Howe et al., 2008). To collate all of the information so that access is facile, logical, and user-friendly is an enormous undertaking, well beyond the capabilities of manual curation, so utilizes a range of computational tools that include annotation (Howe et al., 2008), text mining (Singhal et al., 2016), visualization, and querying (Welter et al., 2014), although accuracy remains a challenge. One example of the complications of annotation is the human gene *CDKN2A* (Cyclin Dependent Kinase Inhibitor 2A) that has 10 synonyms in the literature, one of which, *p14*, is also a synonym for 5 other genes (Howe et al., 2008). Limitations in annotation are regarded as one of the key reasons for the poor reproducibility of high-throughput gene expression studies (Rung and Brazma, 2013).

Responsibility for maintaining and updating databases with revised information is frequently unclear. Genetic screening for severe, orphan childhood recessive diseases revealed that a large proportion of literature-annotated disease mutations warehoused in publicly accessible databases, such as the *Human Gene Mutation Database and Online Mendelian Inheritance in Man* were "incorrect, incomplete, or common polymorphisms" (Bell et al., 2011). While 12%–13% of such mutations were simply incorrect, a further 74% could be accounted for by simple substitutions, many of which were erroneously annotated as disease mutations. As noted by the authors, without an accurate and authoritative database, progress toward prevention, diagnosis and treatment of these diseases will be severely hampered (Bell et al., 2011). The reality is that curators of many databases wrestle with keeping them updated and accurate (e.g., Omenn et al., 2015; Rupp et al., 2016; Weichenberger et al., 2017).

The costs associated with curating and maintaining the multitude of databases are significant, such that after a while many become "stale" despite still being accessible, and it can be challenging to ascertain when they were last updated. Even KEGG, a highly regarded and widely used database, has had to resort to a subscription and licensing model to supplement the grants and maintain viability.

1.5.3 Cross-Database Integration

Most genetic associations with human disease occur in gene regulatory regions rather than the protein-coding regions, so extrapolation of GWAS findings to specific genes, proteins, or networks is frequently absent. It would be beneficial to link different datasets to develop a more complete picture of how the genetic variant results in an alteration to the phenotype. However, even at the level of gene expression microarrays, it is generally accepted that only data using a common platform can be integrated reliably in a quantitative manner (Chen et al., 2007). Integrating multiple datasets to quantitatively relate genomic, transcriptomic, and proteomic signatures that have used different sample preparation methods, technical approaches and procedures, analytical tools, and annotation methods, is extremely challenging.

There are several important considerations to bear in mind. One complication is that the effects of gene regulation, like the diseases themselves, are often tissue or even cell-selective, and involve tissue/cell-specific molecular networks (Lukk et al., 2010; Melé et al., 2015; Ni et al., 2016; Price et al., 2015). However, identifying the actual cells that are relevant, aside from identifying appropriate cell-specific data in the database, is not always clear-cut. As exemplified for coronary artery disease (CAD) by Price et al. (2015), it might be necessary to consider liver/hepatocytes involved in lipoprotein metabolism, cells of the immune system implicated in the development of CAD, cells of the blood, such as platelets and leukocytes, aside from the more obvious endothelial and smooth muscle cells of the artery wall itself.

While several thousand genes show tissue-preferential expression, only about 200 genes are expressed exclusively in a given tissue (Melé et al., 2015). The primary separation is between blood and solid tissues, and between solid tissues, the brain is the most distinct, within which the cerebellum is the most clearly differentiated (Melé et al., 2015). Interestingly solid tissue cell lines show similar gene expression patterns to each other and are distinct from their tissues of origin, but with close similarity to blood cell lines (Lukk et al., 2010), suggesting they are not good models of human tissues. Melé et al. (2015) identified 1993 genes that globally change expression with age, 753 with tissue-specific sex-biased expression, and 31 with tissue-specific ethnicity-based expression, predominantly in the skin. Failure to consider these factors when attempting to correlate gene expression to molecular networks that regulate the phenotype can obviously lead to erroneous conclusions.

1.5.4 Analytical Tools

These large datasets bring unique analytical and statistical challenges (Fan et al., 2014). The plethora of analytical tools for these large datasets is overwhelming, is constantly being revised, updated, and changed, discussion of which could fill a book. What is disturbing is the extent to which the chosen analytical methods can impact the results and conclusions (Barrows et al., 2010; Clooney et al., 2016). For example, an analysis of the microbiome in human stool samples found that the chosen analytical method was responsible for more variance in gene expression than differences between the species of microbiota (Clooney et al., 2016). Updated analytical tools lead to different interpretations than those used when the data were first generated and deposited, but there is little incentive to go back and reanalyze and correct historical data. There is a concern that multiple analytical methods can be employed until the answer that is sought can be found, analogous to *p*-hacking to find a statistical test that shows a "significant effect" in other types of experiments (Simmons et al., 2011).

1.6 THE REPRODUCIBILITY PROBLEM

Research is defined as "the systematic investigation into and study of materials and sources in order to establish facts and reach new conclusions" (https://en.oxforddictionaries.com/definition/research). A more precise definition applicable to biomedical research "is the broad area of science that involves the investigation of the biological process and the causes of disease through careful experimentation, observation, laboratory work, analysis, and testing" (CBRA, 2017) to which the authors would add "to identify the molecular causes of disease in order to identify drug targets that can be used to develop safe and efficacious therapeutics to benefit patients."

For a variety of reasons, it has been historically unusual for researchers to submit a study for peer review that is an attempt (or failed attempt) to repeat one already published in the literature and equally unusual for a journal to publish one (Buntin et al., 2011; Dirnagl and Lauritzeb, 2010; Fanelli, 2012; Ioannidis, 2005; Ioannidis and Trikalinos, 2005; Song et al., 2013; ter Riet et al., 2012; Tsilidis et al., 2013). The reproducibility of the original finding is therefore not commonly reported even when it undergoes evaluation. This is referred to as the "file drawer" effect and represents a publication bias where "journals are filled with the 5% of the studies that show Type 1 errors, while the file drawers are filled with the 95% of the studies that show nonsignificant results" (Rosenthal, 1979). This bias runs the risk of publishing the 1 study out of 10 that showed a desired effect, and ignoring the other 9—both by a single investigator as well as other researchers who tried to reproduce the study (Begley, 2012).

The publication of negative outcomes occurs most frequently under a limited number of scenarios, for example, when:

- The new finding is potentially important: for example, a new disease mechanism or putative drug target, a critical methodology (Obokata et al., 2014a,b) or when a drug candidate enters Phase IIa clinical studies. In these instances, the stakes are high and external scrutiny becomes far greater, and where the inability to reproduce findings in preclinical research, often in animal models, leads to failed translation in the clinical setting—generating considerable concerns especially in the biopharmaceutical sector (Section 1.9).
- A systematic review of the type pioneered by the Cochrane Library, or a metaanalysis of multiple studies on a topic are compiled and analyzed, and can lead to conclusions that differ from many of the individual component studies and influence patient standard of care.

- The *Proteus Phenomenon*. It has been observed that the more extreme a research finding, the more likely it is to be statistically significant and published quickly (Ioannidis and Trikalinos, 2005). As discussed in Chapter 2, exploratory studies that are underpowered are more susceptible for producing outlying results that are not true and are difficult to reproduce. Consequently, further studies addressing the same topic might yield results in the same direction, but of a much lower magnitude, and hence less provocative and less attractive for publication. Since the initial publication represents results that are disproportionate to reality, it is equally possible that a follow up study could find the opposite results. Highly contradictory outcomes are more attractive for publication as they foster interest in determining the "right" answer, and potentially increase journal Impact Factors (Drucker, 2016). This bias in publication might, in large part, underlie the frequency with which dramatic findings in key areas of topical interest are frequently followed by equally dramatic findings in the opposite direction—an event termed the *Proteus phenomenon*, named after the Greek god able to assume multiple forms (Ioannidis and Trikalinos, 2005). Further studies then fall in between these two extremes, a regression to the mean identical to that described in clinical trials.

  While examples of the *Proteus phenomenon* have been published (Ioannidis, 2006; Ioannidis and Trikalinos, 2005; Pfeiffer et al., 2011), most of these relate to genetic markers in molecular medicine—a relatively easy target since genetic associations are low in the reproducibility hierarchy. What is much less clear is the extent to which this is a widespread phenomenon, its full impact on issues of reproducibility, and whether or not it is a result of publication bias, sloppy preliminary (exploratory) studies, or a combination of issues. Ioannidis and Trikalinos (2005) warned that the increased availability of large databases for interrogation by different groups simultaneously might result in increased contradictory publications occurring almost contemporaneously and increasing confusion, although whether that occurred in the ensuing decade has not been reported.

On a broad scale, the validity and reproducibility of data emerging from preclinical biomedical research has become a cause for concern (Begley and Ioannidis, 2015; Collins and Tabak, 2014; Dolgin, 2014; Jarvis and Williams, 2016; Prinz et al., 2011) irrespective of whether these data originate in academia or industry. Consequently, for the better part of the past decade, there have been numerous articles in the research literature and mainstream media on the shortcomings in the quality control processes in scientific publishing. These include peer review (Chapter 5.2) and scientific self-correction (Alberts et al., 2015; Kraus, 2014), both of which have been deemed inadequate in their current forms (Begley and Ellis, 2012; Estes, 2012; Flier, 2017; Ioannidis, 2005; Prinz et al., 2011). Separately, concerns regarding the relevance of modern day biomedical research to human disease (Horrabin, 2003) and a lack of efficiency in its execution (Chalmers et al., 2014; Macleod et al., 2014) have led to estimates that 50% of research, preclinical and otherwise, cannot be reproduced (Button et al., 2013; Goodman et al., 2016; Horton, 2015). Together these phenomena have given rise to concerns about the sustainability of the current biomedical research model in an era of relative economic austerity (Alberts et al., 2014; Balch et al., 2015).

Triaging the landmark Ioannidis, Prinz et al., and Begley and Ellis articles, The (Economist, 2013a) published a high profile, widely read, and widely cited report under the rubric, "Unreliable Research" that focused on the issue of reproducibility. This article also highlighted additional concerns regarding peer review of research articles (Chapter 5.2–5.4), and the very low standards of scrutiny and output in the fringes of the scientific literature (Beall, 2012; Bohannon, 2013; Butler, 2013), which has undergone an exponential increase over the period 2010–14 representing 25%–30% of all published articles (Chapter 5.6.4) (Beall, 2016; Bohannon, 2015; Shen and Björk, 2015). The relevance of fringe research publication to the core rubric of science is small, but its overt visibility due to illogical, often Pollyannaish claims that are supported by minimal and/or low quality data is attractive to the mainstream media, and newsworthy, leading to its widespread dissemination. The worst types of fringe publications are those in so-called "predatory" journals, defined as "counterfeit journals to exploit the open-access model in which the author pays" (Beall, 2012). Predatory publishers generally lack transparency and any notion of peer review, and are considered as dishonest. Beall notes, "They aim to dupe researchers, especially those inexperienced in scholarly communication….[by setting up]… websites that closely resemble those of legitimate online publishers, and publish journals of questionable and downright low quality. Many purport to be headquartered in the United States, United Kingdom, Canada or Australia but really hail from Pakistan, India or Nigeria. Some predatory publishers spam researchers, soliciting manuscripts but fail to mention the required author fee. Later, after the paper is accepted and published, the authors are invoiced for fees, typically US $1,800. Because the scientists are often asked to sign over their copyright to the work as part of the submission process (against the spirit of open access) they feel unable to withdraw the paper and send it elsewhere." A recent development in the predatory publisher sphere is the disappearance of Beall's controversial blog site at the beginning of 2017 (Chawla, 2017). Beall noted that "he was worn out by publishers who threatened him and harassed his colleagues" (Gillis, 2017) and by "intense pressure" from his university about the list that led to him "fearing for my job." This led Beall (2017) to document the pressures brought to bear on his activities by irked predatory publishers and, in some instances, colleagues in the library sciences (Beall, 2017). This article also includes Beall's personal concerns regarding open access publishing and the current state of scientific publishing—a "once-proud scholarly publishing industry is in a state of rapid decline.. [with]... predatory publishers pos[ing] the biggest threat to science since the Inquisition."

Beall's list has been replaced by a "more transparent" subscription-based list, Cabell's Journal Blacklist (https://www.cabells.com/about-blacklist) that uses 65 criteria to determine whether a journal is "deceptive", this term replacing predatory as a descriptor (Gillis, 2017).

Major efforts are ongoing to find effective and logical remedies to the basic reproducibility problem, that are focused on good research practices involving improvements in experimental design, analysis, and disclosure and reporting (Begley and Ioannidis, 2015; Collins and Tabak, 2014; Curtis et al., 2015; Jarvis and Williams, 2016; Mullane et al., 2015; Wadman, 2013) together with improvements in *Classical Peer Review* (*CPR*; Chapter 5.3.2.1) that are intended to identify research that is credible, distinguishing it from that which is not.

1.6.1 Factors Contributing to the Reproducibility Problem

Among the factors proposed to contribute to the reproducibility problem are:

1. inadequate standards of training and mentoring of new generations of scientists (Flier, 2017);
2. an inability to repeat an experiment due to a scarcity of resources (Alberts et al., 2014; Balch et al., 2015; Goodstein, 1995) or adequate experimental detail (Vasilevsky et al., 2013);
3. institutional culture and pressure–or lack thereof (Chapter 6.6.2; NASEM, 2017);

4. distorted reward systems for researchers—often termed "perverse incentives"—that lead to hubris and fraud (Alberts et al., 2015; Begley and Ioannidis, 2015; Ioannidis, 2014; Smaldino and McElreath, 2016); and
5. detrimental research practices which, in many respects, parallel the ongoing ethical malaise in 21st century society (Bishop, 2013; Mullane and Williams, 2015; Redman, 2015).

1.6.2 Human Contributions to the Reproducibility Problem

Changes in cultural norms in biomedical research in the 21st century frequently reflect personal agendas that are prioritized to the absolute detriment of the greater good, eroding the commitment, inquisitiveness, and altruism that has been the cornerstone of the biomedical research culture that has sustained its success (Begley and Ioannidis, 2015; Ioannidis, 2014; Kraus, 2014). This has led to detrimental research practices that are enabled by the "perverse incentives" (Alberts et al., 2014; Begley and Ioannidis, 2015) provided by the structure of 21st academic biomedical research, the main currencies of which are citations and grants. These dictate the outcomes of grant applications and career advancement such that Richard Harris, the author of *Rigor Mortis: How Sloppy Science Creates Worthless Cures, Crushes Hope, and Wastes Billions* has noted (Harris, 2017, p. 3) that "scientists often face a stark choice: they can do what's best for medical advancement by adhering to rigorous standards of science, or they can do what they perceive is necessary to maintain a career in the hypercompetitive environment of academic research."

Historically, scientific research has been enabled by philanthropists and by initiatives in federal research funding (Chapter 6; Triggle and Williams, 2015) after World War II. In the United States, this was part of the *Endless Frontier* (Bush, 1945) and in the United Kingdom, Butler's Education Act (Barber, 2014), both of which enabled access to higher education for a broader portion of the population. Thus, from the 1950s to the early 1970s, these initiatives provided funding to biomedical researchers with an expectation of infinite career growth, while also catalyzing the tools of molecular biology discovered by Watson and Crick (DNA), Milstein and Kohler (Monoclonal Antibodies), Mullis (PCR—polymerase chain reaction), and Boyer and Cohen (DNA cloning). These techniques aided in increasing the precision and, when accurate, the relevance of research outcomes, with the potential to revolutionize drug discovery. The flip side of the coin was that these sophisticated tools tended to become the focus of research activities with questions about human health becoming incidental or irrelevant (Horrabin, 2003). This prompted a reductionist, hypothesis-driven but poorly validated approach to biomedical research that is continually superseded by new tools with even greater precision and promise, but that, until recently, moved farther and farther away from the patients they were intended to help.

Two contrasting events in the 1980s dramatically changed the biomedical research landscape. First, the Bayh-Dole Act of 1980 (Boettiger and Bennett, 2006; Markel, 2013) gave universities ownership of their discoveries made using federal funds, with the obligation that the university should seek to license and commercialize such inventions. Second, the promises of the tools of molecular biology were beginning to be realized, with the first biologic products and monoclonal antibodies reaching the marketplace. This triggered interest from Venture Capitalists to invest in a fledgling biotechnology industry (Chapter 6.4.1.1). The two events together also provided an environment that allowed individual scientists to retain their research interests and, often, their academic positions, while having the opportunity to become entrepreneurs, starting their own companies, making themselves and their institutions wealthier (Kotecki, 2016; Vasagar, 2001). Inevitably, this dual role has created biases—perceived or real—in data generation, evaluation, analysis, and sharing, with inherent conflicts of interest that compromise the integrity and value of information that is put into the public domain. There are also many interesting tales of rogue entrepreneurial scientists involved with insider stock trading and swindles that were supported by fraudulent, misinterpreted or nonexistent data. These have resulted in fines and, in some rare instances, prison sentences (Hodgson, 2016; Interlandi, 2006; Nutt, 2016; Prud'Homme, 2004).

Increased competition for funding and positions has, however resulted in aspects of biomedical research becoming self-serving, involving a substantial flow of taxpayer funding via federal grants to support researchers to conduct research, the main purpose of which appears to be to consolidate a position from which to apply for more grants, *ad infinitum, ad nauseam*. In some instances, the science becomes secondary to the pursuit of personal gain and career advancement whereby some researchers build a biotechnology company (Shimasaki, 2014) or aid in the commercialization of academic research which, by virtue of the stakeholders and the funding process, have been viewed as little more than Ponzi schemes (Mirowski, 2012).

While the onus to ensure integrity and maintain high scientific standards rests primarily with the individual researcher, nonetheless, institutes and the scientific community at large also have key roles to play. The NASEM (2017) report highlights the pressures many academic researchers face where an increasing number solely depend on grant funding for salaries (with all the attendant vagaries of federal support), and are required to maintain a publication record in high profile journals; and where the growing number of graduate and postdoctoral students are competing for an ever-diminishing supply of faculty positions. The report stresses the need for the scientist to "be free from pressures and influences that can bias research results" and compromise objectivity. Separately it may be argued that there is a mandate for the scientific community to recognize, acknowledge, and take responsibility for improving those aspects of the current biomedical research culture that reflect poorly on its societal role and that have the potential to further undermine its credibility with the public and result in cuts in funding (Alberts et al., 2014; Kornfeld and Titus, 2016; Nature Medicine Editorial, 2016). Although the shortcomings in reproducibility more often than not involve honest mistakes—faulty design, biased analysis, and the inherent vagaries of biomedical research (Flier, 2017; McClain, 2013)—this is frequently not obvious to the public. However, society clearly can understand deliberate fraud and there is a perception that it is the fraud that affects societal attitudes to science rather than shortcomings in design and analysis. Thus, as noted by Broad and Wade (1983) more than 30 years ago, "no matter how small the percentage of scientists who might be fakers of data, it require[s] only one case to surface every few months or so for the public credibility of science to be severely damaged." In many respects, outright fraud is merely the tip of an iceberg of the multiple reasons for failed data reproduction. The real concern is how much of the iceberg is due to lack of training, lack of authentication/validation of experimental materials, and lack of insight and knowledge that leads to poor experimental design and analysis, selective reporting, etc., and how much of it is due to more insidious issues, such as deliberate fraud, plagiarism, intrinsic incompetence, and hubris (Alberts et al., 2014; Begley and Ioannidis, 2015; Ioannidis, 2014; Nisbet and Markowitz, 2014; Redman, 2015; Tharyan, 2012).

An additional factor that confounds risk minimization is that many scientists cannot conceive that a colleague in their particular field of research would engage in any activities that are contrary to the pursuit of good science. This includes the routine use of substandard design, execution, and analysis of experiments, or indulging in plagiarism, or outright fraud. Instead, unacceptable behaviors are downplayed or even ignored, thus undermining the bedrock of biomedical research (Stroebe et al., 2012). Accordingly, the biomedical research community must acknowledge the possibility of fraud, data manipulation, and the malicious intent of some authors. Indeed, deliberate fraud (Pulverer, 2016) is thought to be far more widespread than what the "research elites" that manage scientific publication and funding are currently prepared to admit (Kornfeld and Titus, 2016; Wilson, 2016).

Scientific progress is based on the notion that false premises are corrected by the emergence of robust and compelling new data that not only contradict the "state of the art" but aid in reconciling the past and the present by identifying honest misconceptions. This of course assumes that all misconceptions are the product of misapprehension, not flawed design, analysis, or interpretation. Given the discussion earlier, this can no longer be considered as a given in a society where ethics are undervalued or ignored (Bishop, 2013) and where the consequences for errant behavior are arbitrary, and often minimal (Kornfeld and Titus, 2016; Triggle and Williams, 2015).

1.6.2.1 Litigation in Basic Biomedical Research

Another change in biomedical research, again reflective of 21st century society, is the trend for litigation whereby researchers accused of fraud or other forms of scientific misconduct seek legal recourse. Unfortunately, this can dissuade institutions, funding bodies, and whistleblowers from engaging in prosecuting fraud and, indeed, even seeking to uncover fraud, and therefore to fail to take an appropriate course of action. Fear of retribution, being accused of sharing culpability for the fraud, and/or being sued can make it hard for rogue researchers to be identified publicly, resulting in questionable publications retaining, unchallenged, their currency and influence. Complaints from whistleblowers—usually unidentified individuals—are routinely dismissed as "a difference of opinion"—and can compromise careers, especially when the whistleblower is junior to the person who is the subject of their complaint (Goldberg, 2015; Nutt, 2016). Instead, the onus is often on the whistleblower. Some whistleblowers have had a major impact in improving institutional research standards, but many are harassed by the administration at their institution or have their concerns repeatedly dismissed despite having spent considerable time and personal resources in order to gather data to prove their case, while others are simply ignored (Gross, 2016; Yong et al., 2014).

Even when papers are retracted, the authors often continue to "win" grants and publish new research. Often, all that is required to retain a modicum of credibility is a claim by the principal investigator or laboratory head that the false data were included "by mistake." This is normally accepted, especially by publishers who fear litigation if they were to "blacklist" an author. Like big business, accused individuals and their institutions will often pay a fine or offer a short statement of retraction to avoid prosecution while not admitting "any liability or wrongdoing." Instead of being the subject of professional or criminal investigation, they are given a slap on the wrist, for example, being barred from applying for grants for 3 years (Kornfeld and Titus, 2016; Nutt, 2016). However, the arbitrary and capricious nature of penalizing fraudulent activities is evidenced by the laboratory of fired NIH neurologist, Allen Braun where all the researchers have been barred from publishing their data, including graduate students and postdocs who need the publications to embark on their careers, demonstrating that collateral damage can extend to associates not accused of any misdeeds (Couzin-Frankel, 2017). Contrast that outcome with the notorious case of John Darsee, a cardiologist at Emory and Harvard who fabricated much of the data contained in over 100 publications with 47 coauthors. While Darsee was fired and the coauthors could lay claim to being hoodwinked by his activities, nevertheless a review of 18 of his retracted papers (Stewart and Feder, 1987) revealed a catalog of serious flaws (besides whether the data were valid), including reuse of historical control data in multiple articles without even acknowledging the fact; publication of the same data in different journals under different titles (with four coauthors in common and where Darsee was neither the first nor the last author); multiple cases of inconsistent values between the text and figures and standard deviations that did not match the datasets; issues so egregious that the coauthors must have been aware, and in some cases actually defended Darsee (Stewart and Feder, 1987), and where there is no evidence that the coauthors were ever brought to task over the incident.

The lack of significant penalties for scientific fraud also extends into the corporate world of biopharmaceuticals where the management of a number of companies are widely known to have selectively reported clinical data that presented the company in a favorable light, while ignoring negative data to enhance stock value. Much of this information is murky in origin, open to interpretation, and certainly gist for corporate lawsuits against those making such assertions. One such entity is the "stealth" clinical diagnostic company, Theranos, that has been the subject of numerous articles in the mainstream media describing the lack of transparency in its technology. Historically, Theranos submitted very little, if any, of its purported breakthrough technology for peer review (Bilton, 2016a; Carreyrou, 2015; Ioannidis, 2015) with much of their research apparently being "doctored," having failed quality-control checks (Bilton, 2016b; Carreyrou, 2016). Indeed, Theranos is reported to have actually set up a secret company to buy commercial clinical devices in order to run "fake 'demonstrations tests' for prospective investors and business partners" while pretending to showcase its own technology (Weaver, 2017). As a result, the US Centers for Medicare and Medicaid Services, in addition to revoking the license of the company to operate a clinical testing laboratory in California because of unsafe practices that posed an "immediate jeopardy to patient health and safety," banned the company's founder from the blood-testing business. But only for 2 years: a questionably lenient penalty (Flam, 2016; Scott, 2016) that may be superseded by numerous investor lawsuits (Kossoff, 2017).

Because of the risk of litigation, many questionable research findings have sat in the publication cloud, unmolested, despite unspoken (and sometimes open) criticism, and skepticism regarding their provenance. Some of these articles will have been unfairly maligned while others that are truly fraudulent may have maintained a baleful influence. This miasma has ameliorated only somewhat with the advent of the website, *RetractionWatch* (http://retractionwatch.com).

1.6.3 The Impact of the Internet on Publishing and Disseminating Information

With the advent and spread of the Internet, the process of scientific publication has irrevocably changed, becoming faster, more transparent, and interactive. Identifying and recruiting reviewers has similarly become a rapid, semiautomated process. This has allowed the *CPR* process to accept papers for publication within days of submission and their publication online within days of acceptance. Software development has enabled the emergence of *postpublication peer review* (PPPR; Chapter 5.3.2.3) and *postpublication commentary* (*PPC*; Chapter 5.3.2.4). These, while well intended, are insufficiently well curated such that they enable aspects of "vigilante science" that serve no useful purpose (Blatt, 2015). PPPR and *PPC* are thus highly vulnerable to "'trolling'—the posting of disruptive or malicious comments" (Stoye, 2015). PPPR and *PPC* have had limited impact despite extensive proselytization by their advocates that they are a quantal improvement compared to the traditional peer review process.

Nevertheless, the Internet publishing model, together with the proliferation of research papers submitted and published, contrasts markedly with the era of paper publishing with its relatively slow peer review process and perhaps, a greater and more thoughful degree of due diligence. Internet publishing has also contributed to more papers having a shorter "shelf-life" with many rapidly disappearing "without trace" in the virtual sea of the 1.4–2 million papers that are indexed in PubMed *each year*.

Similarly, the large number of journals with low levels of peer review standards commensurate with their low, or nonexistent Impact Factor offers multiple alternative forums for dissemination such that a submitted manuscript will eventually find a journal that will accept it, a trend in publication described by Peres-Neto (2016) as "where rather than if." Separately there are the predatory journals, many of which have published nothing, and which lurk in cyberspace, awaiting the naive researcher only too happy to be parted from their grant money, especially with a publication at the end of the interaction. These journals are of dubious merit for authors who use them who are, as noted, naïve or work in an environment (possibly one involving the scientific culture of a complete country; Economist 2013b) that lacks the appropriate ethical standards and judgment necessary to recognize their lack of value. Although many researchers working in institutions in countries with a long and distinguished record of research publication may regard the predatory platform as irrelevant to their activities, the fact that these journals exist and have proliferated attests to the existence of a market, and one that is growing exponentially. An eightfold growth in the number of papers published in the predatory literature over the period 2010–14 (Shen and Björk, 2015) has increased the contribution of this sector to an estimated 20% of the total published scientific literature in 2015 (Ware and Mabe, 2015).

1.6.3.1 High and Low Profile Journals

Another component of the reproducibility problem is the large number of papers published in the biomedical research literature that are rarely cited; instead they are "sitting in a wasteland of silence, attracting no attention whatsoever" (Davis, 2011). As a result, the findings reported in these papers are rarely subjected to independent verification. This may be because they reflect a research area that is a scientific backwater or because they are published in a relatively obscure journal (Schmidt, 2014). On the other hand, it may be because they are too far ahead of their time or because they show a hypothesis to be false and thus terminate an interesting and easily fundable avenue of research. These explanations do not imply that there is anything wrong with the work. However, a lack of independent verification may be because the research community assumes the data are false for unknown reasons and choses to ignore the publication. This assumption may or may not be correct and may be the result of the journal itself having a low Impact Factor (an example of the publishing sector vicious circle), resulting in a readership bias. In contrast, papers that are published in high profile, high Impact Factor journals like *Cell*, *Nature*, or *Science* attract interest and have inherent "credibility" owing to the "marque" of the journal, so readers are more inclined to assume that the findings are likely to be correct. These inspire attempts to reproduce the studies which, when they cannot be reproduced, immediately have a high profile which is why high Impact Factor journals have high retraction rates. Fang and Casadevall (2011) have reported a positive correlation between the journal Impact Factor, the visibility of a journal, and the number of retractions. This may seem counterintuitive, since one would expect the high Impact Factor journals to be publishing papers that are more likely to be correct although their peer review mechanisms appear no better than those in lower impact journals. Instead, high profile journals are more likely to publish cutting edge and impactful research—research that attracts more attention—and, consequently more attempts to reproduce the findings than research published elsewhere (Bishop, 2012; Fang and Casadevall, 2011). This is a perception facilitated by the public relations efforts of the high impact journals to ensure that their output is avidly awaited and disseminated by the mainstream media, sometimes even before the paper documenting the research is officially published.

The corollary of this is that there is a greater chance of "getting away" with publishing false or fabricated findings, if an author opts for a lower impact journal (and also focuses on more mundane research issues) as few will read the work, fewer will regard it as important, and yet fewer still will attempt to replicate the findings.

1.7 THE LITERATURE ON REPRODUCIBILITY IN BIOMEDICAL RESEARCH

1.7.1 An Age-Old Concern

Concerns related to reproducibility in biomedical research are far from new, despite the apparent surprise (Economist, 2013a) that greeted the papers from Prinz et al. (2011) and Begley and Ellis (2012). For example, the reproducibility of published findings formed part of the feasibility assessment for initiating a new project, and was *de rigueur* throughout the pharmaceutical industry for many decades before the Prinz et al. (2011) paper was published.

When original research findings could not be reproduced, it was a standard operating procedure to contact the original authors to resolve any potential oversights or methodological disconnects. This discussion often led to a satisfactory resolution of the problem by changing reagents or protocols or clarifying procedures. When the cause of failed reproducibility remained elusive or unresolvable, the outcome of the study which failed to reproduce the original findings was very rarely published but its existence became well known in both academia and industry by way of informal research networks (chatter and gossip), long before the advent of the Internet. Thus, when the research sector was small, controversy was rarely a secret, and the chance of unverifiable findings being accepted as fact was lower than it is today—in large part because the number of scientific publications was only a fraction of today's output.

1.7.2 Concerns in the 21st Century

Prior to the Prinz et al. paper, concerns with data reproducibility had been expressed by Ioannidis (2005) in his seminal article "Why most published research findings are false" on experimental reproducibility that addressed the probability that the majority of findings in research were false positives and has been accessed well over a million times. This article was discussed in two articles in the mainstream media, one entitled "Lies, Damned Lies, and Medical Science" (Freedman, 2010) in the *Atlantic Monthly* and the other, "The Truth Wears Off" (Lehrer, 2010) in the *New Yorker*.

The Freedman piece focused on the metaresearch being conducted by Ioannidis and his colleagues discussing many types of bias (including data selection and data analysis, exciting rather than plausible theories, etc.) that distorted research outcomes and led to the publications that were "pervasively flawed…misleading, exaggerated, and often flat out wrong" and that frequently contradicted previous reports in the peer-reviewed scientific literature.

The Lehrer report also cited the contributions of Ioannidis et al. to the reproducibility debate extending the discussion to the phenomenon known as the "decline effect" which has been discussed in detail (Section 1.4.3). This term had been originally used in the psychological sciences to describe how early reports of evidence for significant extrasensory perception (ESP) in experiments on the phenomenon of "psychic" powers and has since been used as a general term to describe a situation where initial experimental results are highly impressive, for example, significant at $p < 0.05$, etc., but with time become less significant as investigators attempt to reproduce the original findings.

Thus, the novelty in the Prinz et al. (2011) paper was its specific focus on *numerous instances* of a failure to reproduce findings from activities conducted in a single research center that was dedicated solely to target validation as well as its high profile and graphically colorful venue, *Nature Reviews Drug Discovery* (Jarvis and Williams, 2016).

An inevitable question, given the widespread response—and, in some instances, outrage—to the Prinz et al., and Begley and Ellis papers, is whether scientific misconduct is on the increase or simply reflects the fact that the means to detect its occurrence has become more efficient via the Internet. Data suggest that the approximate 1% incidence rate for scientific misconduct in published articles has remained unchanged for more than a decade (Blatt, 2015).

1.7.3 Reproducibility in the 21st Century: Origins, Scope and Momentum

1.7.3.1 Why Most Published Research Findings are False (Ioannidis, 2005)

Most published findings in biomedical research are positive, reflecting a publication bias in favor of authors submitting, and reviewers accepting, positive versus negative findings (Dirnagl and Lauritzeb, 2010; Fanelli, 2012; Ioannidis, 2005; Ioannidis and Trikalinos, 2005; ter Riet et al., 2012; Song et al., 2013).

In this article, Ioannidis identified a number of factors that contribute to the false positive rate including: the statistical power of a study; the level of statistical significance; bias as represented by the selective or distorted reporting of data (described by Babbage in the 19th century as "trimming"; Gross, 2016); prejudice and financial interests; and effect size where the tested relationships were greater in number and had not all been preselected as endpoints. As a result, this leads to a greater flexibility in "designs, definitions, outcomes, and analytical modes" which can be interpreted in terms of experiments becoming infinitely variable—changing in their substance and endpoints from the conceptualization of the study to its interpretation for a variety of reasons including a lack of rigor and/or knowhow and/or bias/honesty.

Couched in language and equations that, despite their intended transparency, were challenging to the statistically naïve, especially those unfamiliar with the concepts of Bayesian statistics, Ioannidis' paper focused on the aspects of statistical issues, particularly the relevance of the "ill-founded strategy of claiming conclusive research findings solely on the basis of a single study assessed by formal statistical significance, typically for a *p*-value less than 0.05." With the further revelation that "Research is not most appropriately represented and summarized by *p*-values" many biologists who were taught the infallibility of the *p* value understandably became nervous and confused by this, and subsequent papers from the prolific Ioannidis. Whether data are sufficiently "good" to justify that the *p* value is a different issue and is discussed further in Chapter 2 along with the *null hypothesis.*

1.7.3.2 Believe it or not: How Much can we Rely on Published Data on Potential Drug Targets? (Prinz et al., 2011)

Prinz et al. (2011) was the first of the two papers from the pharmaceutical industry that served as catalysts to highlight concerns regarding reproducibility. It originated from the German pharmaceutical company, Bayer HealthCare and reported that 65% (43/69) of findings published in the literature identifying targets in the areas of oncology, women's health, and cardiovascular disease could not be reproduced when subjected to internal validation efforts. This evaluation was extended to an internal Bayer survey of 23 colleagues across different scientific disciplines, the results of which indicated that only 20%–25% of the published data used by the Bayer project teams to validate projects could be reproduced in house. While citing an "unspoken rule" in venture capital circles (https://lifescivc.com/2011/03/academic-bias-biotech-failures/) that "at least 50% of published studies, even those in top-tier academic journals, can't be repeated with the same conclusions by an industrial laboratory," Prinz et al., did not conclude that the original findings were incorrect or ambiguous due to deficiencies in the reported experimental parameters (e.g., cell line authenticity, assay formats). Rather they emphasized inappropriate experimental design and analysis (deficiencies in null hypothesis testing, inherently low prestudy probabilities of observed results actually being true), associated with small sample sizes that reflected the pervasive "publish or perish" scientific environment, as the primary contributing factors. The original papers also appear to have overlooked the almost ubiquitous lack of blinding and randomization in the preclinical research area (Hirst et al., 2014; Chapter 2). Of additional note given the high profile of the Prinz et al., report, the data on which it was based did not identify the specific targets involved thus making the findings uncorroboratable while the primary publications being assessed for reproducibility were also not identified (Jarvis and Williams, 2016), making it impossible to verify the findings.

1.7.3.3 Drug Development: Raise Standards for Preclinical Cancer Research (Begley and Ellis, 2012)

A second industry report on reproducibility originated from the Californian biopharmaceutical company, Amgen and the MD Anderson Cancer Center in Houston, Texas, a major US cancer research institute. This paper focused exclusively on "landmark studies" in preclinical cancer research. Of these, the findings in only 6 of these studies in a cohort of 53 could be reproduced in house. While this was concerning in itself, the responses received in following up with the original authors generated additional, if not greater, concern. Thus, Begley and Ellis stated that "To address these concerns, when findings could not be reproduced, an attempt was made to contact the original authors, discuss the discrepant findings, exchange reagents and repeat experiments under the authors' direction, occasionally even in the laboratory of the original investigator. These investigators were all competent, well-meaning scientists who truly wanted to make advances in cancer research. In studies for which findings could be reproduced, authors had paid close attention to controls, reagents, investigator bias, and describing the complete data set. For results that could not be reproduced, however, data were not routinely analyzed by investigators blinded to the experimental versus control groups. Investigators frequently presented the results of one experiment, such as a single Western-blot analysis. They sometimes said they presented specific experiments that supported their underlying hypothesis, but that were not reflective of the entire data set. There are no guidelines that require all data sets to be reported in a paper; often, original data are removed during the peer review and publication process." In a subsequent meeting "with the lead scientist of one of the problematic studies" that was summarized by Begley; "We went through the paper line by line, figure by figure," ..... "I explained that we re-did their experiment 50 times and never got their result. He said they'd done it six times and got this result once, but put it in the paper because it made the best story. It's very disillusioning" (Begley, 2012). Begley provided further background detail on the Amgen reproducibility efforts to Harris (Begley, 2017b) noting that "On about twenty occasions we actually sent [company] scientists to the host laboratory and watched them perform experiments themselves." These were blinded studies that "most of the time ...failed."

The general conclusions in the Begley and Ellis paper were twofold: (1) that reproducible research tended to reflect a high level of attention to controls, reagent quality, and descriptions of complete datasets and; (2) research that could not be reproduced often lacked detailed descriptions of experimental methodologies and involved data sets that were representative in nature precluding testing of reproducibility due to vagueness.

As in the Prinz et al. (2011) report, the data in the Begley and Ellis paper were again largely unverifiable with none of the original papers being identified due to the execution of confidentiality agreements with some of the authors of the original reports (Nature, 2012), a situation that had not been viewed favorably (Gorski, 2012; Jarvis and Williams, 2016). This concern however can be viewed as a little disingenuous. Unless one can argue that Prinz, and Begley and Ellis fabricated their findings, which is unlikely, it is a little harsh to criticize them. After all, any criticism of this kind would argue that any study in which the identity of the subject matter or materials is anonymized must be regarded as illegitimate. Since the anonymity of human subjects is *de rigueur* in all ethically designed and reported clinical trials, then the acceptance of such criticism would argue for the abandonment of the current model for clinical research—which some have advocated (Grove, 2011)—a viewpoint that has garnered little in the way of support (Gorski, 2011; Lowe, 2011).

1.7.3.4 Reproducibility: An Academic Viewpoint

Prompted by the Prinz et al., and Begley and Ellis papers, Mobley et al. (2013) conducted an anonymous survey of academic researchers at the MD Anderson Cancer Center to determine the frequency and potential causes of nonreproducible results. From this survey, they reported that the problem of data reproducibility was well known in academia with some 50% of respondents having encountered at least one incident of irreproducibility in their career up to the date of the survey. Many were unable to identify its cause. When contacting the original authors to resolve their findings, responding researchers were met with "a less than 'collegial' interaction" from the original authors of whom "almost half responded negatively or indifferently." While individual responses to the survey were provided, the data in this paper were somewhat unrepresentative with only a 16% (434/2692) response rate (Section 1.12.1).

1.7.3.5 *Reproducibility in the Psychological Sciences*

An additional milestone in the evolving literature was the Reproducibility Initiative (RI) conducted by the *Open Science Collaboration* (OSC, 2015), a large-scale collaborative effort over the period November 2011–December 2014, to estimate the reproducibility of psychological science findings. This study, also known as *Reproducibility Project: Psychology* or *OSC2015* set the goal of replicating (e.g., reproducing) the findings from 100 published papers in the discipline of psychology that were published from 2008 onward in three leading journals in the field, *Psychological Science, Journal of Personality and Social Psychology,* and the *Journal of Experimental Psychology: Learning, Memory, and Cognition.* Reproducibility was assessed on the basis of significance (reported p values), effect sizes, subjective assessments of expert teams, and metaanalyses. While the reader is directed to the original report for full details (OSC, 2015) and discussion in Chapter 5.12–5.14, the salient outcomes of the initiative were that of the 97% of the original studies that reported significant results ($p < 0.05$), only 36% of the repeated studies obtained significance. Thus 65% of the findings could not be reproduced (according to the impartial arbiter of statistical significance) even though some of the original authors were involved in the attempted reproduction effort. This appeared to indicate that reproducibility in the psychological science is low (Gibson, 2012; Gilbert et al., 2016; Stroebe et al., 2012) and led to the conclusion that statistically significant outcomes have an extraordinarily poor level of credibility given their poor level of reproducibility. There is no agreed reason for this, with flaws in design, analysis and basic scientific knowledge, and know-how all identified as culprits. This is discussed further in Chapter 5.12.

1.7.3.6 *Researcher Awareness of Fraud*

Four years before the Mobley report, Fanelli (2009) had conducted a metaanalysis of 21 surveys that had asked scientists about their experiences related to misconduct in science. The results—which Fanelli characterized as "conservative" given the sensitivity of the questions being asked—led to approximately 2% of scientists to admit having "fabricated, falsified, or modified data…at least once" with another 34% admitting to "questionable research practices." Of additional relevance was that 3 years prior to the Fanelli study, in 2006, the NIH Office of Research Integrity (ORI) had funded a Gallup survey looking into research integrity. The final version of the survey results was not released until 2008 and "estimated that 1.5% of all research conducted each year would be fraudulent," (Wells, 2008) consistent with Fanelli's later findings. Given that these findings were based on admission of guilt, this may be the tip of yet another iceberg.

Ten years later in a subsequent paper, *How to Make More Published Research True*, Ioannidis (2014) proposed that false positives in the biomedical literature would diminish if there was an increase in collaborative research, and an improved culture of reproducibility and replication with improved statistical methodology and user literacy with more stringent statistical thresholds, and improved study design standards for peer review, reporting, and dissemination of research. Added to this was the elimination of "nonmeritocratic practices" (Ioannidis, 2014) also known as "perverse incentives," improvements in the training of the scientific workforce and modifications in the reward system for science, with the latter being increasingly identified as dysfunctional and in need of restructuring (Alberts et al., 2014; Begley and Ioannidis, 2015; Ioannidis and Khoury, 2014; Kraus, 2014).

In addressing necessary changes in the historically nepotistic grant-funded ecosystem (Ioannidis, 2014), the adoption of which may take many years, especially when citation-based incentives uniformly select for bad science (Smaldino and McElreath, 2016), the absence of clear consequences and incentives to change entrenched behaviors (Fishburn, 2014; Nardone, 2008) represents a major shortcoming that no amount of wishful thinking (or additional grant funding) will overcome. In yet another paper, Ioannidis (2016a) distinguished clinical from preclinical or discovery research, terming the latter as "blue-sky" and "speculative" that unlike clinical research lacks prespecified deliverables but is nonetheless valuable. This is an important distinction as both the psychological and clinical sciences are often viewed as benchmarks against which preclinical research and its reproducibility should be marked. This has led to the viewpoint that preclinical experimentation should seek to emulate the gold standard of randomized clinical trials (RCTs; Henderson et al., 2013; Muhlhausler et al., 2013), another issue that is discussed in Chapter 2.

It is worthwhile noting that preclinical research is not always "blue sky" or exploratory, a quality that is certainly absent from the preclinical research dossier collated to support an IND (McGonigle and Williams, 2014), but should include properly powered, confirmatory studies. Finally, a point worth repeating *ad infinitum* and which is discussed in Chapter 2 is that the design of the experiment is a key to a good outcome. Deciding how to analyze an experiment *after* it has been completed is bad science (Chapter 3). Harking, hypothesizing after the results are known (Kerr, 1998; Motulsky, 2014), where multiple hypotheses are considered after review of the data and one is selected as though it were the basis for the study, is another source of irreproducibility since it not only introduces bias, but fails to account statistically for the multiple comparisons.

1.8 IS THERE A REPRODUCIBILITY CRISIS?

Approximately 4 years after the initial flurry of papers on reproducibility, a *Nature* survey involving 1,576 researchers found that more than 70% of respondents had tried and failed to reproduce an experiment (Baker, 2016b). And of this albeit self-selected and small cohort, 52% noted there was a "significant" crisis with another 38% noting a "slight" crisis, a view that is perhaps more in line with Gorski's, that "Reproducibility is… a chronic problem…. not a crisis" (Gorski, 2016). In responding to a question as to the cause of problems with reproducibility, more than 60% cited the pressure to publish and selective reporting which one respondent ascribed to the need to compete for grants coupled with an increased bureaucratic burden that "takes way from time spent doing and designing research." This is interesting given 21st century expectations for biomedical researchers, summarized by Faulkes (2016): "When I was in grad school, I had to write a paper and publish it. Now, people are suggesting that I also pre-register my experiments; curate and upload all my raw data (which may be in non-standard or proprietary formats); deposit pre-prints; publish the actual paper in a peer-reviewed journal (because that's not going away); promote it through social media; upload it into sites like Academia.edu or ResearchGate; update my publication information in databases like ORCiD, ImpactStory, and institutional measures; and watch for comments on post-publication peer review sites like PubPeer and engage with them as necessary."

Among the approaches suggested to improve reproducibility were: "more robust experimental design; better statistics; and better mentorship." Journal checklists (Chapter 5.11) received a 69% endorsement while 80% of respondents thought "funders and publishers should do more to improve reproducibility." This indicates that researchers are asking for help and lacking in the initiative and necessary know how to ensure that their own research is adequate—again reflecting disconnects in their training and mentoring.

1.9 TROUBLE AT THE LABORATORY?

An additional source of accessible, expert, and at times, provocative commentaries on topics related to reproducibility that include scientific publishing, peer review, and the integrity of the clinical trial process are certain publications in the mainstream media. These include the *Guardian* newspaper in the United Kingdom, and the *Economist, New York Times, Wall Street Journal, Forbes, Atlantic Magazine, Wired, Slate,* and *Vox*. Setting the tenor of these contributions was Colquhoun's frequently cited polemic in the *Guardian*,"Publish-or-perish: Peer review and the corruption of science" (Colquhoun, 2011) along with commentary by Goldacre in his book, *Bad Pharma* (Goldacre, 2012) and in his column/blog, *Bad Science* (http://www.badscience.net) that focuses on shortcomings in clinical trial execution in the pharmaceutical industry (with an often devastating and journalistically entertaining bottom line). Concomitant with this coverage has been an explosion of articles "researching research" (Ioannidis, 2016b) also termed metaresearch or metascience (Munafò et al., 2017) facilitated by both alternative (*Public Library of Science (PLoS), PeerJ, eLife*) and traditional (*Nature, Science, Cell, PNAS, Lancet, etc.*) publishing venues as well as raw data access in formal databases. These have further highlighted reproducibility problems, and have proposed remedies—including the various *Reproducibility Initiatives* (Morrison, 2014; OSC, 2015; Van Noorden, 2014) and checklists for journals and grant submissions (Jarvis and Williams, 2016; Nature Editorial, 2016).

Additionally, there has been considerable discussion of the role of peer review, contested, long standing, still unresolved, and perhaps irresolvable (Colquhoun, 2011; Dzeng, 2014; Horrabin, 2001; Horton, 2000; Rennie, 1986; Rennie et al., 2003; Schekman, 2013; Smith, 2006; Souder 2011; van Dalen and Henkens, 2012), a topic that is covered further in Chapter 5.2.

1.10 RETRACTIONS

When a previously published effect cannot be repeated resulting in a *negative finding,* historically it has been very difficult—if not impossible—for a researcher to find a peer reviewed forum where this work will be accepted for publication. This is a consequence of the widespread bias in the peer review process toward positive findings (Buntin et al., 2011; Dirnagl and Lauritzeb, 2010; Fanelli, 2012; Ioannidis, 2005; Matosin et al., 2014; Pusztai et al., 2013; Song et al., 2013; ter Riet et al., 2012; Tsilidis et al., 2013; Young et al., 2008). Publication of contrary findings, a correction, or a retraction in the original journal in which the flawed finding was published provides a logical opportunity to be "complete and honest, and clearly articulate(s) what things…wrong" setting the record straight, making the investigator whole and avoiding the issue of citation penalties (Lu et al., 2013). When the originating author is not the source of the subsequent failure to reproduce, the publication of the failed attempt to reproduce the finding can provide an opportunity for open discussion between interested parties and to avoid other researchers investing their time and resources trying to reproduce erroneous findings (Gewin, 2014). However, negative results have rarely been considered interesting or worthy of publication and certainly do not add to the "grant worthiness" of a biomedical researcher, except when it is related to a groundbreaking discovery and the *Proteus phenomenon* (Section 1.5).

The number of retractions—the public withdrawal of a published research paper—has increased markedly over the past decade (Boston, 2015; Steen, 2011a,b; Steen et al., 2013; Van Noorden, 2011), although it is still relatively small with estimates that vary from 1 in 10,000 (van der Vet and Nijveen, 2016), 1.4 per 100,000 (Marcus and Oransky, 2014) and "a tiny fraction of all published papers, perhaps several hundred of the one million published yearly" (Flier, 2017). While the majority of retractions were thought to be attributable to honest mistakes (Steen, 2011a), some two-thirds involve active misconduct that includes plagiarism, image manipulation, and other types of fraud leaving only 25%–33% being characterized as "honest mistakes" (Fanelli, 2009; Fang et al., 2012; Gewin, 2014).

Retraction, self or imposed, for example, a journal expression of concern or an outright rejection imposed by an Editor, is viewed as "the worst outcome of publication" (Souder, 2011). Thus, authors are not motivated to retract research papers because of the potential impact on their tenure, funding prospects, or prestige (Marcus and Oransky, 2014). In such instances, other scientists can express concerns via open blog sites like *PubPeer*. The lack of peer criticism on these sites however, leads to what has been termed "vigilante science" (Blatt, 2015) and dismissed accordingly. Concerned researchers and the editors of scientific journals can contact the authors and/or their institution when issues occur regarding papers with data that is suspected to be fraudulent, plagiarized, or cannot be reproduced, three very different degrees of concern. Taking these approaches, resolution can be achieved by the original journal publishing an expression of concern regarding a study or, when they have satisfactorily investigated a particular case, unilaterally retracting the paper.

When the authors and/or their institution are intransigent (Marcus and Oransky, 2014), negotiations can often become protracted and may involve legal complications (Lowe, 2011; *RetractionWatch*, 2016). This has consequences, especially for the researchers involved since it may result in loss of citations, not only for the retracted work but also the prior work of the "wronged" author (Lu et al., 2013) in the case of refutation or plagiarism. This is an occurrence that has been described as being "consistent with the Bayesian intuition that the market inferred their work was mediocre all along" (Azoulay et al., 2015).

1.10.1 *RetractionWatch*

A major resource in documenting irreproducible research, much of it leading to retractions of suspect papers that neither the authors nor the journals were motivated to do, is the website *RetractionWatch* launched by The Center for Scientific Integrity in 2010 (Carey, 2015; Gewin, 2014; Marcus and Oransky, 2011, 2014; Steen, 2011a,b).

RetractionWatch reports almost on a daily basis, new instances of suspected plagiarism, image manipulation, and fraud and the ongoing efforts to resolve these, all of which represents compulsive reading that is avidly covered in the mainstream media (Boston, 2015;

Carey, 2015). *RetractionWatch* has also collated a list of the more egregious offenders, the *Retraction Watch Leaderboard* (http://retractionwatch.com/the-retraction-watch-leaderboard/), where scientists receive scores for their retractions, and also publishes a yearly *Top Ten* list of offenders (Marcus et al., 2015). At the time of writing (Spring of 2017), two anesthesiologists, Fujii and Boldt, occupy the top positions on the Leaderboard with a combined 277 retractions. The retraction scores for these individuals were: Yoshitaka Fujii from Japan (Carlisle, 2012; Marcus and Oransky, 2015) leading with 183 retractions and Joachim Boldt from Germany (Wise, 2013) with 94. They were followed by Peter Chen, a Taiwanese physicist with 60, Diederik Stapel, a social psychologist from the Netherlands with 58 retractions (Stroebe et al., 2012; Verfaellie and McGwin, 2011), and Shigeaki Kato, a Japanese endocrinologist with 38 retractions. Another anesthesiologist, Scott S. Reuben (Bornemann-Cimenti et al., 2016; Borrell, 2009) from the United States was listed at position 9 with 25 retractions and was one of the unusual instances where a jail term was the outcome for his misdeeds (Kornfeld and Titus, 2016). Conclusions based on the data contained in the combination of retracted papers have had a major impact on the practice of anesthesia and also the postoperative treatment of surgically-related pain. It has been estimated that Rueben's fraudulent papers affected the treatment of millions of patients, and facilitated the sales of prescription COX-2 NSAIDs to the tune of billions of dollars (Borrell, 2009).

In a derisive letter to the Editor, Kranke et al. (2000) had noted in the title how "incredibly nice!" the data in Fujii's papers were. A belated *posthoc* analysis of a set of these papers indicated that the odds of some of them being actually experimentally derived were in the order of 10^{-33} which was viewed as "a hideously small number" (Carlisle, 2012). Despite the Kranke et al., letter, the 48 articles cited in the letter were not retracted and Fujii was able to publish another 11 papers in the same journal in the ensuing 10 years before a new editor took appropriate action (Stroebe et al., 2012).

Salient questions to explore are why the field of anesthesia is so susceptible to egregious fraud and numb to its consequences and whether the incidence of fraud in the area has increased in recent years. In considering the latter question, the editor of one of the journals involved, Shafer (2011) argued that the serial frauds perpetrated by Boldt and Reuben were far from new and that their personal misconduct dated back for over a decade, but failed to address this question. What was new however was the discovery which was facilitated by access to multiple sources of information available via the Internet. Conversely, others have wondered whether the field of anesthesiology is more vulnerable to research fraud because the individuals who self-select for this medical specialty may be more solitary by nature and receive less peer oversight than other medical subdisciplines.

As discussed further in Chapter 5.2, the peer review process has not always kept pace with changes in the scientific culture with an antiquated, entrenched code of conduct (Ioannidis, 2014) more appropriate to a bygone era than the cultural and ethical mores of the 21st century (Redman, 2015). The latter have been described in terms of a "Cheating Culture" (Callahan, 2004) that reflects "Norms [that] may arise within an organization that give implicit permission for unethical misconduct. A cheating culture exists when enough people are breaking the rules that there is a perception that 'everybody' is corrupt and there is no clear imperative for ethical behavior. In extreme instances, there may be the belief that one cannot be competitive by following formal rules and that cheating is the key to success" (Stone, 2005), analogous to the arguments made in sports by athletes who take performance enhancing drugs. This would suggest that the necessary trust in others which has been assumed to be the core of scientific behavior and peer review (Kraus, 2014) has disappeared with the "trust me model that is no longer considered appropriate in corporate life nor in government" being extinct (Begley and Ioannidis, 2015).

Another variable deemed to be a major cause of fraud is a lack of basic rigor, competence, and expertize in the peer review process that is reflected in the finding that many of the retracted anesthesia papers lacked basic documentation including the mandatory Institutional Review Board (IRB) approval necessary to conduct studies in humans thus removing them from the administrative and peer physician radar.

1.10.2 Continued Citation of Retracted Publications

Retraction of inaccurate or fraudulent publications does not resolve the perpetuation of the inaccuracies that continue to be used and cited in secondary studies by other researchers. Of 82 retracted articles published between 1973 and 1987, there were 733 citations occurring more than 6 months after their retraction, a fall of only 18% compared to the preretraction period, with only 2.9% citations mentioning the retraction (Pfeifer and Snodgrass, 1990). Similarly examining 235 retracted articles from the period 1966 to 1996 that received 2,034 postretraction citations, only approximately 7% acknowledged the retraction, while all other citations used them with positive connotations (Budd et al., 1999). Moreover, most citations occurred in the Introduction or Discussion sections of the secondary publications, inferring that the retracted studies impacted the hypothesis being tested or the interpretation of the results (Budd et al., 1999).

Such studies make clear that retracted publications are not being identified appropriately and removed from the literary canon, and that their continued use could contribute to issues of reproducibility. A long-standing concern has been the lack of transparent information on retracted works, and inconsistencies in the format (Pfeifer and Snodgrass, 1990), while there had been no reliable source of such information. Even as the *RetractionWatch* Leaderboard (http://retractionwatch.com/the-retraction-watch-leaderboard/) notes, there are inconsistencies in the number of retractions between different sources, and a Medline search of 233 retracted articles revealed that 22% gave no mention of the retraction (Decullier et al., 2013). While steps to improve identification of retracted articles are improving, it is taking longer to issue retractions (Fang et al., 2012), perhaps due to a lengthy investigative process, after which it can take up to 3 years for retraction notices to appear on *PubMed* (Decullier et al., 2014).

It is particularly troubling when citation of retracted studies is used to justify clinical studies and place patients in harm's way. Evaluating 180 retracted clinical papers, Steen (2011c) found 851 citations that induced prospective clinical trials treating patients. While over 28,000 patients were enrolled and 9,189 treated in studies directly related to the retractions, some 400,000 and 70,501 patients were enrolled and treated, respectively, in secondary studies "that drew ideas or inspiration from the primary study." While some patients were likely harmed, it is difficult to discern how many patients were compromised after the original study was retracted, since the design and execution of clinical trials takes significant time. Steen (2011c) cites one particular example where studying the combination of chemoembolization and radiofrequency ablation in patients with liver cancer that was found to be fraudulent (Cheng et al., 2008), still spawned 144 publications on the faulted procedure after the retraction was published, even if they did not all cite the Cheng article. This demonstrates the insidious nature of retracted studies. They can perpetuate erroneous science placing patients at risk, through secondary continuation of bogus ideas that are hard to trace back to their fraudulent beginnings but become entrenched in the literature.

1.10.3 The Spectrum of Irreproducibility

The spectrum of irreproducibility is the product of behaviors that extend from calculated, often serial, fraud, termed pathological (Pulverer, 2016) at one end to the "slippery slope" of erroneous design and analysis that are the product of ignorance or a misguided or poorly thought through hypothesis at the other (Curtis et al., 2015; Mullane et al., 2015). Whether such behaviors constitute fraud depends on whether the author deliberately did things they knew were wrong. Unfortunately, the degree of ignorance and hubris in the sector means that the incidence of malpractice that is not strictly speaking fraud is probably immense.

1.10.4 Research Misconduct

A metaanalysis of 146 reports of scientific misconduct from the US Federal Office of Research Integrity (ORI) that covered the period 1992–2003, by Kornfeld (2012) concluded that individual research misconduct could be ascribed both to environmental factors—pressure to publish and perverse incentives (Alberts et al., 2014; Ioannidis, 2014) that included academic and/or financial rewards—and individual psychological traits with the latter being divided into the following categories:

- *the desperate*, whose fear of failure overcame a personal code of conduct;
- *the perfectionist*, for whom any failure is a catastrophe;
- *the ethically challenged*, who succumb to temptation;
- *the grandiose*, who believe that their superior judgment did not require verification; and
- *the sociopath*, who is totally absent a conscience.

Scientists both in academia and industry can be sorted into these aforementioned categories, with the caveat that given the multiple interactions and interdependencies of applied research and its deliverables, there are far more checks and balances and peer and management oversight in industry (as compared with academia) that can lead to the rapid identification of research misconduct resulting in more tangible consequences, for example, demotion, reassignment, or termination for cause.

1.10.5 Fraud

Issues with fraud in biomedical research were the subject of US Congressional hearings as far back as 1981 when key witnesses, described as "über establishment" figures in the biomedical research community, generated consternation when they described the problem of scientific fraud as being "exaggerated" and "not a matter of general societal concern." This led the Chairman of the hearings, one Congressman Albert Gore (of climate change, hanging chads and Nobel Prize fame) to express perplexity since the community spokespersons considered "the problem of fraud.... a private one.... that should be dealt with by informal codes of the scientific community... and was not an important ethical problem.. [to].. worry those...charged with the public trust" (Gross, 2016). While this attitude is still a prominent one evidenced by many researchers in biomedical research (who fiercely defend their perceived independence), and their institutions, it is not shared by Begley, Ioannidis, the anonymous sages at the *Economist*, or two NIH scientists, Walter Stewart and Ned Feder. Stewart and Feder who were harassed by the hierarchy at the NIH for their investigation of "scientific misdeeds" until they ultimately left. The reader is referred to Gross' engaging and insightful accounts of this and examples of scientific misconduct in the 1970s and 1980s for additional detail (Gross, 2016).

In defining scientific fraud as "serious misconduct with intent to deceive.... the very antithesis of ethical behavior in science," Goodstein (1995) presciently noted that because of increasing competition for scarce funding resources, fraud in science would increase. This viewpoint was revisited by Triggle and Miller (2002) and some 2 decades later by Alberts et al. (2014) and documented by Steen et al. (2013) and others. Goodstein further stated that while "Science is self-correcting....[since fraud]...will eventually be discovered and rejected..[this].. does not protect us against fraud, because injecting falsehoods into the body of science is never the purpose of those who perpetuate fraud." The expectation that science continues to be effectively self-correcting and, as a result, makes fraud pointless, illustrates the difficulty facing "meticulous scientists" (Lemaitre, 2016), who describe themselves as "rigorously honest," making it difficult for them to conceptualize why others, especially those individuals documented in the following sections, would engage in serial fraud that destroys their careers.

1.10.6 Notable Examples of Fraud—Biomedical Researchers Behaving Badly

When articles are retracted in the area of biomedical research on the basis of demonstrable fraud, they are often subjected to deep investigation and hindsight. This has the potential to serve as a helpful catalyst for efforts to find practical solutions to rectify the problem (Bartlett, 2015). While there are many instances of fraud documented on the *RetractionWatch* website (http://retractionwatch.com), the following five were selected as landmarks for reasons that will hopefully be self-evident (Table 1.1). In reviewing the literature on fraud in biomedical research, the interested reader will find other egregious examples including that of the cardiologist, Darsee (Kochan and Budd, 1992) who is number 24 in the list of the 30 researchers with the most retractions on the *RetractionWatch* leaderboard with 17 retractions (http://retractionwatch.com/the-retraction-watch-leaderboard/). At the time of his serial frauds, Darsee outraged the scientific community; over 30 years later, his transgressions may be viewed as tame compared with the 23 other researchers above him on the list, a sign of a continuing and progressive lapse in ethical standards? Or the transparency enabled by the Internet?

1.10.6.1 MMR Vaccine: Andrew Wakefield—Royal Free Hospital

Following from an initial paper that proposed a link between the measles virus and Crohn's disease (Wakefield et al., 1993), Wakefield and coworkers subsequently published a paper in *Lancet* in 1998 (Wakefield et al., 1998) suggesting a link between the use of measles, mumps, and rubella (MMR) vaccine and Crohn's disease. The "findings" of this association could not be reproduced and were in time independently refuted by the Department of Health and

TABLE 1.1 Serial Fraudsters

Investigator	Topic	Detail	Societal consequences	Resolution
Wakefield	MMR vaccination causing Crohn's Disease, "autistic enterocolitis, autism"	Irreproducible "bogus data" supporting vaccine-induced autism pathology Conflicts of interest involving litigation against MMR vaccine manufacturers and Newco to develop diagnostics and treatment for "autistic enterocolitis"	Reduction in MMR vaccine use from 92% in 1995 to 84% or less in 2002. Measles outbreaks in England, Wales, and California.	Retraction of original paper. Investigator struck off medical register.
Woo-Suk	Human embryo-derived stem cell lines for personalized therapeutics	Data on 9/11 of cells fabricated with ethical lapses in the collection of human eggs. Fraud, embezzlement of research funds, and bioethics		Suspect papers retracted. Patent issued for NT-1 stem cell line. Investigator dismissed from Seoul National University and now running South Korean Institute making genetically identical copies of dogs for pet owners
Stapel	Human psychology studies	Investigator planned hypotheses, methods, collection, and outcomes of his experiments and then pretended to run the experiments to gather "too good to be true" results. Denied collaborators and students access to fraudulent "raw data."	Led to field of social psychology being viewed as unique in its ability to manipulate/create data to conduct research fraud. "Career-killing behavior" for colleagues and students	58 suspect papers retracted. Investigator dismissed. Questioning of core validity of social psychology research—for example, a science with "fuzzy" endpoints like the "absurd hypothesis that listening to a Beatles song could make you 1.5 years younger". Autobiography—*Ontsporing* (*Derailed*) (2012) reflecting on circumstances and motivation for fraud.

Obokata	Creation of pluripotent STEM cells using in vitro stress manipulation	Within 4 months 133 attempts to replicate in seven labs failed. STAP cells not a genetic match with mice of origin probably "normal" embryonic stem cells investigator plagiarized text and manipulated images in the papers. Extraordinary claims not supported by extraordinary data.	Contribution to suicide of supervisor. Questioning of rigor of peer review at Nature.	Retraction of suspect papers 7 months after publication. Investigator resigned from RIKEN Institute. Authored book *Ano Hi (That Day)* (2016) claiming she was "framed" and implicating mentor in fraud. Set up "STAP HOPE PAGE" website in 2016 with instructions for making STAP cells.
Potti	Microarray analysis of human tumors to derive a drug response signature to predict the patient response to chemotherapy	Forensic bioinformatics assessment identified "careless, inexplicable errors" in Potti findings. Concerns also expressed within Duke by student dismissed as a "difference of opinion."	Microarray drug response signature used to design and conduct clinical trials in 117 patients after concerns raised. *60 Minutes* (March 5, 2012)—"one of the biggest medical research frauds ever—one that deceived dying patients, the best medical journals, and a great university" http://www.cbsnews.com/news/deception-at-duke-fraud-in-cancer-care/.	Investigator resigned from Duke. Resolution of fraud took 7 years. 2016 lawsuit accusing Duke University of engaging in a civil conspiracy compromising clinical trials in cancer patients.

Medical Research Council in the United Kingdom (Siva, 2010) and a Japanese research group (Iizuka et al., 2000). Nonetheless, the unfounded MMR vaccine association was additionally extended to other bowel disorders and to autism with the "discovery" of a new syndrome described as "autistic enterocolitis" (Deer, 2010), an autism that was thought by its discoverers to be worthy of a Nobel Prize (a thought that is unusually high on the list of delusional career aspirations of those engaged in overt scientific fraud). It was also the basis of a speculative lawsuit, the supportive data for which was highly suspect (Deer, 2010). The original Wakefield et al., paper was ultimately retracted in 2010 (Editors of the Lancet, 2010) on the basis of false claims "that children were 'consecutively referred' and that investigations were 'approved' by the local ethics committee." Major conflicts of interest were also identified (Goodlee et al., 2011) that included Wakefield's litigation against MMR vaccine manufacturers (Deer, 2011a) and the founding of a new company to develop diagnostics and treatment for "autistic enterocolitis" (Deer, 2011b). A subsequent retrospective cohort study (Jain et al., 2015) involving 95,727 children some of whom had siblings with what was now termed autism spectrum disorders (ASD) concluded that the "receipt of the MMR vaccine was not associated with increased risk of ASD, regardless of whether older siblings had ASD. These findings indicate no harmful association between MMR vaccine receipt and ASD even among children already at higher risk for ASD."

Despite these refutations of the "findings," the Wakefield paper led to a grass roots anti-vaccination movement that resulted in a reduction in MMR vaccine use, from a level of 92% in 1995 to 84% or less in 2002 reducing herd immunity to a level well below the 90%–95% required to protect the entire population. In 1998, the year that the original autism-association article was published, 56 measles cases were reported in the United Kingdom. A decade later, in 2008, measles had become endemic in England and Wales with 1,348 cases and 2 confirmed deaths (Thomas, 2010; Hiltzik, 2014). A similar measles outbreak in California in early 2015 was also ascribed to a lack of, or incomplete, vaccination (Majumder et al., 2015). Despite the unusual weight of scientific evidence against the MMR-autism association and the fact that Wakefield had been "struck off" as a licensed physician in the United Kingdom, he made a movie entitled "*Vaxxed; From Cover Up to Catastrophe*" that represented his original premise, and was controversially shown at the Tribeca Film Festival in March, 2016 (Hoffman, 2016; Senapathy, 2016).

When a solidly refuted research finding is championed by the uninformed—often celebrities who lack scientific training but who provide an imprimatur of faux credibility to the topic (Tarkan, 2016)—the outcome is often a pernicious cult that is detrimental to public well-being. The fact that the perpetrator is often immune from the criminal prosecution that might be anticipated given the negative impact of the fraud reinforces the belief of acolytes that the perpetrator is right and that the system is wrong despite "the energy, emotion, and money ..[being],.. diverted away from efforts to understand the real causes of autism" (Goodlee et al., 2011) and the associated suffering and death. If they did what they were accused of doing and have not been convicted, how could the accusation have any merit?

1.10.6.2 Embryonic Stem Cells: Hwang Woo-Suk—Seoul National University

A stem cell researcher in South Korea, Hwang published papers in 2004 in which he claimed to have extracted 11 stem cell lines from human embryos that had the potential to be used as personalized therapies (Hwang et al., 2004, 2005). The data on nine of these proved to have been fabricated with ethical lapses in the collection of the human eggs used (Cyranoski, 2004; Cyranoski and Check, 2005). These papers were eventually retracted with Hwang being charged with fraud, embezzlement of research funds, and bioethics violations. Despite this, he has continued his research in stem cell science and currently heads an institute in South Korea that makes genetically identical copies of dogs for pet owners (Cyranoski, 2014a) based on work conducted concurrently with the retracted stem cell work (Lee et al., 2005). While ethical issues remain, the scientific status of the disputed papers appears to be unresolved with patents being issued after the retractions for at least one of the claimed stem cell lines, NT-1 (Cyranoski, 2014a).

1.10.6.3 Environment and Human Behavior: Diederik Stapel—Tilburg University

A "Wunderkind" social psychologist at Tilburg University in the Netherlands, Stapel currently holds the number 3 position on the *RetractionWatch* Leaderboard with 58 retractions, having fabricated many of his publications in the field of social psychology, for example, human behavior. One such fabrication was a prominent paper in *Science* that suggested that environmental untidiness led to racism (Stapel and Lindenberg, 2011).

Stapel prefabricated entire experiments (Stroebe et al., 2012; Verfaellie and McGwin, 2011) with his *modus operandi* being to construct the hypotheses, methods, data collection, and outcomes of his experiments and then pretend to run the experiments on his own at local schools. He would then fabricate the data and provide these to apparently unsuspecting colleagues and students for further analysis (Jump, 2011; Stroebe et al., 2012). When these individuals requested access to the raw data they were routinely rebuffed. Stapel's fraud was finally revealed by whistleblowers at Tilburg University and led to his suspension, censure, and ultimate dismissal (Stroebe et al., 2012). In an interim report, the University stated that his behavior had caused "severe damage to young people at the beginning of their careers, as well as to the general confidence in science, in particular social psychology" (Levelt Committee, 2011). Many of the scientists being mentored by Stapel evidently completed their theses without ever doing any actual experiments. His colleagues in the field of social psychology underwent a bout of major soul searching as to why his "too good to be true" fraudulent results (Jump, 2011; Stroebe et al., 2012) went undetected for so long (Verfaellie and McGwin, 2011) and eventually ascribed it to "a general culture of careless, selective, and uncritical handling of research and data." (Bhattacharjee, 2013). Some colleagues were also indignant that the final report from Tilburg University read as if Stapel's fraud was a phenomenon unique to the field of social psychology with one individual noting that "there are no grounds for concluding either that research fraud is any more common in social psychology than other disciplines or that its editorial processes are particularly poor at detecting it" (Gibson, 2012; Stroebe et al., 2012). Another colleague noted that "to understand fraud, we should think about how it begins and escalates, not how it ends. By the time such fraud is exposed, bad choices that would usually lead to only minor transgressions have escalated into outright career-killing behaviour" (Crocker, 2011).

In a follow-up article entitled *The Mind of a Con Man* that involved extensive interviews with Stapel, Bhattacharjee (2013) noted that "Stapel did not deny that his deceit was driven by ambition. But it was more complicated than that, he told me. He insisted that he loved social psychology but had been frustrated by the messiness of experimental data, which rarely led to clear conclusions. His lifelong obsession with elegance and order, he said, led him to concoct sexy results that journals found attractive." "It was a quest for aesthetics, for beauty—instead of the truth." While Stapel's misdeeds have, due to their magnitude, tended to occupy center stage in recent reports of fraud in the social psychology sciences, others including Dirk Smeesters, Lawrence Senna (Yong, 2012), and Marc Hauser (Wade, 2010), have been similarly suspected or proven responsible for manipulating data, both human and animal. This has led to widely-held concerns that psychology, because of its reliance on self-reporting from subjects, is a discipline with soft, "fuzzy" endpoints. This viewpoint is reinforced by outcomes in social psychology like those reported in a study in which evidence was generated (Simmons et al., 2011) for the "absurd hypothesis that listening to a Beatles song could make you 1.5 years younger" (Estes, 2012).

A take home from this example of egregious fraud, however is not nuanced. If it is easy to get away with fraud in a particular discipline because the practitioners of that particular discipline are poorly trained in the scientific method, gullible, complacently uncritical, and naïve, this does not excuse the fraud. Fraud is fraud. The issue for the discipline, however, is how to acquire and implement the necessary vigilance, know-how and expertize, as exemplified by rigorous reporting requirements and thorough peer review, that would ensure that fraud cannot walk through the front door of subject publications, unmolested, in the future, while not destroying the core aims of scientific inquiry (Woodward and Goodstein, 1996).

1.10.6.4 *Stimulus-Triggered Acquisition of Pluripotency (STAP): Haruko Obokata—RIKEN Institute*

Two papers published in *Nature* reporting the reprogramming of mammalian somatic cells using a stressor, for example, transient low-pH—citric acid, to generate pluripotent stem cells (Obokata et al., 2014a,b), were viewed as a major breakthrough in the "hot" field of stem cell research in January of 2014 (De Los Angeles et al., 2015; Goodyear, 2016; Normile and Vogel, 2014; Rasko and Power, 2015). However, by April 2014 as the result of numerous failed attempts—some 133 in 7 laboratories—to reproduce these findings, RIKEN the host institution, found Obokata guilty of research misconduct. By July 2014 these papers were retracted.

Of especial note in this instance of fraud was the short time period, some 6 months between the publication of these papers and their retraction that no doubt reflected the potential importance of the finding and an increased awareness and concern regarding the importance of the failed replication. There was also a positive side in that the detection of the fraud was a welcome instance of the scientific community successfully self-correcting.

Like Stapel, Obokata was considered a "star" both in her area of stem cell research and as a symbol of Japan's *rikejo* ("science women"; Hongo, 2014) with inevitable rumors of a Nobel Prize for her breakthrough STAP research. When the fraud was uncovered and the outcomes reported at the now obligatory forum of a press conference, Obokata "apologized for many things that day. She apologized for insufficient efforts, ill-preparedness, and unskillfulness, for errors of methodology, and sloppy data management. They were all, she said, benevolent mistakes, due to her youth and inexperience. But she denied fabricating her results" (Rasko and Power, 2015). It was also found that the STAP cells did not genetically match the mice from which they originated indicating they were probably "normal" embryonic stem cells and that Obokata had plagiarized text and manipulated images in the papers. As final closure, when Obokata was provided the opportunity by RIKEN to repeat her own experiments, she was unable to do so (Hongo, 2014). Nonetheless, in 2016, Obokata launched a website "STAP HOPE PAGE" (Jiji, 2016) to help researchers reproduce her findings.

In a more personal context, her case raised issues as to whether aggressive reporting in the scientific and mainstream media violated her human rights (Stemwedel, 2015) and whether the latter was a factor in the suicide of her supervisor, Yoshiki Sasai (Cyranoski, 2014b).

An additional wrinkle in the STAP saga was that one of the coauthors of the original papers (Obokata et al., 2014a,b) was Charles Vacanti, at one time the Chairman of Anesthesiology at Brigham and Women's Hospital in Boston who claimed that he and his brother had originated the STAP concept (Goodyear, 2016). Obokata had conducted postgraduate work in Vacanti's laboratory before joining RIKEN and even after the STAP fraud issues emerged, Vacanti still claimed, without providing any evidence (Goodyear, 2016; Rasko and Power, 2015), that he could create STAP cells, this despite the continued and consistent failure of others to do so (De Los Angeles et al., 2015).

1.10.6.5 Microarray Genetic Analysis for Personalized Cancer Treatment: Anil Potti—Duke University Medical Center

In 2006 Potti, together with his supervisor Joseph Nevins, published a paper in *Nature Medicine* (Potti et al., 2006) claiming that a microarray analysis of human tumors could be used to derive a drug response signature that could predict the individual patient response to chemotherapy. The discovery of a reliable "omics-based predictive diagnostic was important in that it provided a means to design individual patient treatment and thus avoid the trial and error approach common with treating cancer." Because of the importance of these findings and the failure of previous predictive tests, two biostatisticians at the MD Anderson Cancer Center, Baggerly and Coombes, used forensic bioinformatics to assess the Potti findings and found errors "some....careless...[and]..others... inexplicable" (Kolata, 2011). These concerns were initially published in a note in *Nature* (Coombes et al., 2007) and later in greater detail in the *Annals of Applied Statistics* (Baggerly and Coombes, 2009). Despite the initial *Nature* note, the Potti findings were used in 2008 as the basis to design and conduct clinical trials in 117 cancer patients. These trials were suspended in 2009 following the publication of the full Baggerly and Coombes paper that led to the retraction of several high-profile papers by Potti. The concurrent identification of overt misrepresentations in Potti's CV together with 11 cases of malpractice (http://retractionwatch.com/2015/05/01/malpractice-case-against-duke-anil-potti-settled/) also resulted in his resignation from Duke. A patient lawsuit accusing Duke University of engaging in a civil conspiracy (Goldberg, 2015) was reportedly resolved (Ramkumar, 2015).

Internal concerns regarding Potti's research were documented in an email in March, 2008 by a concerned medical student in the Nevins laboratory, Bradford Perez. This was downplayed by the Duke Administration as a "difference of opinion" and only came to light as a result of an in-depth report from the *Cancer Letter* in 2015 (Goldberg, 2015). Unlike the rapid resolution of the Obokata STAP fraud, that conducted by Potti took 7 years to resolve (http://retractionwatch.com/2015/11/07/its-official-anil-potti-faked-data-say-feds/). Additional detail on the Potti case is available in the NASEM (2017) report pp. 234–240.

1.10.7 Fraud as an Opportunity for Reflection and a Learning Moment?

Of the five egregious examples discussed (Table 1.1.), the retractions in the preclinical research conducted by Stapel, Hwang, and Obokata have, at least in the short term, had little direct impact on patient care although they have had a major impact on research directions in the areas of cloning, social psychology, and stem cell research. Conversely, the papers published by Wakefield and by Potti directly affected the practice of medicine, in the former instance leading to the establishment of an antivaccination movement, the actions of which have put others at unnecessary risk and led to a controversial mandatory vaccination law in California (Salmon et al., 2015) while the latter compromised clinical trials in cancer patients.

It is also noteworthy that the number of papers that are ultimately retracted is immaterial, with the single case of Wakefield doing more harm to patients and society than the 183 that were retracted in the case of Fujii.

There is a natural inclination to downplay the problem of fraud by ignoring it or treating it as an aberration. A far better reaction is to assess whether there are lessons to be learnt from the fraudsters themselves and the reactions of their institutions and funding organizations, lessons that will aid in tangibly improving the reproducibility, fidelity, quality, and relevance of biomedical research. This is certainly the viewpoint that was captured under the rubric "Can a Longtime Fraud Help Fix Science?" in the blog by Bartlett (2015) and was shared by a number of authors including Kakluk (2009), Borsboom and Wagenmakers (2013), Witkoski (2014), and Berry (quoted in Goldberg, 2015).

For additional insights in understanding and dealing with scientific misconduct and fraud in biomedical research together with recommendations and best practices to address these issues, the reader is referred to the monograph, *The Management of Scientific Integrity within Academic Medical Centers* (Snyder et al., 2015).

1.11 REPRODUCIBILITY AND TRANSLATIONAL MEDICINE

More than half of the failures in Phase II clinical trials of new drug candidates are due to lack of efficacy (Harrison, 2016; Kimmelman and Federico, 2017). Concerns regarding the reproducibility of animal findings are routinely cited as one contributing factor in the poor translation of research findings to the clinical setting where translation is described as "the transfer of new understandings of disease mechanisms gained in the laboratory into the development of new methods for diagnosis, therapy, and prevention and their first testing in humans" (Sung et al., 2003).

So while reproducibility and translation are not synonymous or interchangeable concepts (Jarvis and Williams, 2016), they do overlap. The issues that influence reproducibility—large effect sizes in underpowered studies, inadequate experimental design and analytical methods, use of poorly validated tools and models, etc.—also impact effective translation as they undermine the robustness of the observation, a metric that goes beyond reproducibility. Consequently, even if a finding is reproducible, if it is not of a magnitude that can be meaningful when applied to a clinical setting, and observed in a population with diverse genetic and epigenetic backgrounds, then it has limited utility. A second departure for translatability being differentiated from reproducibility is around the endpoints selected for demonstrating efficacy. In the research setting, it is relatively easy to incorporate either mechanism-related measures to assess compound activity, for example, collagen deposition in a pulmonary fibrosis model, as a proxy for clinically applicable endpoints of lung function, such as forced vital capacity, or to rely on surrogate measures of efficacy—for example, contextual fear conditioning (Comery et al., 2005) or a Morris water maze test (Jian et al., 2016) in transgenic models of Alzheimer's disease—that bear no relationship to the standard clinical measures, such as the Clinical Dementia Rating scale (CDR sum of boxes; O'Bryant et al., 2008) that addresses six functional domains related to memory, orientation, judgment and problem solving, community affairs, home and hobbies, and personal care. Generally the argument is made that there is a correlation or temporal association between the pathology and the measures used in animal models, although such associations alone constitute scientifically weak evidence, as demonstrated in numerous amusing examples on the Internet between per capita consumption of margarine in the United States and the divorce rate in Maine; automobile drivers killed in a collision with a train and US crude oil imports from Norway (http://www.tylervigen.com/spurious-correlations); and *iPhone* sales and individuals dying from falling down stairs (https://hbr.org/2015/06/beware-spurious-correlations). However, more important to this thesis is that even if such studies were reproducible, their translatability to the clinical setting, based on dubious associations, constitutes a major leap of faith.

Predicated on a Phase I clinical trial of the fatty acid amide hydrolase (FAAH)-inhibitor BIA 10-2474 that led to serious neurological complications including one fatality, Kimmelman and Federico (2017) argued that "a lack of emphasis on evidence for the efficacy of drug candidates is all too common in decisions about whether an experimental medicine can be tested in humans" and recommended that the regulatory bodies—including the institutional review boards (IRBs) responsible for approving clinical trial protocols place an emphasis equal to that on compound safety on understanding compound efficacy. They propose "three questions to assess clinical promise":

1. *What is the likelihood that the drug will prove clinically useful?*
   - How have other drugs in the same class or against the same target performed in human trials?
   - How have other drugs addressing the same disease process fared?
2. *Assume the drug works in humans. What is the likelihood of observing the preclinical results?*
   - Are the treatment effects seen in animals large and consistent enough to suggest a tangible benefit to patients?
   - How well do animal models reflect human disease?
3. *Assume the drug does not work in humans. What is the likelihood of observing the preclinical results?*
   - Have effects of random variation and bias been minimized (e.g., by sample sizes, randomization, blinding, dose-response curves, and proper controls)?
   - Do the conditions of the experiment (for instance age of animal models, timing of treatments, and outcomes) match clinical scenarios?
   - Have effects been reproduced in different models and/or in independent laboratories?

Providing answers to these sorts of questions is far from a trivial undertaking and has been likened to "translat..[ing].. a text with the sophistication and depth of Shakespeare using a first-grader's vocabulary and experience, because our knowledge about the functions of most pathways in various cell types, during different developmental stages, and under normal physiological conditions, is still rudimentary and piecemeal" (Zoghabi, 2013).

Nonetheless, such considerations underline the need for both ensuring that the preclinical data used as the basis of moving a new drug candidate into clinical trials be "true" as defined by Goodman et al. (2016), for example, reproducible and generalizable, and also have context in relationship to the reported effects of other known compounds, if any, acting at the same disease target (Kimmelman and Federico, 2017).

1.11.1 Animal Models—Predictive Value in Translational Research

A key—and often the penultimate—step in the preclinical drug discovery process is to demonstrate the efficacy of a drug candidate in an animal model that is thought to reflect aspects of the human disease state that consequently has predictive validity (Drucker, 2016; Groenink et al., 2015; McGonigle and Ruggeri, 2014; NAP, 2015; van der Worp et al., 2010). For example, an animal model of pain where morphine, acting directly via morphine receptors, has been repeatedly shown to reverse pain behavior induced by chemical or physical trauma may be considered as predictive for the testing of new compounds that have the potential to work in similar human pain states. Unfortunately, in this example, the translational models have proven to be far less predictive than preclinical data would suggest, only being valid for compounds that act via an opioid receptor. Newer analgesics that act via one of the many (more than 20) novel mechanisms identified in the past 20 years that show robust and reproducible efficacy in a morphine model of pain yet have consistently failed to produce analgesia in humans, indicates the high level of false positives in preclinical studies (Fairbanks and Goracke-Postle, 2015).

Animal models of human disease states remains a contentious issue throughout the spectrum of preclinical research (Denayer et al., 2014; Greek and Hansen, 2013; Peters et al., 2015; Sams-Dodd, 2006; Vandamme, 2015) with many researchers confusing the inability to predict the translation of the therapeutic activity of NCEs observed in animal models to the clinic with failed reproduction *per se*. Indeed, many NCEs that evidence robust, dose-dependent and replicated activity in preclinical models often fail in the clinic. As noted by Grove (2005) "Humans are incredibly complex biological systems, and working with them has to be subject to safety, legal, and ethical concerns.... The result is wide-scale experimentation with animal models of dubious relevance, whose merit principally lies in their short lifespan."

The fidelity of animal models of disease to the human condition is generally low, irrespective of the therapeutic area, given species differences in biological mechanisms and systems pharmacology between rodents and humans, and also because the mechanistic causality of many diseases remains unknown (Groenink et al., 2015; Llovera et al., 2015). As a result, many animal models, especially transgenics, are exclusively models of presumed and often unproven molecular mechanisms of a disease rather than disease states per se. Additional shortcomings of animal models relate to the putative disease phenotype in the animal which is temporally more acute than that of the actual human disease state. Thus chronic pain in humans reflects a subtle and complex process that may be decades in development with extensive neuronal remodeling. This is in stark contrast with chemically—or surgically-induced animal models of pain that are acute (hours) to subchronic (weeks) in onset and duration and focus on a limited repertoire of known or putative pain-associated targets.

Similarly, Type 2 diabetes (T2D) research is generally modeled in C57BL/6J mice 2–6 months of age as the animals are readily available and cost effective (Drucker, 2016). However, in humans T2D onset occurs in the fifth through ninth decades of life by which time humans have suffered years of low grade tissue inflammation, hypertension, dyslipidemia, fibrosis, weight gain, and impaired glucose tolerance. Since younger animals have greater organ and cellular plasticity and cell proliferation, they present a background that generally makes it easier for putative therapeutics to reverse a disease phenotype than that present in older animals, thus confounding their use in translational research (Drucker, 2016). Furthermore, in genetic models, the knockout or overexpression of a discrete disease-associated target is an artificial, reductionistic monogenic approach with many, if not the majority, of human diseases being multifactorial involving contributions from dozens of genes that have modest effect sizes (Fuchsberger et al., 2016). Gene effects on disease onset often occur in early stages of development and also involve environmental contributions. For instance, in schizophrenia, a disease that involves alterations in early brain development (Kahn et al., 2015), researchers have been looking for a single susceptibility gene (the one disease, one gene mantra) for more than 2 decades. As of 2008, schizophrenia has been associated with 3,608 polymorphisms in 516 different genes (Allen et al., 2008) suggesting that the underlying genetics of the disease are complex and that identifying a major susceptibility gene is unlikely (Farrell et al., 2015). An additional confound of using germline knockouts as translational disease models is their potential "for developmental adaptation and physiological compensation secondary to the loss of the key gene and protein at the earliest stages of development" (Drucker, 2016). Adaptation and compensation can lead to systems redundancy (Doyle et al., 2012) where a gene knockout can be compensated for by an alternative pathway. Thus, preclinical efficacy that supports compound advancement to the clinic can frequently be replicated to an extreme level of precision and still be reflective of a model that is not predictive for the human condition. In fact, the absence of efficacy in the clinic, the lack of translation of an NCE (New Chemical Entity), has become, like that of the preclinical animal models, very reproducible, as discussed in the examples below.

1.11.2 Limitations in Animal Models of Human Disease

The lack of effective therapeutics to treat chronic disease states, for example, AD, T1D, and neuropathic pain, has prompted considerable efforts in basic research to meet "unmet medical needs." Despite this imperative, many potential therapeutics with convincing drug-like properties and appropriate efficacy in animal models that are thought to recapitulate molecular and phenotypic aspects of the human disease have consistently failed in the clinic—prompting Greek and Hansen (2013) to note that in "reductionism-based animal models, the failures have been epic."

1.11.2.1 Amyotrophic Lateral Sclerosis (ALS)

A number of compounds reported in approximately 50 papers published before 2008 were identified as having survival benefit in the standard ALS SOD1^{G93A} mouse model but failed to translate to the clinic and provide human benefit. This led to the conclusion that "the high noise floor of the model and the failure of the selected studies to replicate support the conclusion that the bulk of published studies using the SOD1^{G93A} mouse model may unfortunately be measurements of biological variability due to inappropriate study design" (Scott et al., 2008). Blinding and randomization were conspicuously lacking in this example.

Additional testing of a cohort of approximately 100 potential therapeutics that had been reported preclinically to attenuate ALS symptoms in animal models found that the original preclinical findings could not be replicated (Perrin, 2014). Only 8 of these compounds were advanced to clinical trials, but subsequently failed to show efficacy indicating that the preclinical data used for their advance to the clinic was a reflection of false positive findings. This was attributable to limitations in the ALS mouse model or the design and analysis of studies that used it. Specifically, in ALS patients and some mouse models of the disease, the paralysis caused by deterioration of the neurons innervating skeletal muscle progresses over time. However, in the ALS mouse model where a mutant form of the RNA-binding protein TDP43 was associated with motor neuron loss, protein aggregation, progressive muscle atrophy, and a defective version of the SOD1 gene (which is mutated in 10% of the familial ALS population), this progressive deterioration was not observed (Perrin, 2014). Additionally, muscle deficits in TDP43 mice were mild with animals dying of acute bowel obstruction caused by deterioration of gut smooth muscle rather than the progressive muscle atrophy observed in human ALS (Hatzipetros et al., 2014).

Further analysis of the failure to replicate the original findings identified four factors that increased variability (Perrin, 2014) and that reflected a lack of user consideration of the nuances in using animal models of ALS: (1) a failure to exclude animals whose deaths were unrelated to the disease; (2) not randomizing littermates between control and treatment groups; (3) not accounting for gender since male mice demonstrate ALS symptoms as much as a week before females and die approximately a week earlier—differences that could be construed as a compound-related effect; and (4) the loss of disease phenotype due to multiple copies of the disease-causing gene building up with breeding in a mouse colony that were not passed on in a stable fashion as cells divide leading to subsequent generations having fewer copies of the transgene and a less severe disease phenotype.

More recent studies have focused on TDP-43 misfolding rather than SOD1 dysfunction as a causative factor in ALS. This has highlighted the benefits of attenuating the loss of function and gain of function/dominant-negative toxicity role of TDP-43 in initiating ALS by preventing its misfolding and/or enhancing its clearance, the latter using enhancers of chaperone-dependent TDP-43 folding, and activators of the ubiquitin proteasome system and autophagic pathways (Scotter et al., 2015). While intellectually appealing, it remains to be seen if this concept has any greater success at clinical translatability.

1.11.2.2 Stroke

Considerable efforts have been expended over the past 4 decades to identify effective interventions to reduce mortality and improve outcomes from stroke. These include a variety of small molecule therapeutics acting through targets thought to be directly involved in mediating the excitotoxicity and free radical formation that occurs subsequent to an ischemic event (van der Worp and van Gijn, 2007).

Despite a huge investment of resources, of more than 1,000 putative neuroprotective NCE treatments demonstrating benefit in animal models [mainly in the MCAO (middle cerebral artery occlusion) model in gerbil or rat], 114 of which were examined in clinical protocols in which aspirin and thrombolytic (e.g., alteplase, rTPA) had shown robust efficacy, none were efficacious (O'Collins et al., 2006). Subsequent analysis of these results (Macleod et al., 2009) identified several variables, including timing of NCE administration as well as animal age, comorbidities, and physiological status, as contributing to the disparity between findings from the animal models and the clinical trial outcomes reflecting bias in the preclinical models that resulted in an "overstatement of neuroprotective efficacy" (Sena et al., 2010). That these models are still being used in preclinical stroke research—more than a decade after extensive peer-reviewed metaanalyses of their total lack of translational value was published—attests to laziness or hubris.

1.11.2.2.1 METAANALYSES OF DATA FROM ANIMAL MODELS OF STROKE

In a metaanalysis of animal data (Bath et al., 2009), the clinical failure of the free radical scavenger, NXY-059, in large clinical trials (5,028 patients) in acute ischemic stroke (Diener et al., 2008) was evaluated in the context of the positive preclinical animal model data (Bath et al., 2009). This included reduced infarct volume and motor impairment in experimental stroke models (transient, permanent, and thrombotic) in rodents, rabbits, and primates (Macrae, 2011). Analysis of the data from 585 animals (NXY-treated 332; control, 253) from mice, rats, and marmosets that originated from 12 laboratories which reflected 26 experiments (four of which were unpublished) showed that NXY-059 was neuroprotective in preclinical models that met the established STAIR (Stroke Therapy Academic Industry Roundtable) criteria (Macrae, 2011). There was evidence however, that publication biases were present in the preclinical studies reviewed. Additionally, while spontaneously hypertensive rat (SHR) models of stroke were included in the metaanalysis, NXY-059 was retrospectively found to be effective only in normotensive rats. Another concern was that sample size calculations were absent from all the studies. Thus the discrepancy between the preclinical and clinical data may have resulted from: (1) a lack of relevance of the preclinical data to the human situation; (2) efficacious doses in rats and marmosets not being predictive of the human situation and; (3) issues with brain access of the free radical scavenger.

The metaanalysis concluded that because of the various biases, the preclinical efficacy of NXY-059 may have been overestimated (Bath et al., 2009; Dirnagl and Macleod, 2009; Macleod et al., 2008, 2009). Based on these conclusions, the authors recommended that metaanalysis of all available preclinical data on an NCE be conducted before the initiation of clinical trials.

Additional factors identified in the translational failures in stroke included potential differences between human brain and that of rodents and the initial assessment of novel therapeutics in a stringently controlled preclinical testing environment where the animals used and the laboratory conditions differ markedly from the heterogeneity seen in stroke patients, and the time of compound administration *after* the ischemic episode.

To further address these issues and those raised in the NXY-059 metaanalysis study, a primate (cynomologous macaque) embolic stroke model was used in a blinded crossover study to assess a novel neuroprotectant, Tat-NR2B9c (aka NA-1) which blocks neurotoxic signaling events by uncoupling the postsynaptic protein, PSD-95 (Cook et al., 2012). Primates treated with Tat-NR2B9c *after* the onset of embolic strokes showed reduced stroke numbers and stroke volumes that anticipated the outcomes of the corresponding human trial, ENACT (Evaluating Neuroprotection in Aneurysm Coiling Therapy; Hill et al., 2012), an apparent success in terms of translation.

Another approach to improve the translational process in stroke involved the use of a *preclinical* randomized controlled multicenter trial (pRCT), the design and rigor of which was based on that of a typical Phase III RCT. This study was conducted in six independent research centers using two models of stroke, the cMCAO (permanent distal middle cerebral artery) and fMCAO (transient middle cerebral artery occlusion) in C57BL/6J mice to assess the effects of a CD49d antibody that inhibits leukocyte migration into the brain (Llovera et al., 2015). Standardization of procedures between the six centers followed similar procedures to those used by Crabbe et al. (1999) (Section 1.4.1.1) and involved 315 mice, 81 in the cMCAO model and 174 in the fMCAO model.

Anti-CD49d, given 3 h after stroke induction, consistently decreased leukocyte migration and infarct volume in the cMCAO model that involved small cortical infarcts but not in the fMCAO model where larger lesions were induced. Anti-CD49d had no effect on behavioral outcomes (rotarod test, adhesion removal test) that were confounded by great intercenter variability. These outcomes were concluded to reflect a variety of subtle differences between the two models including: insufficient statistical power in the fMCAO model due to unexpected variability that could have been addressed by increasing group sizes; differences in the time point of assessment due to higher mortality rates in the fMCAO mice; differences in the neuroinflammatory markers between the two models (that were reflected in a twofold higher infiltration of leukocytes in the cMACO versus the fMCAO model); the use of unequal numbers of animals in the two models; an absence of primary data due to mortality or a lack of infarct demarcation; and low performance in the behavioral outcomes. While a major *tour de force* in using a pRCT approach, the Llovera et al. (2015) study still highlighted many of the limitations in rodent animal studies of stroke and their utility in translation. Furthermore, a completed Phase II clinical trial of the anti-CD49d antibody showed no benefits (Clinicaltrials.gov identifier NCT01955707).

1.11.3 Issues in Translatability

In the disease states discussed earlier, all of which are neurological, the ability to predict efficacy in humans based on animal models remains poor (Hartung, 2013; Hobin et al., 2012; Wendler and Wehling, 2012; Wehling, 2009). Attempts to understand the disconnects between preclinical data and the clinical trial outcomes reveal a variety of causes. Some of these, like experimental design and powering and data analysis, and reagent validity are addressed in this monograph. Others remain unknown (Mullane and Williams, 2015) and are therefore insurmountable which makes putative animal models of human disease states less predictive in assessing potential human efficacy.

This is not unique to the field of neurology as numerous examples from other therapeutic areas have questioned the relevance and predictivity of animal models (Groenink et al., 2015; McGonigle and Ruggeri, 2014; NAP, 2015) other than these providing another measure of a pharmacodynamic response to an NCE that is accompanied by a pharmacokinetic component that can aid in dose selection for human testing (Caldwell et al., 2004; Kleiman and Ehlers, 2016).

Vatner (2016) has argued that the shift in research funding from traditional physiology to molecular medicine that has been ongoing for the past 30 years (Jobe et al., 1994) has resulted in a dearth of laboratories with the appropriate physiological expertize, in this particular instance, cardiovascular physiology, along with the many decades of practical experience associated with the discipline. This has deprived biomedical research of a critical aspect, that of integrating molecular with whole animal data, that has confounded the integrative hierarchy of experimentation (Kenakin et al., 2014) and led to the whole animal aspects of physiology being poorly understood and inaccurately represented. As a result, integrative physiology is treated as an incidental to molecular findings (or to paraphrase a one-time candidate for the Nobel Prize in Physiology or Medicine—"compound A binds to target Z in vitro, let's start clinical trials") that has been replaced with "wishful thinking" (Vatner, 2016). Similar concerns regarding the decline of integrative biology had been made over 20 years ago by Jobe et al. (1994), who questioned the relevance of in vitro studies to physiological mechanisms that probably has been a major contributor to the translational shortcomings of the animal models discussed earlier—reducing them to in vivo test tubes via transgenic manipulations.

1.12 CONCLUSIONS

In setting the stage for the remainder of this monograph on reproducibility, the present Chapter has focused primarily on the seminal commentaries, papers, and events that have driven the debate on reproducibility to its current level of visibility, active debate, and attempted resolution. This has necessarily required a focus on fraud in the biomedical sciences—a topic that is often far more accessible and interesting than, for example, the use of Bayesian statistics, cell line authentication, or translational confounds.

Fraud per se is probably a very minor contributor (estimated by various sources as being 0.02% or less of all research activities) to the overall issue of reproducibility in biomedical research and its resolution is more likely to occur via the actions of grant review bodies, institutional standards, and ultimately law enforcement than peer review as these can wield (but frequently do not) the big stick of tangible personal consequences. However, by its very nature, published fraud and related misdemeanors have received increased attention by scientists invested in "research on research" who have found a ready outlet for their considerable body of work in the open access literature. It has also been gleefully blown out of proportion—as much that occurs in the 21st century increasingly tends to be—by the mainstream media avidly seeking content, the more sensational, the better. Indeed, Drucker, (2016) has noted "The media itself has an extraordinary appetite for scientific and medical information, especially stories with a hint of therapeutic relevance. The media beast is insatiable, although even my mother has now learned that most "medical breakthrough stories" featured on the television, radio, in print, or disseminated via the Internet and social media are almost always exaggerated and often frankly incorrect."

In the majority of instances, articles on "research on research" in the form of metaanalyses and also via blogs, where issues on the validity of published research findings are raised, perform an invaluable service that grant review bodies and research institutions fail to acknowledge. They do however give the inappropriate impression that reproducibility is in crisis, and has recently escalated dramatically, when it is in fact a chronic problem made more visible by the increased transparency offered by the Internet.

To document the occurrence of egregious fraud in a monograph that is intended to address the topic of reproducibility may seem to some as unnecessary as it tends to highlight the most negative aspects of the reproducibility issue. However, as noted previously, serial fraud is probably only the tip of the reproducibility iceberg and there needs to be a concerted effort on the part of the biomedical research community as a whole to acknowledge fraud in order to effect meaningful change rather than delegating the responsibility for remedying the situation to journal editors and peer reviewers.

Many scientists consider biomedical research fraud as being at one end—the extreme end—of the reproducibility spectrum with its practitioners, like the five discussed previously as not being worthy of being called a scientist. At the other end of the reproducibility spectrum is the phenomenon of an "honest mistake" where ethical and competent scientists can sometimes, inadvertently, generate bad science. Between these extremes are various forms of investigator bias, overt and unintended that can lead to honest mistakes. Collectively, this leads to a perception by many researchers that 99% of his or her colleagues will subscribe to the same scientific values that they were taught and espouse and would, accordingly, express absolute disbelief (as apparently occurred repeatedly with Stapel and Potti) if they were accused of being involved, intentionally or circumstantially (laboratory members making up or manipulating data) in aspects of fraud (Wilson, 2016).

Like the reproducibility spectrum, fraud is also a semantic continuum that ranges from the egregiously fictitious activities of a Stapel to the selection bias evidenced by one of Begley's unnamed colleagues—no doubt a distinguished cancer researcher—who made the decision to ignore five negative findings in a series of six cancer experiments and instead focus on, and publish, the one finding that was deemed to be "the best" (Begley, 2012). This emulated in intent the behaviors of Stapel, Obokata, and others whose sociopathic priorities, situational and financial, demonstrate a lack of personal responsibility that adds to the erosion of public trust in biomedical research (Alberts et al., 2014; Begley and Ioannidis, 2015) and its integrity (Kaiser, 2014). In the bigger picture, the frauds perpetrated by Wakefield, Hwang, Stapel, Obokata, and Potti can be viewed as examples of extreme personal bias, in the same category as the bias that researchers deal with on a daily basis that differs only in degree from that of overt fraudsters.

While there are grey areas in reproducibility, especially where unintentional bias is involved, biomedical researchers can take simple steps to remove ambiguity in their work and ensure that the research conducted in their laboratories is transparent, relevant, and reproducible. Other unintentional biases unique to modern day science might include the assembly and interrogation of large databases where concerns of keeping them updated, accurate, and well-curated requires significant, long-term investments, while continually modified and refined analytical tools can constantly adjust interpretations of the large datasets, but rarely are the preceding publications revised. The research community needs to "grasp the nettle" if they want to ensure that the continued funding of research remains a high priority for the public and lawmakers (Alberts et al., 2014; Moher et al., 2016). This monograph is intended to aid in that endeavor.

In the remainder of this monograph, the various reproducibility issues are documented together with suggested solutions to improve shortcomings. This includes the provision of substantive guidance on aspects of the design, execution, analysis, and reporting of research findings and the separate but important issue of transparency—if it can't be *seen*, how can anyone know it has been *done*?

In addition, consideration is given to the various mechanisms intended to encourage, evaluate, and judge whether guidance to facilitate reproducibility have been followed. These include the following: (1) the more stringent application, oversight, and proactive support of the prepublication *CPR* peer review process (Chapter 5.2) by supplementing, this with Internet-enabled PPPR (Chapter 5.3.2.3) and *PPC* (Chapter 5.2.3.4) which have the potential, if used correctly, to reinforce the scope, standards, and execution of the process of self-correction (Alberts et al., 2015; Nature, 2013); (2) the use of publication/peer review checklists in concert with more rigorous and detailed guidelines on the requirements of journals for publication (Begley, 2013; Collins and Tabak, 2014; Curtis et al., 2015; Mullane et al., 2015) and; (3) formal reproducibility initiatives (Errington et al., 2014; OSC, 2015; Van Noorden, 2014), the value of which has been questioned (Mullane and Williams, 2017).

Equally important are initiatives focused on improved training and mentoring, represent the best hope for improving reproducibility and include: (1) improving the general standards for the training and mentoring of new generations of scientists (Collins and Tabak, 2014; Flier, 2017; Kornfeld and Titus, 2016; Michael, 2015; Munafò et al., 2017); (2) increasing the focus on logical and transparent design, powering, execution, and analysis of experiments (Chapters 2 and 3); and (3) using properly validated reagents, cell lines, and animal models (Freedman and Inglese, 2014; Freedman et al., 2015; Chapter 2).

The dedicated fraudster will hopefully find little useful in this monograph—other than the possible opportunity to learn about the more recently identified examples of fraud and misrepresentation and, by default, understand how to enhance their efforts in avoiding detection. These individuals certainly will not be deterred by any strategies proposed to diminish their impact on advancing science as opposed to their careers. Instead this monograph is intended as an aid for the other 95 + percent of well-intentioned scientists, who wish to avoid the pitfalls and traps that can ensnare their best efforts, and be able to judge the quality, and hence the reliability and truthfulness of published biomedical research.

1.12.1 A Note on Estimates and Surveys

Throughout this monograph, the authors cite various estimates in the literature and mainstream media of the occurrence of various issues related to reproducibility. These generally range from 0.01 (or less) to 50% and are reported with little in the way of convincing substantiation or authoritative sourcing. This may be due, in part, to the lack of transparency in the process of identifying and proving fraud, the extensive delays in the investigative processes and then further delays in informing journal editors and retractions making their way to *PubMed* or other databases. Also, as mentioned, there is a decidedly grey area between outright pathological fraud and the practices of "sloppy science" with selective reporting, Harking, *p*-hacking, and overt bias in selection, interpretation, and presentation that could be argued as fraudulent practices although the perpetrators clearly would not regard themselves as practicing anything other than the responsible conduct of research (NASEM, 2017). The use of the cited range of anywhere between 0.01%–50% is therefore used with the recognition that while it is broad, it probably underestimates the extent of the problem.

Similarly, several surveys are cited that need to be taken with a grain of salt since they have extremely low response rates that are compounded by the responders self-selecting. For instance, the *Nature* survey on whether there is a reproducibility crisis (Baker, 2016b) that involved 1,576 researchers was, depending on how many biomedical researchers there are in the world, woefully underpowered but nonetheless extensively cited. In the United States, this population is estimated at 70,000–80,000 (Heggeness et al., 2017), while a conservative extrapolation would estimate a global group size as 2–3 times larger making the *Nature* survey group of 1,576 representative of somewhere between 0.66% and 2.25% of the actual biomedical research workforce and thus scientifically questionable.

References

Alberts, B., Kirschner, M.W., Tilghman, S., Varmus, H., 2014. Rescuing US biomedical research from its systemic flaws. Proc. Natl. Acad. Sci. USA 111, 5573–5777.

Alberts, B., Cicerone, R.J., Fienburg, S.F., Kamb, A., McNutt, M., Nerem, R.M., et al., 2015. Self-correction in science at work. Science 348, 1420.

Allen, N.C., Bagade, S., McQueen, M.B., Ioannidis, J.P.A., Kavvoura, F.K., Khoury, M.J., et al., 2008. Systematic meta-analyses and field synopsis of genetic association studies in schizophrenia: the SzGene database. Nat. Genet. 40, 827–834.

Almeida, J.L., Cole, K.D., Plant, A.L., 2016. Standards for cell line authentication and beyond. PLoS Biol. 14, e1002476.

Arrowsmith, C.H., Audia, J.E., Austin, C., Baell, J., Bennett, J., 2015. The promise and peril of chemical probes. Nat. Chem. Biol. 11, 536–541.

Artus, M., van der Windt, D.A., Jordan, K.P., Hay, E.M., 2010. Low back pain symptoms show a similar pattern of improvement following a wide range of primary care treatments: a systematic review of randomized clinical trials. Rheumatology 49, 2346–2356.

Azoulay, P., Bonatti, A., Krieger, J.L., 2015. The Career Effects of Scandal: Evidence from Scientific Retractions. NBER Working Paper No. 21146. Available from: http://www.nber.org/papers/w21146.

Baggerly, K.A., Coombes, K.R., 2009. Deriving chemosensitivity from cell lines: forensic bioinformatics and reproducible research in high-throughput biology. Annal. Appl. Stat. 3, 1309–1334.

Bahrndorff, S., Alemu, T., Alemneh, T., Nielsen, J.L., 2016. The microbiome of animals: implications for conservation biology. Int. J. Genomics 2016, 7.

Baker, M., 2015. Blame it on the antibodies. Nature 521, 274–276.

Baker, M., 2016a. Muddled meanings hamper efforts to fix reproducibility crisis. Nature. Available from: http://www.nature.com/news/muddled-meanings-hamper-efforts-to-fix-reproducibility-crisis-1.20076.

Baker, M., 2016b. Is there a reproducibility crisis? Nature 533, 452–454.

Baker, M., Dolgin, E., 2017. Reproducibility project yields muddy results. Nature 541, 269–270.

Balch, C., Arias-Pulido, H., Banerjee, S., Lancaster, A.K., Clark, K.B., Perilstein, M., et al., 2015. Science and technology consortia in U.S. biomedical research: a paradigm shift in response to unsustainable academic growth. Bioessays 37, 119–122.

Barber, M., 2014. Rab Butler's 1944 act brings free secondary education for all. BBC News School Report, January 17, 2014. Available from: http://www.bbc.co.uk/schoolreport/25751787.

Barrows, N.J., Le Sommer, C., Garcia-Blanco, M.A., Pearson, J.L., 2010. Factors affecting reproducibility between genome-scale siRNA-based screens. J. Biomol. Screening 15, 735–747.

Bartlett, T., 2015. Can a Longtime Fraud Help Fix Science? Chron. Higher Edu. Available from: http://chronicle.com/article/Can-a-Longtime-Fraud-Help-Fix/231061/.

Bath, P.M.W., Gray, L.J., Bath, A.J.G., Buchan, A., Miyata, T., Green, A.R., 2009. On behalf of the NXY-059 Efficacy Meta-analysis in individual Animals with Stroke (NEMAS) investigators. Effects of NXY-059 in experimental stroke: an individual animal meta-analysis. Br. J. Pharmacol. 157, 1157–1171.

Beall, J., 2012. Predatory publishers are corrupting open access. Nature 489, 179.

Beall, J., 2016. Beall's List of Predatory Publishers 2016. Scholary Open Access. Available from: http://scholarlyoa.com/2016/01/05/bealls-list-of-predatory-publishers-2016/.

Beall, J., 2017. What I learned from predatory publishers. Biochem. Med. 27, 273–278.

Begley, C.G., 2013. Reproducibility: six red flags for suspect work. Nature 497, 433–434.

Begley, C.G., 2017a. Quoted in Harris R. 2017. Rigor Mortis: How Sloppy Science Creates Worthless Cures, Crushes Hope, and Wastes Billions. Basic Books, New York, p. 26.

Begley, C.G., 2017b. Quoted in Harris R. 2017. Rigor Mortis: How Sloppy Science Creates Worthless Cures, Crushes Hope, and Wastes Billions. Basic Books, New York, p. 9.

Begley, C.G., Ellis, L.M., 2012. Drug development. Raise standards for preclinical cancer research. Nature 483, 531–533.

Begley, C.G., Ioannidis, J.P.A., 2015. Reproducibility in science: improving the standard for basic and preclinical research. Cir. Res. 116, 116–126.

Begley, S., 2012. In cancer science, many "discoveries" don't hold up. Reuters. Available from: http://www.reuters.com/article/2012/03/28/us-science-cancer-idUSBRE82R12P20120328.

Bell, C.J., Dinwiddie, D.L., Miller, N.A., Hateley, S.L., Ganusova, E.E., et al., 2011. Carrier testing for severe childhood recessive diseases by next-generation sequencing. Sci. Transl. Med. 3, 65ra4.

Bhattacharjee, Y., 2013. The Mind of a Con Man, New York Times Magazine. Available from: http://www.nytimes.com/2013/04/28/magazine/diederik-stapels-audacious-academic-fraud.html?pagewanted=all&_r=1&.

Bilton, N., 2016a. The secret culprit in the Theranos mess. Vanity Fair. Available from: http://www.vanityfair.com/news/2016/05/theranos-silicon-valley-media.

Bilton, N., 2016b. Exclusive: How Elizabeth Holmes's House of Cards came tumbling down. Vanity Fair. Available from: http://www.vanityfair.com/news/2016/09/elizabeth-holmes-theranos-exclusive.

Bishop, D., 2012. Novelty, interest and replicability. Bishop Blog. Available from: http://deevybee.blogspot.co.uk/2012/01/novelty-interest-and-replicability.html.

Bishop, D., 2015. Publishing replication failures: some lessons from history. Bishop Blog. Available from: http://deevybee.blogspot.co.uk/2015/07/publishing-replication-failures-some.html.

Bishop, W.H., 2013. The role of ethics in 21st century organizations. J. Bus. Ethics 118, 635–637.

Blatt, M.R., 2015. Vigilante science. Plant Physiol. 169, 907–909.

Boettiger, S., Bennett, A.B., 2006. Bayh-Dole: if we knew then what we know now. Nat. Biotechnol. 24, 320–323.

Bohannon, J., 2013. Who's afraid of peer review? Science 342, 60–65.

Bohannon, J., 2015. How to hijack a journal. Science 350, 903–905.

Bollen, K., Cacioppo, J.T., Kaplan, R., Krosnick, J., Old, J.L., 2015. Social, Behavioral, and Economic Sciences Perspectives on Robust and Reliable Science, National Science Foundation, Arlington, VA.

Bornemann-Cimenti, H., Szilagyi, I.S., Sandner-Kiesling, A., 2015. Perpetuation of retracted publications using the example of the Scott S. Reuben case: incidences, reasons and possible improvements. Sci. Eng. Ethics 22, 1063–1072.

Borrell, B., 2009. A medical madoff: anesthesiologist faked data in 21 studies. Sci. Amer. Available from: http://www.scientificamerican.com/article/a-medical-madoff-anesthestesiologist-faked-data/.

Borsboom, D., Wagenmakers, E.-J., 2013. Book Review. Derailed: The Rise and Fall of Diederik Stapel. APS Observer. Available from: http://www.psychologicalscience.org/index.php/publications/observer/2013/january-13/derailed-the-rise-and-fall-of-diederik-stapel.html.

Boston, M., 2015. Retracted Scientific Studies: A Growing List. New York Times. Available from: http://www.nytimes.com/interactive/2015/05/28/science/retractions-scientific.

Bothwell, L.E., Greene, J.A., Podolsky, S.H., Jones, D.S., 2016. Assessing the gold standard—lessons from the history of RCTs. N. Engl. J. Med. 374, 2175–2181.

Bradbury, A., Pluckthun, A., 2015. Standardize antibodies used in research. Nature 518, 27–29.

Braude, S.E., 1979. ESP and Psychokinesis: A Philosophical Examination. Temple University Press, Philadelphia, PA, p. 2.

Broad, W., Wade, N., 1983. Betrayers of the Truth. Touchstone/Simon and Shuster, New York, p. 12.

Buntin, M.B., Burke, M.F., Hoaglin, M.C., Blumenthal, D., 2011. The benefits of health information technology: a review of the recent literature shows predominantly positive results. Health Aff. 30, 464–471.

Budd, J.M., Sievert, M., Schultz, T.R., Scoville, C., 1999. Effects of article retraction on citation and practice in medicine. Bull. Med. Libr. Assoc. 87, 437–443.

Bush, V., 1945. Science, the Endless Frontier: A Report to the President. US Government Printing Office, Washington, DC.

Butler, D., 2013. Investigating journals: the dark side of publishing. Nature 495, 433–435.

Button, K.S., Ioannidis, J.P.A., Mokrysz, C., Nosek, B.A., Flint, J., Robinson, E.S.J., et al., 2013. Power failure: why small sample size undermines the reliability of neuroscience. Nat. Rev. Neurosci. 14, 365–376.

California Biomedical Research Association (CBRA), 2017. Fact Sheet What is Biomedical Research? Available from: http://www.ca-biomed.org/pdf/media-kit/fact-sheets/FS-WhatBiomedical.pdf.

Callahan, D., 2004. The Cheating Culture: Why More Americans Are Doing Wrong to Get Ahead. Harcourt, Orlando, FL.

Caldwell, G.W., Masucci, J.A., Yan, Z., Hageman, W., 2004. Allometric scaling of pharmacokinetic parameters in drug discovery: can human CL, Vss and $t_{1/2}$ be predicted from in vivo rat data? Eur. J. Drug Metabol. Pharmacokinet. 29, 133–143.

Carlisle, J.B., 2012. The analysis of 169 randomised controlled trials to test data integrity. Anaesthesia 67, 521–537.

Carey, B., 2015. Science, Now Under Scrutiny Itself. New York Times. Available from: http://www.nytimes.com/2015/06/16/science/retractions-coming-out-from-under-science-rug.html?_r=0\.

Carreyrou, J., 2015. Hot startup theranos has struggled with its blood-test technology. Wall St. J. Available from: http://www.wsj.com/articles/theranos-has-struggled-with-blood-tests-1444881901.

Carreyrou, J., 2016. Theranos whistleblower shook the company—and his family. Wall St. J. Available from: http://www.wsj.com/articles/theranos-has-struggled-with-blood-tests-1444881901.

Casadevall, A., Fang, F.C., 2010. Reproducible science. Infect. Immun. 78, 4792–4795.

Chalmers, I., Bracken, M.B., Djulbegovic, B., Garattini, S., Grant, J., Gülmezoglu, A.M., et al., 2014. How to increase value and reduce waste when research priorities are set. Lancet 383, 156–165.

Chau, C.H., Rixe, O., McLeod, H., Figg, W.D., 2008. Validation of analytical methods for biomarkers employed in drug development. Clin. Cancer Res. 14, 5967–5976.

Chawla, D.S., 2017. Mystery as controversial list of predatory publishers disappears. ScienceInsider. Available from: http://www.sciencemag.org/news/2017/01/mystery-controversial-list-predatory-publishers-disappears.

Chen, J.J., Hsueh, H.M., Delongchamp, R.R., Lin, C.J., Tsai, C.A., 2007. Reproducibility of microarray data: a further analysis of microarray quality control (MAQC) data. BMC Bioinformatics 8, 412.

Cheng, B.Q., Jia, C.Q., Liu, C.T., Fan, W., Wang, Q.L., Zhang, Z.L., Yi, C.H., 2008. Chemoembolization combined with radiofrequency ablation for patients with hepatocellular carcinoma larger than 3 cm: a randomized controlled trial. JAMA 299, 1669–1677.

Clooney, A.G., Fouhy, F., Sleator, R.D., O'Driscoll, A., Stanton, C., Cotter, P.D., Claesson, M.J., 2016. Comparing apples and oranges? Next generation sequencing and its impact on microbiome analysis. PLoS One 11, e0148028.

Colquhoun, D., 2011. Publish-or-perish: peer review and the corruption of science. Guardian, Available from: http://www.theguardian.com/science/2011/sep/05/publish-perish-peer-review-science.

Colquhoun, D., 2014. An investigation of the false discovery rate and the misinterpretation of p-values. R. Soc. Open. Sci. 1, 140216.

Collins, F.S., Tabak, L.A., 2014. Policy: NIH plans to enhance reproducibility. Nature 505, 612–613.

Comery, T.A., Martone, R.L., Aschmies, S., Atchison, K.P., Diamantidis, G., et al., 2005. Acute γ-secretase inhibition improves contextual fear conditioning in the Tg2576 mouse model of Alzheimer's disease. J. Neurosci. 25, 8898–8902.

Cook, D.J., Teves, L., Tymianski, M., 2012. A translational paradigm for the preclinical evaluation of the stroke neuroprotectant Tat-NR2B9c in Gyrencephalic nonhuman primates. Sci. Transl. Med. 4, 154ra133.

Coombes, K.R., Wang, Baggerly, K.A., 2007. Microarrays: retracing steps. Nat. Med. 13, 1226–1227.

Couzin-Frankel, J., 2017. Firing of veteran NIH scientist prompts protests over publication ban. Science. Available from: http://www.sciencemag.org/news/2017/02/firing-veteran-nih-scientist-prompts-protests-over-publication-ban.

Crabbe, J.C., Wahlsten, D., Dudek, B.C., 1999. Genetics of mouse behavior: interactions with lab environment. Science 284, 1670–1672.

Crocker, J., 2011. The road to fraud starts with a single step. Nature 479, 151.

Cryan, J.F., O'Mahoney, S.M., 2011. The microbiome-gut-brain axis: from bowel to behavior. Neurogastroenterol. Motility 23, 187–192.

Curtis, M.J., Bond, R.A., Spina, D., Ahluwalia, A., Alexander, S.P.A., et al., 2015. Experimental design and analysis and their reporting: new guidance for publication in BJP. Br. J. Pharmacol. 172, 3461–3471.

Cyranoski, D., 2004. Korea's stem-cell stars dogged by suspicion of ethical breach. Nature 429, 3.

Cyranoski, D., 2014a. Cloning comeback. Nature 505, 468–471.

Cyranoski, D., 2014b. Stem-cell pioneer blamed media 'bashing' in suicide note. Nat. News. Available from: http://www.nature.com/news/stem-cell-pioneer-blamed-media-bashing-in-suicide-note-1.15715.

Cyranoski, D., Check, E., 2005. Clone star admits lies over eggs. Nature 438, 536–537.

Davis, P., 2011. Quoted in Mandavilli A. Peer review: trial by twitter. Nature 469, 286–287.

Decullier, E., Huot, L., Samson, G., Maisonneuve, H., 2013. Visibility of retractions: a cross-sectional one-year study. BMC Res. Notes 6, 238.

Decullier, E., Huot, L., Maisonneuve, H., 2014. What time lag for a retraction search on PubMed? BMC Res. Notes 7, 395.

De Los Angeles, A., Ferrari, F., Fujiwara, Y., Mathieu, R., Lee, S., Lee, S., et al., 2015. Failure to replicate the STAP cell phenomenon. Nature 525, E6–E9.

Deer, B., 2010. Wakefield's "autistic enterocolitis" under the microscope. BMJ 340, c1127.

Deer, B., 2011a. How the case against the MMR vaccine was fixed. BMJ 342, c5347.

Deer, B., 2011b. How the vaccine crisis was meant to make money. BMJ 342, c5258.

Denayer, T., Stöhr, T., Van Roy, M., 2014. Animal models in translational medicine: validation and prediction. New Horiz. Transl. Med. 2, 5–11.

Diener, H.C., Lees, K.R., Lyden, P., Grotta, J., Davalos, A., Davis, S.M., et al., 2008. NXY-059 for the treatment of acute stroke: pooled analysis of the SAINT I and II Trials. Stroke 39, 1751–1758.

Dinan, T.G., Roman, M., Stilling, R.M., Stanton, C., Cryan, J.F., 2015. Collective unconscious: how gut microbes shape human behavior. J. Psychiat. Res. 63, 1–9.

Dirnagl, U., Lauritzeb, M., 2010. Fighting publication bias: introducing the negative results section. J. Cereb. Blood Flow Metab. 30, 1263–1264.

Dirnagl, U., Macleod, M.R., 2009. Stroke research at a road block: the streets from adversity should be paved with meta-analysis and good laboratory practice. Br. J. Pharmacol. 157, 1154–1156.

Dolgin, E., 2014. Drug discoverers chart path to tackling data irreproducibility. Nat. Rev. Drug Discov. 13, 875–876.

Doyle, A., McGarry, M.P., Lee, N.A., Lee, J.J., 2012. The construction of transgenic and gene knockout/knockin mouse models of human disease. Transgenic Res. 21, 327–349.

Drummond, C., 2009. Replicability is not reproducibility: nor is it good science. Available from: http://www.site.uottawa.ca/ICML09WS/papers/w2.pdf.

Drucker, D.J., 2016. Never waste a good crisis: confronting reproducibility in translational research. Cell Metab 24, 348–360.

Dzeng, E., 2014. How academia and publishing are destroying scientific innovation: a conversation with Sydney Brenner. King's Rev. Available from: http://kingsreview.co.uk/magazine/blog/2014/02/24/how-academia-and-publishing-are-destroying-scientific-innovation-a-conversation-with-sydney-brenner/.

Earp, B.D., Trafimow, D., 2016. Replication, falsification, and the crisis of confidence in social psychology. Front. Psychol. 6, 621, 2015.

Economist, 2013a. Unreliable research. Trouble at the lab. The Economist. Available from: http://www.economist.com/news/briefing/21588057-scientists-think-science-self-correcting-alarming-degree-it-not-trouble.

Economist, 2013b. Looks good on paper. Available from: http://www.economist.com/news/china/21586845-flawed-system-judging-research-leading-academic-fraud-looks-good-paper.

Editors of the Lancet, 2010. Retraction—Ileal-lymphoid-nodular hyperplasia, non-specific colitis, and pervasive developmental disorder in children. Lancet 375, 445.

Engber, D., 2016. Cancer research is broken. Slate. Available from: http://www.slate.com/articles/health_and_science/future_tense/2016/04/biomedicine_facing_a_worse_replication_crisis_than_the_one_plaguing_psychology.html.

Errington, T.M., Iorns, E., Gunn, W., Tan, F.E., Lomax, J., Nosek, B.A., 2014. Science Forum: an open investigation of the reproducibility of cancer biology research. eLife 3, e04333.

Estes, S., 2012. The myth of self-correcting science. Atlantic Magazine. Available from: http://www.theatlantic.com/health/archive/2012/12/the-myth-of-self-correcting-science/266228/.

Ezenwa, V.O., Gerardo, N.M., Inouye, D.W., Medina, M., Xavier, J.B., 2012. Animal behavior and the microbiome. Science 338, 198–199.

Fairbanks, C.A., Goracke-Postle, C.J., 2015. Neurobiological studies of chronic pain and analgesia: rationale and refinements. Eur. J. Pharmacol. 759, 168–181.

Fan, J., Han, F., Liu, H., 2014. Challenges of big data analysis. Natl. Sci. Rev. 1, 293–314.

Fanelli, D., 2009. How many scientists fabricate and falsify research? A systematic review and meta-analysis of survey data. PLoS One 4, e5738.

Fanelli, D., 2012. Negative results are disappearing from most disciplines and countries. Scientometrics 90, 891–904.

Fanelli, D., Ioannidis, J.P.A., 2013. US studies may overestimate effect sizes in softer research. Proc. Natl. Acad. Sci. USA 110, 15031–15036.

Fang, F.C., Casadevall, A., 2011. Retracted science and the retraction index. Infect. Immun. 79, 3855–3859.

Fang, F.C., Steen, R.G., Casadevall, A., 2012. Misconduct accounts for the majority of retracted scientific publications. Proc. Natl. Acad. Sci. USA 109, 17028–17033.

Farrell, M.S., Werge, T., Sklar, P., Owen, M.J., Ophoff, R.A., O'Donovan, M.C., et al., 2015. Evaluating historical candidate genes for schizophrenia. Mol. Psychiatr. 20, 555–562.

Faulkes, Z., 2016. Mission creep in scientific publishing. NeuroDojo. Available from: http://neurodojo.blogspot.com/2016/02/mission-creep-in-scientific-publishing.html?m=1.

Fishburn, C.S., 2014. Repairing reproducibility. SciBx 7. Available from: http://www.nature.com/scibx/journal/v7/n10/full/scibx.2014.275.html.

Fisher, K.H., Wright, V.M., Taylor, A., Zeidler, M.P., Brown, S., 2012. Advances in genome-wide RNAi cellular screens: a case study using the *Drosophila* JAK/STAT pathway. BMC Genomics 13, 506.

Flam, F., 2016. Lesson of theranos: fact-checking alone isn't enough. BloombergView. Available from: http://www.bloomberg.com/view/articles/2016-08-08/lesson-of-theranos-fact-checking-alone-isn-t-enough.

Flier, J.S., 2017. Irreproducibility of published bioscience research: diagnosis, pathogenesis and therapy. Mol. Metab. 6, 2–9.

Freedman, D.H., 2010. Lies, damned lies, and medical science. Atlantic Monthly. Available from: http://www.theatlantic.com/magazine/archive/2010/11/lies-damned-lies-and-medical-science/308269/.

Freedman, L.P., Gibson, M.C., 2015. The impact of preclinical irreproducibility on drug development. Clin. Pharmacol. Ther. 97, 16–18.

Freedman, L.P., Inglese, J., 2014. The increasing urgency for standards in basic biologic research. Cancer Res. 74, 4024–4029.

Freedman, L.P., Cockburn, I.M., Simcoe, T.S., 2015. The economics of reproducibility in preclinical research. PLoS Biol. 13, e1002165.

Fuchsberger, C., Flannick, J., Teslovich, T.M., Mahajan, A., Agarwala, V., Gaulton, K.J., et al., 2016. The genetic architecture of type 2 diabetes. Nature 536, 41–47.

Geraghty, R.J., Capes-Davis, A., Davis, J.M., Downward, J., Freshney, R.I., Knezevic, I., et al., 2014. Guidelines for the use of cell lines in biomedical research. Br. J. Cancer 111, 1021–1046.

Gewin, V., 2014. Retractions: a clean slate. Nature 507, 389–391.

Gibson, S., 2012. Don't tar discipline with Stapel brush. Times Higher Edu. Available from: https://www.timeshighereducation.com/dont-tar-discipline-with-stapel-brush/422194.article.

Gilbert, D.T., King, G., Pettigrew, S., Wilson, T.D., 2016. Comment on "estimating the reproducibility of psychological science". Science 351, 1037a.

Gillis, M., 2017. U.S. company launches a new blacklist of deceptive academic journals. University Affairs. Available from: http://www.universityaffairs.ca/news/news-article/u-s-company-launches-new-blacklist-deceptive-academic-journals/.

Glass, D.J., 2014. Experimental Design for Biologists, second ed. Cold Spring Harbor Press, New York, Cold Spring Harbor.

Global Lipids Genetics Consortium, 2013. Discovery and refinement of loci associated with lipid levels. Nat. Genet. 45, 1274–1283.

Goodlee, F., Smith, J., Marcovitch, H., 2011. Wakefield's article linking MMR vaccine and autism was fraudulent. BMJ 342, c7452.

Goldacre, B., 2012. Bad Pharma. London, Fourth Estate, 2012.

Goldberg, P., 2015. Duke officials silenced med student who reported trouble in Anil Potti's Lab. Cancer Lett. Available from: http://www.cancerletter.com/articles/20150109_1.

Goodman, S.N., Fanelli, D., Ioannidis, J.P.A., 2016. What does research reproducibility mean? Sci. Transl. Med. 8, 342ps12.

Goodstein, D., 1995. Conduct and misconduct in science. Ann. NY Acad. Sci. 775, 31–38.

Goodyear, D., 2016. The stress test. Rivalries, intrigue, and fraud in the world of stem-cell research. New Yorker. Available from: http://www.newyorker.com/magazine/2016/02/29/the-stem-cell-scandal.

Gorski, D., 2011. The wrong way to "open up" clinical trials. Science-Based Medicine. Available from: https://www.sciencebasedmedicine.org/the-wrong-way-to-open-up-clinical-trials/.

Gorski, D., 2012. The problem with preclinical research? Or: a former pharma exec discovers the nature of science. Science-Based Medicine. Available from: https://www.sciencebasedmedicine.org/the-problem-with-preclinical-research/.

Gorski, D., 2016. Is there a reproducibility "crisis" in biomedical science? No, but there is a reproducibility problem. Science-Based Medicine. Available from: https://www.sciencebasedmedicine.org/is-there-a-reproducibility-crisis-in-biomedical-science-no-but-there-is-a-reproducibility-problem/.

Greek, R., Hansen, L.A., 2013. Questions regarding the predictive value of one evolved complex adaptive system for a second: exemplified by the SOD1 mouse. Prog. Biophys. Mol. Biol. 113, 231–253.

Groenink, L., Folkerts, G., Schuurman, H.-J., 2015. European Journal of Pharmacology, special issue on translational value of animal models: introduction. Eur. J. Pharmacol. 759, 1–2.

Gross, G., 2016. Scientific Misconduct. Annu.Rev. Psychol. 67, 693–711.

Grove, A.S., 2005. Efficiency in the health care industries: a view from the outside. JAMA 294, 490–492.

Grove, A., 2011. Rethinking clinical trials. Science 333, 1679.

Harris, R., 2017. Rigor Mortis: How Sloppy Science Creates Worthless Cures, Crushes Hope, and Wastes Billions. Basic Books, New York.

Harrison, R.K., 2016. Phase II and phase III failures: 2013-2015. Nat. Rev. Drug Discov. 15, 817–818.

Hartung, T., 2013. Food for thought, look back in anger—What clinical studies tell us about preclinical work. ALTEX 30, 275–291.

Hatzipetros, T., Bogdanik, L.P., Tassinari, V.R., Kidd, J.D., Moreno, A.J., Davis, C., et al., 2014. C57BL/6J congenic Prp-TDP43A315T mice develop progressive neurodegeneration in the myenteric plexus of the colon without exhibiting key features of ALS. Brain Res. 1584, 59–72.

Heggeness, M.L., Gunsalus, K.T.W., Pacas, J.G., McDowell, G., 2017. The new face of US science. Nature 541, 21–23.

Henderson, V.C., Kimmelman, J., Fergusson, D., Grimshaw, J.M., Hackam, D.G., 2013. Threats to validity in the design and conduct of preclinical efficacy studies: a systematic review of guidelines for in vivo animal experiments. PLoS Med. 10, e1001489.

Hill, M.D., Martin, R.H., Mikulis, D., Wong, J.H., Silver, F.L., ter Brugge, K.G., et al., 2012. Safety and efficacy of NA-1 in patients with iatrogenic stroke after endovascular aneurysm repair (ENACT): a phase 2, randomised, double-blind, placebo-controlled trial. Lancet Neurol. 11, 942–950.

Hiltzik, M., 2014. More on the unsavory history of the vaccine-autism 'link'. LA Times, Available from: http://www.latimes.com/business/hiltzik/la-fi-mh-vaccineautism-link-20140122,0,1151028.story#axzz2rXPki7fV.

Hirst, J.A., Howick, J., Aronson, J.K., Roberts, N., Perera, R., Koshiaris, C., et al., 2014. The need for randomization in animal trials: an overview of systematic reviews. PLoS One 9, e98856.

Hobin, J.A., Deschamps, A.M., Bockman, R., Cohen, S., Dechow, P., Eng, C., et al., 2012. Engaging basic scientists in translational research: identifying opportunities, overcoming obstacles. J. Transl. Med. 10, 72.

Hodgson, J., 2016. When biotech goes bad. Nat. Biotech. 34, 284–291.

Hoffman, J., 2016. Vaxxed review—one-sided film leaves the elephant in the room. Guardian. Available from: https://www.theguardian.com/film/2016/apr/02/vaxxed-from-cover-up-to-catastrophe-review.

Holder, D.J., Marino, M.J., 2017. Enhancing reproducibility: logic in experimental design and execution in pharmacology and drug discovery. Curr. Protocol Pharmacol. 76, A.3G.1–A.3G.26.

Hongo, J., 2014. Timeline: the rise and fall of Haruko Obokata in 2014. Wall St. J., Japan JAPANREALTIME. Available from: http://blogs.wsj.com/japanrealtime/tag/haruko-obokata/.

Horrabin, D.F., 2001. Something rotten at the core of science? Trends Pharmacol. Sci. 22, 51–52.

Horrabin, D., 2003. Modern biomedical research: an internally self-consistent universe with little contact with medical reality? Nat. Rev. Drug Discov. 2, 151–154.

Horton, R., 2000. Genetically modified food: consternation, confusion, and crack-up. Med. J. Aust. 172, 148–149.

Horton, R., 2015. Offline: what is medicine's 5 sigma? Lancet 285, 1380.

Howe, D., Costanzo, M., Fey, P., Gojobori, T., Hannick, L., et al., 2008. The future of biocuration. Nature 455, 47–50.

Hróbjartsson, A., Gøtzsche, P.C., 2010. Placebo interventions for all clinical conditions. Cochrane Database Syst. Rev. 20 (1), CD003974.

Hsu, J.C., Chang, J., Wang, T., Steingrímsson, E., Magnússon, M.K., Bergsteinsdottir, K., 2007. Statistically designing microarrays and microarray experiments to enhance sensitivity and specificity. Brief Bioinform. 8, 22–31.

Hwang, W.S., Ryu, Y.J., Park, J.H., Park, E.S., Lee, E.G., Koo, J.M., et al., 2004. Evidence of a pluripotent human embryonic stem cell line derived from a cloned blastocyst. Science 303, 1669–1674.

Hwang, W.S., Roh, S.I., Lee, B.C., Kang, S.K., Kwon, D.K., Kim, S., et al., 2005. Patient-specific embryonic stem cells derived from human SCNT blastocysts. Science 306, 1777–1783.

Interlandi, J., 2006; An Unwelcome Discovery. New York Times. Available from: http://www.nytimes.com/2006/10/22/magazine/22sciencefraud.html.

Ioannidis, J.P.A., 2005. Why most published research findings are false. PLoS Med. 2, e124.

Ioannidis, J.P.A., 2006. Evolution and translation of research findings: from bench to where? PLoS Clin. Trials 1, e36.

Ioannidis, J.P.A., 2014. How to make more published research true. PLoS Med. 11, e1001747.

Ioannidis, J.P.A., 2015. Stealth research. Is biomedical innovation happening outside the peer-reviewed literature? JAMA 313, 663–664.

Ioannidis, J.P.A., 2016a. Why most clinical research is not useful. PLoS Med. 13, e1002049.

Ioannidis, J.P.A., 2016b. The mass production of redundant, misleading, and conflicted systematic reviews and meta-analyses. Milbank Quart. 94, 485–545.

Ioannidis, J.P.A., Khoury, M.J., 2014. Assessing value in biomedical research: the PQRST of appraisal and reward. JAMA 312, 483–484.

Ioannidis, J.P., Trikalinos, T.A., 2005. Early extreme contradictory estimates may appear in published research: the Proteus phenomenon in molecular genetics research and randomized trials. J. Clin. Epidemiol. 58, 543–549.

Ioannidis, J.P.A., Allison, D.B., Ball, C.A., Coulibaly, I., Cui, X., et al., 2009. Repeatability of published microarray gene expression analyses. Nat. Genetics 41, 149–155.

Iizuka, M., Chiba, M., Yukawa, M., Nakagomi, T., Fukushima, T., Watanabe, S., Nakagomi, O., 2000. Immunohistochemical analysis of the distribution of measles related antigen in the intestinal mucosa in inflammatory bowel disease. Gut 46, 163–169.

Jain, A., Marshall, J., Buikema, A., Bancroft, T., Kelly, J.P., Newschaffer, C.J., 2015. Autism occurrence by MMR vaccine status among US children with older siblings with and without Autism. JAMA 313, 1534–1540, 2015.

Jarvis, M.F., Williams, M., 2016. Irreproducibility in preclinical biomedical research: perceptions, uncertainties and knowledge gaps. Trends Pharmacol. Sci. 37, 290–302.

Jian, C., Zou, D., Liu, X., Meng, L., Huang, J., et al., 2016. Cognitive deficits are ameliorated by reduction in amyloidβ accumulation in Tg2576/p75(NTR+/−) mice. Life Sci. 155, 167–173.

Jiji, 2016. Obokata sticks to guns, launches website with instructions for making STAP cells. Japan Times. Available from: http://www.japantimes.co.jp/news/2016/04/01/national/science-health/obokata-sticks-guns-launches-website-boasting-way-make-stap-cells/#.V3MzeVfKKeN.

Jobe, P.C., Adams-Curtis, L.E., Burks, T.F., Fuller, R.W., Peck, C.C., Ruffolo, R.R., Snead, O.C., et al., 1994. The essential role of integrative bio-medical sciences in protecting and contributing to the health and well-being of our nation. Physiologist 37, 79–86.

Jump, P., 2011. A star's collapse. Times Higher Education. Available from: https://www.insidehighered.com/news/2011/11/28/scholars-analyze-case-massive-research-fraud.

Kaiser, M., 2014. The integrity of science—lost in translation? Best Pract. Res. Clin. Gastroenterol. 28, 339–347.

Kahn, R.S., Sommer, I.E., Murray, R.M., Meyer-Lindenberg, A., Weinberger, D.R., Cannon, T.D., et al., 2015. Schizophrenia. Nat. Rev. Dis. Primers 1, 15067.

Kakluk, P., 2009. The legacy of the Hwang case: research misconduct in biosciences. Sci. Eng. Ethics 15, 545–562.

Kass, R.E., Caffo, B.S., Davidian, M., Meng, X.-L., Yu, B., Reid, N., 2016. Ten simple rules for effective statistical practice. PLoS Comput. Biol. 12, e1004961.

Kenakin, T., Bylund, D.B., Toews, M.L., Mullane, K., Winquist, R.J., Williams, M., 2014. Replicated, replicable and relevant—target engagement and pharmacological experimentation in the 21st Century. Biochem. Pharmacol. 87, 64–77.

Kerr, N.L., 1998. HARKing: hypothesizing after the results are known. Pers. Soc. Psychol. Rev. 2, 196–217.

Kimmelman, J., Federico, C., 2017. Consider drug efficacy before first-in-human trials. Nature 542, 25–27.

Kleiman, R.J., Ehlers, M.D., 2016. Data gaps limit the translational potential of preclinical research. Sci. Transl. Med. 8, 320 ps1.

Kochan, C.A., Budd, J.M., 1992. The persistence of fraud in the literature: the Darsee case. J. Am. Soc. Info. Sci. 43, 488–493.

Kolata, G., 2011. How bright promise in cancer testing fell apart. NY Times. Available from: http://www.nytimes.com/2011/07/08/health/research/08genes.html?_r=0.

Kornfeld, D.S., 2012. Perspective: research misconduct: the search for a remedy. Acad. Med. 87, 877–882.

Kornfeld, D.S., Titus, S.L., 2016. Stop ignoring misconduct. Nature 537, 29–30.

Kossoff, M., 2017. Theranos's latest lawsuit may be its worst yet. Vanity Fair. Available from: http://www.vanityfair.com/news/2017/01/theranoss-latest-lawsuit-may-be-its-worst-yet.

Kotecki, P., 2016. In focus: as Lyrica profits dry up, Northwestern seeks another 'blockbuster' drug. dailynorthwestern.com. Available from: http://dailynorthwestern.com/2016/04/10/in-focus/in-focus-as-lyrica-profits-dry-up-northwestern-seeks-another-blockbuster-drug/.

Kranke, P., Apfel, C.C., Roewer, N., 2000. Reported data on Granisetron and postoperative nausea and vomiting by Fujii et al. are incredibly nice! Anesth. Anal. 90, 1004–1006.

Kraus, W.L., 2014. Editorial: do you see what i see? Quality, reliability, and reproducibility in biomedical research. Mol. Endocrinol. 38, 277–280.

Lau, J., Ioannidis, J.P.A., Terrin, N., Schmid, C.H., Olkin, I., 2006. The case of the misleading funnel plot. BMJ 333, 597–600.

Lee, B.C., Kim, M.K., Jang, G., Oh, H.J., Yuda, F., Kim, H.J., et al., 2005. Dogs cloned from adult somatic cells. Nature 436, 641.

Lehrer, J., 2010. Annals of science. The truth wears off. Is There something wrong with the scientific method? New Yorker. Available from: http://archives.newyorker.com/?i=2010-12-13#folio=052.

Lemaitre, B., 2016. An essay on science and narcissism: how do high-ego personalities drive research in life sciences? brunolemaitre.ch, Switzerland, Lausanne. Available from: https://www.amazon.com/Essay-Science-Narcissism-high-ego-personalities-ebook/dp/B01DS47AN4.

Levelt Committee, 2011. Interim report regarding the breach of scientific integrity by Prof. D. A. Stapel. Tilburg University. Available from: https://www.tilburguniversity.edu/upload/547aa461-6cd1-48cd-801b-61c434a73f79_interim-report.pdf.

Llovera, G., Hofmann, K., Roth, S., Salas-Pérdomo, A., Ferrer-Ferrer, M., Perego, C., et al., 2015. Results of a preclinical randomized controlled multicenter trial (pRCT): anti-CD49d treatment for acute brain ischemia. Sci. Transl. Med. 7, 299ra121.

Longo, D.L., Drazen, J.M., 2016. Data sharing. N. Engl. J. Med. 374, 276–277.

Loscalzo, J., 2012. Irreproducible experimental results: causes, (mis) interpretations, and consequences. Circulation 125, 1211–1214.

Lowe, D., 2011. Andy Grove's idea for opening up clinical trials. In the pipeline. Sci. Transl. Med. Available from: http://blogs.sciencemag.org/pipeline/archives/2011/09/28/andy_groves_idea_for_opening_up_clinical_trials.

Lu, S.F., Jin, G.Z., Uzzi, B., Jones, B., 2013. The retraction penalty: evidence from the web of science. Sci. Rep. 3, 3146.

Lucanic, M., Plummer, W.T., Chen, E., Harke, J., Foulger, A.C., Onken, B., et al., 2017. Impact of genetic background and experimental reproducibility on identifying chemical compounds with robust longevity effects. Nat. Commun. 8, 14256.

Lukk, M., Kapushesky, M., Nikkilä, J., Parkinson, H., Goncalves, A., et al., 2010. A global map of human gene expression. Nat. Biotech. 28, 322–324.

Macleod, M.R., Van der Worp, B., Sena, E.S., Howells, D.W., Dirnagl, U., Donnan, G.A., 2008. Evidence for the efficacy of NXY-059 in experimental focal cerebral ischaemia is confounded by study quality. Stroke 39, 2824–2829.

Macleod, M.R., Fisher, M., O'Collins, V., Sena, E.S., Dirnagl, U., Bath, P.M., et al., 2009. Good laboratory practice: preventing introduction of bias at the bench. J. Cereb. Blood Flow Metab. 29, 221–223.

Macleod, M.R., Michie, S., Roberts, I., Dirnagl, U., Chalmers, I., Ioannidis, J.P.A., et al., 2014. Biomedical research: increasing value, reducing waste. Lancet 383, 101–104, 2014.

Macrae, I.M., 2011. Preclinical stroke research—advantages and disadvantages of the most common rodent models of focal ischaemia. Br. J. Pharmacol. 164, 1062–1078.

Majumder, M.S., Cohn, E.L., Mekaru, S.R., Huston, J.E., Brownstein, J.S., 2015. Substandard vaccination compliance and the 2015 measles outbreak. JAMA Pediatr. 169, 494–495.

Marcus, A., Oransky, I., 2011. Science publishing: the paper is not sacred. Nature 480, 449–450.

Marcus, A., Oransky, I., 2014. What studies of retractions tell us. J. Microbiol. Biol. Edu. 15, 151–154.

Marcus, A., Oransky, I., 2015. How the biggest fabricator in science got caught. Nautilus. Available from: http://nautil.us/issue/24/error/how-the-biggest-fabricator-in-science-got-caught.

Marcus, A., McCook, A., Oransky, I., 2015. The top 10 retractions of 2015. The Scientist. Available from: http://www.the-scientist.com/?articles.view/articleNo/44895/title/The-Top-10-Retractions-of-2015/.

Marino, M., 2014. The use and misuse of statistical methodologies in pharmacology research. Biochem. Pharmacol. 87, 78–92.

Markel, H., 2013. Patents, Profits, and the American People—The Bayh–Dole Act of 1980. N. Engl. J. Med. 369, 794–796.

Mathur, M., 2016. Replication of "Why People are Reluctant to Tempt Fate" by Risen & Gilovich (2008, J. Personal. Social Psychol.) Risen & Gilovich replication writeup.pdf (Version: 1). OSC. Available from: https://osf.io/nwua6/.

Matosin, N., Frank, E., Engel, M., Lum, J.S., Newell, K.A., 2014. Negativity towards negative results: a discussion of the disconnect between scientific worth and scientific culture. Dis. Model. Mech. 7, 171–173.

McClain, S., 2013. Not breaking news: many scientific studies are ultimately proved wrong! Guardian. Available from: http://www.theguardian.com/science/occams-corner/2013/sep/17/scientific-studies-wrong.

McGonigle, P., Ruggeri, B., 2014. Animal models of human disease: challenges in enabling translation. Biochem. Pharmacol. 87, 162–171.

McGonigle, P., Williams, M., 2014. Preclinical pharmacology and toxicology - contributions to the translational interface. Ref Module Biomed Sci. Available from: http://dx.doi.org/10.1016/B978-0-12-801238-3.05242-9.

McGorry, R.W., Webster, B.S., Snook, S.H., Hsiang, S.M., 2000. The relation between pain intensity, disability, and the episodic nature of chronic and recurrent low back pain. Spine 25, 834–841.

Melé, M., Ferreira, P.G., Reverter, F., DeLuca, D.S., Monlong, J., et al., 2015. The human transcriptome across tissues and individuals. Science 348, 660–665.

Menezes, P., Miller, W.C., Wohl, D.A., Adimora, A.A., Leone, P.A., Miller, W.C., Eron, Jr., J.J., 2011. Does HAART efficacy translate to effectiveness? Evidence for a trial effect. PLos One 6 (7), e21824.

Michael, A., 2015. Ask The Chefs: How Can We Improve the Article Review and Submission Process? the scholarly kitchen. Available from: http://scholarlykitchen.sspnet.org/2015/03/26/ask-the-chefs-how-can-we-improve-the-article-review-and-submission-process/.

Mirowski, P., 2012. The Modern Commercialization of Science is a Passel of Ponzi Schemes. Social Epistemol. 26, 285–310.

Mobley, A., Linder, S.K., Braeuer, R., Ellis, L.M., Zwelling, L., 2013. A survey of data reproducibility in cancer research provides insights into our limited ability to translate findings from the laboratory to the clinic. PloS One 8, e6322.

Moher, D., Glasziou, P., Chalmers, I., Nasser, M., Bossuyt, P.M.M., Korevaar, D.A., et al., 2016. Increasing value and reducing waste in biomedical research: who's listening? Lancet 397, 1573–1586.

Morris, A.P., Voight, B.F., Teslovich, T.M., Ferreira, T., Segrè, A.V., et al., 2012. Large-scale association analysis provides insights into the genetic architecture and pathophysiology of type 2 diabetes. Nat. Genet. 44, 981–990.

Morrison, S.J., 2014. Reproducibility project: cancer biology: time to do something about reproducibility. eLife 3, e03981.

Motulsky, H.J., 2014. Common misconceptions about data analysis and statistics. J. Pharmacol. Exp. Ther. 351, 200–205.

Moyé, L.A., Deswal, A., 2002. The fragility of cardiovascular clinical trial results. J. Card. Fail. 8, 247–253.

Muhlhausler, B.S., Bloomfield, F.H., Gillman, M.W., 2013. Whole animal experiments should be more like human randomized controlled trials. PLoS Biol. 11, e1001481.

Mullane, K., Williams, M., 2015. Unknown unknowns in biomedical research: does an inability to deal with ambiguity contribute to issues of irreproducibility? Biochem. Pharmacol. 97, 133–136.

Mullane, K., Williams, M., 2017. Enhancing reproducibility: failures from reproducibility initiatives underline core challenges. Biochem. Pharmacol. 138, 7–18.

Mullane, K., Enna, S.J., Piette, J., Williams, M., 2015. Guidelines for manuscript submission in the peer-reviewed pharmacological literature. Biochem. Pharmacol. 97, 224–239.

Munafò, M.R., Nosek, B.A., Bishop, D.V.M., Button, K.S., Chambers, C.D., et al., 2017. A manifesto for reproducible science. Nat. Hum. Behav. 1 Article no. 0021.

Nardone, R.M., 2008. Curbing rampant cross-contamination and misidentification of cell lines. BioTechniques 45, 221–227.

National Academies Press (NAP), 2015. Institute for Laboratory Animal Research. Round Table on Science and Welfare in Laboratory Animal Use. Reproducibility Issues in Research with Animals and Animal Models: Workshop in Brief. Washington, DC, The National Academies Press. Available from: http://www.nap.edu/catalog/21835/reproducibility-issues-in-research-with-animals-and-animal-models-workshop.

National Academies of Science (NASEM), 2017. Engineering and Medicine. Fostering Integrity in Research. Washington, DC, The National Academies Press. Available from: http://www.nap.edu/21896.

Nature, 2012. Editorial note. Nature 485, 41.

Nature, 2013. Time to talk. Online discussion is an essential aspect of the post-publication review of findings. Nature 502, 593–594.

Nature Editorial, 2016. Repetitive flaws. Nature 529, 256.

Nature Medicine Editorial, 2016. Take the long view. Nat. Med. 22, 1.

Neimark, J., 2014. The dirty little secret of cancer research. Discover. Available from: http://discovermagazine.com/2014/nov/20-trial-and-error.

Neimark, J., 2015. Line of attack. Science 347, 938–940.

Ni, J., Koyuturk, M., Tong, H., Haines, J., Xu, R., Zhang, X., 2016. Disease gene prioritization by integrating tissue-specific molecular networks using a robust multi-network model. BMC Informatics 17, 453.

Nisbet, M., Markowitz, E.M., 2014. Understanding public opinion in debates over biomedical research: looking beyond political partisanship to focus on beliefs about science and society. PLoS One 9, e88473.

Normile, D., Vogel, G., 2014. STAP cells succumb to pressure. Science 344, 1215–1216.

Nosek, B.A., Errington, T.M., 2017. Making sense of replications. eLife 6, e23383.

Nutt, A.E., 2016. The scientist nearly went to jail for making up data. Washington Post. Available from: https://www.washingtonpost.com/news/speaking-of-science/wp/2016/04/01/when-scientists-lie-about-their-research-should-they-go-to-jail/?utm_term=.6bf002c03709.

O'Bryant, S.E., Waring, S.C., Cullum, C.M., Hall, J., Lacritz, L., et al., 2008. Staging dementia using clinical dementia rating scale sum of boxes scores: a Texas Alzheimer's research consortium study. Arch. Neurol. 65, 1091–1095.

O'Collins, V.E., Macleod, M.R., Donnan, G.A., Horky, L.L., van der Worp, B.H., Howells, D.W., 2006. 1,026 experimental treatments in acute stroke. Ann. Neurol. 59, 467–477.

Obokata, H., Wakayama, T., Sasai, Y., Kojima, K., Vacanti, M.P., Niwa, H., et al., 2014a. Retracted: stimulus-triggered fate conversion of somatic cells into pluripotency. Nature 505, 641–647.

Obokata, H., Sasai, Y., Niwa, H., Kadota, M., Andrabi, M., Takata, N., et al., 2014b. Retracted: bidirectional developmental potential in reprogrammed cells with acquired pluripotency. Nature 505, 676–680.

Omenn, G.S., Lane, L., Lundberg, E.K., Beavis, R.C., Nesvizhskii, A.I., Deutsch, E.W., 2015. Metrics for the human proteome project 2015: progress on the human proteome and guidelines for high-confidence protein identification. J. Proteome Res. 14, 3452–3460.

OSC (Open Science Collaboration), 2015. Estimating the reproducibility of psychological science. Science 349, aac4716.

Panagiotou, O.A., Willer, C.J., Hirschhorn, J.N., Ioannidis, J.P.A., 2013. The power of meta-analysis in genome-wide association studies. Annu. Rev. Genomics Hum. Genet. 14, 441–465.

Pattinson, D., 2012. Plos One launches reproducibility initiative. Available from: http://blogs.plos.org/everyone/2012/08/14/plos-one-launches-reproducibility-initiative/.

Peres-Neto, P.R., 2016. Will technology trample peer review in ecology? Ongoing issues and potential solutions. Okios 125, 3–9.

Perrin, S., 2014. Preclinical research: make mouse studies work. Nature 507, 423–425.

Peters, S.M., Pothuizen, H.H.J., Spruijt, B.M., 2015. Ethological concepts enhance the translational value of animal models. Eur. J. Pharmacol. 759, 42–50.

Piwowar, H.A., Vision, T.J., Whitlock, M.C., 2011. Data archiving is a good investment. Nature 473, 285.

Pfeifer, M.P., Snodgrass, G.L., 1990. The continued use of retracted, invalid scientific literature. JAMA 263, 1420–1423.

Pfeiffer, T., Bertram, L., Ioannidis, J.P.A., 2011. Quantifying selective reporting and the proteus phenomenon for multiple datasets with similar bias. PLoS One 6, e18362.

Potti, A., Dressman, H.K., Bild, A., Riedel, R.F., Chan, G., Sayer, R., et al., 2006. Retracted: genomic signatures to guide the use of chemotherapeutics. Nat. Med. 12, 1294–1300.

Price, A.L., Spencer, C.C.A., Donnelly, P., 2015. Progress and promise in understanding the genetic basis of common diseases. Proc. R. Soc. B 282, 20151684.

Prinz, F., Schlange, T., Asadullah, K., 2011. Believe it or not: how much can we rely on published data on potential drug targets? Nat. Rev. Drug Discov. 10, 712–713.

Prud'Homme, A., 2004. The cell game. Sam Waksal's Fast Money and False Promises—and the Fate of Imclone's Cancer Drug. Harper Business, New York.

Pulverer, B., quoted in Meadows A. "Research Mechanics", OA, Ethics, and More: Three Chefs Musings on APE 2016. The Scholarly Kitchen. Available from: https://scholarlykitchen.sspnet.org/2016/02/03/research-mechanics-oa-ethics-and-more-three-chefs-musings-on-ape-2016/.

Pusztai, L., Hatzis, C., Andre, F., 2013. Reproducibility of research and preclinical validation: problems and solutions. Nat. Rev. Clin. Oncol. 10, 720–724.

Ramkumar, A., 2015. Duke lawsuit involving cancer patients linked to Anil Potti settled. Duke Chronicle. Available from: http://www.dukechronicle.com/article/2015/05/duke-lawsuit-involving-cancer-patients-linked-anil-potti-settled.

Rasko, J., Power, C., 2015. What pushes scientists to lie? The disturbing but familiar story of Haruko Obokata. Available from: http://www.theguardian.com/science/2015/feb/18/haruko-obokata-stap-cells-controversy-scientists-lie.

Redman, B.K., 2015. Are the Biomedical Sciences Sliding Toward Institutional Corruption? And Why Didn't We Notice It? Edmond, J. Safra Working Papers, No. 59. Harvard University, 2015. Available from: http://papers.ssrn.com/sol3/papers.cfm?abstract_id=2585141.

Rennie, D., 1986. Guarding the guardians A conference on editorial peer review. JAMA 256, 2391–2392.

Rennie, D., Flanagin, A., Smith, R., Smith, J., 2003. Fifth international congress on peer review and biomedical publication. Call for research. JAMA 289, 1438.

Retraction Watch, 2016. Lawsuit against Ole Miss for rescinded Sarkar job offer dismissed; briefs filed in PubPeer case. RetractionWatch. Available from: http://retractionwatch.com/2016/01/20/judge-dismissed-lawsuit-against-ole-miss-for-rescinded-offer/.

Rogers, G.B., Kozlowska, J., Keeble, J., Metcalfe, K., Fao, M., Dowd, S.E., et al., 2014. Functional divergence in gastrointestinal microbiota in physically-separated genetically identical mice. Sci. Rep. 4, 5437.

Rosenthal, R., 1979. The file drawer problem and tolerance for null results. Psychol. Bull. 86, 638–641.

Roth, K.A., Cox, A.E., 2015. Science isn't science if it isn't reproducible. Am. J. Pathol. 185, 2–3.

Rung, J., Brazma, A., 2013. Reuse of public genome-wide gene expression data. Nat. Rev. Genet. 14, 89–99.

Rupp, B., Wlodawer, A., Minor, W., Helliwell, J.R., Jaskolski, M., 2016. Correcting the record of structural publications requires joint effort of the community and journal editors. FEBS J. 283, 4452–4457.

Ruxton, G.D., Colegrave, N., 2011. Experimental Design in the Life Sciences. Oxford University Press, Oxford, UK.

Salmon, D.A., McIntyre, C.R., Omer, S.B., 2015. Making mandatory vaccination truly compulsory: well intentioned but ill conceived. Lancet Infect. Dis. 15, 872–873.

Sams-Dodd, F., 2006. Strategies to optimize the validity of disease models in the drug discovery process. Drug Discov. Today 11, 355–362.

Schekman, R., 2013. How journals like Nature, Cell and Science are damaging science. Guardian, Available from: http://www.theguardian.com/commentisfree/2013/dec/09/how-journals-nature-science-cell-damage-science.

Schmidt, C.W., 2014. Research wranglers: initiative to improve reproducibility of study findings. Environ. Health Perspec. 122, A188–A191.

Schmidt, S., 2009. Shall we really do it again? The powerful concept of replication is neglected in the social sciences. Rev. Gen. Psychol. 13, 90–100.

Scott, M., 2016. Everything you need to know about the Theranos saga so far. Wired. Available from: http://www.wired.com/2016/05/everything-need-know-theranos-saga-far/.

Scott, S., Kranz, J.E., Cole, J., Lincecum, J.M., Thompson, K., et al., 2008. Design, power, and interpretation of studies in the standard murine model of ALS. Amyotroph. Lateral Scler. 9, 4–15.

Scotter, E.L., Chen, H.J., Shaw, C.E., 2015. TDP-43 proteinopathy and ALS: insights into disease mechanisms and therapeutic targets. Neurotherapeutics 12, 352.

Sen, S., 2011. Francis Galton and regression to the mean. Significance (The Royal Statistical Society), pp. 124–126.

Sena, E.S., van der Worp, H.B., Bath, P.M.W., Howells, D.W., Macleod, M.R., 2010. Publication bias in reports of animal stroke studies leads to major overstatement of efficacy. PLoS Biol. 8, e1000344.

Serikawa, K., 2015. Baseball, regression to the mean, and avoiding potential clinical trial biases. Available from: https://kyleserikawa.com/2015/05/18/baseball-regression-to-the-mean-and-avoiding-potential-clinical-trial-biases/.

Senapathy, K., 2016. No Andrew Wakefield, You're Not Being Censored And You Don't Deserve Due Process. Forbes.com Opinion. Available from: http://www.forbes.com/sites/kavinsenapathy/2016/03/28/no-andrew-wakefield-youre-not-being-censored-and-you-dont-deserve-due-process/#6715cbc0225d.

Shafer, S., 2011. Research Fraud in Anesthesia. American Society of Anesthesiologists Newsletter. Available from: http://www.asahq.org/resources/publications/newsletter-articles/2011/may2011/research-fraud-in-anesthesia.

Shen, C., Björk, B.-C., 2015. Predatory open access: a longitudinal study of article volumes and market characteristics. BMC Med. 13, 230.

Shimasaki, C., 2014. Biotechnology Entrepreneurship. Starting, Managing, and Leading Biotech Companies. Elsevier Academic, Waltham, MA.

Simmons, J.P., Nelson, L.D., Simonsohn, U., 2011. False-positive psychology: undisclosed flexibility in data collection and analysis allows presenting anything as significant. Psychol. Sci. 22, 1359–1366.

Singhal, A., Leaman, R., Catlett, N., Lemberger, T., McEntyre, J., Polson, et al., 2016. Pressing needs of biomedical text mining in biocuration and beyond: opportunities and challenges. Database 2016.

Siva, N., 2010. Wakefield's first try. Slate. Available from: http://www.slate.com/articles/health_and_science/medical_examiner/2010/06/wakefields_first_try.html.

Smaldino, P.E., McElreath, R., 2016. The natural selection of bad science. R. Sci. Open Sci. 3, 160384.

Smith, R., 2006. Peer review: a flawed process at the heart of science and journals. J. R. Soc. Med. 99, 178–182.

Snyder, P.J., Mayes, L.C., Smith, W.E., 2015. The Management of Scientific Integrity within Academic Medical Centers. Academic Press, San Diego.

Song, F., Hooper, L., Loke, Y.K., 2013. Publication bias: what is it? How do we measure it? How do we avoid it? Open Access J. Clin. Trials 2013, 71–81.

Sorge, R.E., Martin, L.J., Isbester, K.A., Sotocinal, S.G., Rosen, R., et al., 2014. Olfactory exposure to males, including men, causes stress and related analgesia in rodents. Nat. Methods 11, 629–632.

Souder, L., 2011. The ethics of scholarly peer review: a review of the literature. Learned Pub. 24, 55–74.

Stapel, D.A., Lindenberg, S., 2011. Retracted: coping with chaos: how disordered contexts promote stereotyping and discrimination. Science 332, 251–253.

Steen, R.G., 2011a. Retractions in the scientific literature: do authors deliberately commit research fraud? J. Med. Ethics 37, 113–117.

Steen, R.G., 2011b. Retractions in the scientific literature: is the incidence of research fraud increasing? J. Med. Ethics 37, 249–253.

Steen, R.G., 2011c. Retractions in the medical literature: how many patients are put at risk by flawed research? J. Med. Ethics 37, 688–692.

Steen, R.G., Casadevall, A., Fang, F.C., 2013. Why has the number of scientific retractions increased? PLoS One 8, e68397.

Stemwedel, J.D., 2015. Is aggressive science reporting a human rights violation? Forbes. Available from: http://www.forbes.com/sites/janetstemwedel/2015/08/29/is-aggressive-science-reporting-a-human-rights-violation/#4fa718257488.

Stewart, W.W., Feder, N., 1987. The integrity of the scientific literature. Nature 325, 207–214.

Stone, A., 2005. The cheating culture. BusinessWeek Archives. Available from: http://www.bloomberg.com/bw/stories/2005-06-20/the-cheating-culture.

Stoye, E., 2015. Post publication peer review comes of age. Chemistry World. Available from: http://www.rsc.org/chemistryworld/2015/01/post-publication-peer-review-stap-comes-age.

Stroebe, W., Postmes, T., Spears, R., 2012. Scientific misconduct and the myth of self-correction in science. Persp. Psychol Sci. 7, 670–688.

Sung, N.S., Crowley, Jr., W.F., Genel, M., Salber, P., Sandy, L., et al., 2003. Central challenges facing the national clinical research enterprise. JAMA 289, 1278–1287.

Suresh, K.P., 2011. An overview of randomization techniques: an unbiased assessment of outcome in clinical research. J. Hum. Reprod. Sci. 4, 8–11.

Tarkan, L., 2016. Why Robert De Niro Promoted—then Pulled—Anti-Vaccine Documentary. Fortune. Available from: http://fortune.com/2016/03/29/robert-de-niro-anti-vaccine-documentary/.

ter Riet, G., Korevaar, D.A., Leenaars, M., Sterk, P.J., Van Noorden, C.J.F., et al., 2012. Publication bias in laboratory animal research: a survey on magnitude, drivers, consequences and potential solutions. PLoS One 7, e43404.

Tharyan, P., 2012. Criminals in the citadel and deceit all along the watchtower: irresponsibility, fraud, and complicity in the search for scientific truth. Mens Sana Monogr. 10, 158–180.

Thomas, J., 2010. Paranoia Strikes Deep: MMR Vaccine and Autism. Psychiatric Times. Available from: http://www.psychiatrictimes.com/autism/%E2%80%9Cparanoia-strikes-deep%E2%80%9D-mmr-vaccine-and-autism#sthash.PDAqrm2v.dpuf.

Torrente, A., Lukk, M., Xue, V., Parkinson, H., Rung, J., Brazma, A., 2016. Identification of cancer related genes using a comprehensive map of human gene expression. PLoS One 11 (6), e0157484.

Triggle, D.J., Miller, K.W., 2002. Doctoral education: another tragedy of the commons? Am. J. Pharm. Edu. 66, 287–294.

Triggle, C.R., Williams, M., 2015. Challenges in the biomedical research enterprise in the 21st century: antecedents in the writings of David Triggle. Biochem. Pharmacol. 98, 342–359.

Tsilidis, K.K., Panagiotou, O.A., Sena, E.S., Aretouli, E., Evangelou, E., Howells, D.W., et al., 2013. Evaluation of excess significance bias in animal studies of neurological diseases. PLoS Biol. 11, e1001609.

van Dalen, H.P., Henkens, K., 2012. Intended and unintended consequences of a publish-or-perish culture: a worldwide survey. J. Am. Soc. Inform. Sci. Technol. 63, 1282–1293.

Van der Staay, F.J., Steckler, T., 2002. The fallacy of behavioral phenotyping without standardization. Genes Brain Behav. 1, 9–13.

van der Vet, P.E., Nijveen, H., 2016. Propagation of errors in citation networks: a study involving the entire citation network of a widely cited paper published in, and later retracted from, the journal Nature. Res. Integr. Peer Rev. 1, 3.

van der Worp, H.B., van Gijn, J., 2007. Clinical practice. Acute ischemic stroke. N. Engl. J. Med. 357, 572–579.

van der Worp, H.B., Howells, D.W., Sena, E.S., Porritt, M.J., Rewell, S., O'Collins, V., et al., 2010. Can animal models of disease reliably inform human studies? PLoS Med. 7, e1000245.

Vanden Berghe, T., Hulpiau, P., Martens, L., Vandenbroucke, R.E., Van Wonterghem, E., Perry, S.W., et al., 2015. Passenger mutations confound interpretation of all genetically modified congenic mice. Immunity 42, 200–209.

Van Noorden, R., 2011. Science publishing: the trouble with retractions. Nature 478, 26–28.

Van Noorden, R., 2014. Parasite test shows where validation studies can go wrong. Nature. Available from: http://www.nature.com/news/parasite-test-shows-where-validation-studies-can-go-wrong-1.16527.

Vandamme, T.F., 2015. Rodent models for human diseases. Eur. J. Pharmacol. 759, 84–89.

Vasagar, J., 2001. Rise of the wealthy Oxford scientists. Guardian. Available from: https://www.theguardian.com/uk/2001/apr/21/highereducation.education.

Vasilevsky, N.A., Brush, M.H., Paddock, H., Ponting, L., Tripathy, S.J., LaRocca, G.M., Haendel, M., 2013. On the reproducibility of science: unique identification of research resources in the biomedical literature. Peer J. 1, e148.

Vatner, S.F., 2016. Why so few new cardiovascular drugs translate to the clinics. Circ. Res. 119, 714–717.

Verfaellie, M., McGwin, J., 2011. The case of Diederik Stapel. Psychological Science Agenda, American Psychological Association. Available from: http://www.apa.org/science/about/psa/2011/12/diederik-stapel.aspx.

Wade, N., 2010. Harvard Finds Scientist Guilty of Misconduct. New York Times. Available from: http://www.nytimes.com/2010/08/21/education/21harvard.html?_r=0.

Wadman, M., 2013. NIH mulls rules for validating key results. Nature 500, 14–16.

Wakefield, A.J., Pittilo, R.M., Sim, R., Cosby, S.L., Stephenson, J.R., Dhillon, A.P., Pounder, R.E., 1993. Evidence of persistent measles virus infection in Crohn's disease. J. Med. Virol. 39, 345–353.

Wakefield, A.J., Murch, S.H., Anthony, A., Linnell, J., Casson, D.M., Malik, M., et al., 1998. RETRACTED: Ileal-lymphoid-nodular hyperplasia, non-specific colitis, and pervasive developmental disorder in children. Lancet 351, 637–641.

Ware, M., Mabe, M., 2015. STM Report, fourth ed. Available from: http://www.stm-assoc.org/2015_02_20_STM_Report_2015.pdf.

Weaver, C., 2017. Theranos secretly bought outside lab gear and ran fake tests, court filings allege. Wall Street J. Available from: https://www.wsj.com/articles/theranos-secretly-bought-outside-lab-gear-ran-fake-tests-court-filings-1492794470.

Wehling, M., 2009. Assessing the translatability of drug projects: what needs to be scored to predict success? Nat. Rev. Drug Discov. 8, 541–546.

Weichenberger, C.X., Pozharski, E., Rupp, B., 2017. Twilight reloaded: the peptide experience. Acta Cryst. D73, 211–222.

Wells, J.A., 2008. Final Report: Observing and Reporting Suspected Misconduct in Biomedical Research Gallup/Office of Research Integrity. Available from: http://ori.hhs.gov/sites/default/files/gallup_finalreport.pdf.

Welter, D., MacArthur, J., Morales, J., Burdett, T., Hall, P., et al., 2014. The NHGRI GWAS catalog, a curated resource for SNP-trait associations. Nucl. Acids Res. 42, D1001–D1006.

Wendler, A., Wehling, M., 2012. Translatability scoring in drug development: eight case studies. J. Transl. Med. 10, 39.

Wilbanks, J., Friend, S.H., 2016. First, design for data sharing. Nat. Biotech. 34, 377–379.

Wilson, W.A., 2016. Scientific regress. First Things. Available from: http://www.firstthings.com/article/2016/05/scientific-regress.

Wise, J., 2013. Boldt: the great pretender. BMJ 346, f1738.

Witkoski, T., 2014. From the archives of scientific fraud—Diederik Stapel. Psychology Gone Wrong. The Dark Sides of Science and Therapy. Available from: https://forbiddenpsychology.wordpress.com/2014/06/26/from-the-archives-of-scientific-fraud-diederik-stapel/.

Woodward, J., Goodstein, D., 1996. Conduct, misconduct, and the structure of science. Amer Sci 84, 468–478.

Young, N.S., Ioannidis, J.P.A., Al-Ubaydi, O., 2008. Why current publication practices may distort science. PLoS Med. 5, 1418–1422.

Yong, E., 2012. Uncertainty shrouds psychologist's resignation. Nature. Available from: http://www.nature.com/news/uncertainty-shrouds-psychologist-s-resignation-1.10968.

Yong, E., Ledford, H., Van Noorden, R., 2014. Research ethics: 3 ways to blow the whistle. Nature 503, 454–457.

Zoghabi, H.Y., 2013. The basics of translation. Science 339, 250.

CHAPTER

2

Experimental Planning and Execution

Kevin Mullane, Michael J. Curtis, Michael Williams

OUTLINE

Research in the Biomedical Sciences. http://dx.doi.org/10.1016/B978-0-12-804725-5.00002-1

2.1 INTRODUCTION

In the burgeoning literature on reproducibility, the inability to reproduce research findings has been primarily - if not exclusively - attributed to the misuse of statistical tests, naïve inference/extrapolation from animal models, and/or fraud. The present chapter argues however, that the current reproducibility crisis is as equally ascribable to poor experimental design and execution as to the misuse of statistics. Accordingly, the type of statistical analysis—within reason—can be irrelevant when the provenance of the input data is questionable—a classical GIGO (garbage in, garbage out) scenario.

The data generated from an experiment is a function of the experimental design, its execution, and analysis. Unless an experiment is designed in the context of a specific question that clearly defines the anticipated outcome(s), its usefulness will be compromised to the point that it may lack any value.

Over the past 2 decades, the types of experiments that can be done to understand cellular, tissue, and whole organism functions have increased exponentially along with increasingly sophisticated technologies, including single-cell sequencing (SCS) (Grun and van Oudenaarde, 2015; Wang and Song, 2017), optogenetics (Kim et al., 2017), and polychromatic flow cytometry (PFC) (Chattopadhyay et al., 2014), that can interrogate cellular genotypes and phenotypes at the single-cell level and complement Big Data–based omics disciplines. These types of studies have increased both the level of sophistication at which single cells can be interrogated and the complexity of experimentation, not the least of which is significant cell-to-cell heterogenity within a tissue (Pettit et al., 2014) that can have important implications in terms of drug responses, for example, resistance to chemotherapy (Waclaw et al., 2015).

Many of these newer approaches require the creation of new guidelines for experimental design where the techniques used historically, for example, powering, randomization, frequentist statistics, and analysis, are inappropriate in their present forms, such that the standards related to defining and ensuring reproducibility are in flux. With this in mind, the current chapter is focused on more established issues and resolutions in experimental planning and execution.

2.2 HYPOTHESIS GENERATION

A hypothesis is "a posited explanation held up for falsification by subsequent experimentation" (Glass, 2014) with "falsification" being used to denote Popper's dictum (Popper, 1959) rather than deliberate researcher fraud where "fabricate" is the accurate descriptor.

Popper's dictum states that confirmatory evidence cannot prove an assertion/claim to be true but that contradictory evidence can make the assertion/claim invalid. The goal of an experiment therefore, is to show that the hypothesis is false and, if this is done convincingly, the hypothesis should be abandoned. Repeated failures to falsify the hypothesis will eventually lead to its acceptance as a theorem. This has been further elaborated in a quote ascribed to Einstein that "No amount of experimentation can ever prove me right; a single experiment can prove me wrong" and can be examined experimentally via the use of the null hypothesis (https://en.wikipedia.org/wiki/Falsifiability).

2.2.1 Falsification

The concept of falsification is illustrated in the statement that "all swans are white" which was invalidated by the unexpected discovery of the rare Australian black swan. This was the topic of Taleb's bestselling, erudite musings on the impact of highly improbable events which, based on history, can "never happen" but surprisingly do happen, often with dire consequences (Taleb, 2010). Like real world statistical usage (Ashton, 2012), hypothesis generation is not necessarily a formal part of either a research initiative or a manuscript submitted for publication.

2.2.2 Hypothesis not Needed

In some research disciplines including chemistry (Jogalekar, 2014) where the main activity is synthesizing molecules, and in "big data" disciplines like genome wide (GWAS; Bush and Moore, 2012) and phenome-wide (Phe-WAS; Bush et al., 2016) association studies, high throughput screens to identify compounds (Biesecker, 2013; Glass, 2014), and large scale cell-based phenotypic screens (Finkbeiner et al., 2015), the reason for running the experiments is hypothesis-free or hypothesis-generating, and the data are often complex. For this reason, falsification can be an irrelevant approach, with hypotheses being developed after the experimental data have been generated (Howitt and Wilson, 2014).

However, in the majority of biomedical research activities, hypothesis testing is a useful approach and is a key part of experimental design. Nevertheless, restating the hypothesis representing "the formulation of a scientific idea," "an expectation of the outcome of the enquiry" (Medawar, 1996), or merely the provision of a logical reason for doing an experiment, however "blue sky" (Ioannidis, 2016), can remove the mystique associated with the hypothesis concept and a researcher, especially a graduate student, may feel relieved to be on less hallowed ground.

In biomedical research, there are generally four frameworks for conducting an experiment, or, given the need to reproduce experimental findings, a *series* of experiments as described here.

1. To understand differences between "normal" and "diseased" tissues in order to understand disease causality. Tissue from diseased patients and from animal models that have been manipulated genetically, chemically, or surgically can be used to understand disease mechanisms and identify new drug targets.
2. To evaluate the effects of a research compound, a new chemical entity (NCE) or a known drug—a drug in this instance being a clinically approved compound that is used in patients for the treatment of a disease—on a biological system. This can involve the validation of a putative disease target or the assessment of NCEs as hits or leads for a drug discovery project.
3. To develop new technologies that can facilitate systems biology–based interrogation of human and nonhuman cell systems to discover new mediators and pathways.
4. To develop new assay systems that facilitate frameworks 1–3.

2.2.3 The Null Hypothesis

A null hypothesis, or more formally, the null hypothesis significance testing procedure (NHSTP) is based on the premise that there is no significant difference between two or more data sets in an experiment or series of experiments. Experiments that disprove or "falsify" a null hypothesis create the basis for establishing that there is a difference, thus allowing rejection of the null hypothesis.

The null hypothesis is another experimental concept that is not especially well taught to graduate students. Accordingly, the recent banning of its use by one psychology journal due to issues with *p*-values, for example, "the $p < 0.05$ bar is too easy to pass and ... an excuse for lower quality research" (Trafimow and Marks, 2015) is confusing—especially in the absence of any validated replacement making it impossible for an author to claim that any "difference" between two means constitutes an "effect". Coupled with the copious literature on the misuse of the *p*-value (Baker, 2016a; Colquhoun, 2014; Demidenko, 2016; Halsey et al., 2015; Ioannidis, 2005; Lovell, 2013; Motulsky, 2014; Nuzzo, 2014; Siegfried, 2010), the new graduate with minimal practical experience or training in statistics, as well as the established experimenter who uses legacy statistical routines that have served them well (i.e., have not been an obstacle in their ability to obtain grants and publish papers) in their career to date, when faced with what appears to be the wholesale dissolution of two seminal statistical concepts, the *p*-value and the null hypothesis, are left to watch statisticians endlessly debate the situation without delineating—in a cohesive and responsible manner—alternatives that are viable, accessible, and understandable to nonexperts and that are also useful in the real world of experimentation (Ashton, 2012).

In response to such concerns, the American Statistical Association (ASA) recently issued a statement on *p*-values (Wasserstein and Lazar, 2016) together with a series of principles (Table 2.1). While comprehensive it is not necessarily as useful as it could be to those without an extensive background in statistics. It states what *p*-values actually mean, and warns that just because $p < 0.05$ (or 0.01 or indeed 0.0001), this does not indicate that an effect is real or that an hypothesis is proven. Furthermore, the statement notes that statistical significance does not necessarily mean biological significance.

TABLE 2.1 ASA Statement on Statistical Significance and *p*-Values (Wasserstein and Lazar, 2016)

ASA principle	Supporting text
1. *p*-values can indicate how incompatible the data are with a specified statistical model.	A *p*-value provides one approach for summarizing the incompatibility between a particular set of data and a proposed model for the data using the Null hypothesis. The most common context is a model, constructed under a set of assumptions, together with a so-called "null hypothesis." Often the null hypothesis postulates the absence of an effect, such as no difference between two groups, or the absence of a relationship between a factor and an outcome. The smaller the *p*-value, the greater the statistical incompatibility of the data with the null hypothesis, if the underlying assumptions used to calculate the *p*-value hold. This incompatibility can be interpreted as casting doubt on or providing evidence against the null hypothesis or the underlying assumptions.
2. *p*-values do not measure the probability that the studied hypothesis is true, or the probability that the data were produced by random chance alone.	Researchers often wish to turn a *p*-value into a statement regarding the truth of a null hypothesis, or about the probability that random chance produced the observed data. The *p*-value is neither. It is a statement about data in relation to a specified hypothetical explanation, not a statement about the explanation itself.

TABLE 2.1 ASA Statement on Statistical Significance and *p*-Values (Wasserstein and Lazar, 2016) (*cont.*)

ASA principle	Supporting text
3. Scientific conclusions and business or policy decisions should not be based only on whether a *p*-value passes a specific threshold.	Practices that reduce data analysis or scientific inference to mechanical "bright-line" rules (such as "$p < 0.05$") for justifying scientific claims or conclusions can lead to erroneous beliefs and poor decision making. Researchers should bring many contextual factors into play to derive scientific inferences, including the design of a study, the quality of the measurements, the external evidence for the phenomenon under study, and the validity of assumptions that underlie the data analysis. Pragmatic considerations often require binary, "yes-no" decisions, but this does not mean that *p*-values alone can ensure that a decision is correct or incorrect. The widespread use of "statistical significance" (generally interpreted as $p < 0.05$) as a license for making a claim of a scientific finding (or implied truth) leads to considerable distortion of the scientific process.
4. Proper inference requires full reporting and transparency.	*p*-values and related analyses should not be reported selectively. Conducting multiple analyses of the data and reporting only those with certain *p*-values (typically those passing a significance threshold) renders the reported *p*-values essentially uninterpretable. Cherry-picking promising findings, also known by such terms as data dredging, significance chasing, significance questing, selective inference, and "*P*-hacking," leads to a spurious excess of statistically significant results in the published literature and should be vigorously avoided. One need not formally carry out multiple statistical tests for this problem to arise. Whenever a researcher chooses what to present based on statistical results, valid interpretation of those results is severely compromised if the reader is not informed of in the context of that choice and its basis. Researchers should disclose the number of hypotheses explored during the study, all data collection decisions, all statistical analyses conducted, and all *p*-values computed. Valid scientific conclusions based on *p*-values and related statistics cannot be drawn without at least knowing how many and which analyses were conducted, and how those analyses (including *p*-values) were selected for reporting.
5. A *p*-value, or statistical significance, does not measure the size of an effect or the importance of a result.	Statistical significance is not equivalent to scientific, human, or economic significance. Smaller *p*-values do not necessarily imply the presence of larger or more important effects, and larger *p*-values do not imply a lack of importance or even lack of effect. Any effect, no matter how tiny, can produce a small *p*-value if the sample size or measurement precision is high enough, and large effects may produce unimpressive *p*-values if the sample size is small or measurements are imprecise. Similarly, identical estimated effects will have different *p*-values if the precision of the estimates differs.
6. By itself, a *p*-value does not provide a good measure of evidence regarding a model or hypothesis.	Researchers should recognize that a *p*-value without context or other evidence provides limited information. For example, a *p*-value near 0.05 taken by itself offers only weak evidence against the null hypothesis. Likewise, a relatively large *p*-value does not imply evidence in favor of the null hypothesis; many other hypotheses may be equally or more consistent with the observed data. For these reasons, data analysis should not end with the calculation of a *p*-value when other approaches are appropriate and feasible.

It is however, necessary, and indeed unavoidable, that in biomedical research the *p*-value must be used in a binary manner to identify an effect. The binary inference is generated by the context (i.e., the standard question—is there an effect?), and results in a value that is dependent on the experimental design and the critical questions that inform the design (see Chapter 3). These include: (1) what value of *p* is necessary in order to provide a degree of confidence that the probability of making a correct statement is logical based on the data? (2) How big a group size is required (and is feasible given ethical and funding constraints) to generate outcomes that reduce the chance of making a false inference due to under powering? (3) What types of positive and negative controls are required to minimize the risk of fallacious inference? All these points come under the umbrella of best experimental practice based on experience.

The ASA statement further notes that in view of the prevalent misuses of and misconceptions concerning *p*-values, some statisticians prefer to supplement or even replace *p*-values with other approaches. These include: methods that emphasize estimation over testing, such as confidence, credibility, or prediction intervals; Bayesian methods; alternative measures of evidence, such as likelihood ratios or Bayes Factors (van Ravenzwaaij and Ioannidis 2017); and other approaches, such as effect sizes, decision-theoretic modeling and false discovery rates. All these measures and approaches rely on further assumptions, but they may more directly address the size of an effect (and its associated uncertainty; Lovell, 2013) or whether the hypothesis is correct (Chapter 3).

Bayesian statistics (Eddy, 2004; Puga et al., 2015; Goodman et al., 2016) are being increasingly discussed in the biomedical research community as an alternative to traditional frequentist statistical inference, but are not being used because they rely on a priori knowledge of the true population mean and variance. They also fail to provide an alternative to the *p*-value and accordingly do not provide a better answer to binary questions regarding as to whether there is an effect or not. With a Bayesian approach, the potential plausibility of an experimental outcome, rather than in its potential frequency make it "comparatively easy for observers to incorporate what they know about the world into their conclusions, and to calculate how probabilities change as new evidence arises," a viewpoint not widely appreciated by statisticians due to issues with subjectivity (Nuzzo, 2014) and subjective probabilities (Colquhoun, 2014) and the enhanced possibility of bias.

2.3 EXPERIMENTAL PLANNING

Experiments can be divided into three distinct types, preliminary, exploratory or hypothesis generating, and confirmatory or hypothesis testing. Of these, confirmatory/hypothesis testing are routinely conducted using a null hypothesis-based approach (Tukey, 1980). Before initiating an experiment, it is essential that a researcher has a clear idea as to what question is being asked, how the biological system being used is mechanistically/phenotypically appropriate to provide the type of data that can answer the question being posed, and whether the system is suitable, in terms of its robustness and readout characteristics—its signal-to-noise ratio or "noise"—to provide data that is reproducible and real (Goodman et al., 2016).

When planning an experiment, it is important that the researcher creates a written outline providing the rationale and context behind the experiment and what the expectations are for the outcomes. While this usually involves the postulation of a hypothesis, this is often preceded by a series of *exploratory* or *hypothesis-generating* experiments that help in defining the hypothesis and the conditions required to test it. Sometimes these exploratory experiments are preceded by a series of *preliminary* experiments where the researcher evaluates a number of different questions to establish a baseline to better inform the design and execution of the exploratory experiments (Lovell, 2013). Preliminary experiments are usually not adequately powered or rigorously designed, as they are used to provide a rapid and inexpensive means to see if an experimental signal can be produced and whether the data obtained is worthy of additional follow-up efforts. Once a convincing set of exploratory data is in hand, the researcher can then proceed to *confirmatory* or *hypothesis confirmation* experiments that are rigorously designed, powered, randomized, and blinded and are designed for peer-reviewed publication.

The key considerations in initiating a series of experiments are:

- What is (are) the specific question(s) being asked?
- What are the types of data that would inform answers to the questions raised?
- What type of experiment is being conducted: "preliminary," "exploratory," or "confirmatory?"
- What are the outcomes that might be expected?
- What is the observed variation in the data obtained and hence the group size required to test the hypothesis?
- What are the types of checks and balances, for example, controls and alternative test systems, that would be required to rule out alternative interpretations of the outcomes?

2.3.1 Preliminary Experiments

A preliminary approach to experimentation involves small sample sizes usually with multiple variables in terms of experimental conditions that are conducted with a minimum of expense to establish whether there is any reason to continue to explore an idea. Often rather than being a waste of time, a preliminary experiment can help in refining the question(s) being asked while at the same time providing a rough comparison between different experimental conditions, for example, different buffers, pH, and temperatures, or to determine the sample size necessary for testing the hypothesis. An example of a preliminary experiment would be "what happens if I treat a cell or animal with compound *X*?" where compound *X* has already been shown to affect a key disease target or phenotypic readout that has relevance to the research area of the investigator. Given that a series of preliminary experiments generally utilize too few samples to permit statistical analysis and are viewed as "fishing expeditions" undertaken and evaluated on the basis of the experience/intuition of the investigator, such experiments are used to address questions that are insufficiently well structured to constitute a bona fide hypothesis. Preliminary experiments may fail to yield any useful information due to inadequate group sizes, a lack of controls, randomization and blinding, or a poor signal-to-noise ratio. In conducting preliminary experiments the design of experiment (DOE) methodology is often used to optimize experimental conditions, where multiple factors are varied rather than the "one factor at a time" (OFAT) approach, which is less efficient (Lovell, 2013).

When the outcome is consistent with the initial investigator "hunch," the preliminary findings can be replaced by data from subsequent, more formal experiments that are conducted with adequate resources and greater rigor (experimental design and execution) and powering. Accordingly, such data are unlikely to be submitted for (or accepted via) peer review, unless they are a necessary part of the experimental narrative for the development of additional follow-up studies reported in a paper. On their own, they are usually *unpublishable* which means that if they evade peer review yet appear in print this will further confound reproducibility.

2.3.2 Exploratory Experiments

While often confused with preliminary experiments, exploratory experiments are intended to explore a hypothesis (or hypotheses) with sufficient rigor and flexibility to establish a robust basis for subsequent confirmatory studies. While the majority of these are quietly forgotten when they do not provide the desired outcome—those that are successful are characterized as useful in "generating robust pathophysiological theories of disease." (Kimmelman et al., 2014) and should be adequately powered with a priori defined experimental endpoints and statistical analyses so that they can be communicated meaningfully, and independently reproduced.

2.3.3 Confirmatory Experiments

Once a hypothesis or research idea has been formulated and undergone feasibility assessment in a series of exploratory studies, more formal and substantive confirmatory or hypothesis-testing experiments are undertaken. These provide a gold standard for testing the validity of a hypothesis and use best standards in terms of clearly defined outcomes and use "rigid and prespecified designs, a priori stated hypotheses, prolonged durations, and the most clinically relevant assays and endpoints available" (Henderson et al., 2013; Kimmelman et al., 2014; Muhlhausler et al., 2013), which involve clearly defined endpoints with appropriate powering, randomization, and blinding and that have been reproduced in a series of independently conducted experiments that can occur either in the lab where the hypothesis originated or via crowdsourcing initiatives, such as The Pipeline Project, where "a nonadverserial replication (e.g., reproduction) process" occurs before publication (see Chapter 5.12.5).

2.3.4 Relationship of Intervention and Effect: A Need to Avoid Premature and Erroneous Associations

Over 30 years ago, with a Nobel Prize and a Knighthood still to come, Sir James Black (1986) warned of concluding that because a compound is known to block a specific enzyme or target, and is then shown to block another response (say cellular motility or contraction), that the same enzyme/target mediates the observed response phenotype because of the circularity in defining the compound in terms of its target (e.g., inhibitor of enzyme *X*) and then using the activities of the compound, in turn, to define the target. However, "when 20 or more quite different chemicals with different physical and pharmacological properties" (Black, 1986) show commonality between their effects on the target and the observed response, then it is reasonable to conclude that the two are related. Moreover, as warned by Bunnage et al. (2013) "There are many widely used chemical probes that do not meet generally accepted potency and selection criteria, and the conclusions made from their use are suspect"; a point reinforced by Arrowsmith et al. (2015) that is often not appreciated by many researchers. This is discussed in greater detail in Section 2.5.2.

2.3.5 A Reductionist Approach can Lead to Erroneous Assumptions About Targets and Therapeutics

Most drugs fail Phase II clinical trials due to a lack of efficacy, indicating that the underlying therapeutic hypothesis driving target selection was in error, and that the assays chosen to select the compounds lacked a strong, relevant, disease-associated foundation. Inappropriate interpretations using suspect tool compounds is part of the wider concern regarding the reductionist approach to science and the focus on molecular components rather than systems. Black (2010) made a useful analogy, writing "Components are to systems as words are to poems and pigments to paintings. The decomposition of poems and paintings into words and pigments is not reversible."

Black (2010) also commented on how every physiological system is a balance between stimulation and inhibition, where "each control arm is subject to some kind of feedback control about which our reductionist efforts have left us even more ignorant." He envisaged a system of "convergent control" that is more nuanced than a bioinformatics-derived systems biology pathway, exemplified by "I imagine a growth factor giving a stem cell, say, not a command, but a piece of advice, such as 'Other things being equal, you should start dividing'! The other equal things are other chemical messengers, which have to impinge on the cell at the same time to achieve its activation." This advice–consent arrangement, controlled by chemical convergence, means that suppressing one component might not create the desired activity, except under rigidly defined conditions that rarely translate to clinical utility. Certainly cancer treatment has become dependent on a cocktail of drugs that target different components of a tumor system to achieve an overall suppression of growth. High blood pressure is often treated with a mix of antihypertensive drugs, where even though the targets are known (e.g., components of the renin–angiotensin system, β-adrenoceptors, and voltage-dependent calcium channels), agreement on how they work and interact is lacking.

The high clinical failure rate of therapeutics discovered using specific targets based on hypothetical mechanisms of disease has led to the reemergence of disease-related phenotypic assays as a key element in drug discovery (Mungall et al., 2017; Wagner and Schreiber, 2016; Zheng et al., 2013). Citing several successful recent examples where phenotypic assays led to success, Vincent et al. (2015) highlighted three critical components that need to be considered carefully, namely, the *assay system*, the *stimulus* needed to provoke a disease-like phenotype, and the *relevance* of the endpoints measured. Where possible it is proposed that phenotypic screens should utilize assays with clear links to disease, for example, using patient-derived iPS cells containing disease-associated mutations, which develop a spontaneous phenotype due to the inherent disease-causing elements within the cell(s), negating the need for chemical or physical provocation, and express a functional manifestation of the disease that has clinical relevance rather than a biological response, such as a change in gene expression (Vincent et al., 2015). While such assays are often unavailable, Black's drug discoveries (1986, 2010) did not depend on patient-derived cells, but instead adopted tissue-based assays that used clear phenotypic endpoints directly related to the human physiological process under investigation, rather than reflective of a particular disease state. Moreover, a clinically relevant (measurable) endpoint might be a few steps removed from the compound–target interaction, thereby having the potential to be modified by other factors, such that a more direct reflection of the compound–target interaction might be necessary for early assays and development of structure–activity relationships, with subsequent studies moving toward more clinically defined measures. However, "leaps of faith" between biology and disease must be minimized before therapeutics are advanced into clinical trials.

The renewed interest in, and value of, phenotypic analyses is also evidenced by the emergence of phenome-wide association studies, or PheWAS, as a complementary approach to the genome-wide association studies (GWAS) of the last 20 years. While GWAS have been successful at identifying genetic polymorphisms associated with a variety of diseases (Visscher et al., 2017), it has become apparent that most diseases are polygeneic, with each component contributing only a minor effect, and GWAS have been exceedingly poor at determining which variants are clinically and biologically meaningful (Hebbring, 2013). Khoury (2013) has noted that "GWAS have many limitations, such as their inability to fully explain the genetic/familial risk of common diseases; the inability to assess rare genetic variants; the small effect sizes of most associations; the difficulty in figuring out true causal associations; and the poor ability of findings to predict disease risk." Indeed, the limitations of GWAS led Boyle et al., (2017) to propose that since "essentially all genes expressed in relevant tissues affect traits" rather than describing these as polygenic, they are in fact, more appropriately described as omnigenic. This in turn led Weiss (2017) to consider whether GWAS (and Big Data) has been a hoax and "a drain on public resources and, perhaps worse, on the public trust in science" due to redundancies in genetic causation that result from duplication and the complexity of the pathways involved and an almost infinite number of contributing sites on the genome. He posited that until GWAS was tried it was not possible to anticipate how useful it would be in mapping complex traits, but while it has proved useful, its continued use is not justifiable since "complex traits are not genetic in the usual sense of being due to tractable, replicable genetic causation."

PheWAS utilize electronic medical records and epidemiological studies to interrogate the genetic variants across a range of human diseases or traits. By testing for associations of one or multiple genetic variants across a broad range of clinical phenotypes, the functional impact of the genetic change can be assessed. As approximately 17% of genes or gene regions are associated with more than one trait or disease (Sivakumaran et al., 2011), this method can determine common elements and pathways between certain diseases, providing risk associations and highlighting common biomarkers, and also identifying drug targets and drug repurposing opportunities (Bush et al., 2016). For example, about 80% of rheumatoid arthritis patients who are seronegative for rheumatoid factor have fibromyalgia (Doss et al., 2017), and while GWAS have previously identified 160 disease phenotypes that map to the major histocompatibility region on Chromosome 6, many of which have been confirmed by PheWAS together with the identification of novel associations with 8 new diseases (Liu et al., 2016).

While PheWAS offer a new dimension of assessing patients and treatment strategies, there are several hurdles that need to be overcome before concerns over validity and reproducibility can be laid to rest (Bush et al., 2016; Hebbring, 2013). In the United States, most patient medical records utilize International Classification of Diseases (ICD) codes for billing and procedures, and it is these codes (or their alternatives) that are used to define phenotypes. ICD codes have several limitations, for example, they have been updated from ICD-9-CM to ICD-10. However, most historical records use the older codes; they have a degree of redundancy, as exemplified by the 496 different codes for tuberculosis that differ primarily based on the site of infection (Bush et al., 2016); the quality of the data and the manner in which they are collected can be highly variable; and importantly, the codes are used to define a basis for a diagnostic test, not the final diagnosis. Many tests—bone scans, mental acuity, and cancer evaluations—occur later in life, after multiple comorbidities might have influenced the phenotype, and disentangling these associations to determine which are causative can be challenging. A statistically significant threshold for the multiple comparisons, analogous to that defined for GWAS after years of nonreproducible studies, has not yet been established. Finally, although touted as an unbiased search for associations, in reality the starting point for most PheWAS depends on the association of a previously identified genetic variant with a particular disease phenotype (Bush et al., 2016; Roden, 2017).

Despite these caveats and the current clinical focus, PheWAS approaches will in time replace GWAS to inform preclinical studies in terms of understanding the phenotypic consequences of subtle gain- or loss-of-function mutations; the contributions of novel pathways to lead compound phenotypes (Grainger, 2013), including safety (Pirmohamed et al., 2015); and development of new generations of animal models that will better reflect the human disease state.

2.3.6 Conflating Studies: Exploratory With Preliminary, Confirmatory With IND Enabling

As part of their delineation of exploratory and confirmatory studies, Kimmelman et al. (2014) define the former as involving "small sample sizes with the flexibility to employ multiple methodologies at the molecular, cellular, and whole animal levels" that "may or may not employ inferential statistics" and which "include tests of an intervention's efficacy against disease in live animals as a way of validating the pathophysiological theories (efficacy studies)." This confuses exploratory studies, which represent the vast majority of preclinical research efforts, with preliminary findings that are unpublishable in the absence of the findings being subjected (or subjectable) to statistical analysis. Additionally, these authors extend the "clinical relevance" of preclinical studies to include "large sample sizes" that "aim less at elaborating theories or mechanisms of a drug's action than rigorously testing a drug's clinical potential and restricting the advance of ineffective interventions advanced into clinical testing." The "confirmatory" designation used by Kimmelman et al. (2014) envisages these studies in the context of drug discovery as being translationally predictive rather than hypothesis testing in the traditional sense of exploratory versus confirmatory (Tukey, 1980).

Putting aside caveats on the translatability of animal models that have been discussed in Chapter 1.11, this designation of "confirmatory" studies is, at least in intent, equivalent in scope to the datasets required to support an Investigational New Drug (IND) application (Federico et al., 2014; Khanna, 2012; McGonigle and Williams, 2014), which almost always lies outside the scope, experience and resources required for conducting experiments focused on conventional preclinical hypothesis-testing paradigms. Irrespective of how rigorously this concept of a preclinical confirmatory experiment parallels that of a clinical study, an IND package not only assesses compound efficacy in an animal model, but also extends to defining the physicochemical, pharmacokinetic, and toxicological properties of an IND candidate.

Nonetheless, Mogul and Macleod (2017) take Kimmelman at al.'s translationally predictive experiment to other levels of clinical relevance with the confirmatory study becoming a "preclinical trial," a final "impeccable" confirmatory study without which they controversially advocate "no publication without confirmation." Elements of the preclinical trial would involve providing a protocol for the proposed confirmatory study, its key outcome measures, and a plan for statistical analysis. The study would: (1) adhere to the highest standards of rigor in design, analysis, and reporting; (2) be held to a higher standard of statistical significance with $p < 0.01$ rather than $p < 0.05$, an approach that has been advocated by Colquhoun (2014) and Johnson (2013), both of whom have recommended $p \leq 0.001$; (3) be performed by an independent laboratory or consortium; and (4) increase the number of animals used sixfold.

The concept of performing clinical trial-like, pivotal efficacy studies as a paradigm to restrict "the advance of ineffective interventions..... into clinical trials" (Kimmelman et al., 2014), given the current issues regarding the unresolved predictive limitations of animal studies of human disease (Landis et al., 2012; McGonigle and Williams, 2014) and pharmacokinetic issues (Caldwell et al., 2004; Fan and de Lannoy, 2014; Ferl et al., 2016; Kleiman and Ehlers, 2016), while laudable, is simplistic and naïve and would predicate studies that would normally be far outside the limited resources of an academic laboratory, placing an additional burden on increasingly limited research funds.

Mogul and Macleod (2017) further describe plans for a generalizability study, generalizability being defined as "the persistence of an effect in settings different from and outside of an experimental framework" (Goodman et al., 2016), that would necessarily involve a multicenter consortium. They fail to note, however, that consortia-based preclinical trials have already been evaluated in the area of stroke and lack robust success (Chapter 1.11.2.2.1).

A major issue with these proposals is their cost. Mogul and Macleod (2017) suggest that "government funders and industry partners, which have spent billions of dollars on disappointing clinical trials, would be prepared to shift resources to support such an improved system," a utopian expectation against the backdrop of the failed multicenter preclinical stroke initiatives and other demands on available research funding. Another approach would be to reduce waste and increase efficiency in research (Glasziou et al., 2014; Ioannidis et al., 2014). The cost of irreproducible research activities in the United States is estimated to be as much as \$28bn a year (Freedman et al., 2015a). Increased efficiencies, for example, the use of software-based approaches to improve the design of preclinical in vivo studies may be a useful first step (Laajala et al., 2016). However, finding ways to reduce waste have yet to be shown to be either actionable or useful (Moher et al., 2016) much like the always well-intended but virtually impossible to implement initiatives in cutting bureaucratic waste of all types.

2.4 ADDRESSING THE BASICS IN EXPERIMENTAL DESIGN

2.4.1 Sample Sizes

One of the most contentious issues in biomedical research is the appropriate sample size required in an experiment to avoid false positives and false negatives that would compromise its reproducibility (Colquhoun, 2014; Ioannidis, 2005; Marino, 2014; Motulsky, 2014).

The concept of sample size (denoted as "n") refers to the number of independent values in an individual experimental group. Thus, if there are two experimental groups, for example, control and experimental manipulation, each comprising 5 mice, and 1 tissue sample is prepared from each mouse to populate 1 lane of a western blot, $n = 5$ values will be generated for each of the 2 groups ($n = 5$/group, $n = 10$/study). If each tissue sample is divided into 3 aliquots this will result in 15 samples per group. However, this does not mean that the n value is 15 per group as the 3 aliquots are triplicates that can be used to assess reproducibility within the assay (in this case, a western blot) and the technical dexterity of the investigator, but only 1 value from the triplicate, usually the mean of the triplicate, should be used in the data analysis. Sample size determination, therefore, represents the number of independent samples necessary to power a study, not the intraexperimental replicates within a study. The example $n = 5$ experiment needs to be performed a minimum of 3 times to demonstrate reproducibility and is the format typically reported in the peer-reviewed literature (see Section 2.10.1).

To calculate the sample size requires consideration of four parameters: (1) The anticipated size of the experimental effect; (2) desired significance level; (3) anticipated

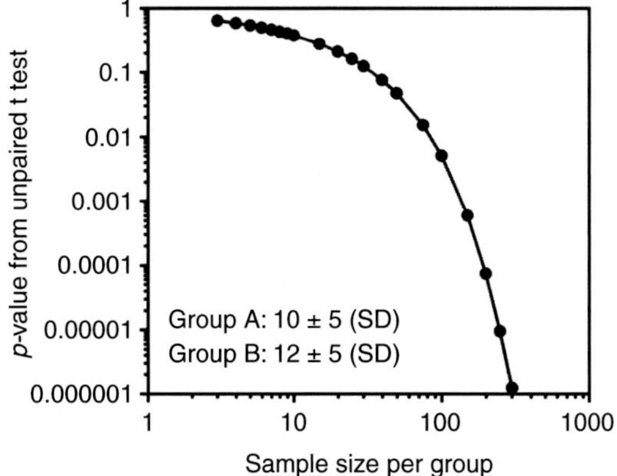

FIGURE 2.1 ***p*-values depend upon sample size.** This graph shows *p*-values computed by unpaired *t* tests comparing two sets of simulated data. The means of the two samples are 10 and 12 with a difference of 2. The S.D. of each sample is 5. Given these parameters, a group size of approximately 80 would be required to obtain a *p*-value of < 0.05. A simulated t test for various sample sizes is plotted on the *x*-axis and shows the *p*-value is dependent on sample size. Both axes are plotted on a logarithmic scale. For the example presented, even though the means and standard deviations are identical for each simulated experiment, the *p*-values are not identical. With $n = 3$ in each group, the *p*-value is 0.65. When $n = 300$, the *p*-value is 0.000001. *Source: From Motulsky, H.J., 2014. Common misconceptions about data analysis and statistics. J. Pharmacol. Exp. Ther. 351, 200–205, used with permission of ASPET.*

variability in the data; (4) the power required for the study, topics that are dealt within Chapter 3.

Knowledge of this information allows determination of a minimum sample size necessary to detect an effect at a predetermined *p*-value (e.g., < 0.05) using power analysis (see Chapter 3.5.2).

Some additional issues worthy of discussion in relation to group sizes follow.

2.4.1.1 The Effect of Sample Size on the p-Value and the emerging argument for effect sizes.

In Fig. 2.1, the effect of sample size on the *p*-value is demonstrated using a computer simulation. In this example, which takes an extreme situation given the minimal difference between the means and the large standard deviation for each group, 50% for A and 42% for B, with $n = 3$ in each group, the *p*-value is 0.65. When *n* is increased to 300 in each group, the *p*-value becomes 0.000001 (Motulsky, 2014). In changing the *p* bar, the risk of false negative findings (Type 2 error) is increased since, if the *p* bar is set at < 0.001 then a *p*-value of < 0.01 will be "not significant." Colquhoun (2014) has argued that because of this, the effect size should be used to denote effect rather than the *p*-value, a position advocated by Sullivan and Feinn (2012) and Lovell (2013) that necessitates the use of confidence intervals (CIs) rather than a *p*-value since CIs provide more information by giving an estimate of the size of the effect including the mean difference and its variability, the latter via the width of the 95% CI (Lovell, 2013). Furthermore, a two-sided 95% CI can also serve as the test of a hypothesis "because if the interval does not straddle zero, this means that the null hypothesis of zero is rejected at $p = 0.05$" (Lovell, 2013).

Increasing the group size from 3 to 5 has considerable benefits in addition to reducing the scope for error described earlier (Fig. 2.1). Increasing the group size to 7, 12, or 300 has further benefits but proportionately less. Thus the improvement in increasing the group size from $n = 3$ to $n = 5$, as a rule of thumb, is sufficiently large that its benefit clearly outweighs its costs (in terms of time and resources). The *British Journal of Pharmacology* (Curtis et al., 2015) has set the minimally acceptable experimental group size to $n = 5$.

2.4.1.2 Surrogate Readouts With Reduced Variability to Reduce Group Size

In an ideal world, the design of an experiment should be informed by its feasibility. In contrast to late stage pivotal clinical trials that typically involve 1,000–3,000 patients, preclinical studies are restricted by technical, financial, and ethical issues, the latter focusing on reducing unnecessary animal usage. There is a school of thought that the animals used in biomedical research can be replaced by computer simulations where the efficacy and toxicity of new drug candidates can be assessed by optimizing the interactions of new drug candidates with their presumed targets (Badyal et al., 2009). While major advances have been made in using alternatives to animals to understand human disease and drug targets to facilitate drug discovery, computational simulations of compound-target interactions and systems biology-based approaches to compound effects have remained adjunctive to, rather than replacing, animals, and numerous high visibility biosimulation initiatives have disappeared without trace (http://www.bio-itworld.com/newsitems/2006/june/06-15-06-entelos/). For more than 50 years, scientists have focused on the *3Rs initiative*—replacement, reduction, refinement—to reduce animal usage in research (Burden et al., 2015; Hooijmans et al., 2010; McGrath et al., 2015; https://www.nc3rs.org.uk/the-3rs). There is however, an inherent conflict in the agenda of the *3Rs initiative* minimizing the use of animals in research and that of biomedical researchers whose intent is to understand disease pathophysiology to develop therapeutics. This results in situations where ethical constraints result in the approval of studies that are underpowered to the extent that they are unable to provide any useful data. In such instances, the ethical recourse is not to run the experiment rather than waste the limited number of animals that an Institutional Animal Care and Use Committee (IACUC; Jones-Bolin, 2012) approves in a research protocol. This does not preclude logical efforts to reduce the number of animals required for a given study. If the power analysis indicates that 30 or more animals/group are required to precisely measure the experimental effect one solution is to seek a surrogate readout for the variable of interest. This has been a common procedure in the drug safety arena, when evaluating the adverse drug reaction (ADR) potential of NCEs to evoke the potentially lethal syndrome, *torsades de pointes* (TdP) (Lee et al., 2010; Pugsley et al., 2008, 2009). Variables, such as potency in blocking the ion current, IK_r, or the dose of compound producing a 10% prolongation in the QT interval of the ECG are affected in a reproducible way by most compounds with a TdP liability, with statistical significance obtained with manageable group sizes owing to this robust reproducibility (Pugsley et al., 2008).

2.4.1.3 Guidelines for Animal Usage

A number of initiatives to establish minimal standards for preclinical experimentation are covered in Chapter 5.10.1. The most cited of these are the *ARRIVE* guidelines that consist of a checklist of 20 items for animal research (Kilkenny et al., 2009, 2010; McGrath et al., 2010), and the Gold Standard Publication Checklist (GSPC; Hooijmans et al., 2010, 2011) incorporating aspects of the RCT approach into preclinical studies to aid in reducing false positive and negative rates (Henderson et al., 2013; Kimmelman et al., 2014; Muhlhausler et al., 2013). While these guidelines have become very visible in preclinical biomedical research, they are rarely followed as researchers often refer to them and other guidelines *after* a study is completed and being prepared for publication, rather than at the outset in planning the study design (Avey et al., 2016; Baker et al., 2014; Smith et al., 2016), and this oversight is rarely noticed as part of the peer review process (Baker et al., 2014).

2.4.2 Experienced Statistical Input

As noted in Chapter 3, statistical methods are a mandatory part of experimental design, and not something to be addressed at the manuscript writing or submission stage.

In the preclinical research setting, formalized input from a trained statistician is, despite its intrinsic value, for the most part, absent or avoided. On the one hand, this reflects a shortage of statisticians in preclinical research and on the other, a prejudice that is reflected in the quote attributed to the Nobel physicist, Rutherford "If your experiment needs a statistician, you need a better experiment" (and extended in the concept of the "bloody obvious" test by Kitchen (1987)) and issues with the dogmatic approach exhibited by many statisticians that precludes their input being perceived as having any practical use (Ashton, 2012).

Nonetheless, because of a growing awareness of investigator bias, intended or unconscious, there is an increasing recognition in preclinical research regarding the value of the input from a trained statistician either as an advisor or as a core member of a research team (Basken, 2017; Collins and Tabak, 2014; Gore and Stanley, 2015; Peers and Ceuppens, 2012; Peers et al., 2014) to ensure that the experimental design and the means of analysis have been clearly defined and, to the extent possible, are free of bias, before an experiment is initiated. This can enhance both the credibility of a study and its reporting (Adams-Huet and Ahn, 2009). When statistical input cannot be accessed, the article *Ten Simple Rules for Effective Statistical Analysis* (Kass et al., 2016) provides important practical insights. Indeed, the NIH (Collins and Tabak, 2014; Wadman, 2013) and some mainstream journals have increased their emphasis on the need for a priori statistical planning (Nature, 2013; Van Noorden, 2014).

In 2014, *Science* established a seven-member statistics board of reviewing editors (SBoRE) to monitor and provide guidance—if required—on the statistical analysis used in submitted manuscripts. In welcoming this as "a step forward," Ioannidis noted "that for the majority of scientific papers nowadays statistical review is more essential than expert review" (Van Noorden, 2014), implying that journals should impose even greater emphasis on statistics in peer review.

Many journals do not have ready access to experimental design and statistical analysis expertise other than that offered by peer reviewers (Gore and Stanley, 2015).

Additionally in the current system, whether the findings from a preclinical study are true (i.e., can be reproduced), remains a matter of chance, as much of biomedical research proliferates without the application (by authors) or the validation (by peer reviewers) of best experimental practices in statistics or indeed reproduction by other investigators. Flaws in design and analysis, often the result of inadequate training and mentorship, are responsible for many of the failures in reproducibility that cannot be rectified by an *a posteriori* statistical analysis (Mullane and Williams, 2017).

2.5 PRECLINICAL EXPERIMENTAL SYSTEMS

A biochemistry laboratory will generally have all the necessary instrumentation, reagents and experimenter expertise, and experience to prepare in vitro drug target samples (enzyme, cell lines, cell membranes, mitochondria, etc.) or cell-based disease models and measure their responses to manipulation/interrogation (Horvath et al., 2016), while a pharmacology or physiology laboratory will be well versed in, and equipped for, experimentation involving tissue preparation and animal use. With time a laboratory will need to acquire new technology that are outside its present scope of expertise to aid in advancing its research goals. In some instances, the cost and time delays necessary to set up new equipment and train researchers in its use may be sufficiently prohibitive that it is easier to either collaborate with another laboratory already having the expertise or contract the work out to a contract research laboratory (CRO); (e.g., Batelle, SRI, Eurofins Cerep, etc.) provided the latter has a proven track record of quality, cost effectiveness, and a timely turnaround.

2.5.1 The Assay Capability Tool (ACT)

Gore and Stanley (2015) reported on an assay capability tool (ACT) developed by Pfizer, and informed by the *British Journal of Pharmacology* guidelines for publication (Curtis et al., 2015) that is intended to aid in selection of appropriate experimental approaches in the preclinical arena. This is based on 13 questions encompassing essential experimental design and analysis strategies. The ACT documents assay strengths, weaknesses, and precision and the level of confidence in their use (Table 2.2) and can be used to guide the planning of experiments once an assay approach has been selected and its validity, specificity, sensitivity, and stability assessed (Glass, 2014).

TABLE 2.2 The Assay Capability Tool (ACT)—Gore and Stanley (2015)

Question to consider	Importance
Aligning assay capability with research objectives	
Q1: Are the scientific objectives for running the assay recorded in a protocol/SOP?	The scientific questions to be answered, the measurements to be obtained and analyzed along with their required precision (as defined by, e.g., a standard error or confidence limits) must be stated in the protocol/standard operating procedure (SOP) to prevent data dredging and misinterpretation of the results.
Q2: What will a successful assay outcome look like in order to guide decision making?	Prespecifying decision criteria leads to crisp decisions and ensures unbiased interpretation of results. State the primary endpoint and state the minimum response or effect required. As all assay results include inherent uncertainty, it is also necessary to state the level of uncertainty that can be tolerated for acceptable decision making.
Q3: Is the experimental design, as described in the protocol/SOP, aligned loosely with the objectives?	The design and conduct should be addressed in light of the objectives. Once the objectives and definitions of success are defined, the assay must be designed so that the analysis can deliver the objectives. Consultation with a statistician is highly recommended.
Enabling assay capability by managing variation	
Q4: Is assay development and validation fully documented?	Describe the work done in order to verify that the assay is fit for purpose. Identify key learnings/issues/concerns arising from experiments done during assay development. Assay developers should document validation runs using positive and negative controls and standard compounds to provide benchmarks and reassurance to the users of the resulting data.

TABLE 2.2 The Assay Capability Tool (ACT)—Gore and Stanley (2015) *(cont.)*

Question to consider	Importance
Q5: Have the sources of variability present in the assay been explored?	All assays exhibit variability and it is important to know what the sources of variability are and their relative sizes. The major sources of variation and the statistical methods that will be used for their control should be summarized in the assay protocol/SOP. Understanding and controlling the sources of variability in an assay are critical to achieving the required precision as captured in the standard errors and confidence intervals for the key endpoints.
Q6: Is the proposed sample size/level of replication fit for purpose?	An assay that enables a crisp decision requires sufficient, but not excessive, precision. Sample size should always be based on what is known about the assay's variability in the laboratory where it will be run and the quantitative definition of what a successful assay outcome will look like. *Relying on historical precedent or published values should not be the default Strategy.*
Q7: Is there a comprehensive protocol/SOP detailing study objectives, key endpoints, experimental design, methods of analysis, and a timetable of activities?	A comprehensive assay protocol/SOP supports efficient decisions by specifying the methods to be used to control variation (e.g., randomization, blocking, use of covariates, and blinding). It helps to ensure uniformity in assay execution resulting in assay results that are reproducible and comparable from one run to another. It promotes transparency by documenting the actual conditions of the assay.
Q8: How is assay performance monitored over time? What is the plan for reacting to signs of instability?	Repeated assay use should be tracked to detect changing conditions that may affect the interpretation of the results and to understand the natural variability in the assay. Quality control (QC) charts are useful to monitor the consistency of controls or standards over time. Ongoing monitoring is necessary to understand any changes and their implications for interpretation of the results and to trigger remediation when necessary.
Objectivity in assay conduct	
Q9: Are inclusion/exclusion criteria for the assay specified in the protocol/SOP?	Criteria for the inclusion/exclusion of animals, cells, plates, etc. in an assay should be predefined and clearly stated in the protocol/SOP. This ensures all the appropriate data are collected and eliminates selection bias.
Q10: Is the management of subjectivity in data collection and reporting defined in the protocol/SOP?	There is a need to ensure that the scientist remains unaware of the treatment applied to the experimental unit. Even when the assay measurement is obtained automatically without human intervention there is possibility for bias. The use of randomization and blinding is highly recommended. Studies of a long duration should be blocked to ensure that no bias is introduced by changing conditions over time.
Q11: If the raw data are processed (e.g., by summarization or normalization) prior to analysis, is the method for doing this specified in the study protocol/SOP?	Methods of processing raw data prior to statistical analysis should be clearly stated in the assay protocol/SOP. For example, is it the raw response data, change from baseline or log transformed data that are to be analyzed; or are the raw data summarized into an area under the curve or average? This ensures that assay methods and results can be reproduced and validated.
Q12: Are rules for treating data as outliers in the analysis specified in the protocol/SOP?	Rules for treating data as outliers should be clearly stated in the assay protocol/SOP. Rules should be in place for the removal of individual data points, whole animals/plates and dose groups as required. This ensures all the appropriate data are analyzed and eliminates selection bias.
Q13: Is the analysis specified in the study protocol/SOP? Is it fit for purpose?	The statistical analysis must reflect the study design and assay objectives. Inappropriate statistical analyses can result in misleading conclusions and a false sense of precision. Consultation with a statistician is highly recommended.

The ACT list (Table 2.2) addresses three major areas:

1. *Aligning assay capability with research objectives* that includes an SOP (Standard Operating Procedure) that defines assay precision; prespecified decision criteria including the objective and primary endpoints of the study to prevent data dredging.
2. *Enabling assay capability by managing variation.* This includes verifying that the assay is fit for the purpose intended, a topic that is discussed further in the next section; developing an SOP for the methods used in an assay to minimize the influence of unwanted variability by randomization, experimenter blinding, the choice of appropriate statistical tests and sample sizes, and reproduction of the assay using quality control (QC) measures (Baker, 2016b) to identify conditions that may affect interpretation of the results and require remediation (e.g., microtiter plate drift; Harrison and Hammock, 1988).
3. *Objectivity in assay conduct.* This covers the predefined criteria for the inclusion/exclusion of samples (animals, cells, etc.) to avoid selection bias, the processing of data prior to statistical analysis, statistical testing, outliers and randomization, and blinding to avoid bias.

2.5.2 Validation of Reagents for Experimentation

A key issue in ensuring reproducibility is to ensure that all reagents used in a research system are validated and authenticated so that a researcher can be confident in their provenance. The concept of validation also extends to how the reagents are used. Research compounds being used as tools to interrogate a system must not only be chemically authenticated, but must be used at concentrations or doses that retain their target selectivity *if* that selectivity has been robustly established (Arrowsmith et al., 2015). Research funding bodies and publications have not historically required proof of reagent validation, a situation that is, however, changing.

2.5.2.1 Compound Authentication

All the materials used in an experiment must be qualitatively and quantitatively assayed for purity and structural fidelity using certified standards (http://www.sigmaaldrich.com/analytical-chromatography/analytical-standards.html?gclid=CNvknY7Mxs8CFQaUaQodZUUPWQ). Currently reagents are assumed to have been validated by the manufacturer (e.g., Millipore Sigma), although this presumption necessitates that the researcher exercise caution in their choice of supplier. For chemicals/compounds that are used for screening and as reference standards, the age of the sample must also be taken into account. If it is obtained from a reputable commercial vendor, its structure and purity are routinely provided and assumed by the investigator to be correct. If the sample is obtained from a colleague, is an "old" sample, or a source of unknown provenance, a recent chemical analysis using mass, IR, or NMR spectrometry is necessary. Given the inconvenience of doing this, unless the cost is prohibitive, it is probably easier to reorder the sample from a reputable supplier.

In the area of HTS to identify compound "hits" for a disease target of interest, there have been many instances where the "active" material representing the last few milligrams in a sample vial stored for 20 or 30 years fails to replicate its original activity. Thus, when the sample is resynthesized according to its original structure and its use fails to reproduce the original finding, this may be due to there being an unknown and now often unknowable product of decomposition of the original compound.

2.5.2.2 Compound Validation and Characterization

The essential properties of a compound are its target specificity, potency, efficacy (agonist, antagonist, or allosteric modulator), selectivity, target engagement, and bioavailability. The first four properties reflect the *pharmacodynamic* (PD) properties of the compound, its effect on the target, cellular transduction cascade, tissue function, system function, and whole animal to which it is exposed. The term specificity is used here to denote a defined molecular pharmacological property, for example, the muscarinic antagonistic properties of a compound that are mediated by binding to the orthosteric ligand recognition site of a muscarinic receptor. This is distinct from *nonspecific* or *functional* muscarinic antagonism whereby a compound blocks a downstream molecular target involved in muscarinic signaling. A selective muscarinic antagonist has only one property—muscarinic antagonism. Almost all compounds lose selectivity when their concentration is increased, so almost all compounds and drugs are best described as relatively selective (with the property qualified by information about the concentration range over which the selectivity pertains). Target engagement (Kenakin et al., 2014) which involves the duration (or *residence time*) of a binary compound–target complex that is a function of the conformational dynamics of target macromolecules that affect drug binding and dissociation rather than the binding affinity of a compound per se, and can dictate in vivo pharmacological activity (Copeland, 2016) and selectivity (Willemsen-Seegers et al., 2017). Bioavailability is the effect of the test system, tissue or animal, on the compound and is part of the *pharmacokinetic* (PK) profile of the compound. If the compound has specificity, and additionally binds selectively to its intended target (over a defined concentration range) but has poor to no bioavailability due to a lack of solubility, high first past metabolism in the liver or high plasma protein binding, it is unlikely to manifest any meaningful PD profile. Another key property associated with a compound is its *therapeutic index*, the ratio of the concentration/dose of a compound that produces a desired effect in relation to its propensity to produce unwanted side effects. These important concepts are covered in highly accessible detail in Kenakin's *A Pharmacology Primer: Techniques for More Effective and Strategic Drug Discovery* 4th *Edn.* (Kenakin, 2014).

Of particular relevance to the present discussion is the inappropriate use of compounds as research tools (Arrowsmith et al., 2015). Many compounds, agonists, antagonists, and allosteric modulators that are used to characterize a biological system involving a receptor, an enzyme, a signaling pathway, etc., can have questionable selectivity. When used, such compounds may have been only partially characterized—against a single target or target family—or deployed at a concentration that is orders of magnitude greater than the concentration at which it has been reported to be selectively active. Arrowsmith et al. (2015) cite examples of both. LY294002 was reported as a selective PI3 kinase inhibitor ($Ki = 1.4\ \mu M$) in 1994. A decade later it was known to interact with many other cellular targets in same concentration range. Given the availability of more selective PI3 kinase inhibitors, LY294002 is therefore not recommended as a probe for the PI3 kinase, yet it is still widely used for this purpose. A second example is 3-Deazaneplanocin A (DZNep), a 3-deazaadenosine analog reported as a potent inhibitor of *S*-adenosyl homocysteine hydrolase ($Ki = 50$ pM) in 1986. In 2007 at a concentration 1 million times greater than that required to inhibit SAH hydrolase in vitro, DZNep was reported to reduce methylation of histone H3K27, the substrate of EZH2 methyltransferase. While the effect of DZNep was ascribed to downregulation of EZH2 methyltransferase expression, it was implied in the literature that DZNep acted as a catalytic inhibitor of EZH2 methyltransferase and its observed effects were ascribed to this mechanism.

Another example of the need for caution in using chemical probes to define drug target associated signaling pathways are inhibitors of the human protein kinase family that comprises over 500 members. Many of these, like the prototypic inhibitor, staurosporine (Tanramluk et al., 2009) are "too nonspecific for useful conclusions to be made" (Bain et al., 2007).

The tyrosine kinase/Raf kinase inhibitor, sorafenib, which is approved for the treatment of renal cell and hepatocellular carcinomas, was originally designated as a selective B-Raf kinase inhibitor (in vitro activity in the range 260–540 nM). It was subsequently found to be active at a number of other kinases as these were discovered or became available as screens and included: RET (2 nM); FLT3 (13 nM); VEFGR (7–28 nM); KIT (16–18 nM); FMS (29 nM); PDGFRβ (37 nM); and ABL1 (T135I) (160 nM; Davis et al., 2011; Karaman et al., 2008). Additionally, sorafenib was also reported to be a potent (*K*i = 56 nM) 5-HT_{2B} receptor antagonist (Lin et al., 2012). Given its lack of selectivity, the use of sorafanib as a probe to determine the involvement of Raf kinase in a pathway is illogical as, for the same reason, is the use of SP600125 and AS601245 as selective JNK inhibitors (Bain et al., 2007).

In the context of ubiquitous targets like the protein kinases where selectivity of tool compounds versus other protein kinases is often minimal (Bain et al., 2003, 2007; Uitdehaag et al., 2012; Davis et al., 2011; Karaman et al., 2008), it is important to be even more cautious in concluding associations and mechanisms based on responses to individual compounds. Measures other than just IC_{50} values (the concentration causing 50% enzyme inhibition) are suggested to provide a more accurate portrayal of kinase inhibitor selectivity, including on- and off-rates and the residence time of the inhibitor with the target (Willemsen-Seegers et al., 2017), and certain scoring methods including window (WS) and ranking (RS) scores (Bosc et al., 2017). However, since many compounds target the ATP binding site of the kinase (Adams and Lee, 2007), there are inherent limitations on selectivity and it is necessary to employ orthogonal approaches to testing hypothesized associations between measured kinase inhibition and cellular responses. Such approaches include:

- The rank order of a number of available inhibitors with differing potencies must match at the specific kinase and on the response (Black, 1986; Titov and Liu, 2012). While one caveat might be that some compounds could, for example, exhibit low cellular penetration to decrease their apparent potency, this possibility can be tested.
- Use of inactive enantiomers or closely related chemical analogs that lack activity on both the target and response, to indicate the effect is not due to a non-specific (e.g. physicochemical) interaction.
- Selective knockdown of the target using siRNA (Bartz and Jackson, 2005), shRNA (Cheung et al., 2011) or CRISPR-Cas9 (Moore, 2015) approaches to reproduce the effect of the inhibitor. Moreover, targeted knockdown can sensitize the cell to the effect of chemical inhibition (Nijman, 2015; Titov and Liu, 2012).
- Overexpression of the compound target can also confer resistance (decrease in compound potency; Nijman, 2015; Titov and Liu, 2012).
- Creating a mutation in the compound binding site region of the kinase that does not affect activity but renders the mutant enzyme resistant to the inhibitor, as reported for SB 203580 induced inhibition of p38α MAPK (Eyers et al., 1999).
- Targeting other components of the pathway to mimic the response. For example, upstream of p38α are two activating kinases, MKK3 and MKK6, as well as an MKK-independent activation mechanism involving transforming growth factor

activated protein kinase 1 (TAK1)-binding protein; while downstream substrates for p38α include MAPKAPK2 and MAPKAPK3 (Adams and Lee, 2007).

While the present discussion has focused on protein kinases, similar activities should be utilized to help validate any prospective drug target, provided it is encoded by a gene.

Finally, as briefly mentioned earlier, compound selectivity is a critical function of the concentration used. The tendency to use a concentration of 50 μM of Compound A that is a selective antagonist for signaling pathway target X at a concentration of 20 nM in an effort to increase the extent of blockade to effect a change in phenotype is misguided and invalidates any interpretation of the ensuing results. Indeed, the ablation of a cell-based response by 50 μM Compound A may be, as likely, due to cellular toxicity as inhibition of target X.

2.5.2.3 *Validating a Concentration/Dose Response Curve*

The concentration/dose response curve is a key element of biomedical research and drug discovery. If an experimental outcome in a cell, tissue, or animal fails to show a robust, quantifiable concentration- or dose-dependence, colloquially the DRC, then the relevance of the agent under evaluation being related to an interaction with a discrete molecular target is questionable (Kenakin, 2014; Rang, 2006). When assessing the effect of a drug, drug candidate or lead compound on a particular biological phenotype and relating this to discrete cell membrane and intracellular events—the systems biology of the phenotype—it is critical that a DRC (Chapter 4.6.1) be established that conforms to the Law of Mass Action. This can then be used to characterize the pharmacology of the phenotypic response and provide a credible basis for the biological relevance of the effect. In many studies, effects that show no concentration/dose response or a U-shaped DRC are used without comment or concern as the basis of an intensive and often expensive evaluation of the signaling pathways involved in the initial response which leads to questionable validity or relevance. This is frequently technology- rather than intellectually-enabled - and has correspondingly subjective value.

2.5.2.4 *Cell Line Authentication*

The immortalized animal and human cell lines routinely used in biomedical research are usually obtained from established cell culture banks, for example, ATCC, (American Type Culture Collection), CellBank Australia, ECACC (European Collection of Animal Cell Cultures), UKSCB (UK Stem Cell Bank), etc., where authentication is conducted on a routine basis. Cells obtained from historical collections present in an investigator's laboratory freezer or from other investigators may have been maintained for years—if not decades—as monolayers and passaged. As a result, they may have become pleiomorphic, with a drift in their genetic fidelity due to the accumulation of genetic variations, copy number variants (CNVs) and transcriptional aberrations that compromise their cellular phenotype (Almeida et al., 2016; Geraghty et al., 2014; Neimark, 2014, 2015). Additionally, over time, cell lines may become mislabeled or contaminated with other cell lines, bacteria, mycoplasma, fungi, and viruses (Freedman et al., 2015b; Nardone, 2008) further complicating the cellular phenotype with the result that, in cancer research, "a fifth to a third or more of cell lines tested were mistakenly identified—with researchers unwittingly studying the wrong cancers, slowing progress toward new treatments and wasting precious time and money" (Marcus, 2012). While the problem of misidentification or contamination of cell lines in research has been well known for nearly 50 years, researchers appear reluctant to verify the identity of their cell lines (Nardone, 2008; Neimark, 2015) perhaps because large bodies of established research may have been based on the use of incorrectly defined cell lines. A striking example of the consequences of not authenticating cell lines is that of the bladder cancer cell line, KU7. This cell line has been widely use since 1984. Some 3 decades later, using short tandem repeat (STR) DNA analysis, KU7 was found to be "an exact match" with the cervical adenocarcinoma HeLa cell line leading to the recommendation that research using KU7 cells should appropriately be ascribed to HeLa cell lines (Jager et al., 2013). Other examples of "imposter" cell lines include: the MDA-MB-435 cell line widely used to study breast cancer that was identified as an M14 melanoma cell line (Rae et al., 2007); the uterine endometrial cell line, hTERT-EEC that was shown to have breast cancer origins; and the KAT (Kentucky Ain Thyroid) cancer cell line that was found to be contaminated with melanoma and colon cancer cells (Neimark, 2014). These examples and others, led to the formation of the International Cell Line Authentication Committee (ICLAC; Capes-Davis et al., 2010). This currently lists 475 cell lines in its database, of which, 438 are designated as misidentified with "no known authentic stock" (http://iclac.org/databases/cross-contaminations/) with 24% being contaminated with the ubiquitous HeLa cervical cancer cell line that has little resemblance or relevance to its origin (Masters, 2002). Indeed, some 18%–36% of the cell lines used in biomedical research may be cross-contaminated (Hughes et al., 2007) with rates of 25%–85% being reported in China (Ye et al., 2015) such that one in six researchers (and a greater number in China) use cells that are contaminated or misidentified and hence potentially irrelevant. This can lead to whole areas of research being undermined because of a blatant disregard for what have become well-known issues. For example, the MDA-MB-435 cell breast cancer line was unambiguously documented as an imposter in 2007 (Rae et al., 2007), yet an additional 247 articles using this cell line were published over the period 2008–14 (Prasad and Gopalan, 2015). Since different cancer lines respond differently to drugs (Wilding and Bodmer, 2014), the use of imposter cell lines in drug discovery can lead to compounds with activity at the wrong targets being advanced to clinical trials as in the case of the bexarotene and vemurafenib cases cited by Neimark (2014).

Since cell line authentication using STR analysis with PCR to detect contamination using guidelines from the *British Journal of Cancer* (Geraghty et al., 2014) and ASN standards (Almeida et al., 2016) costs between $50 and $150 per cell line (https://www.scienceexchange.com/services/cell-line-authentication?page=1), this begs the obvious question as to why 73% of researchers in a recent survey (Freedman et al., 2015c) chose to jeopardize their research—and in some instances their careers (Jager et al., 2013)—by not using this robust, facile, and inexpensive test.

2.5.2.5 Antibody Validation

Like cell lines, the use of incompletely or not at all characterized antibodies has been highlighted as yet another avoidable contribution to the inability to reproduce research findings (Baker, 2015). The magnitude of the problem was indicated by 112 researchers (Bradbury and Plückthun, 2015) calling for an international collaboration for antibody standardization which appears to be reflected in the International Working Group on Antibody Validation (IWGAV; Uhlen et al., 2016).

Antibodies, due to their specificity, high selectivity, affinity, and avidity, are uniquely powerful tools to modulate proteins of interest and to identify the presence of protein, the latter via the use of western blots. ENCODE (Encyclopedia of DNA Elements) reported that approximately 25% of 246 histone-modifying antibodies failed to show their purported activity, with only 41% exhibiting 100% selectivity, with 4 being specific, but not for their intended target (Egelhofer et al., 2011). Parseghian (2013) found that these results were the consequence of "Hitchhiker antigens" that expressed antibody complementarity-determining regions (CDRs) that originated from the presence of dead cells and debris in the large-scale bioreactors used to make the antibodies. Parseghian further noted that while clinical grade antibody production incorporated methods to reduce process-specific impurities, cost pressures on research antibody production limited their quality.

Less than 50% of commercial antibodies tested in 2008 were found to be useful in assessing the tissue distribution of their target protein (Berglund et al., 2008) while a more recent survey of methodological reporting in the published biomedical literature (Vasilevsky et al., 2013) found that only 44% of the antibodies used were validated by unique identification. Of the 2 million commercially available antibodies, it has been estimated that these represent 0.25–0.5 million unique "core" antibodies with the remainder being relabeled and sold by different vendors with unknown quality control standards. As Voskuil (2014) has noted, "QC data on the [antibody] product sheet … were generated many years ago thus keeping the sales going, while the actual antibody that generated these data may have sold out and has been replaced by successive other batches (from different animals) and the current batch on sale may no longer be able to generate such data at all." Research Resource Identifiers (RRIDs) represent a facile means to assign a unique tag to each antibody (Chawla, 2015). However, Simson (2016) has pointed out that the relabeling of antibodies for commercial sale results in the same antibody being available from multiple sources, negating or diminishing the value of the RRID.

Commercial antibodies, in addition to having the potential of being mislabeled and thus lacking activity at their purported target or losing activity due to mishandling (Simson, 2016) are often not validated as selective, exhibiting cross reactivity, binding to multiple targets that contain epitopes similar to the intended target protein. Each of these limitations may show interbatch variability. Also, when used in the wrong application (Bordeaux et al., 2010; Roncador et al., 2016; Voskuil, 2014), for example, an antibody selective for a native protein in a flow cytometry study can undergo denaturation when used in a western blot assay (Uhlen et al., 2016), the antibody may fail to recognize the target protein. These issues can be further amplified when antibodies originate with unknown provenance, e.g., from academic research groups, eventually becoming commercial entities.

To resolve these problems, Bradbury and Plückthun (2015) have advocated the cloning of antibodies from B cells, and following their sequencing producing these recombinantly. The resultant recombinant antibodies (rAbs) would then be characterized and validated, with their sequence serving as a unique identifier. An rAb could be faithfully produced independently of any individual animal response to the antigen-multiple antibodies with different affinities for the target protein plus other antibodies unrelated to the antigen. Researchers would then be able to choose precisely which validated rAb is most suitable for their research needs rather than having to engage in a trial and error revalidation process in screening non-rAbs. In theory, rABs would be less prone to cross reactivity and thus more selective in their actions and more appropriate for their selected application (Baker, 2015). The downside may be the estimated $1 billion required to produce characterized rABs to target the primary products of the 20,000 genes in the human genome (Bradbury and Plückthun, 2015).

Additional antibody concerns reported by Baker (2015) include one researcher who wasted 2 years and $500,000 on an antibody that did not recognize the protein of interest. Despite another researcher unforgettably noting that "I wasn't trained that you had to validate antibodies; I was just trained that you ordered them," a recent survey (Freedman et al., 2016) reported that 70% of respondents validated their antibodies—the converse of the 73% who fail to authenticate cell lines (Freedman et al., 2015c). It was however noted that 43% of researchers with 5 or fewer years of experience did not validate their antibodies due to perceived time constraints. In evaluating 13 anti-ERβ (estrogen receptor β) antibodies, Andersson et al. (2017) found that only one, the monoclonal PPZ0506 which is rarely used specifically targeted ERβ. Using PPZ0506 to profile protein expression in 44 normal and 21 malignant human tissues, the authors found no evidence for expression of ERβ protein in normal or cancerous human breast but found ERβ protein to be present "in testis, ovary, lymphoid cells, granulosa cell tumors, and a subset of malignant melanoma and thyroid cancers." This expression pattern contradicts a large body of published research and "highlights how inadequately validated antibodies can lead an exciting field astray," in this instance, no improved endocrine therapies despite over two decades of research.

Both the European Antibody Network (Roncador et al., 2016) and the IWGAV (Uhlen et al., 2016) have presented solutions to antibody validation with the latter proposing five "conceptual pillars" to guide antibody validation in specific research applications. These are:

- *Genetic strategies*: Measuring the anticipated signal in control cells or tissues in which the target gene has been knocked out or down using CRISPR/Cas9 or RNAi.
- *Orthogonal strategies*: Using an antibody-independent method for quantification across diverse sample sets and then examining the correlation between the antibody-based and antibody-independent quantifications.
- *Independent antibody strategies*: Using two or more independent antibodies that recognize different epitopes on the same target protein and confirming specificity via comparative and quantitative analysis.
- *Expression of tagged proteins*: Modifying the endogenous target gene to add sequences for an affinity tag or a fluorescent protein. Correlating the signal from the tagged protein with antibody-based detection methods.
- *Immunocapture followed by mass spectrometry (MS)*: Isolating a protein from solution via target-specific antibody binding coupled with MS analysis to identify proteins that interact directly with the purified antibody as well as proteins that may form a complex with the target protein.

Using these approaches alone or in combination to validate antibodies is anticipated to establish best practices in antibody validation making this a robust process.

2.5.2.6 RNA Interference (RNAi) Validation

RNA interference (RNAi) assays are another essential tool to detect critical gene products and their relationship/involvement to the cellular phenotype and ultimately to disease processes. Short interfering RNA (siRNA) is employed in cell culture to inhibit gene expression, while short hairpin RNA (shRNA) can be genomically integrated to provide a heritable source of RNAi. These are favored techniques as they have the potential to inhibit the expression of any gene, including those for "undrugable targets" (Mohr et al., 2014). At the experimental level, RNAi utilizes intrinsic cellular mechanisms that process endogenous microRNAs (miRNA) involving the RNA-induced silencing complex (RISC). However, as noted by Fellmann and Lowe (2014), gene suppression using these techniques "is somewhat unpredictable and often not as efficient as desired," and that "not all designed sequences are target specific." As a result, large shRNA libraries need to be screened to identify highly potent sequences, but the result is often that less efficient sequences are used with increased risk for off-target effects as higher concentrations are required (Mohr et al., 2014). "Off-target" effects that give rise to false positive and false negative results include sequence homology of the RNAi with nontargeted transcripts; overwhelming the intrinsic RNA processing systems (e.g., competing for binding to the RISC) such that endogenous miRNAs are affected and normal cellular processes are disrupted; and direct toxic effects of introducing large amounts of RNA or their delivery vectors to a cell, which can activate the immune system (Fellmann and Lowe, 2014; Kaelin, 2012). This requires that RNAi findings be corroborated with other techniques. Similar validation concerns exist for DNA constructs (Akama-Garren et al., 2016).

To minimize these off-target confounds, it is necessary to use multiple unrelated shRNA/siRNA sequences that target the same gene, as well as control, nonsilencing forms as negative controls (Forloni et al., 2017). For example, 3–4 different siRNAs for the same target might be pooled, and if successful, the pooled siRNAs are then evaluated individually to look for consistent target knock-down. If only one or two of the sequences account for the activity, then it is likely a nonspecific effect. As an alternative to this deconvolution approach, a different pooled library of siRNAs with different structures obtained from another vendor are used against the identified target for confirmation.

RNAi assays can be used to perform unbiased loss-of-function screens on a genome-wide scale to identify targets and pathways (Bartz and Jackson, 2005), although they may be replaced in the future by assays using CRISPRi libraries that are more specific (Moore, 2015). RNAi has also been used to compare cell-lines expressing cancer-causing genetic mutations with isogenic wild-type control cells, to identify potentially vulnerable targets and pathways specific for the cancer cells (Ashworth and Bernards, 2010; Iorns et al., 2007). In order to minimize these off-target confounds, it is necessary to use multiple unrelated shRNA/siRNA sequences that target the same gene, as well as control, nonsilencing forms as negative controls (Forloni et al., 2017).

2.5.2.7 *Animal and Model Validation*

A disease model or the animal species chosen for use in a research project is either validated, invalidated (shown to be misleading), or not yet evaluated. A valid animal species, or a model of a human disease, is one in which all drugs with specific effects in humans have the same effects (or a similar rank order of potency on a biomarker used as a surrogate for the variable or disease of interest) in the model or animal, and no drug that lacks the effect in humans is found to have an effect in the animal or model. Proof of validity, certain only when there are no false positives or negatives, is rare. This means that selection of model or species is normally based on incomplete information. The validity of species and models in biomedical research has therefore remained a subject of active debate (Arends et al., 2016; McGonigle and Ruggeri, 2014; Schuurman et al., 2015; Varga et al., 2010). When a disease state can be treated by a large number of drugs, and there is also a cohort of drugs that are known to be ineffective, these may serve as positive and negative controls to test the validity of a new model of the targeted disease on the basis of whether or not the positive and negative controls have their anticipated effects. A disease with existing treatment modalities, however, is unlikely to be one for which new treatments are urgently sought. A new model for a disease for which there are no known drugs is a challenge to validate because there are no positive controls.

The phenotype of an established animal model (e.g., a coronary ligation model of sudden cardiac death) is usually stable, but when cell lines are used, or if the model is a transgenic, there is a risk of passenger mutations that function in the background (Vanden Berghe et al., 2015) and contributions from both the microbiome of the animal (Bahrndorff et al., 2016; Dinan et al., 2015; Ezenwa et al., 2012) and the environment (Beura et al., 2016) and breeding errors, all of which may influence outcomes.

One example of this that is discussed in Chapter 1.11.2.1 is the unappreciated loss of the ALS disease phenotype in mice due to a build-up of multiple copies of the disease-causing gene that contributed, along with poor experimental design (lack of blinding or randomization) to multiple compounds showing a favorable profile preclinically but uniformly lacking activity in the clinic (Perrin, 2014).

2.5.2.8 *Equipment Validation/Maintenance*

In addition to the reagents used in an assay, the equipment used to assess the readouts also requires validation. This is usually achieved by ensuring that centrifuges, microtiter plate readers, animal activity chambers, microscopes and other image analyzers, blood pressure transducers etc. are maintained on the schedule recommended by their manufacturers such that they perform to specification. In times of financial hardship, equipment maintenance often becomes a discretionary expenditure (Baker, 2016b). This can result in micropipettes, pressure transducers or pH meters that are not accurately calibrated or a centrifuge that runs at 18,700 rpm while indicating the 21,000 rpm that is required for the experimental protocol because of a faulty speed sensor or a run temperature of 12°C instead of 4°C because of a defective temperature probe. Equipment that is obviously dysfunctional, for example, is broken, is paradoxically preferable to a piece of equipment that works inconsistently but is perceived as working normally. As a result, efforts are in place to improve quality assurance (QA) systems that can ensure and document that laboratory equipment, materials as well as data records are functional, up to date and complete. However, these best research practices remain voluntary with many researchers still viewing them as an unjustified constraint on their activities (Baker, 2016b).

2.5.2.9 *Final Thoughts on Validating Experiments*

Validating a system requires a test run using an experimental design that would be the starting point for a series of preliminary experiments to optimize the system. In doing a test run the following questions can be addressed: are the selected reagents—including cells and animals—fit for purpose, allowing measurement of the required endpoints? For instance, is the specific activity of a radioligand sufficient to produce a reliable signal, that is, is the signal to noise ratio acceptable? Are the data generated quantitative and are they amenable to analysis using the statistical test selected? Does the experimental readout show appropriate specificity, sensitivity, and stability? If the answer to any of these questions is negative, the test run conditions will need to be optimized by tweaking the variables involved.

2.5.2.10 *Animal Care*

The various animal care guidelines (Chapter 5.10) provide recommendations in planning, executing (including time parameters), and reporting research (including information on the species, strain, gender, weight, and source of animals used in a study) even when animals are used as a resource for the preparation of cells and tissues. It is also important that the care, housing, and humane use of the animals follow the US Guide for the Care and Use of Laboratory Animals (NRC, 2011) and that evidence be provided in the methods section indicating that the study has been approved by the Institutional Animal Care and Use Committee (IACUC) for the originating institution (Jones-Bolin, 2012). This information should include documentation on food and water availability, the type of food, housing temperature, humidity and light/dark cycles, caretaker/ investigator gender, and details of the compound/drug administration regimen (including dose, frequency, and route of administration), and, where appropriate, the use of anesthesia and/or analgesics. While these conditions often appear to be a formality, they can have a major impact on experimental outcomes and may explain discrepancies in results between different laboratories, including the failure to replicate studies.

2.6 TRANSPARENCY AND ACCOUNTABILITY: DOCUMENTING EXPERIMENTAL DETAIL

An increasingly important aspect of addressing the issue of reproducibility is the accurate recording of experimental methodologies and the raw data in a form that is accessible for those "skilled in the art." Traditionally, a laboratory notebook, written or electronic, is viewed as a complete and independent record of a researcher's laboratory activities with its content being immediately understandable to any other researcher in the research area. This is unfortunately infrequently the case such that the raw data—if indeed there is any in a notebook or electronic spreadsheet—often disappears along with its author making efforts to reproduce key experiments a mystery worthy of Sherlock Holmes (Engber, 2016). Baker (2016b) cites apocryphal instances of researchers "scribbling data on paper towels, repeating experiments without running controls and guessing at details months after an experiment" rather than formal notebooks that are kept current and audited (Williams et al., 2008).

With the emergence of Open Data Peer Review (ODPR) and open data policies (Chapter 5.3.2.6) that have been increasingly mandated by federal and private research funding bodies (Hahnel, 2015), the voluntary nature of the approach to laboratory QC will no doubt change. Additionally, some journals are requesting that raw experimental data be submitted together with manuscripts for ODPR. In situations where plagiarism and/or fraud is suspected in a submitted manuscript (and where a frequent response is an inability to "locate the data"—the scientific version of "the dog ate my homework"), this can lead to rapid resolution and rejection or retraction of an article depending on its publication status. At the present time it was found that "Most data sharing policies did not provide specific guidance on the practices that ensure data is maximally available and reusable." (Vasilevsky et al., 2017). In the more positive context of the "big data" revolution (Luo et al., 2016; Taglang and Jackson, 2016) raw data sets, public and private, especially those from omics disciplines can be mined by researchers to more precisely model complex diseases (Alyass et al., 2015) while population data sets collected in areas with public health issues can be used to identify and help fight global epidemics (Marr, 2015).

2.7 PHARMACOKINETICS

As in vivo experiments in animal models usually represent the last step in evaluating a disease mechanism/compound before an IND application for approval of NCE administration to human subjects and patients, it is an imperative to assess pharmacokinetics. This is partly to help inform evaluation of the potential efficacy and safety of an NCE based on data on the amount of compound present in the plasma rather than the dose, and partly to help predict human dosing using allometric scaling (Caldwell et al., 2004; Fan and de Lannoy, 2014; Ferl et al., 2016; Kleiman and Ehlers, 2016). Little emphasis is placed on determining the pharmacokinetic properties of an NCE, such that many preclinical experiments conducted in the absence of this information will be a waste of time and resources. Further detail on pharmacokinetics is beyond the scope of the current monograph and the reader is referred to Tozer and Rowland (2015) and Rosenbaum (2016) for additional information on this key topic.

2.8 BIAS IN EXPERIMENTAL DESIGN

Bias in experimentation can be unconscious and unintentional, due to inattention, lack of adequate forethought, flawed research practices, or deliberate (e.g., omission of data that does not fit the preselected hypothesis or selective reporting). An example of unconscious bias would be measuring the size of an implanted tumor in a mouse from the experimentally treated group as smaller than that in the control group because the researcher is not blinded (Vesterinen et al., 2010) and knows which group is which (Loscalzo, 2012).

Investigator bias includes: *Ignorance bias* a term to describe violations in experimental design and analysis that should have been learnt in graduate school. This includes undertaking a study that is not adequately powered, or is analyzed with inappropriate statistical methods (Button et al., 2013; Henderson et al., 2013; Ioannidis, 2005; Motulsky, 2014; Sena et al., 2010). Studies conducted in this way are unlikely to be reproducible when appropriately-designed and analyzed follow-up studies are conducted (Button et al., 2013; Henderson et al., 2013; Kilkenny et al., 2009). Unfortunately, the outcome of biased research, if published before others have the chance to publish unbiased findings, is the establishment of "priority" (Pfeiffer et al., 2011) that makes the subsequent studies the ones that require justification for their outcomes if these are different to the "priority" study. Another example of ignorance bias is the design (and publication) of studies using the minimal number of animals necessary for software to detect significance by Student's *t*-test, typically a minimum n of 3 as discussed in Chapter 3, and Curtis et al. (2015).

Design bias reflects critical features in planning an experiment that undermines its ability to test a hypothesis. This can include a design that is intended to support rather than refute a hypothesis, or is based on preconceived ideas or current dogma (e.g., the central role of amyloid in Alzheimer's disease).

In general, design is biased due to a lack of consideration of the null hypothesis; failure to incorporate relevant controls, and reference standards (positive and negative); and a reliance on single data points (endpoint, time point, or concentration/dose). A lack of randomization is a major reason for bias, and its effect is exacerbated in nonblinded experiments; failing to define endpoints, primary or secondary, until after the experiment is done, rather than a priori, and not incorporating a statistical penalty if multiple endpoints are recorded, is another example of design bias.

Misrepresentation bias that includes the selective reporting of positive rather than negative data, is an issue covered in Chapter 5 with *entrepreneurial/biotech bias* being a variation on the theme of author bias where data are reported in a form that is frequently more beneficial in enhancing the value of a company than disclosing objective science.

2.9 EXPERIMENTAL IMPLEMENTATION

The design of an experiment depends on the hypothesis and the number of variables being tested.

2.9.1 Selecting Groups for a Hypothesis-Testing/Confirmatory Study

The number of groups (as distinct from the number of samples per group) to be included in a study should be sufficient and necessary to allow testing of the hypothesis. If the hypothesis being tested involves the effects of a NCE or drug then appropriate concentrations should be selected that reflect the specificity and selectivity of the drug/NCE. When specificity and/or selectivity are uncertain then full concentration/dose-response studies will be required, with estimation of the key parameters (slope, maximum and IC_{50}/EC_{50} value) that are necessary for determining the specificity and the nature of antagonism (competitive, noncompetitive/allosteric). When selectivity is uncertain, specificity tests on a range of putative molecular targets (multiple bioassays or binding assays) are necessary.

When the readout involves multiple doses/concentrations, or multiple interventions/NCEs/ drugs at a single concentration/dose, with a requirement for single or multiple control groups (vehicle controls, positive and negative controls), there is scope for a very complex design, and, accordingly, a complex analysis. Usually a large, tactically unwieldy study can be broken into smaller parts, each involving a maximum of five groups and as few as possible sets of analysis on multiple ancillary readouts where the variables may be dependent or independent of the primary readout. An inevitability in hypothesis testing is that a simple question will be made complex by an imperative to build in as many readouts as possible—multiple variables that demand multivariate analysis. This imperative is often driven by the perception that complex data sets render the resultant manuscript more appealing to reviewers of high JIF journals. A general rule is that the more ways and more times a single piece of data (a number) is statistically analyzed, the less reliable the significance (p-value) that is generated. Indeed, multiple testing requires use of a "correction for multiple comparisons."

2.9.2 Experimental Tactics: Blinding, Randomization and Controls

Once a decision has been made on groups to be included in a study, a researcher must blind, randomize, and control the study to reduce bias (Bebarta et al., 2003; Hirst et al., 2014; Holman et al., 2015; Ruxton and Colgrave, 2011).

2.9.2.1 *Blinding*

Blinding can be easily achieved by the investigator asking a colleague to relabel the container storing the compound (and relabeling the identical container storing the vehicle), or the identical stock bottles of concentrated compound and of vehicle that will be used for pipetting test samples. The identity of the different groups of experimental materials is recorded in a laboratory notebook or database that is kept by the individual who undertook the blinding, not the investigator undertaking the experiment.

Unblinding of an experiment should not occur until the data have been analyzed. In some instances, the color or smell of the test sample, or the obviousness of the actual or anticipated effect of the test substance may make it difficult to blind the investigator. The color problem can be overcome by wrapping test materials in aluminum foil to obscure the color. Smell may require the use of a respirator mask. When a compound has an overtly visible effect (e.g., a large increase in heart rate), then a positive control can be used such that when the study is blinded the investigator will not be able to differentiate between the compound being tested and the positive control.

2.9.2.2 *Randomization*

Randomization of the participants in treatment groups involves not only animals, but tissues and cell lines. Its intent is to reduce bias (Suresh, 2011) without making the process totally haphazard (http://www.statisticshowto.com/randomization-experimental-design/). There are various ways to ensure randomization: choosing numbers from a container (unrestricted or simple random sampling); restricted randomization where a participant is allocated to a treatment group, while maintaining a balance across treatment groups (permuted block randomization); and where participants are stratified (stratified random sampling). Kim and Shin (2014) provide further background on the randomization process, while software, such as Research Randomizer (https://www.randomizer.org) and NCSS Staistical Software (http://ncss.wpengine.netdna-cdn.com/wp-content/themes/ncss/pdf/Procedures/NCSS/Randomization_Lists.pdf) greatly facilitate the process.

2.9.2.3 *Controls*

An important component of an experiment is its validation using appropriate controls. If a particular experiment does not produce an effect using a widely-used reference standard, for example, morphine in an animal model of pain, then the experiment is invalid irrespective of a positive analgesic outcome produced by any experimental treatment that is the actual subject of the study. There are three types of controls: vehicle, positive, and negative.

2.9.2.3.1 VEHICLE CONTROLS

Irrespective of whether an experiment is conducted in vitro or in vivo, any compounds that are used will need to be dissolved in a vehicle.

The vehicle acts a solvent and may itself have biological activity (Kelava et al., 2011). Common vehicles in biomedical experimentation include: water, saline, ethanol, dimethylsufoxide (DMSO), and propylene glycol (PG). Of these, DMSO is a very effective solvent but has to be used at concentrations below those that will perturb the cell membrane and produce biological effects on its own, for example, apoptosis (Qi et al., 2008), independent of the compound it is solubilizing (Kelava et al., 2011; Santos et al., 2003). In fact, there is a large literature on the therapeutic uses of DMSO (Walker, 1993). The final concentration of DMSO in an experiment should be 0.5% or less. (Hall et al., 2014; Jamalzadeh et al., 2016). Ethanol and PG may also disrupt biological systems, independent of the compound they are solubilizing. The potential for vehicle/solvent effects means that all experiments should have a vehicle/solvent control where a set of wells in a microtiter plate, test tube, or a group of animals are treated with the vehicle alone at the same concentration as that in the experimental solutions. All experimental samples should use the same vehicle at the same concentration. If this is not possible, for example, when a compound will come out of solution if the DMSO percentage drops below 1% and a mixture of PG and ethanol needs to be used instead, the control group (and all the test groups) should, if possible, include both solvents.

2.9.2.3.2 POSITIVE CONTROLS

Positive controls can be used to validate an experiment. For example, any experiment examining a novel analgesic agent should include morphine at a concentration/dose that is known to be effective. If morphine has no effect, the experiment is not valid and cannot be used. When morphine does work, it serves as a useful reference point for the potency/efficacy of the experimental agent under study adding additional value and perspective on the experiment.

2.9.2.3.3 NEGATIVE CONTROLS

A negative control is typically a known compound or drug that, based on its known PD/PK properties, should not produce any effect on functional outcomes in the system under study. Indeed, if possible it is useful to include, as part of the validation, a compound that has the opposite effect of that sought in the test agents for instance an active and inactive (or less active) enantiomer pair. For example, in an animal test to assess the efficacy of a new compound as an antianxiety agent, diazepam would serve as a positive control while a negative control would be Ro15-4513, an inverse agonist of the diazepam receptor which produces anxiety. In the absence of a validated negative control, the next best thing is to evaluate a series of compounds that differ in their structure activity profiles.

2.10 EXPERIMENTAL ANALYSIS

At face value, analysis of the data resulting from an experiment should be a straightforward matter, given that the analysis should follow the design, and the design should take into account the method of analysis.

2.10.1 From an Experiment to the Experiments

The topic of this chapter has been the planning and execution of an experiment. An experiment is not however what is reported in a publication. Instead, the results in a paper submitted for peer review represent the cumulative outcome of *reproducing* an experiment a sufficient number of times to confirm the result, as codified in most studies by the outcome of statistical analysis. In a research notebook, this is typically documented in terms of Experiments 1–5, each run independently (*not* five experiments run in parallel at the same time). From each experiment a data point is generated for a particular intervention, for example, an IC_{50} value that, when reproduced within that experiment provides a mean ± the standard deviation of the result. In combining data from the five reproduced experiments, the five means should be averaged to provide the mean ± the standard deviation that is then reported in a publication (Williams et al., 2008). In some instances, experiments are excluded from the pooled data for a variety of valid reasons. For example, the wrong reagents were used, the antibody was subsequently found to have the wrong (or no) selectivity, the equipment was not properly calibrated, or there was a power outage. Exclusion criteria can then be used to minimize the risk of biased data, that is, values that fall more than 2 SD from the mean as in Dixon's *Q*-test (Dean and Dixon, 1951). This and the reasons for excluding data other than a *Q*-test should always be documented in the experimental record (but is often not the case) to allow other investigators to understand how the data that is finally reported was arrived at. For an IND, each experiment that is included in a data set is mandatorily numbered and its source clearly identified in terms of a notebook number and page numbers, or its electronic equivalent. The investigator needs to annotate which experiments have been included in the reported IC_{50} value as it is often far from obvious to the third party "skilled in the art" when examining the raw data which sets of data were used to derive the final number reported.

2.11 INCREASING THE PROBABILITY OF GENERATING MEANINGFUL RESULTS

While science is a creative endeavor, there is a need for standardization in all aspects of the process—preparation of interventions and processing of the samples and materials, as well as design and analysis—to improve the outcomes. At all times the goal is to reduce unwanted sources of variability, whether this be due to contamination/degradation of reagents, nonvalidated cell lines and animal models, the uncertain provenance and specificity of antibodies, a lack of appropriate controls, baseline variations, background noise, etc, as well as false data analysis. It is also about technique; in a randomized study, suboptimal know-how and process will cause error that is spread evenly across the groups, potentially masking real intervention effects (Type 2 error). If design violations are part of the study (no randomization or blinding), the errors will vary *between* groups potentially contributing to false positive (Type 1 error) or false negative findings. Generating meaningful results therefore requires acknowledging these possibilities and accounting for them. Accordingly:

2.11.1 Dos

- Plan the experiment(s) with an idea to inform a simple and logical hypothesis.
- Evaluate the hypothesis as simply and directly as possible, avoiding overelaboration that leads to the generation of complex data sets that are challenging to analyze and interpret.
- Design the study to include as few groups as possible, but with a number that is sufficient to provide positive and negative controls if these are available (as noted, a novel target may have no known positive or negative ligands).

- Design the study by defining, a priori, how the data will be processed and analyzed statistically. The threshold *p*-value that denotes reliable statistical significance should be stated and not varied in the results.
- Test for a Gaussian distribution in accumulated controls. If the experiment is novel and there are no historical controls, the data set should be examined to assess the relationship between the mean values and the standard errors. If the standard errors are proportional to the mean (i.e., when the mean is 10, the standard error is 1, but when the mean is 100, the standard error is 10) this means the data are not Gaussian but, most likely, log Gaussian, such that if all values are log transformed (the log number is taken) and the mean and the SEM of the log values are calculated, the SEM values will be roughly constant and thus independent of the mean values. If the standard error varies between groups in a nonobvious way (i.e., not in proportion to dose/concentration or effect) do not assume the data are Gaussian, as no transform would be appropriate, meaning that nonparametric statistics should be used (not ANOVA or *t*-type tests) (Curtis et al., 2015).
- Blind the researchers involved in conducting and analyzing the study.
- Randomize the study.
- Use power analysis to determine how much extra power would be obtained by increasing the group size from a starting point of a minimum of $n = 3–5$ per group.
- Design the study to include as many samples as is feasible in terms of time, cost, and animal usage considerations not as few as can be justified by power analysis as the threshold minimum, below which obtaining significance is impossible.
- Determine the slope and EC_{50}/IC_{50} value (and maximum) of a dose/concentration response curve, and how these are affected by compounds, such that substantially more information regarding compound action can be provided, rather than binary (effect/no effect) data. Using a compound at any concentration, let alone 10 times the EC_{50}/IC_{50} value for a defined target, cannot guarantee either that the intended target has been modulated, or that other unintended targets have not been engaged to confuse the outcomes.
- Ensure that the laboratory staff are trained and remain up to date with changes in all relevant technology while networking with colleagues (Baker, 2016c).

2.11.2 Dont's

- Vary the *p*-value used to signify significance. "Effects" that are qualified by $p < 0.01$, $p < 0.001$, etc., in addition to $p < 0.05$, with no explanation for the use of multiple *p*-values or discussion of the meaning of the variation of the *p*-values.
- State that the smaller the *p*-value, the stronger the effect or the more believable the effect. The former is simply false, as the *p*-value gives no indication of the effect size.
- Vary the type of statistical test in a study from one data set to the next without explanation.
- Add more samples (increasing numbers) in a step-by-step process (unblinded), with repeated statistical reanalysis and further additions to group sizes until the *p*-value becomes significant (or until the futility of the exercise is apparent).

- Add numbers to a sample from a pool of historical controls to generate a large "n" for the control group, thereby conferring bogus statistical power (an approach favored by biotechs in fund raising campaigns).
- Test a range of transformations (such as log, cosine, square root, etc.) until the p-value becomes significant (as opposed to interrogating a large number of control and test data to determine which distribution best fits the data), then declaring this to be the appropriate transform for future use (Johnston et al., 1983).
- P-Hack: "When researchers collect or select data or statistical analyses until nonsignificant results become significant" (Head et al., 2015) using "Common practices that include: conducting analyses midway through experiments to decide whether to continue collecting data; recording many response variables and deciding which to report post-analysis, deciding whether to include or drop outliers post-analysis, excluding, combining, or splitting treatment groups post-analysis, including or excluding covariates post-analysis, and stopping data exploration if an analysis yields a significant p-value."

References

Adams, J.L., Lee, J.C., 2007. Progress in achieving proof of concept for p38 kinase inhibitors. In: Metcalf, B.W., Dillon, S. (Eds.), Target Validation in Drug Discovery. Academic Press, pp. 175–198.

Adams-Huet, B., Ahn, C., 2009. Bridging clinical investigators and statisticians: writing the statistical methodology for a research proposal. J. Invest. Med. 57, 818–824.

Akama-Garren, E.H., Joshi, N.S., Tammela, T., Chang, G.P., Bethany, L., Wagner, B.L., Lee, D.-Y., et al., 2016. A modular assembly platform for rapid generation of DNA constructs. Sci. Rep. 6, 16836.

Almeida, J.L., Cole, K.D., Plant, A.L., 2016. Standards for cell line authentication and beyond. PLoS Biol. 14, e1002476.

Alyass, A., Turcotte, M., Meyre, D., 2015. From big data analysis to personalized medicine for all: challenges and opportunities. BMC Med. Genomics 8, 33.

Andersson, S., Sundberg, M., Pristovsek, N., Ibrahim, A., Jonsson, P., Katona, B., et al., 2017. Insufficient antibody validation challenges oestrogen receptor beta research. Nature Commun 8, 15840.

Arends, M.J., White, E.S., Whitelaw, C.B.A. (Eds.), 2016. Models of human disease. J. Pathol. 238, 135–367.

Arrowsmith, C.H., Audia, J.E., Austin, C., Baell, J., Bennett, J., Blagg, J., et al., 2015. The promise and peril of chemical probes. Nat. Chem. Biol. 11, 536–541.

Ashton, J.C., 2012. When biostatistics is a neo-inductionist barrier to science. Br. J. Pharmacol. 167, 1389–1390.

Ashworth, A., Bernards, R., 2010. Using functional genetics to understand breast cancer biology. Cold Spring Harb. Perspect. Biol. 2, a003327.

Avey, M.T., Moher, D., Sullivan, K.J., Fergusson, D., Griffin, G., Grimshaw, J.M., et al., 2016. The devil is in the details: incomplete reporting in preclinical animal research. PLoS One 11, e0166733.

Badyal, D.K., Modgill, V., Kaur, J., 2009. Computer simulation models are implementable as replacements for animal experiments. Altern. Lab. Anim. 37, 191–195.

Bahrndorff, S., Alemu, T., Alemneh, T., Nielsen, J.L., 2016. The microbiome of animals: implications for conservation biology. Int. J. Genomics 2016, 5304028.

Bain, J., McLauchlan, H., Elliott, M., Cohen, P., 2003. The specificities of protein kinase inhibitors: an update. Biochem. J. 371, 199–204.

Bain, J., Plater, L., Elliott, M., Shapiro, N., Hastie, C.J., et al., 2007. The selectivity of protein kinase inhibitors: a further update. Biochem. J. 408, 297–315.

Baker, M., 2015. Blame it on the antibodies. Nature 521, 274–276.

Baker, M., 2016a. Statisticians issue warning over misuse of P values. Nature 531, 151.

Baker, M., 2016b. How quality control could save your science. Nature 529, 456–458.

Baker, M., 2016c. Seek out stronger science. Nature 537, 703–704.

Baker, D., Lidster, K., Sottomayor, A., Amor, S., 2014. Two years later: journals are not yet enforcing the ARRIVE Guidelines on reporting standards for pre-clinical animal studies. PLoS Biol. 12, e1001756.

Bartz, S., Jackson, A.L., 2005. How will RNAi facilitate drug development? Sci. STKE 295, pe39.

Basken, P., 2017. A new theory on how researchers can solve the reproducibility crisis: do the math. Chron. Higher Edu., Available from: http://www.chronicle.com/article/A-New-Theory-on-How/240470.

Bebarta, V., Luyten, D., Heard, K., 2003. Emergency medicine animal research: does use of randomization and blinding affect the results? Acad. Emerg. Med. 10, 684–687.

Berglund, L., Björling, E., Oksvold, P., Fagerberg, L., Asplund, A., Szigyarto, C.A., et al., 2008. A genecentric Human Protein Atlas for expression profiles based on antibodies. Mol. Cell. Proteomics 7, 2019–2027.

Beura, L.K., Hamilton, S.E., Bi, K., Schenkel, J.M., Odumade, O.A., Casey, L.A., et al., 2016. Normalizing the environment recapitulates adult human immune traits in laboratory mice. Nature 532, 512–516.

Biesecker, L.G., 2013. Hypothesis-generating research and predictive medicine. Genome Res. 23, 1051–1053.

Black, J.W., 1986. Pharmacology: analysis and exploration. Brit. Med. J. 293, 252–255.

Black, J., 2010. A life in new drug research. Brit. J. Pharmacol. 160 (Suppl. 1), S15–S25.

Bordeaux, J., Welsh, A., Agarwal, S., Killiam, E., Baquero, M., Hanna, J., et al., 2010. Antibody validation. Biotechniques 48, 197–209.

Bosc, N., Meyer, C., Bonnet, P., 2017. The use of novel selectivity metrics in kinase research. BMC Bioinform. 18, 17.

Boyle, E.A., Li, Y.I., Pritchard, J.K., 2017. An expanded view of complex traits: from polygenic to omnigenic. Cell 169, 1177–1186.

Bradbury, A., Plückthun, A., 2015. Reproducibility: standardize antibodies used in research. Nature 518, 27–29.

Bunnage, M.E., Piatnitski Chekler, E.L., Jones, L.H., 2013. Target validation using chemical probes. Nat. Chem. Biol. 9, 195–199.

Burden, N., Chapman, K., Sewell, F., Robinson, V., 2015. Pioneering better science through the 3Rs: an introduction to the national centre for the replacement, refinement, and reduction of animals in research (NC3Rs). J. Am. Assoc. Lab. Med. Sci. 54, 198–208.

Bush, W.S., Moore, J.H., 2012. Chapter 11: genome-wide association studies. PLoS Comput. Biol. 8, e1002822.

Bush, W.S., Oetjens, M.T., Crawford, D.C., 2016. Unravelling the human genome-phenome relationship using phenome-wide association studies. Nat. Rev. Genet. 17, 129–145.

Button, K.S., Ioannidis, J.P., Mokrysz, C., Nosek, B.A., Flint, J., et al., 2013. Power failure: why small sample size undermines the reliability of neuroscience. Nat. Rev. Neurosci. 14, 365–376.

Caldwell, G.W., Masucci, J.A., Yan, Z., Hageman, W., 2004. Allometric scaling of pharmacokinetic parameters in drug discovery: can human CL, Vss and $t_{1/2}$ be predicted from in vivo rat data? Eur. J. Drug Metabol. Pharmacokinet. 29, 133–143.

Capes-Davis, A., Theodosopoulos, G., Atkin, I., Drexler, H.G., Kohara, A., MacLeod, R.A., et al., 2010. Check your cultures! A list of cross-contaminated or misidentified cell lines. Int. J. Cancer 127, 1–8.

Chattopadhyay, P.K., Gierahn, T.M., Roederer, M., Love, J.C., 2014. Single-cell technologies for monitoring immune systems. Nat. Immunol. 15, 128–135.

Chawla, D.S., 2015. Researchers argue for standard format to cite lab resources. Nature, Available from: http://www.nature.com/news/researchers-argue-for-standard-format-to-cite-lab-resources-1.17652.

Cheung, H.W., Cowley, G.S., Weir, B.A., Boehm, J.S., Rusin, S., Scott, J.A., et al., 2011. Systematic investigation of genetic vulnerabilities across cancer cell lines reveals lineage-specific dependencies in ovarian cancer. Proc. Natl. Acad. Sci. USA 108, 12372–12377.

Collins, F.S., Tabak, L.A., 2014. Policy: NIH plans to enhance reproducibility. Nature 505, 612–613.

Colquhoun, D., 2014. An investigation of the false discovery rate and the misinterpretation of p-values. R. Soc. Open. Sci. 1, 140216.

Copeland, R.A., 2016. The drug–target residence time model: a 10-year retrospective. Nat. Rev. Drug Discov. 15, 87–95.

Curtis, M.J., Bond, R.A., Spina, D., Ahluwalia, A., Alexander, S.P.A., Giembycz, M.A., et al., 2015. Experimental design and analysis and their reporting: new guidance for publication in BJP. Br. J. Pharmacol. 172, 2671–2674.

Davis, M.J., Hunt, J.P., Herrgard, S., Ciceri, P., 2011. Wodicka comprehensive analysis of kinase inhibitor selectivity. Nat. Biotechnol. 29, 1046–1051.

Dean, R.B., Dixon, W.J., 1951. Simplified statistics for small numbers of observations. Anal. Chem. 23, 636–638.

Demidenko, E., 2016. The *p*-value you can't buy. Am. Stat. 70, 33–38.

Dinan, T.G., Roman, M., Stilling, R.M., Stanton, C., Cryan, J.F., 2015. Collective unconscious: how gut microbes shape human behavior. J. Psychiat. Res. 63, 1–9.

Doss, J., Mo, H., Carroll, R.J., Crofford, L.J., Denny, J.C., 2017. Phenome-wide association study of rheumatoid arthritis subgroups identifies association between seronegative disease and fibromyalgia. Arthrit. Rheum. 69, 291–300.

Eddy, S.R., 2004. What is Bayesian statistics? Nat. Biotechnol. 22, 1177–1178.

Egelhofer, T.A., Minoda, A., Klugman, S., Lee, K., Kolasinska-Zwierz, P., et al., 2011. An assessment of histone-modification antibody quality. Nat. Struct. Mol. Biol. 18, 91–93.

Engber, D., 2016. Cancer research is broken. Slate April 19, 2016. Available from: http://www.slate.com/articles/health_and_science/future_tense/2016/04/biomedicine_facing_a_worse_replication_crisis_than_the_one_plaguing_psychology.html.

Eyers, P.A., van den IJssel, P., Quinlan, R.A., Goedert, M., Cohen, P., 1999. Use of a drug-resistant mutant of stress-activated protein kinase 2a/p38 to validate the in vivo specificity of SB 203580. FEBS Lett. 451, 191–196.

Ezenwa, V.O., Gerardo, N.M., Inouye, D.W., Medina, M., Xavier, J.B., 2012. Animal behavior and the microbiome. Science 338, 198–199.

Fan, J., de Lannoy, I.A.M., 2014. Pharmacokinetics. Biochem. Pharmacol. 87, 93–120.

Federico, C.A., Carlisle, B., Kimmelman, J., Fergusson, D.A., 2014. Late, never or non-existent: the inaccessibility of preclinical evidence for new drugs. Br. J. Pharmacol. 171, 4247–4254.

Fellmann, C., Lowe, S.W., 2014. Stable RNA interference rules for silencing. Nat. Cell. Biol. 16, 10–18.

Ferl, G.Z., Theil, F.-P., Wong, H., 2016. Physiologically based pharmacokinetic models of small molecules and therapeutic antibodies: a mini-review on fundamental concepts and applications. Biopharmaceut. Drug. Disp. 37, 75–92.

Finkbeiner, S., Frumkin, M., Kassner, P.D., 2015. Cell-based screening: extracting meaning from complex data. Neuron 86, 160–174.

Forloni, M., Ho, T., Sun, L., Wajapeyee, N., 2017. Large-scale RNA interference screening to identify transcriptional regulators of a tumor suppressor gene. (Eukaryotic transcriptional and post-transcriptional gene expression regulation.). Methods Mol. Biol. 1507, 261–268.

Freedman, L.P., Cockburn, I.M., Simcoe, T.S., 2015a. The economics of reproducibility in preclinical research. PLoS Biol. 13, e1002165.

Freedman, L.P., Gibson, M.C., Bradbury, A.R.M., Buchberg, A.M., Davis, D., Dolled-Filhart, M.P., et al., 2016. The need for improved education and training in research antibody usage and validation practices. BioTechniques 61, 16–18.

Freedman, L.P., Gibson, M.C., Ethier, S.P., Soule, H.R., Neve, R.M., Reid, Y.A., 2015b. Reproducibility: changing the policies and culture of cell line authentication. Nat. Methods 12, 493–497.

Freedman, L.P., Gibson, M.C., Wisman, R., Ethier, S.P., Soule, H.R., Neve, R.M., Reid, Y.A., Never, R.M., 2015c. The culture of cell culture practices and authentication—results from a 2015 survey. BioTechniques 59, 189–192.

Geraghty, R.J., Capes-Davis, A., Davis, J.M., Downward, J., Freshney, R.I., Knezevic, I., et al., 2014. Guidelines for the use of cell lines in biomedical research. Br. J. Cancer. 111, 1021–1046.

Glass, D.J., 2014. Experimental Design for Biologists, Second ed. Cold Spring Harbor Press, New York.

Glasziou, P., Altman, D.G., Bossuyt, P., Boutron, I., Clarke, M., Julious, S., 2014. Reducing waste from incomplete or unusable reports of biomedical research. Lancet 383, 267–276.

Gore, K., Stanley, P., 2015. Helping to drive the robustness of preclinical research—the assay capability tool. Pharma. Res. Perspect. 3, e00162.

Goodman, S.N., Fanelli, D., Ioannidis, J.P.A., 2016. What does research reproducibility mean? Sci. Transl. Med. 8, 342ps12.

Grainger, D., 2013. Industry Voices: PheWAS—the tool that's revolutionizing drug development that you've likely never heard of. FierceBiotech., Available from: http://www.fiercebiotech.com/biotech/industry-voices-phewas-tool-s-revolutionizing-drug-development-you-ve-likely-never-heard-of.

Grun, D., van Oudenaarde, A., 2015. Design and analysis of single-cell sequencing experiments. Cell 163, 799–810.

Hahnel, M., 2015. The year of open data mandates. figshare.com. Available from: https://figshare.com/blog/2015_The_year_of_open_data_mandates/143.

Hall, M.D., Telma, K.A., Chang, K.-E., Lee, T.D., Madigan, J.P., Lloyd, J.R., et al., 2014. Say no to DMSO: dimethylsulfoxide inactivates cisplatin, carboplatin and other platinum complexes. Cancer Res. 74, 3913–3922.

Halsey, L.G., Curran-Everett, D., Vowler, S.L., Drummond, G.B., 2015. The fickle *P* value generates irreproducible results. Nat. Methods 12, 179–185.

Harrison, R.O., Hammock, B.D., 1988. Location dependent biases in automatic 96-well microplate readers. J. Assoc. Off. Anal. Chem. 171, 981–987.

Head, M.L., Holman, L., Lanfear, R., Kahn, A.T., Jennions, M.D., 2015. The extent and consequences of P-hacking in science. PLoS Biol. 13, e1002106.

Hebbring, S.J., 2013. The challenges, advantages and future of phenome-wide association studies. Immunology. 141, 157–165.

Henderson, V.C., Kimmelman, J., Fergusson, D., Grimshaw, J.M., Hackam, D.G., 2013. Threats to validity in the design and conduct of preclinical efficacy studies: a systematic review of guidelines for in vivo animal experiments. PLoS Med. 10, e1001489.

Hirst, J.A., Howick, J., Aronson, J.K., Roberts, N., Perera, R., et al., 2014. The need for randomization in animal trials: an overview of systematic reviews. PLoS One 9, e98856.

Holman, L., Head, M.L., Lanfear, R., Jennions, M.D., 2015. Evidence of experimental bias in the life sciences: why we need blind data recording. PLoS Biol. 13, e1002190.

Hooijmans, C.R., de Vries, R., Leenaars, M., Curfs, J., Ritskes-Hoitinga, M., 2011. Improving planning, design, reporting and scientific quality of animal experiments by using the Gold Standard Publication Checklist, in addition to the ARRIVE guidelines. Br. J. Pharmacol. 162, 1259–1260.

Hooijmans, C.R., Leenaars, M., Ritskes-Hoitinga, M., 2010. A gold standard publication checklist to improve the quality of animal studies, to fully integrate the Three Rs, and to make systematic reviews more feasible. Altern. Lab. Anim. 38, 167–182.

Horvath, P., Aulner, N., Bickle, M., Davies, A.M., Del Nery, E., Ebner, D., et al., 2016. Screening out irrelevant cell-based models of disease. Nat. Rev. Drug Discov. 15, 751–769.

Howitt, S.M., Wilson, A.N., 2014. Revisiting "is the scientific paper a fraud?". EMBO Rep. 15, 481–484.

Hughes, P., Marshall, D., Reid, Y., Parkes, H., Gelber, C., 2007. The costs of using unauthenticated, over-passaged cell lines: how much more data do we need? Biotechniques 43, 575–582.

Ioannidis, J.P.A., 2005. Why most published research findings are false. PLoS Med. 2, e124.

Ioannidis, J.P.A., 2016. Why most clinical research is not useful. PLoS Med. 13, e1002049.

Ioannidis, J.P.A., Greenland, S., Hlatky, M.A., Khoury, M.J., Macleod, M.R., et al., 2014. Research: increasing value, reducing waste 2: increasing value and reducing waste in research design, conduct, and analysis. Lancet. 383, 166–175.

Iorns, E., Lord, C.J., Turner, N., Ashworth, A., 2007. Utilizing RNA interference to enhance cancer drug discovery. Nat. Rev. Drug. Discov. 6, 556–568.

Jager, W., Horiguchi, Y., Shah, J., Hyashi, T., Awrey, S., Gust, K.M., et al., 2013. Hiding in plain view: genetic profiling reveals decades old cross contamination of bladder cancer cell line KU7 with HeLa. J. Urol. 190, 1404–1409.

Jamalzadeh, J., Ghafoori, H., Sariri, R., Rabuti, H., Nasirzade, J., et al., 2016. Cytotoxic effects of some common organic solvents on MCF-7, RAW-264. 7 and human umbilical vein endothelial cells. Avicenna. J. Med. Biochem. 4, e33453.

Jogalekar, A., 2014. Falsification and its discontents. The curious wavefunction. Sci. Am.https://blogs.scientificamerican.com/the-curious-wavefunction/falsification-and-its-discontents/.

Johnson, V.E., 2013. Revised standards for statistical evidence. Proc. Natl. Acad. Sci. USA 110, 19313–19317.

Johnston, K.M., MacLeod, B.A., Walker, M.J., 1983. Responses to ligation of a coronary artery in conscious rats and the actions of antiarrhythmics. Can. J. Physiol. Pharmacol. 61, 1340–1353.

Jones-Bolin, S., 2012. Guidelines for the care and use of laboratory animals in biomedical research. Curr. Protoc. Pharmacol. 4B, A.4B.1–A.4B.9.

Kaelin, Jr., W.G., 2012. Molecular biology. Use and abuse of RNAi to study mammalian gene function. Science 337, 421–422.

Karaman, M.W., Herrgard, S., Treiber, D.K., Gallant, P., Atteridge, C.E., Campbell, B.T., et al., 2008. A quantitative analysis of kinase inhibitor selectivity. Nat. Biotechnol. 26, 127–132.

Kass, R.E., Caffo, B.S., Davidian, M., Meng, X.-L., Yu, B., Reid, N., 2016. Ten simple rules for effective statistical practice. PLoS Comput. Biol. 12, e1004961.

Kelava, R., Cacar, I., Culo, F., 2011. Biological actions of drug solvents. Period Biol. 113, 311–320.

Kenakin, T., 2014. A Pharmacology Primer: Techniques for More Effective and Strategic Drug Discovery, fourth ed. Elsevier Academic Press, San Diego, CA.

Kenakin, T., Bylund, D.B., Toews, M.L., Mullane, K., Winquist, R.J., Williams, M., 2014. Replicated, replicable and relevant-target engagement and pharmacological experimentation in the 21st century. Biochem. Pharmacol. 87, 64–77.

Khanna, I., 2012. Drug discovery in pharmaceutical industry: productivity challenges and trends. Drug Discov. Today. 17, 1088–1102.

Kilkenny, C., Browne, W.J., Cuthill, I.C., Emerson, M., Altman, D.G., 2010. Animal research: reporting in vivo experiments: the ARRIVE guidelines. Br. J. Pharmacol. 160, 1577–1579.

Kilkenny, C., Parsons, N., Kadyszewski, E., Festing, M.F.W., Cuthill, I.C., et al., 2009. Survey of the quality of experimental design, statistical analysis and reporting of research using animals. PLoS One 4, e7824.

Kim, C.K., Adhikari, A., Deisseroth, K., 2017. Integration of optogenetics with complementary methodologies in systems neuroscience. Nat. Rev. Neurosci. 18, 222–235.

Kim, J., Shin, W., 2014. How to do random allocation (randomization). Clin. Orthop. Surg. 6, 103–109.

Kimmelman, J., Mogil, J.S., Dirnagl, U., 2014. Distinguishing between exploratory and confirmatory preclinical research will improve translation. PLoS Biol. 12, e1001863.

Kitchen, I., 1987. Statistics and pharmacology: the bloody obvious test. Trends Pharmacol. Sci. 8, 252–253.

Khoury, M.J., 2013. Public health impact of genome-wide association studies: glass half full or half empty? centers for disease control and prevention. Genomics and Health Impact Blog, Available from: https://blogs.cdc.gov/genomics/2013/08/01/public-health-impact/.

Kleiman, R.J., Ehlers, M.D., 2016. Data gaps limit the translational potential of preclinical research. Sci. Transl. Med. 8, 320ps1.

Laajala, T.D., Jumppanen, M., Huhtaniemi, R., Fey, V., Kaur, A., Knuuttila, M., et al., 2016. Optimized design and analysis of preclinical intervention studies in vivo. Sci. Rep. 6, 30723.

Landis, S.C., Amara, S.G., Asadullah, K., Austin, C.P., Blumenstein, R., Bradley, E.W., et al., 2012. A call for transparent reporting to optimize the predictive value of preclinical research. Nature 490, 187–191.

Lee, N., Authier, S., Pugsley, M.K., Curtis, M.J., 2010. The continuing evolution of torsades de pointes liability testing methods: is there an end in sight? Toxicol. Appl. Pharmacol. 243, 146–153.

Lin, X., Huang, X.-P., Chen, G., Whaley, R., Peng, S., Wang, Y., et al., 2012. Life beyond kinases: structure-based discovery of sorafenib as nanomolar antagonist of 5-HT receptors. J. Med. Chem. 55, 5749–5759.

Liu, J., Ye, Z., Mayer, J.G., Hoch, B.A., Green, C., Rolak, L., et al., 2016. Phenome-wide association study maps new diseases to the major histocompatibility complex region. J. Med. Genet. 53, 681–689.

Loscalzo, J., 2012. Experimental irreproducibility: causes, (mis)interpretations, and consequences. Circulation 125, 1211–1214.

Lovell, D.P., 2013. Biological importance and statistical significance. J. Agric. Food Chem. 61, 8340–8348.

Luo, J., Wu, M., Gopukumar, D., Zhao, Y., 2016. Big data application in biomedical research and health care: a literature review. Biomed. Inform. Insights 8, 1–10.

Marcus, A.D., 2012. Lab mistakes hobble cancer studies but scientists slow to take remedies. Wall St. J.http://www.wsj.com/articles/SB10001424052970204571404577257513760102538.

Marino, M., 2014. The use and misuse of statistical methodologies in pharmacology research. Biochem. Pharmacol. 87, 78–92.

Marr, B., 2015. How big data is changing healthcare. Forbeshttp://www.forbes.com/sites/bernardmarr/2015/04/21/how-big-data-is-changing-healthcare/#320c927732d9.

Masters, J.R., 2002. HeLa cells 50 years on: the good, the bad and the ugly. Nat. Rev. Cancer 2, 315–319.

McGonigle, P., Ruggeri, B., 2014. Animal models of human disease: challenges in enabling translation. Biochem. Pharmacol. 87, 162–171.

McGonigle, P., Williams, M., 2014. Preclinical pharmacology and toxicology—contributions to the translational interface. Ref. Module Biomed. Sci. Available from: http://dx.doi.org/10.1016/B978-0-12-801238-3.05242-9.

McGrath, J.C., Drummond, G.B., McLachlan, E.M., Kilkenny, C., Wainwright, C.L., 2010. Guidelines for reporting experiments involving animals: the ARRIVE guidelines. Br. J. Pharmacol. 160, 1573–1576.

McGrath, J.C., McLachlan, E.M., Zeller, R., 2015. Transparency in research involving animals: the Basel declaration and new principles for reporting research in BJP manuscripts. Br. J. Pharmacol. 172, 2427–2432.

Medawar, P., 1996. Is the scientific paper a fraud? In: Medawar, P., Gould, S.J. (Eds.), The Strange Case of the Spotted Mice: and Other Classic Essays on Science. Oxford University Press, UK, pp. 33–39.

Mogul, J.S., Macleod, M.R., 2017. No publication without confirmation. Nature 542, 409–411.

Moher, D., Glasziou, P., Chalmers, I., Nasser, M., Bossuyt, P.M., Korevaar, D.A., et al., 2016. Increasing value and reducing waste in biomedical research: who's listening. Lancet 387, 1573–1586.

Mohr, S.E., Smith, J.A., Shamu, C.E., Neumüller, R.A., Perrimon, N., 2014. RNAi screening comes of age: improved techniques and complementary approaches. Nat. Rev. Mol. Cell. Biol. 15, 591–600.

Moore, J.D., 2015. The impact of CRISPR-Cas9 on target identification and validation. Drug Discov. Today 20, 450–457.

Motulsky, H.J., 2014. Common misconceptions about data analysis and statistics. J. Pharmacol. Exp. Ther. 351, 200–205.

Muhlhausler, B.S., Bloomfield, F.H., Gillman, M.W., 2013. Whole animal experiments should be more like human randomized controlled trials. PLoS Biol. 11, e1001481.

Mullane, K., Williams, M., 2017. Enhancing reproducibility: failures from reproducibility initiatives underline core challenges. Biochem. Pharmacol. 138, 7–18.

Mungall, C.J., McMurry, J.A., Köhler, S., Balhoff, J.P., Borromeo, C., et al., 2017. The Monarch initiative: an integrative data and analytic platform connecting phenotypes to genotypes across species. Nucleic Acids Res. 45, D712–D722.

Nardone, R.M., 2008. Curbing rampant cross-contamination and misidentification of cell lines. BioTechniques 45, 221–227.

National Research Council (NRC), 2011. Guide for the Care and Use of Laboratory Animals, eighth ed. National Academies Press, Washington DC.

Nature, 2013. Announcement: reducing our irreproducibility. Nature 496, 398.

Neimark, J., 2014. The dirty little secret of cancer research. Discoverhttp://discovermagazine.com/2014/nov/20-trial-and-error.

Neimark, J., 2015. Line of attack. Science 347, 938–940.

Nijman, S.M.B., 2015. Functional genomics to uncover drug mechanism of action. Nature Chem. Biol. 11, 942–948.

Nuzzo, R., 2014. Statistical errors. Nature 506, 150–152.

Parseghian, M.H., 2013. Hitchhiker antigens: inconsistent ChiP results, questionable immunohistology data, and poor antibody performance may have a common factor. Biochem. Cell. Biol. 91, 378–394.

Peers, I.S., Ceuppens, P.R., 2012. In search of preclinical robustness. Nat. Rev. Drug Discov. 11, 733–734.

Peers, I.S., South, M.C., Ceuppens, P.R., Bright, J.D., Pilling, E., 2014. Can you trust your animal study data? Nat. Rev. Drug Discov. 13, 560.

Perrin, S., 2014. Preclinical research: make mouse studies work. Nature 507, 423–425.

Pettit, J.-B., Tomer, R., Achim, K., Richardson, S., Azizi, L., Marioni, J., 2014. Identifying cell types from spatially referenced single-cell expression datasets. PLoS Comput. Biol. 10, e1003824.

Pfeiffer, T., Bertram, L., Ioannidis, J.P.A., 2011. Quantifying selective reporting and the Proteus Phenomenon for multiple datasets with similar bias. PLoS One 6, e18362.

Pirmohamed, M., Ostrov, D.A., Park, K.B., 2015. New genetic findings lead the way to a better understanding of fundamental mechanisms of drug hypersensitivity. J. Allergy Clin. Immunol. 136, 236–244.

Popper, K., 1959. The Logic of Scientific Discovery. Hutchinson, London.

Prasad, V.V.T.S., Gopalan, R.O.G., 2015. Continued use of MDA-MB-435, a melanoma cell line, as a model for human breast cancer, even in year, 2014. NPJ Breast Cancer 1, 15002.

Puga, J.L., Krzywinski, M., Altman, N., 2015. Points of significance: Bayesian statistics. Nat. Methods 12L, 377–378.

Pugsley, M.K., Authier, S., Curtis, M.J., 2008. Principles of safety pharmacology. Br. J. Pharmacol. 154, 1382–1399.

Pugsley, M.K., Authier, S., Towart, R., Gallacher, D.J., Curtis, M.J., 2009. Beyond the safety assessment of drug-mediated changes in the QT interval...what next? J. Pharmacol. Toxicol. Methods 60, 24–27.

Qi, W., Ding, D., Salvi, R.J., 2008. Cytotoxic effects of dimethyl sulphoxide (DMSO) on cochlear organotypic cultures. Hear Res. 236, 52–60.

Rae, J.M., Creighton, C.J., Meck, J.M., Haddad, B.R., Johnson, M.D., 2007. MDA-MB-435 cells are derived from M14 melanoma cells—a loss for breast cancer, but a boon for melanoma research. Breast Cancer Res. Treat. 104, 13–19.

Rang, H.P., 2006. The receptor concept: pharmacology's big idea. Br. J. Pharmacol. 137 (Suppl. 1), S9–S16.

Roden, D.M., 2017. Phenome-wide association studies: a new method for functional genomics in humans. J Physiol.doi: 10.1113/JP273122.

Roncador, G., Engel, P., Maestre, L., Anderson, A.P., Cordell, J.L., Cragg, M.S., et al., 2016. The European antibody network's practical guide to finding and validating suitable antibodies for research. MAbs 8, 27–36.

Rosenbaum, S.E., 2016. Basic Pharmacokinetics and Pharmacodynamics: An Integrated Textbook and Computer Simulations, second ed. Wiley, Hoboken, NJ.

Ruxton, G., Colgrave, N., 2011. Experimental Design for the Life Sciences, third ed. Oxford University Press, Oxford.

Santos, N.C., Figueria-Coelho, J., Martins-Silva, J., Saldanha, C., 2003. Multidisciplinary utilization of dimethyl sulfoxide: pharmacological, cellular, and molecular aspects. Biochem. Pharmacol. 65, 1035–1041.

Schuurman, H.J., Folkerts, G., Groenink, L. (Eds.), 2015. Translational value of animal models. Eur J Pharmacol. 759, 1–356.

Sena, E.S., van der Worp, H.B., Bath, P.M.W., Howells, D.W., Macleod, M.R., 2010. Publication bias in reports of animal stroke atudies leads to major overstatement of efficacy. PLoS Biol. 8, e1000344.

Siegfried, T., 2010. Odds are, it's wrong. ScienceNews 177. Available from: https://www.sciencenews.org/article/odds-are-itswrong?mode=magazine&context=704.

Simson, M., 2016. The relabeling racket: how what we don't know is hurting biomedical research. OWL-legory of the Lab. Available from: https://www.linkedin.com/pulse/relabeling-racket-how-what-we-dont-know-hurting-research-simson-1.

Sivakumaran, S., Agakov, F., Theodoratou, E., Prendergast, J.G., Zgaga, L., et al., 2011. Abundant pleiotropy in human complex diseases and traits. Am. J. Hum. Genet. 89, 607–618.

Smith, M.M., Clarke, E.C., Little, C.B., 2016. Considerations for the design and execution of protocols for animal research and treatment to improve reproducibility and standardization: "DEPART well-prepared and ARRIVE safely". Osteoarthr. Cartil. 25, 354–363.

Sullivan, G.M., Feinn, R., 2012. Using effect size—or why the P value is not enough. J. Grad. Med. Educ. 4, 279–282.

Suresh, K.P., 2011. An overview of randomization techniques: an unbiased assessment of outcome in clinical research. J. Hum. Reprod. Sci. 4, 8–11.

Taglang, G., Jackson, D.B., 2016. Use of "big data" in drug discovery and clinical trials. Gynecol. Oncol. 141, 17–23.

Taleb, N.N., 2010. The Black Swan, second ed. Random House, New York.

Tanramluk, D., Schreyer, A., Pitt, W.R., Blundell, T.L., 2009. On the origins of enzyme inhibitor selectivity and promiscuity: a case study of protein kinase binding to staurosporine. Chem. Biol. Drug Des. 74, 16–24.

Titov, D.V., Liu, J.O., 2012. Identification and validation of protein targets of bioactive small molecules. Bioorganic. Med. Chem. 20, 1902–1909.

Tozer, T.M., Rowland, M., 2015. Essentials of Pharmacokinetics and Pharmacodynamics, second ed. Wolters Kluwer, Lippincott Williams Wilkins, Philadelphia.

Trafimow, D., Marks, M., 2015. Editorial. Basic Appl. Soc. Psych. 37, 1–2.

Tukey, J.W., 1980. We need both exploratory and confirmatory. Am. Stat. 34, 23–25.

Uhlen, M., Bandrowski, A., Carr, S., Edwards, A., Ellenberg, J., Lundberg, E., et al., 2016. A proposal for validation of antibodies. Nat. Methods 13, 823–827.

Uitdehaag, J.C.M., Verkaar, F., Alwan, H., de Man, J., Buijsman, R.C., Zaman, G.J.R., 2012. A guide to picking the most selective kinase inhibitor tool compounds for pharmacological validation of drug targets. Brit. J. Pharmacol. 166, 858–876.

Van Noorden, R., 2014. *Science* joins push to screen statistics in papers. Nature, Available from: http://www.nature.com/news/science-joins-push-to-screen-statistics-in-papers-1.15509.

van Ravenzwaaij, D., Ioannidis, J.P.A., 2017. A simulation study of the strength of evidence in the recommendation of medications based on two trials with statistically significant results. PLoS One 12, e0173184.

Vanden Berghe, T., Hulpiau, P., Martens, L., Vandenbroucke, R.E., Van Wonterghem, E., Perry, S.W., et al., 2015. Passenger mutations confound interpretation of all genetically modified congenic mice. Immunity 42, 200–209.

Varga, O.E., Hansen, A.K., Sandøe, P., Olsson, I.A., 2010. Validating animal models for preclinical research: a scientific and ethical discussion. Altern. Lab. Anim. 38, 245–248.

Vasilevsky, N.A., Brush, M.H., Paddock, H., Ponting, L., Tripathy, S.J., LaRocca, G.M., Haendel M, 2013. On the reproducibility of science: unique identification of research resources in the biomedical literature. PeerJ 1, e148.

Vasilevsky, N.A., Minnier, J., Haendel, M.A., Champieux, R.E., 2017. Reproducible and reusable research: are journal data sharing policies meeting the mark? PeerJ 5, e3208.

Vesterinen, H.M., Sena, E.S., ffrench-Constant, C., Williams, A., Chandran, S., Macleod, M.R., 2010. Improving the translational hit of experimental treatments in multiple sclerosis. Mult. Scler. 16, 1044–1055.

Vincent, F., Loria, P., Pregel, M., Stanton, R., Kitching, L., Nocka, K., et al., 2015. Developing predictive assays: the phenotypic screening "rule of 3". Science Transl. Med. 7, 293ps15.

Visscher, P.M., Wray, N.R., Zhang, Q., Sklar, P., Mark, I., McCarthy, M.I., et al., 2017. 10 years of GWAS discovery: biology, function, and translation. Am. J. Hum. Genet. 101, 5–22.

Voskuil, J., 2014. Commercial antibodies and their validation [version 2; referees: 3 approved]. F1000 Res. 3, 232.

Waclaw, B., Boziv, I., Pittman, M.E., Hruban, R.H., Vogelstein, B., Nowak, M.A., 2015. A spatial model predicts that dispersal and cell turnover limit intratumour heterogeneity. Nature 525, 261–264.

Wadman, M., 2013. NIH mulls rules for validating key results. Nature 500, 14–16.

Wagner, B.K., Schreiber, S.L., 2016. The power of sophisticated phenotypic screening and modern mechanism-of-action methods. Cell Chem. Biol. 23, 3–9.

Walker, M., 1993. DMSO: Nature's Healer. Avery/Penguin Random House, Crawfordsville, IN.

Wang, J., Song, Y., 2017. Single cell sequencing: a distinct new field. Clin. Transl. Med. 6, 10.

Wasserstein, R.L., Lazar, N.A., 2016. The ASA's statement on p-values: context, process, and purpose. Am. Stat. 70, 129–133.

Weiss, K., 2017. The GWAS hoax....or was it a hoax? Is it a hoax? The Mermaid's Tale. Available from: https://ecodevoevo.blogspot.co.za/2017/06/the-gwas-hoaxor-was-it-hoax-is-it-hoax.html.

Wilding, J.L., Bodmer, W.F., 2014. Cancer cell lines for drug discovery and development. Cancer Res. 74, 2377–2384.

Willemsen-Seegers, N., Uitdehaag, J.C.M., Prinsen, M.B.W., de Vetter, J.R.F., de Man, J., et al., 2017. Compound selectivity and target residence time of kinase inhibitors studies with surface plasmon resonance. J. Mol. Biol. 429, 574–586.

Williams, M., Bozyczko-Coyne, D., Dorsey, B., Larsen, S., 2008. Laboratory notebooks and data storage. In: Gallagher, S. (Ed.), Curr Protocol Essential Lab Techniques. Wiley, Hoboken, NJ. 00:2A:A.2A.1–A.2A.28.

Ye, F., Chen, C., Qin, J., Liu, J., Zheng, C., 2015. Genetic profiling reveals an alarming rate of cross-contamination among human cell lines used in China. FASEB J. 29, 4268–4272.

Zheng, W., Thorne, N., McKew, J.C., 2013. Phenotypic screens as a renewed approach for drug discovery. Drug Discov. Today 18, 1067–1073.

CHAPTER

3

Statistical Analysis in Preclinical Biomedical Research

Michael J. Marino

OUTLINE

Research in the Biomedical Sciences. http://dx.doi.org/10.1016/B978-0-12-804725-5.00003-3

3.1 INTRODUCTION—BACKGROUND TO STATISTICS—USE AND MISUSE

A statistic is defined as a piece of information (e.g., a blood pressure reading) drawn or calculated from a sampling of a population which can be used as an estimate of a population parameter (e.g., mean blood pressure). For example, the average height (statistic) of people applying for a driver's license in the state of Ohio (sample) can be used as an estimate of the average height of people in the United States (population parameter). Statistics (not to be confused with the plural of statistic) is the science of quantitative methods that guide data collection, analysis, interpretation, and presentation. Importantly, statistics is not the toolbox that is opened at the end of the experiment to see if the expected result has been obtained. Rather, statistics provide a framework embedded in the scientific method that guides experimental design, execution, and interpretation (Marino, 2014).

With the recent increased focus on reproducibility (Begley and Ioannidis 2015; Jarvis and Williams, 2016), the appropriate use of statistics has been highlighted as a key component of conducting transparent, quality research with any likelihood of generating findings that can be independently replicated (Marino, 2014; Motulsky, 2014). The appropriate use of statistics is not possible without a basic foundational understanding of principles and assumptions that guide the choice of statistical methods. While there clearly are cases of intentional misuse (Simmons et al., 2011), many of the misuses of statistics can be traced to a simple lack of understanding. Unfortunately, as technology-enabled research favors seemingly observational all-or-none outcomes (e.g., a band on a gel is either present or absent) and massive data sets [e.g., genome wide association studies (GWAS)], there has been a shift away from traditional statistical education in the biological sciences. This trend, combined with the ubiquitous availability of powerful and user-friendly statistical software, has amplified the problem of statistical misuse.

3.1.1 Interface of Experimental Design and Hypothesis

In Chapter 2, the importance of careful experimental design has been described, with one takeaway message being that if the intent of the experiment is to reach a conclusion, then the question should be stated as a formal hypothesis which is based on the planned statistical comparisons and takes into account all of the variables. Researchers may consciously strive to design experiments with this view in mind, however many ignore the importance of planning the statistical comparisons *before* running the study. There are at least three reasons to do this work as part of the initial experimental design, the details of which will be discussed in the remainder of this chapter. It is important to emphasize that the choice of statistic is dictated by the study design, and is therefore knowable a priori. For example, a simple experiment designed to test the effect of a drug on rodent body weight may take the form of a single comparison between a test compound group and a vehicle control group. However, if a researcher is interested in a time course of the effect, the design becomes a repeated measures multiple comparison requiring an ANOVA and likely a larger sample size to account for the statistical power that is lost in the multiple comparisons. The latter point raises the second reason to incorporate statistics into the experimental design, namely that the sample size (the n value) should be chosen to provide the appropriate statistical power such that a conclusion may be drawn with confidence. To start with a predetermined n value, and add additional subjects as the study progresses after interim analysis is a common misuse of statistics that greatly increases the false positive rate (Kairalla et al., 2012; Motulsky, 2014). The final reason to make the statistical test part of the initial design is that to do otherwise allows for tests to be chosen or data transformations applied in order to obtain significance. For example, in a case where a nonparametric test is indicated, using a parametric test will sometimes produce a p-value deemed acceptable (i.e., supporting the investigator's hypothesis). This is clearly an unacceptable practice, as the assumptions of the parametric test are not met. Furthermore, this repeated testing is a form of multiple comparison which leads to a reduction in statistical power and once again increases the probability of false positives.

3.1.2 Sampling and Sample Bias

At the core of any experiment lies the concept of sampling. As discussed earlier in the definition of the term statistic, when designing an experiment, the researcher is attempting to measure a statistic that estimates the actual parameter of an entire population. For example, the researcher may be interested in estimating the average height of people in the United States. With the current population of the United States of America approaching 320 million people, measuring each person's height at a rate of 1 person every minute would take over 6 years during which time, the population will have increased by another 17 million. Therefore, it is physically impossible to measure the height of every individual and obtain the true population parameter—even with the determination to do so. The only option is to take a representative sample of the population and measure the height of individuals in this sample and the variability in the sample as an estimate of the population parameter. For example, a random choice of 10 individuals from each of the 50 states would provide a sample of 500 heights that will provide an estimate of the distribution of heights in the 320 million. Larger samples, and more representative samples (i.e., random selection across the entire population) will produce more accurate estimates (Fig. 3.1).

Sample bias occurs when the sample is not representative of the population. This can occur either through errors inherent in the experimental design, or may be caused by some hidden assumption in the design that may require additional experimental work to uncover. To remain with the example of estimating the height of people in the United States of America, the random selection of individuals from each state described earlier should provide a representative sample. However, obtaining even this relatively large number of measurements would still result in a small sample ($<$0.02% of the population). Given the challenges, the researcher may be tempted to choose other methods in the interest of tractability. Choosing the local middle school student body, or the local basketball team as a sample would yield an obvious sampling bias that would respectively under or overestimate the population parameter regardless of the sample size. However, even a seemingly sound strategy of running a newspaper ad in each state asking for volunteers could yield a bias as it is selectively sampling newspaper readers that are likely to volunteer, a group that may or may not share other characteristics (e.g., diet, lifestyle, or age) that could impact on the outcome. In laboratory research, there tends to be a more obvious choice of sample, for example, a set of 12 in-bred C57/BL6 mice as a sample of the population of all C57/BL6 mice. However even in this simple case, it is necessary to pay careful attention to factors, such as choice of vendor, location of breeding, age, sex, weight, etc., all of which could impact on the experimental outcome.

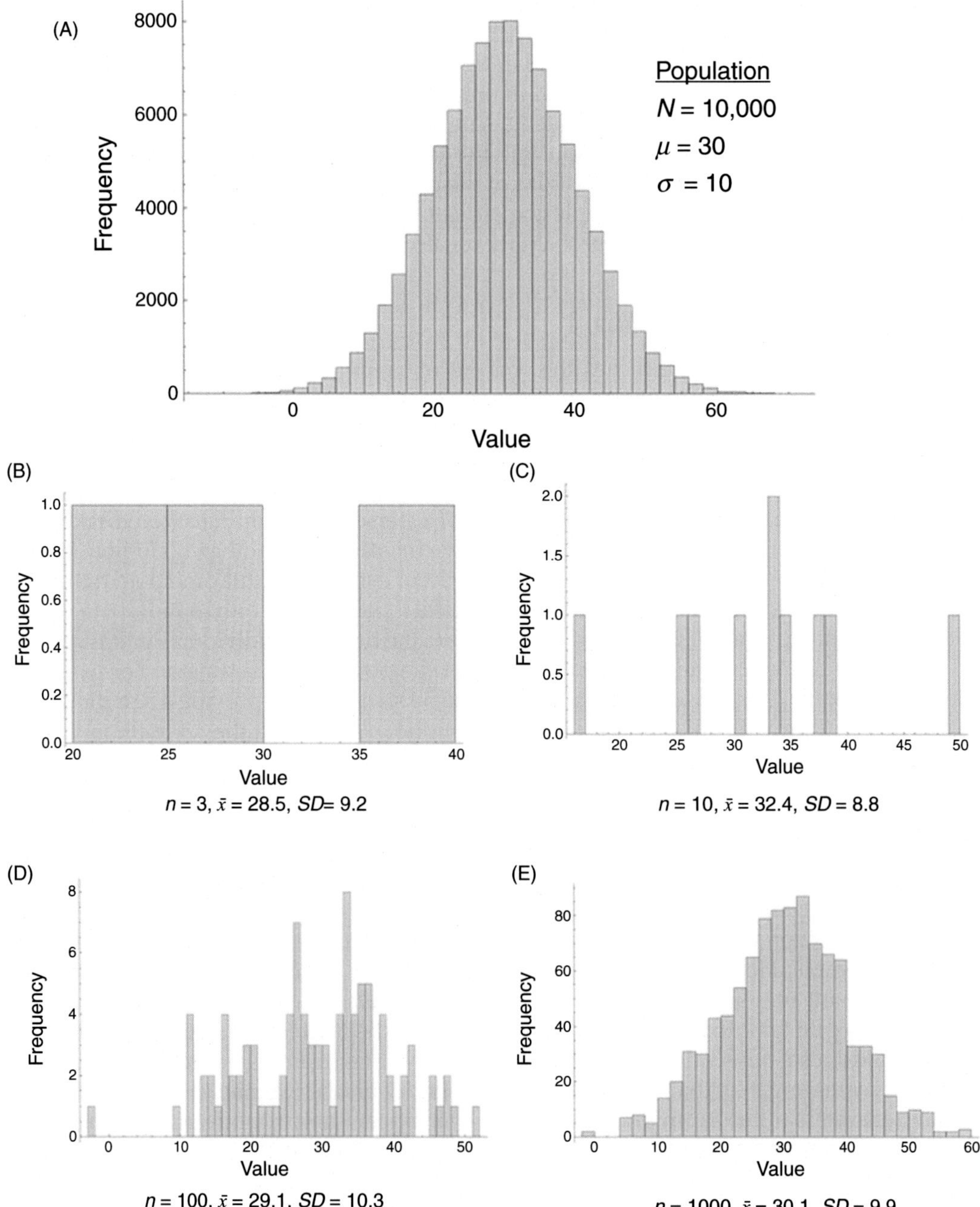

FIGURE 3.1 **Simulation of drawing samples from a population.** (A) The theoretical population composed of 10,000 random numbers normally distributed with a mean of 30 and a standard deviation of 10. Note that these values represent the actual population parameters that are estimated by the sample. (B–E) samples randomly drawn from the population in (A) with sample sizes of 3 (B), 10 (C), 100 (D), and 1000 (E). Note that the larger sample size provides a better estimate of the population mean as well as the shape of the distribution.

Sampling bias is a form of experimental error that is a type of nonsampling error. Nonsampling errors are not the result of chance, but are actual errors in study design, and sampling bias is one of the most common sources of nonsampling error. The earlier example of choosing a middle school or basketball team for the sake of experimental tractability demonstrates how a seemingly random selection can in fact be made nonrandom by the introduction of an experimental design flaw. A more relevant example might be encountered when culturing neurons and choosing to only use culture plates that reach a particular cell density. If that cell density can be mapped to a particular initial cell number and length of time in culture, then the criteria should be based on those parameters. If plates treated identically exhibit different cell densities it may be indicative of some other differentiating factor that could impact on the outcome measure. Other sources of nonsampling error include poorly calibrated equipment and other systematic errors arising from experimental technique.

The difference between the value of the actual population parameter and the value of the measured statistic is termed the sampling error. Sampling error cannot be exactly determined unless the whole population can be measured to determine the parameter. For most day-to-day experimental design and statistical evaluation, it is sufficient to understand that the sampling error exists. Sampling error occurs by chance as samples are drawn from the population and it cannot be avoided.

3.1.3 Descriptive Statistics

Once a sample has been obtained, it is necessary to derive statistics from the sample that summarize the distribution. Two types of values are needed, a measure of central tendency that indicates the approximate center of the sample distribution, and a measure of dispersion that indicates the degree to which individual members of the sample set depart from the central value (Fig. 3.2).

3.1.3.1 Measures of Central Tendency

Measures of central tendency represent a distribution of values by a single typical value. Knowing that the sample statistic can only approximate the population parameter, it should be apparent that a single number cannot represent the entire distribution. However, a measure of central tendency will provide a characteristic of the sample population that is a typical value and is useful when provided with additional statistics, such as a measure of dispersion discussed later.

3.1.3.1.1 MEAN

The commonly used arithmetic mean, or average, is simply the sum of all observations divided by the number of observations (Table 3.1). When thinking in terms of a frequency histogram, the mean can be visualized as the center of mass of the histogram, or the point at which the histogram would balance. While the mean is commonly used, it is not always the most appropriate choice for characterizing a distribution as it is highly sensitive to outliers (Fig. 3.2). It is also important to keep in mind that the arithmetic mean, is just one type of Pythagorean mean, and may not be the most appropriate choice based on the nature of the data being summarized. For example, in cases where the sample values are better interpreted as products (e.g., rates of growth over time), the geometric mean is a better choice (Table 3.1).

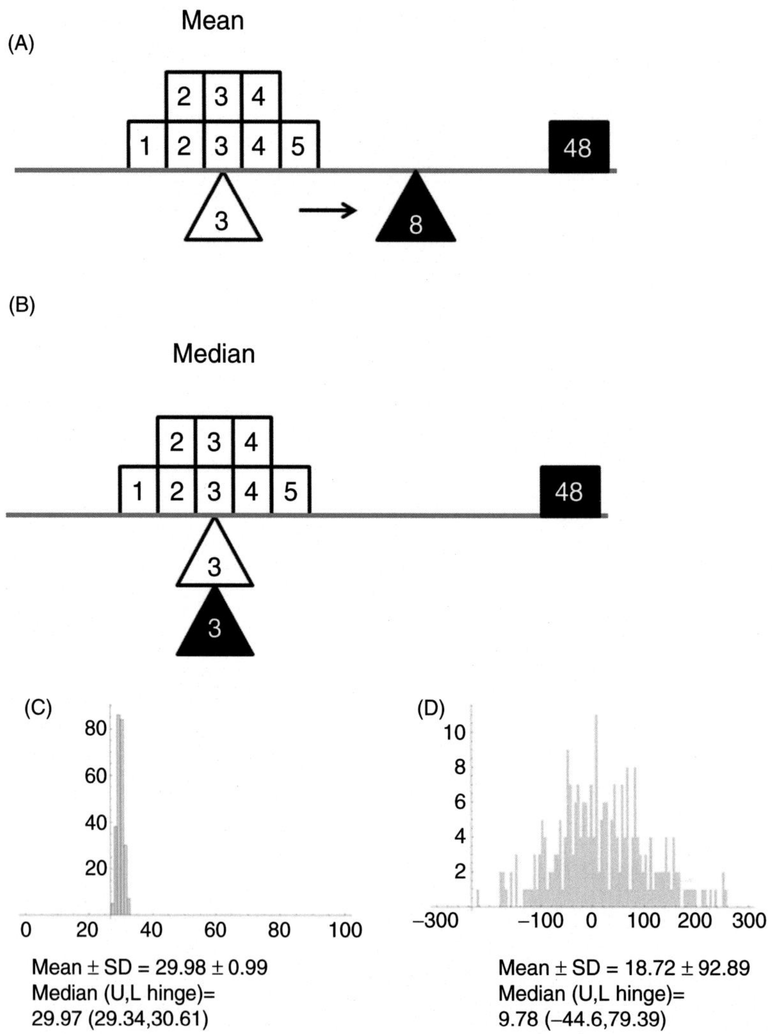

FIGURE 3.2 **A graphical representation of the common measures of central tendency and dispersion.** (A) The mean can be thought of as the balancing point of the histogram. While it is a reasonable indicator of the center of the sample, it is highly sensitive to outliers as depicted by the large shift with the addition of the outlier value 48. (B) The median represents the middle of the distribution, and it is highly resistant to outliers. This makes the median the best choice when dealing with nonnormally distributed data. (C) A simulated sample of 250 values with a mean of 30 and a standard deviation of 1. (D) A simulated sample of 250 values with a mean of 30 and a standard deviation of 100. Note the difference in the *x*-axis, precision of the statistics, and the U and L hinge values (25th and 75th percentiles, respectively).

3.1.3.1.2 MEDIAN

The median, defined as the middle term in a data set arranged in rank order, is the measure of central tendency best suited for describing a data set containing outliers, or where the data is not normally distributed (i.e., follows a normal or Gaussian distribution). The median is determined by rank ordering all the members in a data set and choosing the middle term for which 50% of the values lay above or below. This method provides a single number for data

TABLE 3.1 Measures of Central Tendency and Dispersion Along With Computational Methods

Central tendency		
	Arithmetic mean	$\bar{x} = \frac{\sum_{i=1}^{n} x_i}{n}$
	Geometric mean	$\bar{x} = \left(\prod_{i=1}^{n} x_i\right)^{\frac{1}{n}}$
	Median	Middle term in rank ordered dataset
	Mode	Most common term in dataset
Dispersion		
	Range	(Maximum–Minimum)
	Sample variance	$s^2 = \frac{\sum_{i=1}^{n} (\bar{x} - x_i)}{n-1}$
	Standard deviation	$s = \sqrt{s^2}$
	Upper and lower hinges	Analogous to median taken at 25% and 75% point in rank ordered dataset, respectively

See text for additional details.

sets containing an odd number of values, and a pair of numbers that must be averaged in order to determine the median for a data set containing an even number of values. For example, the data set (1, 3, 7) has a median value of 3, while the data set (1, 3, 4, 8) has a median value of $(3 + 4)/2 = 3.5$. While this computational process does not appear to take into account every sampled value in the distribution, the median can be visualized as being the center of the histogram with an equal number of values on either side (Fig. 3.2). Therefore, the location of every value in the histogram comes into play in determining the median.

3.1.3.1.3 MODE

The mode is the most commonly occurring value in the sampled distribution. The mode is not often useful and is mentioned primarily for the sake of completing the list of measures of central tendency. One point to keep in mind regarding the mode is that while any distribution will have a single mean and a single median, it will not necessarily have a single mode. A distribution could have no mode if no values repeat, and more than one mode as in a bimodal distribution.

3.1.3.2 Measures of Dispersion

While calculation, a measure of central tendency provides a convenient summary of the distribution of values in a sample, it does not, describe the distribution beyond the point of central tendency. If, for example, a study is performed to measure ectopic tumor volume after treatment with a potential chemotherapeutic drug and the mean reduction in tumor volume is found to be 50%, it is expected that repeating that same study is likely produce a reduction close to 50%. But how likely? How surprising would an observation of 70% be? Should a second test compound producing a 40% reduction be considered inferior, or would this result be expected as the normal variation around the mean? Including a measure of dispersion which describes the spread of the sample values about the point of central tendency provides a concise and interpretable summary of the sample.

3.1.3.2.1 RANGE

A measure of dispersion describes, how far the data are spread out about the measure of central tendency. The simplest and most intuitive measure is the range, the difference between the largest and smallest value in the sample (Table 3.1). While the range is simple to understand and easy to compute it carries the limitations of being based on only two numbers in the distribution and being highly sensitive to outliers. Therefore, it is almost always better to employ a more robust measure of dispersion, such as the variance or standard deviation.

3.1.3.2.2 VARIANCE

The variance provides a reasonable estimate of the spread in a distribution, and the calculation of variance is relatively intuitive. If the mean represents the center of the sample distribution, the spread in the distribution can be measured by assessing how far each individual value differs from the mean. Squaring these difference values (to avoid a zero sum) and taking the mean difference provides the average squared difference from the mean, or variance. Importantly, there are different computational methods for calculating population and sample variance (Table 3.1). The difference being that when calculating sample variance (which is almost always the case as the entire population is rarely available for study), a correction factor known as Bessel's correction must be applied such that the average is determined with a denominator of $n-1$. The easiest and perhaps most intuitive explanation for this correction factor is that one degree of freedom has been used in calculating the mean, and this is accounted for by subtracting 1 from the sample size. This is the explanation most commonly found in introductory statistics texts, and while not incorrect, it is a simplification of the true reason which has to do with the desire of obtaining an unbiased estimate of the population variance. In general, it is assumed that values in a sample are closer to the sample mean than to the population mean, and the variance would therefore be underestimated without correction. The degree of freedom explanation is sufficient for an intuitive understanding of variance.

3.1.3.2.3 STANDARD DEVIATION

The sample variance provides an unbiased estimate of population variance, but it has units that are squared relative to the units of the mean value making it difficult to interpret. For example, measuring a mean change in neuronal membrane potential and reporting the result as 12 mV ± 2 mV^2 does not intuitively provide an understanding of the precision of measurement. The standard deviation (SD), simply defined as the square root of the variance (Table 3.1) obviates this issue and provides the most common and usually the best choice of a transparent measure of dispersion. In general, there is no correction of bias for the calculation of standard deviation. The sample standard deviation is calculated from the corrected sample variance; however, the square root introduces nonlinearity, and therefore a bias into the calculation.

3.1.3.2.4 NONPARAMETRIC MEASURES OF DISPERSION

For nonparametric datasets (datasets that are not normally distributed) where the median is used as a measure of central tendency, the interquartile ranges or hinges provide a useful nonparametric measure of dispersion. The quartiles are analogous to the median but taken at the 25% cut in the data set. So the lower quartile is the value that has 25% of the data below and 75% of the data above, while the upper quartile has 75% of the data below and 25% of the data above (Table 3.1).

3.1.3.3 The Standard Error of the Mean (SEM)

Despite its common use in biomedical research, the (SEM) is generally misunderstood and used more by convention than for any logical reasons. The SEM is *NOT* a measure of dispersion and does not provide information regarding the scatter of data about the measure of central tendency. In fact, the SEM is typically (and wrongly) chosen because its value will always be smaller than the standard deviation, providing an illusion that the measurements are more precise than they actually are.

The concept of sampling distributions is central to an understanding of the SEM. As indicated earlier, population parameters are actual and exact values that are characteristic of a population. Statistics are estimates of parameters obtained by sampling the population. Therefore, statistics behave as variables and exist in distributions known as a sampling distribution. If a population is repeatedly sampled with a constant samples size, the mean of each sample set would likely be different each time. In other words, when an experiment is run (i.e., sampled) 3 times, it would be unexpected to get the exact same mean value each time despite the fact that the population parameter that is being estimated with each run is an exact value. This collection of sample means derived from multiple runs of the same experiment represent a sampling distribution of values having a mean (the average of the sample means) and standard deviation (the standard deviation about the sample means). Importantly, this is true for any statistic, not just the mean. There is, for example, a sampling distribution of the standard deviations. For this reason, referring to the SEM by the colloquial "standard error" is incorrect for it does not indicate from which statistic the error is derived.

The SEM is a measure of how accurately the population mean has been estimated based on the *central limit theorem* (CLT). A full discussion of the CLT is beyond the scope of this chapter, and it is sufficient to understand that the CLT rigorously describes how sampling distributions behave relative to the actual population. The SEM is calculated by dividing the standard deviation of the mean of the sampling distribution of the means by the square root of the sample size. Therefore, the SEM will decrease as the sample size increases and a more accurate estimate of the mean is obtained. In addition, the SEM will always be less than the standard deviation. Importantly, the SEM is never subjected to a Bessel correction (i.e., $n-1$). This is a common error that unfortunately is rarely noticed by investigators and impossible to correct by reviewers.

Note that a predictable relation exists between the SEM and sample size in that as sample size increases, the SEM will always decrease independent of the true population variance. This is not true of the sample variance which is limited by the true population variance. This point is underscored by comparing the standard deviations of the samples in Fig. 3.1 with their respective SEM values (Table 3.2). Therefore, it should be obvious that the SEM is not in any way a reliable estimate of dispersion. It can be argued that the use of SEM as an

TABLE 3.2 Standard Deviations From the Random Samples Obtained in Fig. 3.1 Compared to the Standard Error of the Mean Values Calculated by Dividing the Standard Deviation by the Square Root of the Sample Size

N	SD	SEM
3	9.2	5.3
10	8.8	2.8
100	10.3	1.0
1000	9.9	0.3

Note the extreme dependence of the SEM on the sample size and how even for small samples it underestimates the population standard deviation which in this simulation is defined as 10.

indication of the accuracy by which the mean has been determined means that using the SEM in a graphic depiction of data (e.g., a bar graph) allows for a visual estimate of significance. While there is some truth to this argument, it is far better to make such determinations based on probability values derived from inferential methods described later. The SEM is not a descriptive statistic and should not be used as one (Barde and Barde, 2012).

3.2 EXPLORATORY RESEARCH AND EXPLORATORY DATA ANALYSIS

The importance of understanding precisely what question is being asked before launching into an experiment was highlighted in Chapter 2. Importantly, if the intent of the study is to reach a conclusion, then the question should be stated as a formal hypothesis which is based on the planned statistical comparisons, takes into account all of the relevant variables, and uses as context what is known about the function of the system under interrogation. To the last point, sometimes very little is known regarding the function of the system under interrogation. In particular, the information needed to determine basic experimental parameters like optimal cell density, age of mice at time of treatment, test compound dose, etc. are often unknown. Furthermore, quantitative information necessary to perform the necessary power analysis (see later), such as the predicted standard deviation in the control group may not be known.

Take as an example the simple question, "Does a test compound identified in a binding screen for a modulatory site on the 5-HT_6 receptor act as a functional allosteric modulator at the 5-HT_6 receptor?" Assume that the method of measuring 5-HT_6 receptor activation in a recombinant system is established, and that the compound is freely available in the amounts required to perform the study. The experimental design requires a great deal more detail. For example, what is the best agonist to use in testing a modulator? Should the endogenous agonist serotonin (5-HT) be used, or would a more selective 5-HT_6 agonist be a better choice? What concentration of agonist is appropriate? An EC_{20} concentration may be appropriate for a positive allosteric modulator, as it would allow for a larger window to measure the potentiation of the agonist effect. On the other hand, and EC_{80} concentration would be a better choice for a negative modulator as it would allow for a larger window to measure inhibition of the agonist effect. As we do not know the mode of action of the test compound, it would seem best to run the compound against a range of agonist concentrations to see if the putative modulator shifts the agonist concentration–response relationship, and if so in which direction. And what about a starting concentration of the test compound? There may be some suggested starting concentration based on a biochemical measure from the literature or a screening assay, but single concentration/dose studies are rarely informative. Therefore, it will be necessary to choose a range of concentrations to build a concentration–response relationship. This can be done by assuming that 6 concentrations centered around the screening binding affinity would provide a reasonable starting point to construct a curve that exhibits saturation and allows a reasonable estimate of compound potency. Assuming that we already have answers to other questions like how long does the incubation period need to be, what is the best cell density for the assay, and that we want to compare 5-HT to the most selective 5-HT_6 agonist available, we have 6 concentrations of the putative modulator to run against 6 concentrations of each agonist resulting in a $2 \times 6 \times 6$ matrix involving 72 comparisons. While modern high throughput methods would allow this sort of study to be completed as stated, as explained in the relevant sections below it would represent $n = 1$, and with the large number of comparisons, the n-value needed to actually reach a confident conclusion would be untenable. In addition, even with an appropriate n, the false positive rate would be higher than acceptable.

As illustrated by the above example, it is very difficult to design an experiment in the absence of data. The necessary data can sometimes be extracted from historical studies either from the same laboratory or from the literature. But when that is not possible, the only alternative is to run a pilot, or exploratory study (Kimmelman et al., 2014; Tukey, 1980). Exploratory studies are small-scale "see what happens" experiments in which the primary objective is estimation rather than hypothesis testing. Pilot studies provide important information on the feasibility of conducting a larger study. They provide a preliminary assessment on equipment performance, the robustness of assay conditions, and/or the ability of the system being interrogated, cells, or animals, to provide information on the magnitude of the resulting effect, and an estimate of its variability. In animal studies, conducting a pilot study is not only good practice, but may be an ethical necessity as it can lead to a reduction in the overall number of animals used in the main study by allowing an informed estimate of power (see later) and provide information on potential adverse events. The data from an exploratory or pilot study should be subjected to exploratory data analysis (Tukey, 1977) to try and understand how to design confirmatory studies, or in some instances to identify a new line of research. While exploratory studies are not typically suitable for publication, including the details of such studies as a rationale for the design of the confirmatory studies is encouraged as it may aid in future replication of the confirmatory work by another laboratory.

To continue with the above example of the putative 5-HT_6 receptor modulator, we need to consider how to get the most relevant information from the fewest manipulations. Exploratory work deals in trends, not confirmation, so we look at the data and want to have as much confidence as possible in what the data will visually show us. Therefore, simple fast studies are best. For the choice of agonist concentration, the EC_{50} value may not be as sensitive as the EC_{20} value for a positive modulator or the EC_{80} value for a negative modulator, but it does provide a concentration in the middle of the linear portion of the concentration response relationship which would be sensitive to shifts in either direction. By choosing to use the EC_{50} concentration for both the endogenous agonist and the highly selective 5-HT_6 agonist, we can still run 6 concentrations of the test compound yielding a $2 \times 1 \times 6$ matrix. Running this in triplicate would only require 36 wells on a plate, although best practice would be to run the 12 wells on 3 separate plates for a true $n = 3$ (see later). The triplicates are important because they provide information on the variability in the assay and allow for potential outliers to be weighted accordingly. Keep in mind that no data are excluded or compared statistically in exploratory studies. The goal is to explore and gather guidance for the confirmatory study.

3.2.1 Analysis of Exploratory Studies

The tool box for *exploratory data analysis (EDA)* is vast and flexible. In his landmark book "Exploratory Data Analysis," Tukey provided a very accessible and comprehensive introduction that emphasizes the need to spend time reviewing raw data to discover what the results actually indicate (Tukey, 1977). While Tukey's book is a somewhat dated reference predating the now ubiquitous laboratory computer, it is a worthwhile investment for anyone interested in understanding their data.

EDA is not well codified and rightly so. It is more of a conceptual approach to handling data than a strict subfield of statistics. The methods that make up the toolbox of EDA can be as simple as generating a scatter plot or as complex a running autocorrelation analysis on a time series. The goal of this section is to outline a few simple tools for visualizing data sets that can lead to insight. While these tools may appear familiar, they are unfortunately used infrequently being supplanted by bar graphs depicting the mean and standard deviation (or worse the SEM) that only reveal what the researcher already knew. It is expected that the reader will discover or develop additional tools uniquely suited to their own work once the value of the EDA approach becomes evident.

3.2.2 Data Visualization Best Practice

The process of creating effective data visualization that conveys scientific information is not automatic. Scientific thought is clear and precise, and any visualization of that information needs to convey that same clarity and precision. This is, of course important in preparing graphs for publication that can stand on their own and transparently convey information without misleading the audience. However, clarity and precision are even more important in the process of EDA where the researcher must be open to learning what the data are trying to tell, and must be very careful not to mislead themselves. There are a few simple principles to keep in mind that effectively guide the researcher toward making the most effective visualizations of their data. For further information on the effective visualization the reader is referred to in the excellent books by Tufte (Tufte, 2001, 2006).

Principle 1: Avoid chart clutter. Chart clutter serves no purpose but to decorate the visualization. Remember the goal is to communicate the relevant information in a clear and precise fashion. Adding images of test tubes or experimental animals, cross hatching of bars, graphical figure legends that are redundant with axis labels, etc. simply add distraction from the main point that you are trying to communicate.

Principle 2: Avoid making the display complex for the sake of design alone. The classic violation of this principle is the graphing of a simple two-dimensional data set in a three-dimensional format (e.g., a 3D bar graph). If there is no information added by the addition of the third dimension it will take more time for the reader to understand the primary point that you are trying to communicate. Gimmicks that make the visualization appear more impressive seldom aid in understanding the underlying data.

Principle 3: Make every pixel count. If there is ink on the page, it should be there for the sake of providing information. Something as simple as unnecessarily filling in a symbol or a bar on a chart can lead to confusion in rapid interpretation of the visualization.

Principle 4: Use color sparingly and effectively. There are very few cases where color is necessary in the visualization of data. Anyone familiar with the Stroop effect in which the subject is asked to rapidly read a list of color words written in a different color (e.g., the word yellow written in red ink) understands how confounding color can be because the brain processes color differently than language (Stroop, 1935). Both relevant and irrelevant information compete for entry into the central processing centers of the brain leading to a slowing in reaction time and an increased propensity toward errors in interpretation.

Principle 5: Optimize comparisons. Using a common scale for comparisons and being careful to plot magnitude of effect in an interpretable fashion is extremely important. Position on a common scale is the most immediately interpretable representation and the use of measures such as length of a line or area of a shape (e.g., bubble graphs) should be avoided. Plotting key comparisons side by side rather than having nonrelevant comparisons interspersed also aids in interpretation

Principle 6: Use effective annotation: Without violation of the above five principles, layering relevant information on the visualization can be essential. Titles, footnotes, explanatory comments, n-values, and summary statistics can all provide important support for interpretation when used effectively.

3.2.3 Visualization of Exploratory Data

With the earlier mentioned principles in mind, it is possible to outline some of the common methods employed in the visualization of data from exploratory studies. Importantly, these methods are designed to allow transparent visualization of data and the identification of apparent trends. These are not intended as a substitute for inferential methods used in confirmatory studies. Keep in mind that these are just a few examples and other methods exist (e.g., stem and leaf plots) or can be developed that may aid in a particular situation. Perhaps the best part of exploration is that it allows for creativity.

3.2.3.1 Bar Charts

To put it bluntly, bar charts are a useless waste of journal space. There is no information in a bar chart that is not immediately apparent (and more easily interpretable) from the simple in text statement of the mean and standard deviation. Despite this fact, very few journal articles are published without the presence of bar charts. While they should be discouraged in publication, they should be absolutely avoided in EDA. To highlight this issue, we will employ the well-known Anscombe (Anscombe, 1973) data sets (Table 3.3). As can be seen in Fig. 3.3, the bar chart provides a useless visualization of the data which obscures key differences that may be critically important in understanding the experiment.

TABLE 3.3 Anscombe Data Sets Used in Fig. 3.3

	Set A	Set B	Set C	Set D
	8.04	9.14	7.46	6.58
	6.95	8.14	6.77	5.76
	7.58	8.74	12.74	7.71
	8.81	8.77	7.11	8.84
	8.33	9.26	7.81	8.47
	9.96	8.1	8.84	7.04
	7.24	6.13	6.08	5.25
	4.26	3.1	5.39	12.5
	10.84	9.13	8.15	5.56
	4.82	7.26	6.42	7.91
	5.68	4.74	5.73	6.89
Mean	7.5	7.5	7.5	7.5
SD	2.03	2.03	2.03	2.03

3.2.3.2 *Scatter Plots*

Other than a direct listing of values, there is no simpler representation of a data set than a scatter plot. Whether plotting observations against an independent variable, or simply plotting individual observations over time in a run sequence plot, the scatter plot can reveal structure in the data that indicate associations and provide information regarding the distribution of variables. Side by side scatter plots can be quite valuable for comparing data sets and indicating biasing outliers that may not be apparent in summary statistics (Fig. 3.3). All graphing software packages and spreadsheets are capable of producing scatter plots and the generation of these simple visual depictions of data should always be the first step in data analysis. Additional information, such as the mean and standard deviation can be layered on to the scatter plot if it provides clarification.

3.2.3.3 *Frequency Histograms*

Frequency histograms are constructed from a table of frequencies by plotting the values obtained from observation on one axis, and the frequency of that observation on the other. Most graphing software provides the capability to generate frequency histograms in a bar format (Fig. 3.3) and functions exist in spreadsheet programs that can be used to generate these plots in a few steps.

3.2.3.4 *Five Number Summaries and Box Plots*

As noted earlier, the mean is the most commonly reported statistic and one that tells the least about the distribution. The five-number summary (Fig. 3.3) provides a concise and accurate description of a data set with information regarding the shape of the distribution using statistics that are easily calculated and resistant to outliers. The five numbers are the highest and lowest values, the median, and the upper and lower quartiles, or hinges. The quartiles

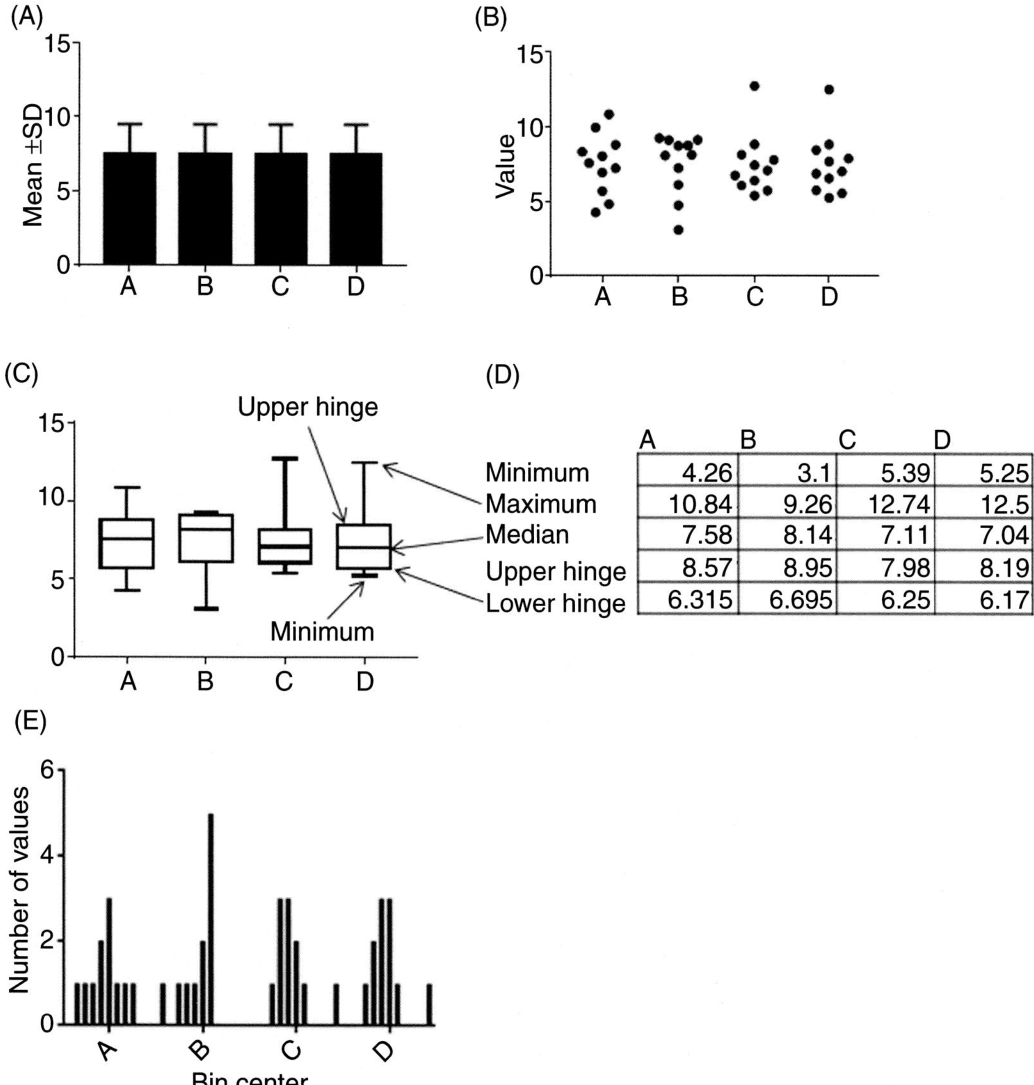

	A	B	C	D
Minimum	4.26	3.1	5.39	5.25
Maximum	10.84	9.26	12.74	12.5
Median	7.58	8.14	7.11	7.04
Upper hinge	8.57	8.95	7.98	8.19
Lower hinge	6.315	6.695	6.25	6.17

FIGURE 3.3 **A comparison of the various ways the data in Table 3.3 can be summarized and visualized.** Bar graphs (A) fail to provide any differentiation between the 4 data sets. Scatter plots (B), Box plots (C), 5 number summaries (D), and histograms (E) all provide a better choice and a full description of the data distribution.

are analogous to the median but taken as the 25% cut in the data set. So the lower quartile is the value that has 25% of the data below and 75% of the data above, while the upper quartile has 75% of the data below and 25% of the data above.

The box plot or stem and whisker plot provides a simple visual representation of the five-number summary (Fig. 3.3). The shape and spread in the distribution is immediately obvious in these plots, and when groups are plotted side by side, overlap in the distributions can be easily assessed. Most graphing packages will produce a box plot and spreadsheet programs can easily be coerced into producing a five-number summary. This type of visualization is concise and provides a great deal of information on the distribution of the samples. The value of box plots in EDA should be obvious, but it should be equally obvious that the use of these plots in publication provide a much better depiction of the data than a bar chart.

3.3 DATA TRANSFORMATION AND NORMALIZATION

While nontransformed data may seem to be closer to the reality or truth being sought in an experiment, there realistically is no reason to trust the scale used in obtaining a measurement (e.g., milliliters, picoamps, scintillation counts) any more or less than a transformed scale (e.g., Log_2 milliliters, picoamps/nanofarad, scintillation counts/second). There are a number of reasons why it may be desirable to transform raw data. First, it may make for a better and more interpretable visualization of the data. For example, an attempt to graph the apparent correlation between brain and body weight across species using a linear scale provides little insight into the relationship between these two measures (Fig. 3.4).

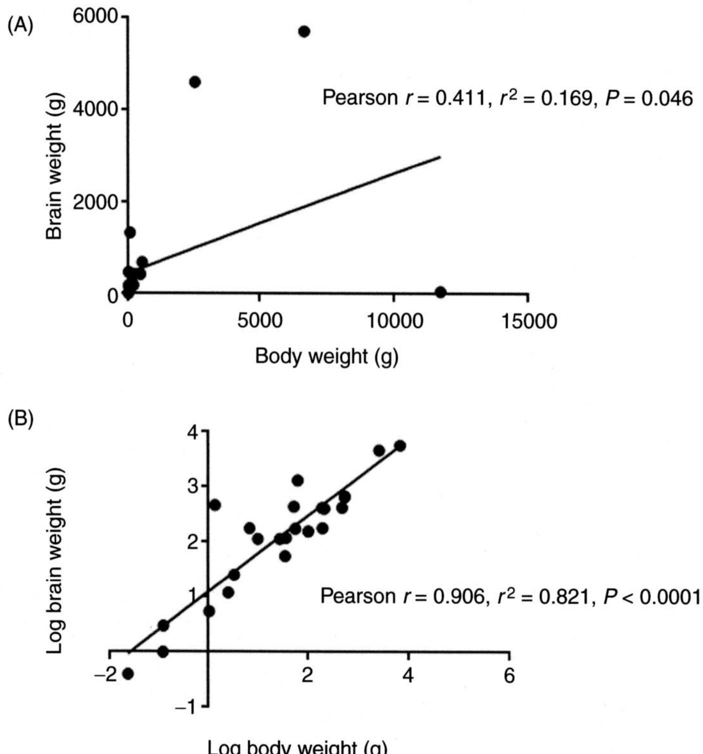

FIGURE 3.4 **Correlation analysis of brain and body weight.** (A) Plotting the raw data provides little insight into any relationship between animals body and brain weight. (B) A simple log transformation of the data reveals an obvious linear correlation that can be quantified using the Pearson coefficient of determination (r^2).

However, application of a simple Log_{10} transformation of the data set reveals a clear relationship that was previously masked by the highly skewed nature of the data (Fig. 3.4). This example also highlights the second reason for data transformation, it may enable a statistical analysis that would not have been possible on the raw data due to the skewed nature of the distribution. For example, the raw brain and body weight data set would not be amenable to a correlation analysis using the *Pearson method* (see later) because the data are obviously not normally distributed, however the log-transformation of the data removes the skew and the resulting data set can be correlated (Fig. 3.4).

The final reason that data might be transformed is to change the units of measurement into units that are either more intuitive, or that enable further analysis. For example, a study could be designed to look at the effect of a compound on the lifespan of the $SOD1^{G93A}$ mouse, a model of amyotrophic lateral sclerosis (ALS; Philips and Rothstein, 2015). If the compound is dosed in the animals drinking water, the data obtained would be in the form of milliliters of water consumed/day. However, the question being asked regarding the number of days the animals lifespan is altered by the compound, so it would be reasonable to take the reciprocal and express the average data in units of days/milliliter. Of course, an additional transformation could be applied based on the concentration of compound in the drinking water in order to express the data in units of days per milligram of compound.

Normalization is a specific type of transformation that can be performed for a number or reasons. As implied by the word itself, normalization may involve an adjustment of the data set designed to transform the sample distribution into a normal distribution. The above example of brain and body mass provides a case where the Log_{10} transformation served as a normalization. Standardization of values such as that performed when calculating z-scores, or more generally the score obtained by the Student's *t*-test represents another form of normalization. The most common use of normalization in biomedical research involves adjusting values measured on different scales, or at widely varying points on a scale, to a common scale. For example, normalizing results to a pretreatment baseline measure.

Take for example a case where a putative reverse transcriptase inhibitor is being tested for HIV antiviral activity in cultured human lymphocytes. Cells are infected with and without HIV stock, and the extent of infection is monitored by measuring a fluorescent marker protein, the expression of which is HIV-dependent. A pretreatment measure of fluorescence provides a baseline measurement from six independent cultures. Cells are then treated overnight with the inhibitor, and fluorescence is again measured the next day, with a decrease in fluorescence indicating a decrease in viral activity. An apparent trend toward a decrease is observed, however, there is a significant degree of variance in both the baseline and the treatment sample. If this variance is correlated across the groups, then normalization to the baseline values would be warranted. This can be easily assessed by running a correlation analysis between the two sets of numbers to determine if a reasonable correlation (i.e., an r^2 value > 0.5) exists (see further for a full explanation of correlation analysis). Other factors, such as cell density may also be correlated with the measures and could also be used as normalization factors. The normalization method may employ a ratio to the baseline mean with values expressed as fold change or percent change, or a subtraction of the baseline mean. Regardless of the choice of normalization method, the reporting of nonnormalized data and the rationale for normalization, in addition to the normalized results is encouraged.

3.4 THE EXPERIMENTAL SUBJECT

A subject is any independent physical entity that can be assigned to a treatment. In an animal experiments the subject would be an individual animal. Extend to in vitro studies, the subject may be a single cell, a reaction in an assay tube, or a well on a culture plate. Samples are derived from the subject. For example, if a tissue biopsy is performed on a mouse, the mouse is the subject and the tissue is the sample. Likewise, in an ELISA assay, the well on the plate is the subject, and the optical absorbance value is the sample. This can be somewhat confusing as there are always two levels of sampling involved in an experiment. The population sample would be the group of mice chosen from the population for study, or in the example of measuring the average height of individuals in the United States, the population sample would be the individuals chosen to be measured. The tissue biopsy from each mouse, or the actual height measured from each individual would be the sample.

Subjects may be sampled multiple times. For instance, in measuring a plasma born marker, we might obtain multiple blood samples from the same animal. Taking multiple measurements from the same subject, whether it be multiple plasma samples from the same rat, multiple wells on the same multiwell plate, or multiple lanes on a gel from the same protein extract, is termed subsampling. Subsampling can be a valuable method to decrease variability by averaging the multiple measurements from the same subject to provide a better estimate for that sample. However, when this is done, it always represents $n = 1$. The reason for this is that most statistical models assume that subjects are sampled independently (see later). Response measurements on the same subject are likely to be more similar to each other than response measurements from a collection of different subjects. For example, measuring the same person's height 3 times would yield numbers much more similar to one another than measuring three different individuals. Thus, subsampling yields measurements that are correlated with each other. Falsely assuming that response measurements are independent leads to underestimation of variability when the measurements are positively correlated and overestimation of variability when measurements are negatively correlated.

3.5 DETERMINING SAMPLE SIZE (n)

The goal of inferential statistics is to quantify the degree of certainty around a conclusion and allow for a decision to be made based on data. As will be discussed further, the number of subjects included in the sample, or n-value, is critically important in avoiding errors in conclusion when applying inferential statistics. At a minimum, a confirmatory study should include $n = 3$ independent subjects for each group. The reasoning behind this is that anything less than three does not allow enough degrees of freedom to calculate both a measure of central tendency and dispersion. However, this only sets a minimal lower limit on the sample size. The only way to accurately quantify certainty and know the likelihood of making an error is to perform a power analysis.

3.5.1 Errors in Statistical Hypothesis Testing—Type I and Type II Errors

Inferential statistics provide a quantitative method to decide if the null hypothesis (H_0) should be rejected. Since H_0 must be either true or false, there are only two possible correct outcomes in an inferential test; correct rejection of H_0 when it is false, and retaining H_0 when it is true. Therefore, there are two possible errors that can be made which have been termed Type I and Type II errors. A type I error occurs when H_0 is incorrectly rejected. This is commonly termed a false positive. A type II error occurs when H_0 is retained when it is in fact false. This error is commonly termed a false negative. From the standpoint of reproducibility, knowing the probability of making a type I or type II error is essential. This probability depends on experimental design and execution, and on the sample size, once again highlighting the importance of power analysis

3.5.2 Power Analysis

The power of a statistical test is defined as the probability that the test will lead to the appropriate rejection of the null hypothesis (H_0) if it is in fact false (Cohen, 1988). The familiar *p*-value is the probability that the obtained data, or more extreme values, could be obtained if H_0 is true. Therefore, the α-value represents the maximum acceptable probability of incorrectly rejecting H_0, or of making a type I error (false positive). The probability of making a type II error, failing to reject the null hypothesis when it is in fact false is termed β. With this in mind, power ($1-\beta$) can be defined as the desired probability that the *p*-value is $< \alpha$ for the expected mean treatment difference, and researchers will typically choose values of 0.8–0.95 as desired.

Statistical power is perhaps the most important concept in statistical analysis relative to the reproducibility of results. In an underpowered experiment, the probability of correctly rejecting the null hypothesis could be as low as 0.5, making the experiment as useful as a coin toss. Unfortunately, few researchers in the biomedical sciences seem to routinely employ power analysis. This is likely due to a lack of understanding of the meaning and importance of statistical power, and by the complexity the required calculations. Most basic statistical packages either fail to include or poorly implement power analysis. It is possible with some effort to produce a spreadsheet that will calculate power values for simple designs based on published descriptions and tables (Cohen, 1992). Fortunately, there are now excellent software packages available (e.g., PASS (http://www.ncss.com), nQuery Advisor (http://www.statsols.com), and G*Power (http://www.gpower.hhu.de)) to calculate power and/or sample size estimates for simple and complicated designs makes power analysis accessible to most scientists (Faul et al., 2007).

Statistical power is influenced by four values:

1. the reliability of results (measured by the SD);
2. the size of the expected effect;
3. the size of the sample (i.e., *n* value); and
4. the predetermined type 1 error rate, more commonly termed the significance level (α).

The type 1 error rate is usually set at an arbitrary conservative value, for example, 0.05. The reliability of results has to be determined from either historical control data or from pilot experiments to estimate the standard deviation in the samples. The size of the expected effect may also be estimated in a pilot study, or may be the subject of an educated guess. These values allow for the calculation of the *n*-value needed to obtain the desired power. As an example, take a simple case where it is hypothesized that a test compound will reduce fasting blood glucose concentrations in a rat model of diabetes mellitus. The null hypothesis is defined as: the test compound produces no effect on fasting blood glucose compared to vehicle treatment, and will be assessed using a standard *t*-test. Historical control data from previous

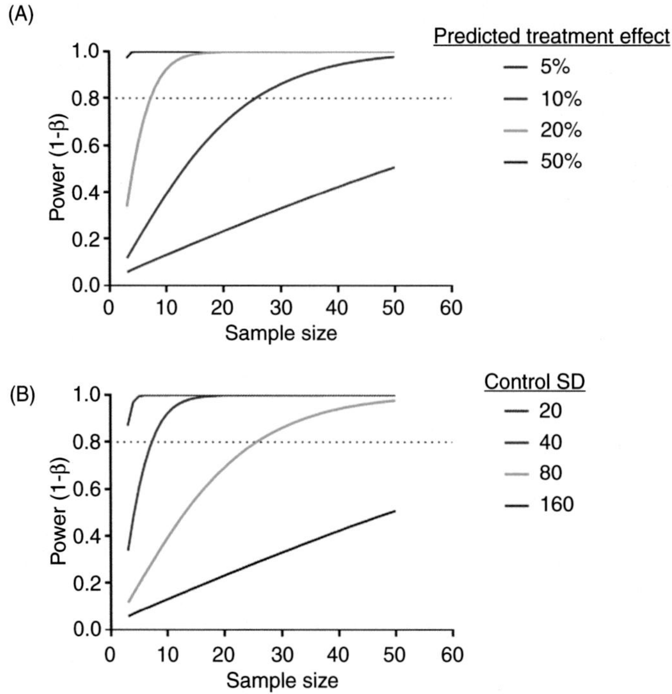

FIGURE 3.5 **Graphical depiction of how statistical power depends on sample size, effect size, and variability.** (A) the graphs show the relationship between sample size and power for a range of predicted treatment effect sizes based on the fasting blood glucose concentration example described in the text. The *dotted line* indicates the 0.80 cut-off for statistical power. The predicted 20% effect shows would require eight animals to have power of greater than or equal to 0.8. (B) Similar graphs to that in part A, but now the effect size is held constant and the standard deviation is varied to show the impact of experimental variability on statistical power.

work provides an estimate of the blood glucose concentration and standard deviation in the proposed vehicle treated group of 320 ± 40 mg/dL (mean ± SD). The expected effect based on a pilot study is a 20% reduction resulting in a predicted fasting blood glucose concentration of 256 mg/dL. Fig. 3.5 shows power curves generated for effect sizes ranging from 5% to 50%. To be confident in an appropriate rejection of the null hypothesis if it is in fact false, a power of ≥0.8 is desired. Therefore, for the expected 20% reduction eight subjects should be included in each group (n = 8 vehicle treated and 8 compound treated rats). It is worth noting that the other way to increase statistical power is to decrease variability (Fig. 3.5) which indicates that if the predicted n-value is too high, improved experimental design may provide an alternate path to a well powered study.

When thinking about statistical power, it is important to keep in mind that statistical significance and biological importance are not the same thing. The statistical significance of a mean difference has to do with the amount of evidence we have that the true difference is not zero (i.e., can we reject the null hypothesis). With a large sample size, even differences that are not biologically important can achieve statistical significance (Fig. 3.5). On the other hand, in a study with low power, the estimated treatment difference may be biologically important if it is real, but may not achieve statistical significance, meaning that we do not have strong evidence that it is real (i.e., nonzero).

3.5.2.1 Misuse of Power Analysis

Post hoc power analysis is a misuse of the method and should always be discouraged (Hoenig and Heisey, 2001; Levine and Ensom, 2001; Thomas, 1997). In light of a statistical test not reaching significance, there are two possible interpretations: (1) the null hypothesis is correct, or (2) the test did not have sufficient power to reject the null hypothesis. A *post hoc* analysis of power is sometimes wrongly used to indicate that null hypothesis is true because there was adequate statistical power to prove otherwise. However, this is a circular argument as power is defined as the probability of correctly rejecting the null hypothesis. If the null hypothesis is actually true, then statistical power is irrelevant. Another misuse that is particularly worrisome is the use of a *post hoc* analysis of power to determine how many subjects to add to the data set in order to achieve significance. As discussed earlier, since power relies only on SD, α, predicted effect, and the n-value, it should be obvious that any effect size will be significant at a given criteria given a high enough n-value (Fig. 3.5). The practice of *post hoc* power analysis and subsequent increase in sample size to achieve significance greatly increases the probability of making a type I error.

3.6 INFERENTIAL OR FREQUENTIST STATISTICS

Moving to the realm of confirmatory studies, or studies designed to allow us to draw a firm conclusion or make a decision through the testing of hypothesis, methods of inference are used to quantify the degree of certainty around the conclusion. Most biomedical researchers will have some familiarity with Frequentist methods of statistical inference. Frequentist methods interpret probability in terms of observations of events in repetitive trials of the same experiment. For example, if an event is observed 8 times in 10 repeated experiments, the probability of that event occurring is estimated at approximately 0.8. The value is approximate as it would require an infinite number of repeated trials to converge on the true probability.

To use a simplified example, assume that the researcher is interested in testing the hypothesis that a test compound increases locomotor activity when administered to rats. The null hypothesis would be stated as the test compound has no effect on locomotor activity. The study is repeated in the appropriate number of rats based on a power analysis, and an appropriate negative control is included resulting in two estimates: the average locomotor activity in the control group, and the average locomotor activity in the compound-treated group. The frequentist wants to answer the question "what is the probability that the compound-treated average could be drawn at random from the same population as the control average if there was, in fact, no difference"? If this probability is low enough (based on a predetermined α level typically set to 0.05) then the null hypothesis can be rejected. Importantly, the frequentist only thinks about the probability of obtaining the average value based on a random selection from the population. There is no probability of the average being correct or probability density function around the population parameter which is being estimated when drawing the sample from the population. This is a main distinction between the way Frequentists use probability and the way Bayesians employ the concept (see later).

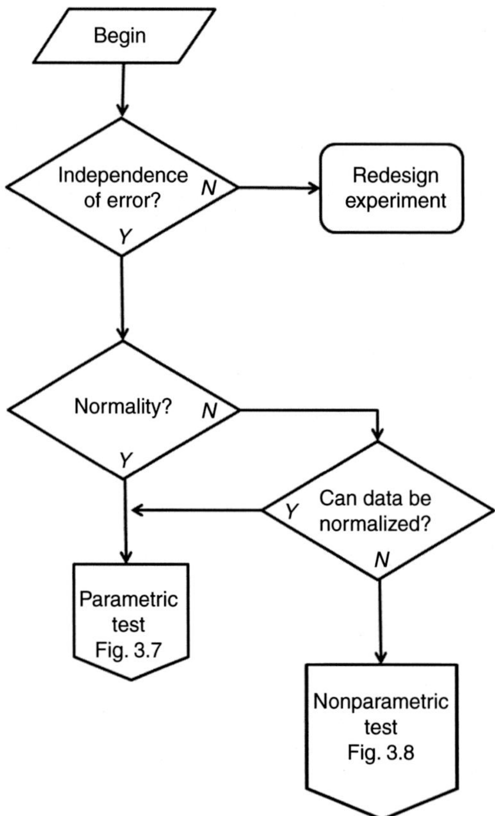

FIGURE 3.6 **A flow diagram depicting the first steps in choosing the appropriate statistical hypothesis test.** This figure covers the assumptions of independence of error and normality. See text for details.

It is beyond the scope of this chapter to go into detail on how the statistical test values are calculated, and the common availability of software to perform these calculations has all but made this level of detail unnecessary. Unfortunately, this ease of computation has also made it easy to choose the wrong statistical test, or to iteratively choose several until one obtains the answer require. This sort of *P*-hacking is inappropriate and will lead to faulty conclusions (see later). It is therefore critically important that the researcher understands how to choose the appropriate statistical test for the study they have designed. This choice is based on the experimental design and the assumptions made by each test, so the assumptions and details of the design can be used in a logical flow diagram to determine the appropriate test (Figs. 3.6–3.8).

3.6.1 The Assumption of Independence of Error

The first, and most critical assumption made by all hypothesis test is that of independence of error (Fig. 3.6). As discussed earlier, sampling and nonsampling errors are the causes for a given sample data point to be different from the population mean. The assumption of independence of error requires that all of the factors contributing to these errors be independent for each data point. Therefore, triplicate measurements from a single 96 well plate or from a single gel for a total of 3 values cannot be used in a statistical test as an $n = 3$ case because the factors contributing to the nonsampling error for each set of triplicates are not independent. This necessitates running the study 3 times, and averaging each of the triplicates which would provide 9 observations, but the triplicate values would be averaged to produce a final $n = 3$. The assumption of independence of error is important for all hypothesis testing and should never be violated.

3.6.2 The Assumption of Normality

Two broad categories of statistical tests can be applied based on the distribution of the sample. Parametric statistical methods (i.e., methods that rely on estimation of parameters) assume that the sample distribution is normally distributed. These statistics are the most commonly employed tests, such as the *t*-test, and ANOVA, however the assumption of normality is rarely considered. A number of quantitative methods exist to test how well a data set can be modeled by a normal distribution (D'Agostino, 1986) and most quality statistical software packages will provide a test of normality. However, it is difficult to test for normality, especially with a relatively small sample size. So often this assumption must be based more on the restrictions imposed by the experimental design. For example, in the earlier described study of locomotor activity, the researcher may choose to place a cutoff level in the study for the sake of tractability. If the researcher decides to halt the study if an animal produces more than 6000 beam brakes in 30 min (with a beam brake being a measure of the animal's movement in the open field, and 6000 indicating an extreme degree of locomotor increase), this will produce a skewed distribution. A compound that produces a large increase in locomotor activity will produce an extreme increase close to the cutoff point with relatively small variance. However, the sampling distribution derived from these values is unlikely to be indicative of the actual distribution, and may be significantly skewed by the cutoff. Therefore, the underlying assumption of normality is unlikely to be met for these data. In cases such as this, a nonparametric test would be the most valid choice (Fig. 3.6).

3.6.3 Factors Regarding the Design and Type of Measurement in Single Comparison Studies

The simple locomotor study described earlier is a case where two conditions are being compared, or what is termed a single comparison of two independent groups. For this type of study, assuming the assumption of normality is met, the *Student's t-test* (Livingston, 2004) is the most appropriate choice (Fig. 3.7), although there are additional test-specific assumptions that must be considered when choosing the final form of the *t*-test. The Student's *t*-test assumes equal variance exists between the two groups. As with the normality assumption, this assumption is often violated measuring effects that are subject to a "ceiling" or "basement." For example, a treatment that moves a necessarily positive value towards zero may also produce a smaller variance by virtue of the fact that the value cannot extend below zero.

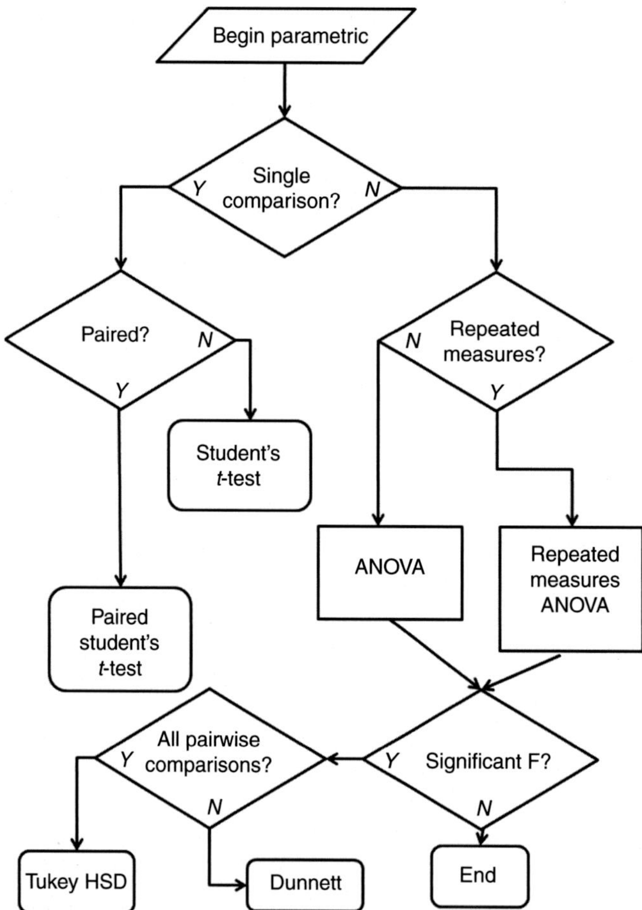

FIGURE 3.7 **Parametric flow diagram.** Following on from Fig. 3.6, a flow diagram depicting the logic used in selecting a parametric hypothesis test. See text for details.

Fortunately, a version of the *t*-test that is robust in the face of unequal variance called *Welch's t-test* can be used in these circumstances.

If the assumption of normality is not met, then the nonparametric *Mann-Whitney U test* (also called the *Wilcoxon rank-sum test*) (Fay and Proschan, 2010; Mann and Whitney, 1947) is the most appropriate choice (Fig. 3.8). The Mann-Whitney U test does not assume that the samples were drawn from normal distributions, but it does assume that the distributions have the same basic shape. The essence of the Mann Whitney U statistic is a comparison of mean ranks, and in doing so the null hypothesis being tested is that the distribution of values is not different. The resultant *p*-value provides the probability that the mean ranks would be found to be different based on chance alone.

If the locomotor study was designed in a way that would measure a baseline locomotor activity in a group of animals, and then the same animals were dosed with the test-compound

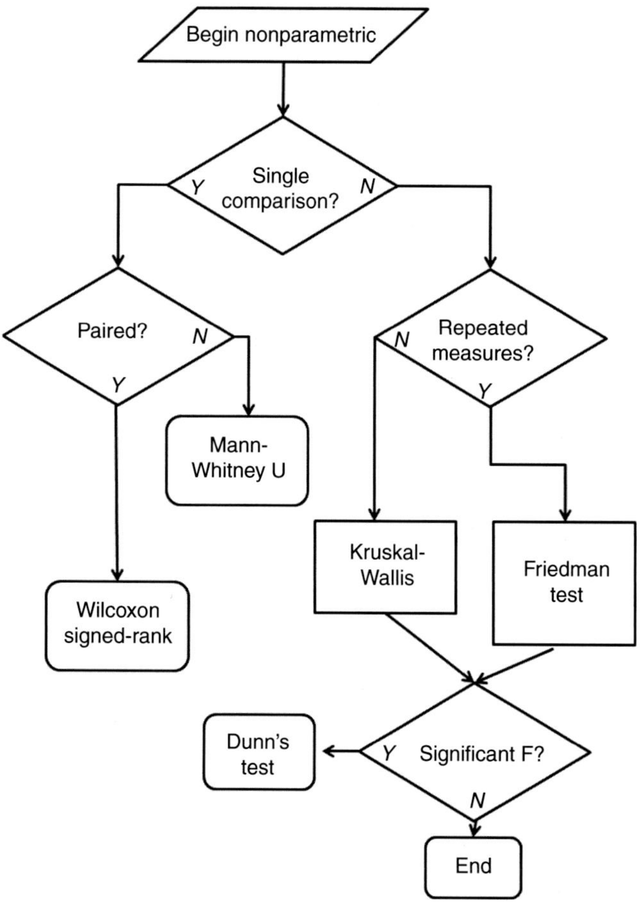

FIGURE 3.8 **Nonparametric flow diagram.** Following on from Fig. 3.6, a flow diagram depicting the logic used in selecting a nonparametric hypothesis test. See text for details.

to determine the locomotor activity after treatment, this becomes a two-group single comparison repeated measures design. This is a type of crossover design in which every subject receives every treatment. In this case, there is reason to pair the values in the first group (baseline or vehicle response) with values in the second group (compound treatment). Pairing assumes that some of the variance is shared between the two groups, and by focusing on the difference between pairs allows for this variance to be blocked. Paired approaches therefore offer greater statistical power.

If the normality assumption is met, the *paired Student's t-test* is the best choice for the two group repeated measures design (Fig. 3.7). While formally similar to the unpaired *t*-test, the paired *t*-test is performed on difference values and assumes that the difference values are normally distributed. Because of the reliance on the difference between pairs, the paired *t*-test does not assume equal variance between the two samples. If the normality assumption is not met, then a nonparametric alternative should be used. The Wilcoxon signed-rank test (Wilcoxon, 1945) is typically the best choice (Fig. 3.8). This test does not assume that difference values are normally distributed; however, it does assume that the distribution of differences is symmetrical about the median.

3.6.4 Multiple Comparisons Between Independent Groups

In studies in which multiple treatments, such as multiple doses of a test compound are employed, the design becomes a multiple comparison. In this case where multiple comparisons are being made, it is important to consider the effect that multiple comparisons have on the power to observe a difference. For example, consider a variation on the locomotor experiment in which three doses of the test compound are administered resulting in 3 comparisons between treatment and control. The first comparison by *t*-test produces a *p*-value of 0.03. This is interpreted as a probability of obtaining this difference based on chance alone is equal to 3%. Therefore, we would expect that if the study was run 100 times, we would obtain this difference 3 times even if there were no real difference between the groups. If we now make the second comparison, it should be apparent that the *p*-value would need to be adjusted because we are now increasing the probability of obtaining a difference by chance alone by making another observation (again from a theoretical population of 100). In other words, the more comparisons that are made, the higher the likelihood is of witnessing the occurrence of a rare event. Therefore, each comparison increases the chance of making a type I error (false positive), and confidence in the results are greatly reduced. It is important that all comparisons made should be planned prior to running the study, all planned comparisons are made, and all comparisons made are reported.

It is possible to compensate for the loss of statistical power produced by making multiple comparisons. The *Bonferroni correction* (Perneger, 1998) is the simplest and most conservative method and consists of simply dividing the α value by the number of comparisons and interpreting the *p*-value against this corrected α value. In the example above with three comparisons, the standard α=0.05 would be divided by 3 making the corrected α= 0.017. This can lead to an overly conservative estimate of significance which increases the possibility of making a type II error (false negative), especially when dealing with large numbers of comparisons. Therefore, the most commonly used approach to multiple comparison designs is to first use an overall test for the difference in means which will only indicate that a difference exists. This is then followed by a *post hoc* test to determine which of the comparisons, if any, achieve significance.

For the parametric case where the normality assumption is met, the comparison of greater than two independent groups without repeated measures is best performed by an *analysis of variance (ANOVA)* (Fisher, 1970) followed by an appropriate *post hoc* test (Fig. 3.7). ANOVA is actually a collection of statistical methods and models the details of which are beyond the scope of this chapter, however it is sufficient to know that ANOVA works by dividing the observed variance into different components (e.g., within groups, between groups). The ANOVA assumes normality in all groups. This is obviously a more critical assumption in ANOVA given that more groups are involved. There is also an assumption of equal variance across groups.

The ANOVA only indicates that there is a difference among the means, but does not indicate which means are different. Assuming the ANOVA allows the null hypothesis to not be retained; subsequent *post hoc* test can be performed to determine which means are significantly different. It is important to note that an overall significance from an ANOVA does not indicate that any of the individual means will be significant on *post hoc* testing. In cases where the overall ANOVA is significant, but the *post hoc* testing does not indicate significance, the interpretation is that treatment did produce a significant effect on the means, however that effect cannot be pinpointed to any one pair of means. This suggests that a better designed follow up study is necessary, so a significant ANOVA with no significance in the *post hoc* test should not be reported as a lack of significance.

The choice of a *post hoc* test is a complex subject. The factors for consideration include whether the comparisons were planned, if all pairwise comparisons are to be made or if comparisons should be restricted to assessing differences from a control group, and how conservative the test will tend to be (i.e., how likely a type II error may be). For planned comparisons between a subset of means, the Bonferroni correction remains the only choice, however as indicated earlier it is a highly conservative test. When making all pairwise comparisons of means, the *Tukey honestly significant test* (Tukey HSD) also known as the *Tukey Kramer method* (Kramer, 1956) is often the most reasonable choice. When comparing multiple groups against a control mean the *Dunnett test* (Dunnett, 1955) is often the appropriate choice (Fig. 3.7).

In the case where the design is a comparison between more than two groups independent groups and the assumption of normality is not met, The *Kruskal-Wallis test* (Kruskal and Wallis, 1952) can be used (Fig. 3.8). It has the same assumptions as the Mann-Whitney U test in that it does not assume that the samples were drawn from normal distributions, but it does assume that the distributions have the same basic shape. As a rank sum test, the Kruskal-Wallis tests against a null hypothesis is testing that the sum of ranks are not different from each other. The resulting p-value is the probability that the difference is due to chance. The Kruskal Wallis test only indicates that a difference exists among the means but does not indicate which means are different. *Post hoc* testing involves the use of the Mann-Whitney U test with Bonferroni correction (Fig. 3.8), sometimes referred to as *Dunn's test*, or more simply termed the *post test* (Daniel, 2000).

3.6.5 Multiple Comparisons with Repeated Measures

The case described earlier for the paired t-test or the Wilcoxon signed-rank test is the simplest form of a repeated measures design. The use of multiple comparisons with repeated measures is very common, especially in studies evaluating the time course of an effect. For example, if the earlier described repeated measures study in which a baseline is measured, and then a measurement is taken after compound treatment in the same animal were extended to include measurements taken every 15 min after dosing for 2 h, each animal will be sampled once for baseline, and 8 times for treatment resulting in 8 comparisons. However, these 9 groups are not independent as they are all obtained from the same set of animals after the same treatment and would be expected to share some variance. This type of study should be analyzed using a one way repeated measures ANOVA (Fig. 3.7).

The repeated measures ANOVA is an extension of the ANOVA that accounts for the shared variance in the groups. The assumptions of the one way repeated measures ANOVA are the same as the ANOVA with the addition of an assumption of sphericity. This is an extension of the equal variance assumption that states that the variance of the differences between all combinations of related groups is equal. Most software packages that provide repeated measured ANOVA will perform tests of sphericity, and these tests should be used as the repeated measures ANOVA is very sensitive to violations of this assumption. Choice of *post hoc* testing is the same as that discussed for multiple comparisons between independent groups by ANOVA (Fig. 3.7).

For situations where the normality assumption is violated in a repeated measures design involving three or more groups, *the Friedman test* (Friedman, 1937), a rank nonparametric version of the analysis of variance can be used (Fig. 3.8). The Friedman test is an extension of the Wilcoxon signed-rank test and carries all of the assumptions of that test described earlier with the additional assumption of sphericity. The null hypothesis for the Friedman test states that all groups have the same median value and the *p*-value is interpreted as the probability that differences in the median can be attributed to chance alone. As with the Wilcoxon signed-rank test, Dunn's test can be used as a *post hoc* analysis to determine which groups are significantly different (Fig. 3.8).

3.7 CORRELATION

Correlation methods allow for the assessment of association between measured variables in a data set. For example, a correlation could be measured between the concentration of a growth factor in the cerebrospinal fluid and the degree of neurodegeneration in a mouse model of Huntington's disease. Misinterpretation of correlation is quite common. The *cum hoc, ergo propter hoc* fallacy (with this, therefore because of this) represents the most serious pitfall in correlation analysis. Correlation between two variables does not imply causality. In the Huntington's model example given earlier, it may be tempting to conclude that the correlation between the growth factor and the neurodegeneration demonstrates a causal relationship. However, the growth factor production may be modulated in response to the neurodegeneration which is actually caused by an unrelated factor. It is equally wrong to assume that a lack of correlation demonstrates a lack of association. Correlation analysis is designed to measure associations fitting a particular model (e.g., linear association). A significant higher order polynomial relation may exist between two variables which would not be detected by common correlation analysis.

Correlation analysis can only be used when assessing association between two measured variables. The use of correlation in an experimental design where one variable is manipulated and another is measured as in a concentration response study is inappropriate. Regression methods, which measure goodness of fit, should not be confused with correlation analysis, which measures association between variables.

Studies in which two variables are independently measured should be designed with a correlation analysis in mind. The range of values covered by both variables should be similar. The methods assume that the two variables are sampled from population distributions (an assumption violated if one of the variables is an independent variable). The output of a correlation analysis will include a correlation coefficient, and a *p*-value. The correlation coefficient ranges from -1 to 1 and indicates the strength and direction of the association. The *p*-value indicates how likely it is that the association could be observed by chance.

In parametric cases, the *Pearson product-moment correlation coefficient* (r) (Rodgers and Nicewander, 1988) is employed to assess association between two variables. The Pearson method assesses linear dependence, and assumes that both variables are sampled from populations which are normally distributed. This assumption is particularly important for small data sets (e.g., $n < 10$). In addition to the requirement for normality, r is also particularly sensitive to outliers.

The output of a correlation analysis using the Pearson method should include the correlation coefficient, a p-value that tests the null hypothesis that $r = 0$, and the coefficient of determination (r^2). Interpretation of the value of r is not strictly quantitative; however, the sign of r indicates the direction of association. In general, the further the r values lay from 0, the stronger the association. The value of r provides a measure of covariance between the two variables normalized to the product of their standard deviations, and squaring r to obtain the coefficient of determination provides a measure of the proportion of the variance that can be accounted for by the association of the two variables. For example, an $\underline{r}^2$ value of 0.8 indicates that 80% of the variance is shared by the two variables.

For nonparametric cases, the *Spearman rank correlation coefficient* (*rs*) (Spearman, 1904) measures association between two variables, but differs from the Pearson product-moment correlation coefficient in that it assesses monotonic (i.e., not necessarily linear) dependence. In practice, *rs* is calculated as r but between the ranked variables. The Spearman method makes no assumptions regarding the distribution of the populations from which the variables are drawn and can assess nonlinear monotonic relationships. Therefore, it is the preferred measure for small data sets, when the assumption of normality is violated or not assessed, or when the relationship appears to be nonlinear. The interpretation of *rs* and the p-value are similar to that discussed above for r. Because ranked values are used in the calculation, the coefficient of determination cannot be calculated for *rs*.

3.8 INTERPRETATION OF INFERENTIAL STATISTICS

The output of a statistical test is all too often reduced to an asterisk on a bar graph. This is unfortunate for at least two reasons. First, there is a great deal of information in parameters like the actual p-value that is obscured by the simplified $P < 0.05$. Second, and more importantly, the practice of reducing the statistical test to a binary readout of significance imbues the $\alpha = 0.05$ with a seemingly magical relevance that it simply does not have. The finding of a $P < 0.05$ does not make an apparent effect real or necessarily biologically relevant any more than $P > 0.05$ indicates a lack of relevance. Significance testing and the resulting p-value are intended to do one thing, control the rate of error, or as one author bluntly put it "the function of significance tests is to prevent you from making a fool of yourself, and not to make unpublishable results publishable" (Colquhoun, 2014).

It is important to understand what the p-value actually represents. As explained in Chapter 2, the experiment is designed around a null hypothesis (H_0). For example, if the goal of the experiment is to determine if a compound of interest produces a given effect (the hypothesis), then H_0: The compound of interest produces no effect different from a vehicle control. The p-value is the probability that the obtained data, or more extreme values, could be obtained if H_0 is true. This is the probability of incorrectly rejecting H_0, or of making a type I error (false positive). This is most definitely not the same thing as the probability that H_0 is false, nor is it a reflection of the probability that the hypothesis being tested is true (i.e., will be reproducible). These two misinterpretations, along with the extreme reliance on p-values alone, have become so pervasive that the American Statistical Association (ASA) has found it necessary to issue a statement on the appropriate use and interpretation of *p*-values (Wasserstein and Lazar, 2016), and some have even suggested a banning of *p*-values and null hypothesis testing from the literature (Trafimow and Marks, 2015).

The fallacy that $P < 0.05$ indicates that H_0 is false and pitfalls in interpretation has been elegantly explained using logical syllogisms (Cohen, 1994). Take, for example, the following flawed syllogism:

> If a food item is a fruit, then it is not a pineapple (a clearly false statement).
> This food item is a pineapple.
> Therefore, this food item is not a fruit (a clearly incorrect conclusion).

The logic breaks down because the main premise is incorrect (pineapples are fruits) even though the logical form is valid. If the syllogism is recast in probabilistic terms, a surprising result is obtained:

> If a food item is a fruit, then it is probably not a pineapple (a true statement given the large number of possible fruits from which the food item can be chosen).
> This food item is a pineapple.
> Therefore, this food item is probably not a fruit (a clearly incorrect conclusion).

The logical form is not valid for probabilistic statements. Casting this in terms of null hypothesis testing the problem becomes apparent.

> If H_0 is true, then the observed result would probably ($P < 0.05$) not occur (i.e., would probably not be a false positive).
> The result did occur.
> Therefore, H_0 is probably not true (a clearly incorrect conclusion).

Frequentists make no assumptions regarding the probability of H_0 being true. The probability of the observed data given H_0 is not the same as the probability of H_0 given the data. This "prior" probability $P(H_0)$ is normally not available, and while it can be assumed in Bayesian methods (see later) it cannot be interpreted based on the *p*-value.

The second fallacy, that the *p*-value is a reflection of the probability, that the hypothesis being tested is true, or that it will be reproducible is more subtle. If $P = 0.05$, then that formally is interpreted as follows: the probability that the results obtained or more extreme results could be obtained if there was, in fact, no difference (i.e., H_0 is true) is 0.05. Therefore, if the same experiment where run 100 times, 5 of those times we would expect to get a result as least as large as the one obtained in the initial experiment even if there was no real difference. However, this does not indicate that the interpretation will only be wrong 5% of the time (Colquhoun, 2014). Recall that statistical power is the probability of correctly rejecting H_0 when it is false (see earlier). The appropriate use of power analysis should lead us to design an experiment with power approximately equal to 0.8. Following Colquhoun (2014) we can consider a result that is real and obtainable in 10% of samples (e.g., a test of a panel of compounds 10% or which are effective). If the population is sampled 1000 times, 100 of those would yield real results, and our statistical test would detect 80 of those as true positives based on the statistical power. Therefore, there would be 20 false negatives. Of the remaining 900 samples, we would expect that 5%, or 45 (based on α =0.05) would show up as false positives with 855 being correctly identified as true negatives. The false discovery rate can be calculated as the total number of false positives (45) divided by the total number of observed positives (45 + 80 =125) giving a value of 36%. Therefore, using an α level of 0.05 and assuming that all of the assumptions of the statistical test are met and the experiment is appropriately powered, positive results are expected to be false 36% of the time.

The earlier discussion of *p*-values raises the question of what can be done. The 36% expected false positive rate is clearly a contributing factor in the apparent crisis of reproducibility (Ioannidis 2005; Ioannidis et al., 2014). However, this should not be interpreted as an indictment of *p*-values or null hypothesis testing which remain valuable tools in that they do control the error rate in the published literature. In addition, what is often missed in discussions of the apparent crisis of reproducibility is that failure to replicate results is part of the scientific process. Rather than providing evidence that the system is broken, reports of failures in replication prove that the system works. If a perfectly designed and executed experiment is expected to yield a false positive 36% of the time, then it seems reasonable that a noticeable number of studies will not reproduce. This, of course, is not an apology for poorly designed experiments or outright fraud. But it does highlight several critical points regarding statistical testing and experimentation:

1. Experiments must be replicated and reproduced. Cost and time are not valid arguments for assuming a greater than 36% chance of being wrong.
2. Experiments should be designed to test a hypothesis in multiple ways using multidisciplinary methods. By building a multidisciplinary case, the confidence in the conclusions is greatly enhanced.
3. Power analysis is not a luxury, but a necessity. The central role of statistical power in calculating the false discovery rate above should make it obvious that power less than 0.8 increase the likelihood of being wrong.
4. Actual *p*-values should be reported. As described earlier, if the *p*-value is ~0.05 the false discovery rate is 36%. A *p*-value of 0.015 (which would often be reported as $P < 0.05$) decreases the false discovery rate to 14%. If the software only reports approximate *p*-values for extreme probabilities (e.g., $P < 0.0001$) then it is reasonable to use these approximations.
5. Avoid the use of the term significant. Many, including the author, have been guilty of applying the term significant to every *p*-value less than 0.05. While it may seem like harmless semantics, the word has unfortunately become synonymous with true or real effect. Based on the discussion earlier, this is obviously misleading. The chosen α-value is arbitrary and does not indicate the probability that the results are correct.

3.9 REPORTING OF RESULTS

The reporting of statistical results has sadly degenerated to a bare minimum treatment with $P < 0.05$ added almost as an afterthought. Using the earlier example where a compound decreasing fasting blood glucose concentrations in a rat model of diabetes mellitus, the results would commonly be reported as "compound X produced a significant decrease in fasting blood glucose ($P < 0.05$, *t*-test)." While this language may be considered acceptable, it is only slightly more informative than stating that a difference was observed. According to the Uniform

Requirements for Manuscripts Submitted to Biomedical Journals (International Committee of Medical Journal Editors, 1997), the author should "describe the statistical methods with enough detail to enable a knowledgeable reader with access to the original data to verify the reported results." The method section should include a listing of the tests performed, and importantly any software used including version numbers. At a minimum, the in-text report should include appropriate measures of central tendency and dispersion (e.g., mean and SD), sample size (n), the calculated test statistic (e.g., t-value), and the probability of observing the effect if H_0 is true (p-value). It is also a good practice to restate the hypothesis being tested as the hypothesis and the statistical test are intimately linked. Therefore, the diabetes example would be better reported as "the ability of compound X to produce a decrease in fasting blood glucose concentration in a rat model of diabetes mellitus was tested and an effect was observed (baseline = 335 ± 42, treatment = 270 ± 58, n = 8 rats/condition, paired t-test t = 2.567, P = 0.0224)."

The format and information reported will vary for each statistical test. Table 3.4 provides a summary of suggested reporting style for some common tests discussed earlier. For further information on appropriate methods of reporting statistical methods, analyses, descriptive statistics, and hypothesis tests, the reader is referred to the SAMPL Guidelines (Lang and Altman, 2015).

TABLE 3.4 Examples of Suggested Reporting Styles for Several Common Hypothesis Tests

Test	Example
Student's t-test	"Treatment with compound X produced a decrease in the measure (vehicle = 335 ± 42, test compound = 270 ± 58 (all mean ± SD), t-test, t =2.567, P =0.0224, n = 8 per condition)." Or "Treatment with compound X did not alter the measure (vehicle = 335 ± 42, test compound = 326 ± 58 (all mean ± SD), t-test, t =0.356, P =0.728, n = 8 per condition)."
Paired Student's t-test	"Treatment with compound X produced a decrease in the measure (baseline = 25.88 ±7.16, treatment = 19.88 ± 5.67 (all mean ± SD), paired t-test, t =3.113, P =0.017, n = 8 per condition)."
ANOVA	"Compound X produced a dose-dependent decrease in the measure (vehicle = 446.2 ± 79.3, 1 nM X = 389.8 ± 81.3, 3 nM X =305.3 ± 55.13, 10 nM X = 218.2 ± 54.3 (all mean ± SD), ANOVA F= 12.62, P < 0.0001, n = 6 per condition. Dunett *post hoc* versus vehicle: P= 0.372 (1 nM X), 0.006 (3 nM X), 0.0001 (10 nM X)" Or "Compound X did not alter the measure at the doses tested (vehicle = 446.2 ± 79.3, 1 nM X = 389.8 ± 81.3, 3 nM X =423 ± 78.9, 10 nM X = 426.5 ± 59.2 (all mean ± SD), ANOVA F= 0.579, P = 0.635, n = 6 per condition.)"
Mann-Whitney U Test	"Compound X increased the response with median (25th,75th percentile) measures in the vehicle group of 219 (201.8,236) and 311 (292,336.8) in the compound X treated group. (Mann–Whitney U = 4, n= 8 per group, P = 0.002)."
Friedman test	"Compound X produced a dose-dependent decrease in the measure (vehicle = 433 (217.5,584.8), 1 nM X = 369.5(234.3,614.8), 3 nM X =306.5 (265,359.3), 10 nM X = 190 (179.3, 230) [(all median (25th, 75th percentile)], Friedman statistic = 9.203, P = 0.018, n = 6 per condition. Dunn's *post hoc* versus vehicle: P > 0.99 (1 nM X), >0.99 (3 nM X), 0.0156 (10 nM X)."

Note the inclusion of measures of central tendency and dispersion, and the use of real p-values when possible. The software used to generate these examples (GraphPad Prism v 7.00) is excellent at providing exact p-values under most circumstances. In some extreme cases, the approximate p-value is all that is available and is appropriate (e.g., P > 0.99 for Dunn's test). For additional information, see Lang and Altman (2015).

3.10 BAYESIAN INFERENCE

As discussed earlier, the frequentist thinks in terms of sampling distributions and the variation of data around fixed parameters. This approach leads to an inferential framework that assess the probability of the existing data being obtained if H_0 is correct. This can be formalized as the conditional probability $P(\text{Data} \mid H_0)$, read as the probability of the data given the null hypothesis (is true). Of course, what the researcher really wants to know is $P(H_0 \mid \text{Data})$, or the related probability that the hypothesis is correct $(1-P(H_0 \mid \text{Data})$ given the data, but as discussed earlier, this information is not available in the frequentist framework.

The Bayesian approach is different at the very basic level of how data are perceived. Data are considered as fixed observations with varying degrees of belief, and parameters are free to vary to fit observations. Parameters are random variables described by probability distributions and bounded by credibility intervals. Under this framework, Bayes theorem can be employed to allow an estimate of the probability of the hypothesis being correct, given the data:

$$P(H \mid \text{Data}) = \frac{P(H)P(\text{Data} \mid H)}{P(\text{Data})}$$

Where P(H) is the probability of the hypothesis being correct before seeing the data, or the prior probability, $P(\text{Data} \mid H)$ is the probability of the data occurring given the hypothesis, P(Data) is the probability of observing the data under any hypothesis, and the $P(H \mid \text{Data})$ is the probability of the hypothesis given the observed data, or the posterior probability. The obvious problem that arises is that the prior probability is not normally known, and therefore must be modeled as a distribution of probabilities that may be updated as additional data becomes available. In practice, the data is combined with the prior probability distribution using Bayes' theorem to give the posterior probability distribution. There are notable benefits of the Bayesian approach including less reliance on assumptions and access to the posterior probability. However, there are notable criticisms to Bayesian inference, the most common focusing on the need to quantify beliefs about the experimental outcome in a prior probability distribution. Since this distribution is rarely known and quantified, the dependence on the prior probability allows for subjectivity in the analysis that makes many scientists, and all frequentists, uneasy. Essentially the subjective nature of the Bayesian methods allow for more "researcher degrees of freedom" and open the door for *P*-hacking (Simmons et al., 2011). Another common criticism is the need for complex calculations necessary to derive the posterior probability distribution. While this criticism is less valid with the availability of fast inexpensive computers and appropriate software, it does mean that much of the calculations are unknown to the scientists. For additional information on Bayesian inference the reader is referred to several introductory manuscripts and monographs (Gelman, 2008; Gelman et al., 1995; van de Schoot and Depaoli, 2014).

3.11 BIG DATA ISSUES

As mentioned at several points in this chapter, technology has evolved to enable the collection of data in high throughput parallel ways that were unimaginable just a few years ago. Unfortunately, our ability to analyze data collected on this scale has lagged behind. The development of "omics" approaches and imaging techniques which can produce thousands and even millions of data points in a single study has produced a situation where the time required to carefully analyze data far exceeds the time needed to acquire the data. As an example, imaging methods can measure variations across the entire brain in a disease population, an ability that has led to a decreased reliance on hypothesis testing and a more look-and-see approach that some have called "connectomics," or "systems biology."

It would seem foolish to simply ignore the massive data sets available through modern techniques. Certainly, there is value in being able to profile variations across the entire brain or the entire genome, however, this sort of study design is a massive multiple comparisons exercise. In the field of genome wide association studies (GWAS) millions of genetic markers measured simultaneously leads to vanishingly small α values with any of the corrections for multiple comparisons describe earlier, a fact that may contribute to low rates of reproducibility (Ioannidis, 2008). The development of bioinformatics approaches, such as pathway analysis (Kelder et al., 2010) should provide valuable tools to explore these large data sets, however the outcome of these sorts of analysis are exploratory and should be considered hypothesis generating rather than hypothesis testing.

3.12 P-HACKING

P-hacking (Simmons et al., 2011) is a broad term that encompasses several types of manipulation that are commonly employed in data analysis despite the fact that they are clear misuses that lead to inappropriate conclusions. Many instances of *P*-hacking can be traced to a simple lack of understanding of the underlying assumptions of the statistical methods (Marino, 2014; Motulsky, 2014). As is stressed throughout this monograph, confirmatory experiments should be planned in every detail (number of groups, number of subjects in each group, exclusion criteria, data transformations, and statistical analysis) prior to actually running the experiment. Any deviation from this practice will lead to *P*-hacking and a higher likelihood of obtaining false positive (i.e., nonreproducible) results. The following outlines some of the most common forms of *P*-hacking along with the reason these practices should be avoided.

3.12.1 Dynamic *n*-Values

As discussed earlier, the *p*-value represents $P(\text{Data} \mid H_0)$. This controls the type I error rate (i.e., with $\alpha = 0.05$, 5% of the time a false positive will be obtained). However, this number is only valid if the type II error rate is controlled by designing an appropriate powered experiment. Assume a study where the actual sample size needed to achieve a power of 0.8 is $n = 15$. Therefore, the practice of starting an experiment with $n = 8$ (approximately half the sample size predicted by power analysis), and checking for significance is flawed for at least two reasons. First, the probability of drawing false negative conclusion is higher than acceptable due to low statistical power (see earlier). Second, the misconception that a significant *p*-value in an underpowered study predicts a significant *p*-value in an appropriately powered study is wrong (Motulsky, 2014). The *p*-value does not depend on the effect size, but rather is heavily influenced by the sample size.

To make matters worse, after checking at $n-8$ and finding no significance, or an apparent "trend" in the data, the researcher decides to add an additional 3 or 4 subjects to each group and check again, a process than can be repeated *ad infinitum*. This leads to a serious and usually ignored problem in that it introduces noncorrected multiple comparison which increases the rate of false positives.

3.12.2 Hypothesizing After the Result is Known (HARKing)

HARKing (Kerr, 1998) is defined as presenting a hypothesis that was developed and based on the results obtained from a study as if it were an a priori hypotheses. This also leads to a form of multiple comparison and increases the false positive rate. HARKing has been highlighted as an issue in the field of brain imaging where the goal of the study is often to first determine a brain region where there is an effect (an essentially exploratory study) and then to estimate the size of the identified effect (a confirmatory study) within the same study (Kriegeskorte et al., 2010). HARKing can occur in any study using any method if the hypothesis being statistically evaluated is not the same as the hypothesis of the experiment was designed to test. For example, if a putative cognitive enhancer is tested in a mouse T-maze assay under the hypothesis that the test compound will improve performance over baseline (pretreatment) results, then the statistical comparison should be performed between treated performance and the performance of animals before they were treated. If a positive control group is included using the standard of care compound donepezil to provide a check on the performance of the assay (i.e., a positive control), and the researcher noticed that the test compound produced a larger improvement in T-maze behavior than that of donepezil, it would be inappropriate to statistically compare the test compound to donepezil and report that the compound represents an improvement over the standard of care compound. Instead, this finding represents an exploratory hypothesis-generating result that forms the basis for a new hypothesis and a new confirmatory study. As discussed throughout this chapter, exploratory studies are of significant value in generating hypothesis. However, a separate confirmatory study is required to estimate the effect size and perform statistical hypothesis testing.

3.12.3 Statistical Roulette

The final, and most egregious example of *P*-hacking is the practice of choosing a statistical test to give the hoped-for result. While seemingly unthinkable, and certainly unethical, the ease at which modern statistical software packages allow data to be reanalyzed has made this type of *P*-hacking all too common. Examples include discovering that a *t*-test provides a *p*-value = 0.06, then trying a paired *t*-test in spite of the fact that the original design was not paired, following an ANOVA with every available *post hoc* test until one choice provides the hoped-for result, or ignoring the results of an ANOVA in favor of making a direct unplanned comparison between two groups using a *t*-test. Many other examples exist, and none are justified. Statistical roulette is basically a form of HARKing and is confounded by the same issues of multiplicities. However, it goes a step further by potentially, and in most cases certainly, using statistical tests in situations where the underlying assumptions are not met. As has been stated previously, and cannot be overemphasized, the statistical test is determined when the hypothesis is generated. Any further analysis of the data is at best exploratory, and if treated as confirmatory represents blatant *P*-hacking which greatly reduces the chance of reproducibility.

3.13 IS THERE A NEED FOR A REAL-WORLD ACCOMMODATION OF THE STATISTICAL IDEAL?

This chapter began with the definition of the term statistic and the field of statistics. Statistics is defined as the science of quantitative methods that guide data collection, interpretation, and presentation. It is a quantitative and mathematically exact framework for "doing science" that provides powerful interpretational tools that are significantly limited by real world assumptions. If these assumptions are not met, then the tools of statistics are rendered useless. Therefore, the statistical framework does limit what can be done in the laboratory, and technological advances have led to a situation where statistics can in some cases be the prime limiting factor. For example, in the field of high throughput screening thousands or even millions of compounds can be tested for a specific biological activity in a single experiment continuing over the course of a few days. It is simply not possible to employ inferential statistics to determine which of the compounds produced a significant biological effect. There are three possible responses in this situation, and each will be evaluated later.

First, the scientist could argue that the statistician is out of touch with the real world, and has not caught up with the needs of the researcher. While this is an extreme and somewhat flawed interpretation of the state of affairs, it is not an uncommon belief. Unfortunately, many researchers faced with this apparent disconnect (although typically on a much smaller scale than a high throughput screening campaign), will opt to place their trust in the technology and use whatever statistical test they are most comfortable with, or whatever statistical test provides the best result. It is hopefully obvious to the reader by now that this approach will lead to a high probability of false positive results.

While the earlier approach puts too much faith in the technology, the second possibility would be to put too much faith in the statistics. Since the statistical framework does not allow inferential statistics to be employed in this situation, the experiment could be scaled down to a more statistically friendly size and run in a way that meets all of the necessary assumptions. So, the researcher could opt to test each compound in independent triplicate runs with appropriate controls and make each comparison using a *t*-test. While no one could argue with the validity of this approach, it is equally flawed. The researcher in this case has hampered scientific progress by not taking advantage of the available technology. In this case, blind faith in statistics is no more useful than ignorance of statistics.

The third option is to use the statistical methods in a way that is in harmony with the technology. Recognizing that the high throughput screening campaign is not a confirmatory study, the research forgets about inferential statistics and turns to EDA. By looking at the distribution of data, extreme values can be identified that may or may not be "significant." The compounds producing these extremes can be singled out for follow up confirmatory studies that are appropriately powered and employ multiple concentrations. Since there are much fewer compounds in the set of extremes, these studies can be appropriately designed for inferential analysis.

It is frustrating, but nevertheless factual that the statistical framework cannot accommodate every experiment that technology allows or that the scientist can conceive. Despite this, statistics remains the best quantitative tool, we have to ensure reproducibility. Rather than rebel against the perceived restrictions that statistics imposes on research, the informed scientist will embrace these restrictions and use them to their advantage at the earliest stages of experimental design.

References

Anscombe, F.J., 1973. Graphs in statistical analysis. Am. Stat. 27, 17–21.

Barde, M.P., Barde, P.J., 2012. What to use to express the variability of data: standard deviation or standard error of mean? Perspect. Clin. Res. 3, 113–116.

Begley, C.G., Ioannidis, J.P., 2015. Reproducibility in science: improving the standard for basic and preclinical research. Circ. Res. 116, 116–126.

Cohen, J., 1988. Statistical Power Analysis for the Behavioral Sciences. Psychology Press, New York.

Cohen, J., 1992. A power primer. Psychol. Bull. 112, 155–159.

Cohen, J., 1994. The earth is round ($p < 0.05$). Am. Psychol. 49, 997–1003.

Colquhoun, D., 2014. An investigation of the false discovery rate and the misinterpretation of p-values. R. Soc. Open Sci. 1, 140216.

D'Agostino, R.B., 1986. Tests for the normal distribution. In: D'Agostino, R.B., Stephens, M.A. (Eds.), Goodness-of-Fit Techniques. Marcel Dekker, New York, pp. 367–420.

Daniel, W.W., 2000. Applied Nonparametric Statistics. Duxbury/Thomson Learning, Pacific Grove, CA.

Dunnett, C.W., 1955. A multiple comparison procedure for comparing several treatments with a control. J. Am. Stat. Assoc. 50, 1096–1121.

Faul, F., Erdfelder, E., Lang, A.G., Buchner, A., 2007. G*Power 3: a flexible statistical power analysis program for the social, behavioral, and biomedical sciences. Behav. Res. Methods 39, 175–191.

Fay, M.P., Proschan, M.A., 2010. Wilcoxon-Mann-Whitney or t-test? On assumptions for hypothesis tests and multiple interpretations of decision rules. Stat. Surv. 4, 1–39.

Fisher, R.A., 1970. Statistical Methods for Research Workers. Oliver and Boyd, Edinburgh.

Friedman, M., 1937. The use of ranks to avoid the assumption of normality implicit in the analysis of variance. J. Am. Stat. Assoc. 32, 675–701.

Gelman, A., 2008. Objections to Bayesian statistics. Bayesian Anal. 3, 445–450.

Gelman, A., Carlin, J.B., Stern, H.S., Rubin, D.B., 1995. Bayesian Data Analysis. Chapman and Hall, Boca Raton, Florida.

Hoenig, J.M., Heisey, D.M., 2001. The abuse of power: the pervasive fallacy of power calculations for data analysis. Am. Stat. 55, 1–6.

International Committee of Medical Journal Editors, 1997. Uniform requirements for manuscripts submitted to biomedical journals. International Committee of Medical Journal Editors. Ann. Intern. Med. 126, 36–47.

Ioannidis, J.P., 2005. Why most published research findings are false. PLoS Med. 2, e124.

Ioannidis, J.P., 2008. Why most discovered true associations are inflated. Epidemiology 19, 640–648.

Ioannidis, J.P., Greenland, S., Hlatky, M.A., Khoury, M.J., Macleod, M.R., Moher, D., Schulz, K.F., Tibshirani, R., 2014. Increasing value and reducing waste in research design, conduct, and analysis. Lancet 383, 166–175.

Jarvis, M.F., Williams, M., 2016. Irreproducibility in preclinical biomedical research: perceptions, uncertainties, and knowledge gaps. Trends Pharmacol. Sci. 37, 290–302.

Kairalla, J.A., Coffey, C.S., Thomann, M.A., Muller, K.E., 2012. Adaptive trial designs: a review of barriers and opportunities. Trials 13, 145.

Kelder, T., Conklin, B.R., Evelo, C.T., Pico, A.R., 2010. Finding the right questions: exploratory pathway analysis to enhance biological discovery in large datasets. PLoS Biol. 8, e1000472.

Kerr, N.L., 1998. HARKing: hypothesizing after the results are known. Pers. Soc. Psychol. Rev. 2, 196–217.

Kimmelman, J., Mogil, J.S., Dirnagl, U., 2014. Distinguishing between exploratory and confirmatory preclinical research will improve translation. PLoS Biol. 12, e1001863.

Kramer, C.Y., 1956. Extension of multiple range tests to group means with unequal numbers of replications. Biometrics 12, 307–310.

Kriegeskorte, N., Lindquist, M.A., Nichols, T.E., Poldrack, R.A., Vul, E., 2010. Everything you never wanted to know about circular analysis, but were afraid to ask. J. Cereb. Blood Flow Metab. 30, 1551–1557.

Kruskal, W.H., Wallis, W.A., 1952. Use of ranks in one-criterion variance analysis. J. Am. Stat. Assoc. 47, 583–621.

Lang, T.A., Altman, D.G., 2015. Basic statistical reporting for articles published in biomedical journals: the "statistical analyses and methods in the published literature" or the SAMPL Guidelines. Int. J. Nurs. Stud. 52, 5–9.

Levine, M., Ensom, M.H., 2001. Post hoc power analysis: an idea whose time has passed? Pharmacotherapy 21, 405–409.

Livingston, E.H., 2004. Who was student and why do we care so much about his t-test? J. Surg. Res. 118, 58–65.

Mann, H.B., Whitney, D.R., 1947. On a test of whether one of two random variables is stochastically larger than the other. Ann. Math. Stat. 18, 50–60.

Marino, M.J., 2014. The use and misuse of statistical methodologies in pharmacology research. Biochem. Pharmacol. 87, 78–92.

Motulsky, H.J., 2014. Common misconceptions about data analysis and statistics. J. Pharmacol. Exp. Ther. 351, 200–205.

Perneger, T.V., 1998. What's wrong with Bonferroni adjustments. BMJ 316, 1236–1238.

Philips, T., Rothstein, J.D., 2015. Rodent models of amyotrophic lateral sclerosis. Curr. Protoc. Pharmacol. 69, 5–21.

Rodgers, J.L., Nicewander, W.A., 1988. Thirteen ways to look at the correlation coefficient. Am. Stat. 42, 59–66.

Simmons, J.P., Nelson, L.D., Simonsohn, U., 2011. False-positive psychology: undisclosed flexibility in data collection and analysis allows presenting anything as significant. Psychol. Sci. 22, 1359–1366.

Spearman, C., 1904. The proof and measurement of association between two things. Am. J. Psychol. 15, 72–101.

Stroop, J.R., 1935. Studies of interference in serial verbal reactions. J. Exp. Psychol. 18, 643–662.

Thomas, L., 1997. Retrospective power analysis. Conserv. Biol. 11, 276–280.

Trafimow, D., Marks, M., 2015. Editorial. Bas. Appl. Soc. Psychol. 37, 1–2.

Tufte, E.R., 2001. The Visual Display of Quantitative Information. Graphics Press, USA.

Tufte, E.R., 2006. Beautiful Evidence. Graphics Press, USA.

Tukey, J.W., 1977. Exploratory Data Analysis. Addison Wesley, Reading, Massachusetts.

Tukey, J.W., 1980. We need both exploratory and confirmatory. Am. Stat. 34, 23–25.

van de Schoot, R., Depaoli, S., 2014. Bayesian analyses: where to start and what to report. Eur. Health Psychol. 16, 75–84.

Wasserstein, R.L., Lazar, N.A., 2016. The ASA's statement on p-values: context, process, and purpose. Am. Stat. 70, 129–133.

Wilcoxon, F., 1945. Individual comparisons by ranking methods. Biometrics Bull. 1, 80–83.

CHAPTER

4

Reporting Results

Michael Williams, Kevin Mullane and Michael J. Curtis

OUTLINE

Research in the Biomedical Sciences. http://dx.doi.org/10.1016/B978-0-12-804725-5.00004-5

4.1 INTRODUCTION

After researchers have planned, executed, reproduced, and objectively analyzed a series experiments related to a stated hypothesis (Chapters 2 and 3), they are usually ready to submit their findings to a peer reviewed journal the scope of which is relevant to the topic of the study. The acceptance and publication of this body of work will enhance the standing of the researcher with their peer research group, demonstrate, with appropriate citations (Ioannidis, 2014), their productivity and relevance to advancing science in their field of research and enhance their ability to receive grants, or in industry, continued project funding.

4.2 PREPARING THE MANUSCRIPT (TABLE 4.1)

The preparation, writing, and dissemination of a body of scientific data in the form of a manuscript involves three key factors: (1) an *hypothesis* on which the premise of the manuscript is based which provides the context for performing a series of experiments; (2) an *experimental data* set that has been appropriately generated, analyzed, reproduced, and collated that provides the basis for the author's conclusions regarding the hypothesis and; (3) the selection of an *appropriate venue* for publication where both the readership and editors of the targeted journal will be genuinely interested and experienced as experts in the subject matter of the manuscript, thus facilitating prompt and appropriate dissemination of the content. Additionally, by selecting an appropriate venue, there is an excellent likelihood that the Editorial Board of the target journal will be able to provide additional expertize and feedback in the peer review process that will both facilitate the review process and potentially aid in improving the quality of the submitted paper (Glasziou et al., 2014). Issues related to the peer review process and reviewer bias are dealt with in detail in Chapter 5.

4.2.1 The Journal Venue—Only the "Best"?

Submitting a paper to what have been termed "luxury" or "prestigious" journals (Garwood, 2014; Schekman 2013) that have a high, double digit, journal Impact Factor (JIF; Chapter 5), can often lead to the author choosing to compromise the content of an article and its intrinsic and consequent value to its target audience in order to fit the perception of the requirements of a luxury journal or to accommodate the reviewers of the article (Eisen, 2014). This is due to an editorial style at luxury journals that is overtly focused on reader and mainstream media impact. This incentivizes authors to "rush into print, cut corners, exaggerate their findings, and overstate the significance of their work" (Alberts et al., 2014). This results in a tendency for the luxury journals to publish primarily papers in "hot areas," even when the paper makes "extraordinary claims in the absence of extraordinary proof" (Truzzi, 1978) as in the case of the STAP retractions in *Nature* (Chapter 1; Goodyear, 2016; Rasko and Power, 2015). At the same time this is associated with a high rejection rate (>90%) for the majority of papers that make less flamboyant claims but are otherwise scientifically worthy. For papers that are accepted by luxury journals, the title can be unilaterally "enhanced" by professional rather than scientific editors to ensure maximal attention, with the abstract also being hyped to increase the number of reader downloads/citations and mainstream media coverage such

TABLE 4.1 Manuscript Sections—IMRAD Format

	Check list	Areas for author attention
Title Page Section 4.11	• Author list, order, and affiliation(s) • Corresponding author • Title • Running title • Keywords • Conflict of interest • Acknowledgement for funding	• COPE considerations for authorship criteria avoiding honorary, guest, courtesy, and ghost authors (Section 4.5.1.1) • Shared first authorship? • Text, figures, and tables consistent with journal guidelines and checklists • Keeping track of individual author contributions • Who actually writes the paper? • Agreement on manuscript content and conclusions • Proof reading of submitted manuscript • Request for specific reviewers OR reviewers to be avoided
Abstract Section 4.9	• Context/background • Major findings and relevance • Biological system/assays used (in vitro/in vivo) • Species • Implications/conclusions	• Abstract is consistent with journal guidance • Accuracy—avoid hyperbole and overt speculation • Abstract consistent with actual data contained in manuscript and conclusions
Introduction Section 4.10	• Background/context • Statement of hypothesis and its importance	• Not a mini-review • Should cite recent key papers relevant to manuscript

Materials and Methods
Section 4.8
	• Detailed information on materials including compounds, animals, cell lines, buffers, and their specific sources
	• Institutional review of experimental protocol statement.
	• Animal source, strain, weight, gender
	• IACUC approval from originating institution
	• Use of *ARRIVE* etc. Guidelines
	• In vivo—pharmacokinetic data used for dosing of reference standard + experimental compounds
	• Compound volume, route, frequency and volume of administration, site, duration of injection, time of day
	• Documented detail and validation of all reagents—sources, purity
	• Reference standards
	• Experimental design (Chapter 2)
	  ◦ Groups and powering
	  ◦ Measures to avoid bias—for example, blinded investigators
	  ◦ Defined endpoints/outcomes
	• Statistical analysis (Chapter 3)
		• Avoid being inaccurate and incomplete
		• Do not omit information for competitive reasons—personal or pecuniary
		• Cell line authentication
		• Antibody validation

(*Continued*)

TABLE 4.1 Manuscript Sections—IMRAD Format (*cont.*)

	Check list	Areas for author attention

Results
Section 4.6
	• Figures and tables presented with sufficiently detailed legends that include key information that can be read as "stand-alone" for example, "reader should be able to understand your work solely by looking at the figures and tables" (Neill, 2007)
	• Discussed in text in order of results
	• Data
	  ◦ Generated using adequate dose/concentration curves
	  ◦ quantitatively analysis and presentation as an IC_{50}, EC_{50}, Kd, pA_2 value etc., reproduced with an estimate of variability (Sem, SD, confidence limits)
		• Controls for vehicle and use of reference standards/ procedures
		• Only results—no methods, included in Figure legends, unless methods-based paper
		• Data independently reproduced
		• Normalized and "representative" data should be clearly justified in methods
		• Appropriate statistical analysis
		• Data not shown/unpublished findings not acceptable
		• Semi-quantitative scoring criteria should be clearly defined
		• Structures of new compounds—or published reference to their published source mandatory
		• Images—quantification, replication, and authenticity should follow emerging guidelines recognizing they may be subject to forensic analysis.

Discussion/ Conclusions
Section 4.7
	• Conclusions to experiment
	  ◦ Top level findings
	    – Relevance to a priori hypothesis of study
	  ◦ Detailed discussion
	    – Limitations of methodologies
	• Placing data in context with literature
	  ◦ Differences and confirmation
	• Framework—"testable hypotheses and observed associations, rather than rigorous proof of cause and effect"
	• Possible next steps/areas for clarification
		• Not a mini-review
		• Avoid speculation and hyperbole, for example, potential "new drug" based on preliminary animal data
		• Do not advertize new experiments in preparation in conclusions

References Section 4.13	• Recommended format	• Adhere to journal guidelines for number • Topical except for seminal articles
Acknowledgements Section 4.12	• Contributors not included as authors • Formal acknowledgment of grant support	• Contributor permissions
Supplementary Materials Section 4.6.4	• Expanded methods section • Specialist content • Videos, photographs.	• Same level of detail as main text • No single observation data points • Accepted by journal often without any peer review
Cover Letter Section 4.14	• Background to submission • Why it should be of interest to the journal • List of potentially conflicted reviewers for editorial consideration	• Ensure that all content reflects the submission and do not carry over from a previous manuscript that was rejected.

that the initial assessment of suitability for publication is "on newsworthiness rather than scientific quality" (Alberts et al., 2014). This has frequently resulted in the abbreviation of the methods section (or its relegation to online only status) to a level where insufficient detail is provided on how to reproduce the experiments used to generate the data reported (Collins and Tabak, 2014; Mullane and Williams, 2017) thus contributing to an inability to reproduce the findings.

4.2.2 The Wrong Journal?

Submitting an original manuscript where 75% or more of the data content involves medicinal chemistry or computer assisted molecular modeling (CAMD) to a biological journal (and *vice versa*), or a primary clinical paper comprised of data from patient cohorts from a clinical trial to a preclinical journal can be an exercise in futility especially with respect to obtaining expert peer review feedback (Lee et al., 2013). This may occur when authors assume that a lack of appropriate expertize at a journal will make it easier for their work to be published. This often backfires when the Editors of a journal reject the submission without review as being inappropriate for their readership. Thus it is probably wiser that authors stay with established journals in their field of expertise and constructively deal with expert and experienced peer review from their actual peer group.

4.3 THE MANUSCRIPT STRUCTURE—IMRAD

The structure of a manuscript varies depending on the journal style. The majority of biomedical research journals generally follow the IMRAD (*Introduction, Methods, Results, And Discussion;* Fig 4.1A) format and flow that is outlined in the "Uniform Requirements for Manuscripts Submitted to Biomedical Journals" (also known as the URM or Vancouver Protocol; ICMJE, 2010; Liumbruno et al., 2013). This format is not arbitrary but instead reflects the process by which the experiments were conceived and conducted. Some journals may reorder these sections, for example, *Cell*—Introduction, Results, Discussion, and Methods (designated as Experimental Procedures) or ignore them. In recent articles (as of October, 2016), *Nature* has no designated sections in its research articles apart from the variable inclusion of a Methods Summary that refers the reader to supplementary materials. Likewise, *Science* has no designated sections, with materials and methods being available only online as supplementary materials (see Chapter 4.6.4). The use of supplementary material sections that are only available online is intended to allow the narrative in the main manuscript to be concisely presented while providing interested readers access to more detail on the methodologies used (Pop and Salzberg, 2015). However, there can be curation issues. While a paper, whether in print or pdf form, is an item of record, a supplementary material file uploaded anywhere but with the publisher of the paper—for example, on a personal website—may have a finite half-life if the website crashes or is closed. Even when the website is immortalized, the paper and the supplementary material are located separately begging the question: what is the item of record—the paper or the paper plus its supplementary material? It is probably better to discontinue the practice of separating the "paper" from its "supplementary material" or better still to include the "supplementary material" actually in the paper.

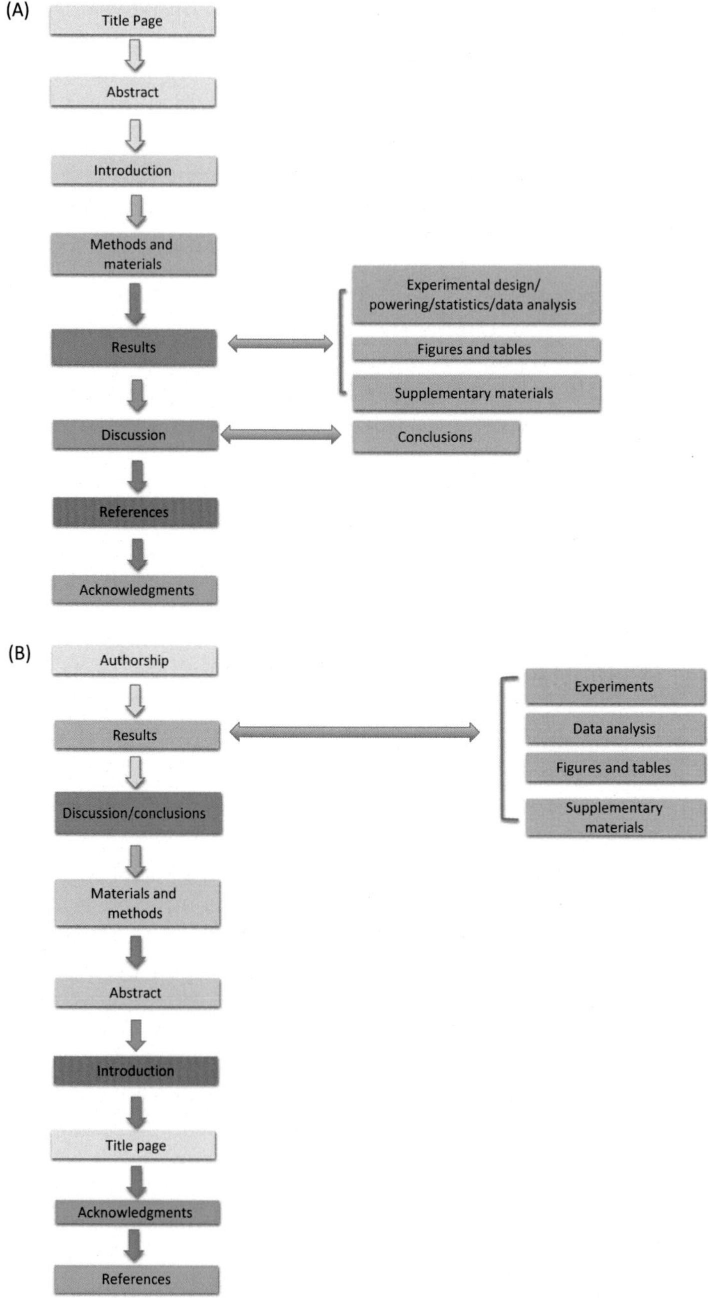

FIGURE 4.1 **IMRAD format and assembling the manuscript.** (A) IMRAD manuscript organization—*Introduction, Methods, Results, And Discussion*. (B) Manuscript writing order (see also Table 4.1).

4.4 INSTRUCTIONS/GUIDELINES FOR AUTHORS

In preparing a manuscript, an imperative for the author(s) is to read and to follow the Instructions/Guidelines for Authors provided by the selected journal. The latter typically describe the style, format, and content of the journal. This can vary greatly from journal to journal (even those from the same publisher) and in some instances the available Instructions/Guidelines are not current such that it is always a good idea to check a recent issue of the journal to ensure that the actual style and content have not changed. Submitting a manuscript that has clearly not followed the journal "house style" and/or where the authors have not read the Instructions/Guidelines for Authors can be a major irritation to the Editorial Office and reviewers. This is especially the case when the actual format of the submitted manuscript is identical to or similar to that of another journal, which often indicates that the manuscript has probably been previously submitted elsewhere and presumably rejected. Most journals will return this type of submission to the authors as an editorial desk decision (e.g. rejection) without peer review.

4.4.1 Guidelines for Style, Grammar, and Accessible Communication

In addition to the documentation and interpretation of the scientific aspects of the paper, the clarity and transparency in its actual writing is a critical element (Spector, 1994). The *lingua franca* of science, irrespective of nationalistic imperatives and concerns, is English that is written by convention in a third party (no "I" or "we") style in the past tense (was). In 2009, 94% of biomedical papers published were in English (Boissier, 2013). Whether "the Queen's" English or American English, the style and grammar in current scientific writing has become increasingly idiosyncratic as the computer has become the major vehicle for text composition (with all the foibles of automatic spell check—for example, clinical trials becomes clinical trails) accompanied by the impact of the Internet (texting, tweeting, emoticons, rap, hyperbole, coprolalia) on communication styles (Section 4.4.1.1), such that the English in a scientific manuscript, especially from researchers whose first language is not English, can be challenging and confusing to read in many instances. This represents an understandable lack of familiarity with written, as contrasted to colloquial, spoken English. In others, it can be a deliberate stylistic choice to the extent that an author takes issue with the archival house style of a journal arguing that writing in the style of the *MailOnline.com* can make the article more "readable," thus improving its value to, and impact on, the reader and that not being allowed to use this format is a form of censorship.

4.4.1.1 New Technology and New Platforms for Communication of Complex Issues

The communication of findings from biomedical research experimentation in the 21st century involves presentation forums from the humble printed page to any number of electronic technologies, including high speed virtual networks, online publications, blogs, etc. that devour and regurgitate the latest findings in research 24 h a day, 7 days a week, etc. with minimal detail as researchers are encouraged "to increase online visibility of their work through visual media and media-sharing platforms, as well as social networking and blogging" (McKee, 2016).

Paradoxically, as researchers "communicate" more, less time is spent in assimilating the information with the actual substance being replaced by the ephemera of communicating.

When actual face-to-face interactions do occur, as at meetings, the widespread use of PowerPoint as the primary vehicle for disseminating information leads to the "dumbing down" of the science due to its hierarchal, bulleted format (Frommer, 2012). PowerPoint, thus promotes "a cognitive style that disrupts and trivializes evidence" (Tufte, 2006) creating "the illusion of understanding and the illusion of control" (Bulmiller, 2010) that compromised the decision-making process in the lead up to the Challenger Shuttle explosion (Tufte, 2006). It has also been argued that PowerPoint is antithetical to critical thinking (Smith, 2015) avoiding the need to internalize an argument in a way that tests its strength and relevance.

Concerns over scientific writing were highlighted in a recent study (Plavén-Sigray et al., 2017) that focused on the readability of over 700,000 abstracts from 122 high-impact scientific journals that were published between 1881 and 2015. This study reported that the readability of science was decreasing, a finding ascribed to a rise in the number of collaborative, multiauthored manuscripts, and an increased use of scientific jargon/"in-group scientific language ("science-ese")"—a reflection of increased scientific specialization. The authors concluded that the decreased readability of science impacted the reproducibility and accessibility of research findings for the dissemination of new approaches to biomedical research, drug discovery and human health is the subject of the widely publicized TED (technology, entertainment, and design) talks. These are intended to "challenge our core beliefs in search of deeper truth, while we celebrate the thinkers, dreamers, and mavericks who dare to offer bold new alternatives via short, powerful 'inspirational' talks (under 18 min)." In TED talks, complex issues are often oversimplified and trivialized with the content being overinterpreted or presented in sound bites. This results in unattainable promises in order to "sell" the content to the audience—an approach viewed as "a recipe for civilizational disaster" (Bratton, 2013). Inevitably, the sound bite nature of these various technological "improvements" in communication affect, in addition to diminishing the process of cognitive thinking, both in producing data (experimental planning, execution and analysis) and in the assessment of data produced by others, also affects attention span and comprehension with many biomedical researchers relying on article abstracts, articles in the mainstream media, and blogs to remain up to date in their research field.

While the present Chapter is intended as a stand-alone guide to effectively reporting research, the reader may find additional benefit in consulting the Chicago Manual of Style (2010) as well as the following papers (Chipperfield et al., 2010; Hoogenboom and Manske, 2012; Liumbruno et al., 2013; Mack, 2013; Neill, 2007; O'Connor and Holmquist, 2009; Shidham et al., 2016; Spector, 1994) for additional insights and guidance on preparing a scientific manuscript. While some aspects of these articles may be contradictory, they are, nonetheless, worthy of consideration as an author prepares a manuscript for submission given that these represent a distillation of many years of practical experience in scientific publishing in all manner of biomedical disciplines.

4.5 ASSEMBLING THE MANUSCRIPT

As noted earlier, a manuscript, that is, ready for submission generally follows the IMRAD guidelines with the addition of a title, abstract, authors, and keywords (Fig. 4.1A). Despite the logical and orderly progression of the IMRAD guidelines as presented in Chipperfield et al.

(2010), a manuscript is more frequently *assembled* and *written* in a different order with the results being the central focus (Fig. 4.1B). Often, when a decision has been made to bring a series of experiments to a conclusion because the results are considered supportive of the original premise of the study, additional scrutiny of the results during the writing process can highlight gaps and inconsistences in logic and content that require additional studies to ensure that the conclusions are robust and have considered, and ruled out, obvious and alternative explanations for the data. The order of the sections below thus follows their order of importance in reporting the research and writing the manuscript rather than the URM guidelines.

4.5.1 Authorship

Authorship of a scientific manuscript is perhaps one of the most important and sensitive issues in science (Dance, 2012; Mack, 2013). It has academic, social, and financial implications (Ioannidis, 2014) that are focused on: (1) appropriately recognizing an individual for their definable contribution to the work contained in the manuscript and; (2) indicating who is responsible and therefore accountable for the published work. To those involved in conceptualizing and designing a series of experiments and generating the data, it can be a highly contentious list, in some instances taking precedence over the actual data content of a manuscript. Thus serious disagreements can occur on the inclusion/exclusion of an individual as an author of a study, the order of authorship, who has final responsibility in writing the manuscript and so on. These issues can often lead to acrimony with disputes in data interpretation, decisions to include authors under a default rubric of "both authors contributed equally to this study" and who is the designation of the corresponding author. The Committee on Publication Ethics (COPE) of The International Committee of Medical Journal Editors (ICMJE; 2010; http://www.icmje.org/recommendations/browse/roles-and-responsibilities/defining-the-role-of-authors-and-contributors.html) recommends that researchers decide on authorship and author order *before* an experiment is even conducted and also suggests that a research group involved in generating the data for the studies revisits the author list regularly as the project evolves (Dance, 2012). Credit for author contributions has been divided into four models: *sequence-determines-credit* approach (SDC) where the order of authors reflects the declining importance of their contribution; *equal contribution norm* (EC) where authors are listed alphabetically to acknowledge similar contributions or to avoid conflict among collaborators; *first-last-author-emphasis norm* (FLAE) where the first author(s) (usually one or two) receives credit for the article and the last author is usually the Principal Investigator (PI) or laboratory chief; and the *percent-contribution-indicated approach* (PCI) where the contribution of each author is quantified and listed in descending order (Tscharntke et al., 2007). This may seem like minutia but it is highly relevant to investigators because their seniority is considered (subjectively) by peers evaluating their applications for research funding, fellowship, and promotions. These issues can be accentuated in "mega author" submissions, that may have 10–20 authors and in "big science" collaborations in the biomedical (Leung et al. 2015; The 1000 Genomes Project Consortium, 2010) and physical (Aad et al., 2015) sciences where the number of authors ranges from the low hundreds to more than 5,000 (Aad et al., 2015). Authorship for such studies has been typically resolved by the inclusion of authors in alphabetical order—the EC model—in some instances irrespective of their contribution or seniority and in others with the more "senior" authors being listed among the first 5 or 10 authors.

Regardless of the model, it is the case that substantive (value-bearing) "credit" is generally given only to the corresponding author (the senior PI) and (if different) the first author (usually the person who did the largest proportion of the laboratory work).

4.5.1.1 Authorship Guidelines

COPE provides criteria and guidelines for the authorship of biomedical research based on the ICMJE guidelines (http://publicationethics.org/files/Authorship_DiscussionDocument.pdf) that are used by many biomedical journals. These state "that an author should meet all four of the following criteria:

- Substantial contributions to the conception or design of the work; or the acquisition, analysis, or interpretation of data for the work;
- Drafting the work or revising it critically for important intellectual content;
- Final approval of the version to be published;
- Agreement to be accountable for all aspects of the work in ensuring that questions related to the accuracy or integrity of any part of the work are appropriately investigated and resolved."

Authorship guidelines have also been addressed in the "Fostering Integrity in Research" report from the National Academies of Sciences, Engineering, and Medicine (NASEM, 2017) with a recommendation that "societies and journals should develop clear disciplinary authorship standards to identify those who have made a significant intellectual contribution as authors that would include contributions to the design or conceptualization of a study, the conduct of research, data analysis or interpretation, or the drafting or revising of a manuscript for intellectual content. Those who engage in these activities should be designated as authors of the reported work, and all authors should approve the final manuscript. In addition to specifying all authors, standards should (1) provide for the identification of one or more authors who assume responsibility for the entire work, (2) require disclosure of all author roles and contributions, and (3) specify that gift or honorary authorship, coercive authorship, ghost authorship, and omitting authors who have met the articulated standards are always unacceptable. Societies and journals should work expeditiously to develop such standards in disciplines that do not already have them." Issues regarding honorary, coercive, and ghost authorship unacceptability are discussed further below.

Despite the clarity of these guidelines (albeit, "substantial" is a subjective notion), they are neither universally appreciated nor always clearly applicable such that there is still significant room for confusion especially when creative, motivational, and intellectual contributions are prioritized, rightly or wrongly, over technical and subjectively debatable "routine" contributions (Liumbruno et al., 2013; Mack, 2013). In the latter context, Wager (2009) has highlighted aspects of the perceived arrogance and hypocrisy of the way that authors deal with "routine" contributions where "Some journals expressly forbid acknowledging individuals such as.... technicians....[as that]....person did not qualify for authorship 'because they were only doing their job' and that because these people were being paid for their services this was reward enough and they should not be listed as authors (which struck me as strange, because presumably the PIs were also 'only doing their job' and were certainly getting paid, yet this did not disqualify them from being listed)" making a logical case for the inclusion of all individuals involved in the research, including technicians.

Fortunately, disagreements over authorship can often be handled by level-headed and objective authors or, if needed, at an institutional level by impartial and uninvolved members of the local research hierarchy before a manuscript is submitted. There is a major difference between presubmission authorship disagreements and postacceptance disputes. The former tend not to involve the journal as most journals require a written statement on authorship inclusion by the corresponding author or all authors on manuscript submission. When authorship changes during peer review, an editorial office is required to seek agreement in writing from the individuals affected and the other authors that they are in agreement with these changes. If not, it can create problems for journals. Indeed when issues of disputed authorship arise (typically when an individual, hitherto unknown to the journal and perhaps a former colleague of the authors, sees a paper and thinks "some of that is MY work" and complains to the journal) either during the peer review process or after publication the authors are required resolve the matter. This can result in prepublication resolution or an erratum or a reprint of the author line postpublication. If the issue cannot be resolved amicably, it can inevitably be referred back to the administrative function of the originating institution to deal which can often lead to the shelving of problematic papers or their retraction.

Other concerns related to authorship involve; (1) *honorary* or *courtesy* authorship where a laboratory head, PI, senior researcher, (or in industry, a line manager equivalent), is automatically included as an author on a paper even when they contributed minimally to the manuscript—what has been described as "an intolerable violation of professional ethics" (Mack, 2013). While this practice has a long standing precedent with the laboratory head, PI, or line manager arguing that they are due recognition for providing the resources (and possibly the original hypothesis), financial, and physical, or in industry, permission, to enable the research to be conducted, journals are increasingly requesting formal documentation of authorship responsibilities with authors being specifically listed as participating in the following categories: research design; conduct of experiments; contribution of reagents (including novel compounds—although this is becoming an increasingly contentious reason for authorship); analyzing data (including statistical analysis); data interpretation; writing or contributing to the writing of the manuscript including revisions: and providing approval for both the submitted manuscript and the final paper. Increasingly, journals that request information on author contributions publish it on the title page or with the references in the published article. Given the complexity and diversity of the contributions required to conduct a series of scientific experiments, some of these categories have been difficult to justify as being objectively tangible contributions, and it is uncertain yet whether "nonplayer" contributions will become regarded as justification for inclusion as an author.

As one example, it has long been customary for a chemist who synthesized a novel "tool compound" or a biologist who had cloned, expressed, and patented a novel drug target and shared each, respectively, with other investigators to be listed as coauthors—rather than acknowledged—on any publication derived from work using the compound or clone. This often resulted in numerous publications—irrespective of the fact that beyond providing the actual materials (which was itself often the primary subject of other publications), these individuals had absolutely no demonstrable contribution to (or often knowledge of) the experiments that had provided the data that were the subject matter of the publication.

A more contentious issue are those papers with *ghost* and/or *guest* authorships. *Ghostwriters* are professional scientific or medical writers in industry or funded by industry who write papers, usually in the clinical area, without having any involvement in the reported studies (Goldacre, 2012). Once completed, key opinion leaders (KoLs) in the subject area of the ghostwritten manuscript are recruited as *guest authors* (and sometimes compensated; Bhattacharjee, 2011; Messerly, 2014; Yahia, 2012) to add their imprimatur and take credit for authorship even though their involvement was often in direct contradiction to the URM guidelines. The practice of guest authorship has transitioned from being considered as scientifically unethical/unacceptable/inappropriate (Bavdekar, 2012; Grassley, 2010; Goldacre, 2012) to being viewed as legal fraud (Flaherty, 2013; Stern and Lemmens, 2011) and has led to legislative oversight (Grassley, 2010) that continues to raise concerns regarding the subversion of the outcomes of research, specifically in clinical trial results, to benefit the sponsoring pharmaceutical companies (Goldacre, 2012, 2016; Ross et al., 2008).

A 2008 survey reported that both honorary and ghost authorship in high impact clinical journals occurred in approximately 21% of published articles (Wislar et al., 2011), an outcome potentially related to pharma-sponsored publication plans to market drugs (Ebrahim et al., 2016; Sismondo, 2009). This led to the suggestion that research funded by the biopharmaceutical industry should not be published in the clinical literature (Smith et al., 2014). While a fairly uncommon practice in the preclinical literature, both ghostwriting, and less overtly, guest authorship does occur with industry-sponsored basic research and can lead to the selective interpretation (or exclusion) of findings.

A final complication of authorship can occur with multicenter research collaborations. These frequently exist not because of any burning desire of the participants to synergistically use their collective expertise and knowledge to interrogate a research problem but rather because funding is available for multicenter research collaborations. One example of these was the EU transnational *Framework Programmes* that began in 1984, part of which were focused on biomedical research in the form of *Networks of Excellence* (NoE) and *Specific Targeted Research Projects* (STRePs). Some manuscripts emanating from these multicentered efforts, while being of sufficient commonality in theme to be funded, tended to suffer from uneven and inconsistent content, for example, reference controls that were different or assayed at different concentrations/doses/time points. When requests were made following peer review of manuscripts for revisions to better integrate the findings, some participants declined further involvement making publication for the other authors problematic.

4.6 RESULTS

The results section represents the core of any scientific manuscript reporting original data and is distinct from the methods, discussion, and conclusions sections, which respectively provide detail on the means used to generate the data and the context for, and relevance of, the results. The results should be clearly described using the past tense in a logical order with the appropriate use of figures and tables. Each figure and each table should be represented by a paragraph in the results section where the data present in each is clearly and concisely presented but not discussed. An increasingly regular event in peer reviewing is authors presenting the results without any reference to the tables or figures, which happens more routinely with review-type articles than papers documenting original research. Another issue is repetition of numerical findings in text and figures or tables (which is confusing and purposeless).

If data are described in the results section of the text (e.g., IC_{50} values), it should be possible for the reader to correlate these directly with the data shown in the figures or tables. Single observations that report the percentage inhibition of a biological response at a single dose or IC_{50} values lacking any indication of replication, for example, are not accompanied with estimates of variability from the mean or 95% confidence intervals, are preliminary ($n = 1$) and are unacceptable in the peer-reviewed literature. Similarly, the citing of data means with errors (SEM or SD) in the text, the source of which cannot be obviously gleaned from the data reported in either the tables or figures is of concern as is data that morphs from lacking significance in the tables or figures to being described as significant in the text.

In addition to the experimental data sets that are the subject of the study being reported, the results section should include control data, for example, the absence of any experimental manipulation, a control for the vehicle(s) used to dissolve a compound, a series of experimental data sets that are the subject of the study being reported and a reference standard(s) (positive and negative controls) to validate the study (Chapter 2).

4.6.1 The Dose/Concentration Response Relationship

Given its central role in biomedical research, the dose/concentration response relationship (hereafter referred to as the DR relationship or a DR curve - DRC) derived from the Law of Mass Action (Bylund and Toews, 2014; Kenakin, 2014; Rang, 2006) requires special consideration in defining what is an acceptable DR relationship and what is not. Ideally, when sufficient amounts of a soluble compound are available and, in the case of in vivo experiments, the compound is sufficiently bioavailable, a logarithmically based dose/concentration response relationship can be obtained (Fig. 4.2). Using a logarithmic scale, in addition to

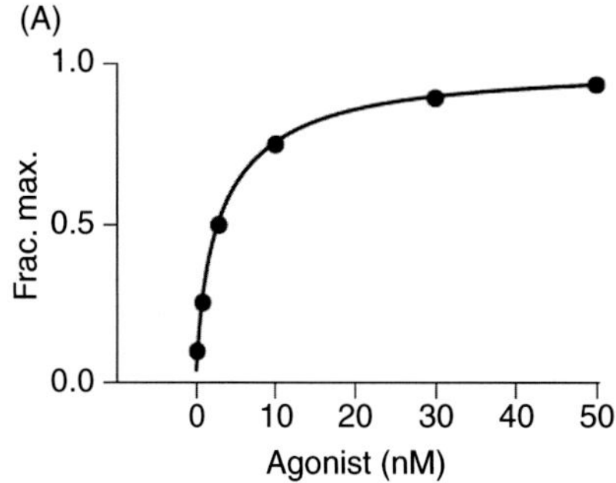

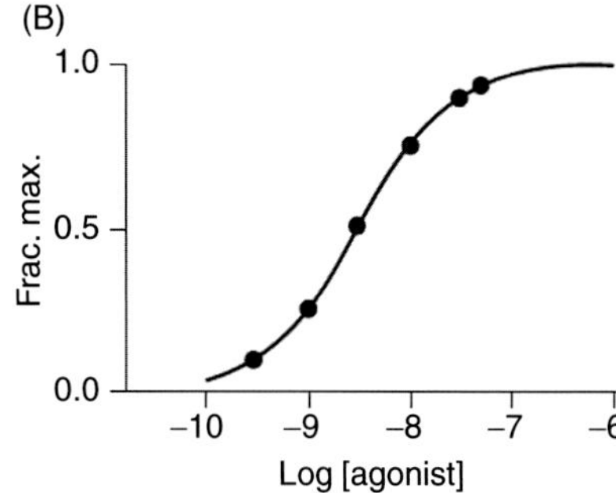

FIGURE 4.2 **Linear versus log dose response (DR) curve.** A dose or concentration response curve (DRC) can provide three key measures: (1) a graded, fractional dose/concentration response to an agonist that reaches a maximum; (2) the maximal number of receptors or response that the agonist can occupy or produce and; (3) a DRC that can be used to define the concentration of the agonist at which a half maximal effect designed as 0.5 or 50% occurs. This is the equilibrium dissociation constant of a compound-receptor complex (K_A) and is a measure of the compound affinity. The K_A value occurs at the midpoint of the curve. Panel A shows a linear DRC with a maximum of 30–50 nM agonist; panel B shows a semilogarithmic DRC with a maximum of approximately 5×10^{-7} M. Estimation of the maximum and the K_A value can be more accurately determined using the semilogarithmic DRC. *Source: Graph from Kenakin, T., 2014. A Pharmacology Primer: Techniques for More Effective and Strategic Drug Discovery, fourth ed. Elsevier Academic Press, San Diego CA, p. 14, used with permission.*

ensuring that the experimental data fit a normal Gaussian distribution to avoid skewing in the data analysis (see Kenakin, 2014; pp. 17–18) permits the evaluation of a broad range of compound concentrations within an experiment via the use of a logical dose/concentration progression. Moreover, it is appropriate as a response is almost always linearly proportional to log concentration over the 30%–70% maximum response range, as predicated by the Law of Mass Action.

In an in vitro assay, for example, a binding assay, biomarker, western blot, etc., a sample is run in triplicate to ensure technical replication within an experiment. This results in a single experimental value ($n = 1$) with a mean and SD for the triplicate observations. When a sample is run in triplicate this generates $n = 1$ not $n = 3$ as some investigators are prone to conclude. Subsequent, *independent* repetitions of the experimental manipulation provide additional independent values for each set of triplicates to provide the group measure.

When assessing the DRC for the effects of a compound at a minimal 6 doses/concentrations these can be measured using linear concentrations/doses, for example, 1–6, 8, 16, 32 nM/mg/kg or logarithmically at 1, 3, 10, 30, 100, 300 nM/mg/kg (Fig. 4.2). Thus the concentration/ dose range in the three examples would be 6 nM or mg/kg, 32 nM or mg/300 nM or mg/kg with the latter providing the broadest data coverage by which to accurately analyze the effects of the compound and derive an accurate IC_{50} or EC_{50} value. If the latter value is 20 nM, then the linear DRC has only two concentrations, one on each side of that bracket the IC_{50} or EC_{50} values. With the log DRC, there are three concentrations on each side of the IC_{50} or EC_{50} value that enhance the accuracy of the measurement.

DRCs should follow the law of mass action and must have a sigmoidal shape, where the response to a compound is related to the degree of occupancy, for example, dose-dependent and where parallel DRCs for mutliple agents indicate that they act at the same site. For further detail see Kenakin (2016). In many instances however, compound effects are often assessed only at a single concentration in an experiment or a DRC may often be generated but not analyzed to yield quantitative metrics, for example, IC_{50} or EC_{50} values, with the authors merely noting that a compound produced an effect over the relevant dose range. Both limit the value of the data. Responses should be compared on the linear portion of the DR curve—with the 50% value point chosen. It is inappropriate to draw a line through the 50% response and identify its intercept on the dose scale, but to calculate the variability in a drug/compound concentration response that produces a 50% response from independently analyzing multiple DR curves using appropriate curve fitting software, for example, Graph Pad. An improved approach is to generate the value of slope, maximum and EC_{50} value for each single DR curve making the estimate of slope more accurate, (Carpenter, 1986). The resultant estimates each have a standard deviation and can be subjected to statistical analysis (Curtis and Walker, 1986). This is illustrated in Fig. 4.3. If the effect of a compound, agonist, antagonist, or allosteric modulator, shows no DR relationship, the author must seriously consider that the effects reported are, for reasons unknown, not the result of the compound engaging a specific target site.

The absence of DR relationship for a compound (other than a well-established reference compound) in a paper, either because a DR evaluation has not been conducted but rather a single dose/concentration evaluated or if there is no apparent dose/concentration response effect (Fig. 4.4) often questions the premise of the study and is grounds for rejection of the manuscript.

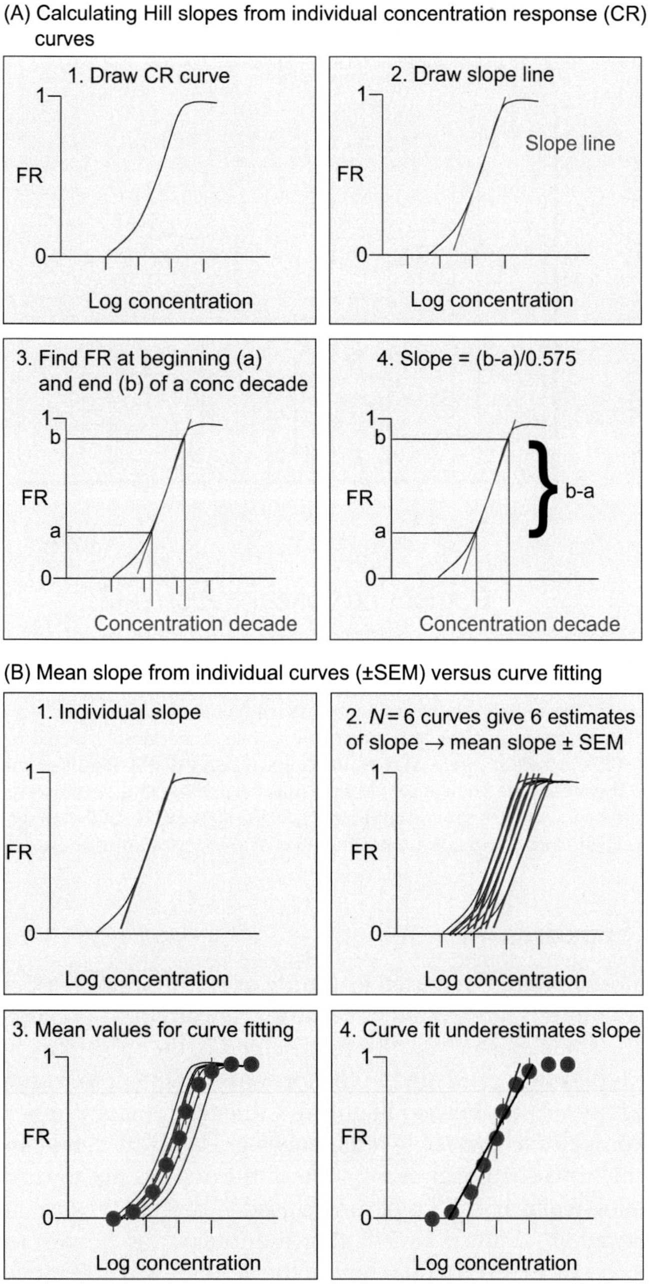

FIGURE 4.3 **Avoiding the underestimate that is inevitable by curve fitting.** (A) The slope of a concentration/dose response relationship defines cooperativity and for a simple 1:1 agonist:receptor interaction, the Hill slope will have a value of 1. The means to find the slope is shown graphically and stepwise in panels 1–4. (B) Individual Hill slopes (*red line* in panel 1) may be averaged and values used to compare concentration/dose response curves in the absence and presence of drug. In panel 2, the family of curves all have almost the same slope, so the mean slope will have a small SEM. They can also be analyzed to give mean and SEM values for the EC_{50} (Curtis and Walker, 1986). An inappropriate analysis is illustrated in panel 3 where the average response to each dose for the 6 curves is plotted. The average response has a slope (*black line* in panel 4) that is shallower than the true Hill slope (*red line* in panel 4, which is the average of the family of Hill slopes in panel 2). When individual curves are used to determine the slope and EC_{50} value this will generate an SEM for each (which is not possible with the averaged curve) and a true estimate of slope (which is underestimated if the mean response values are used) (after Carpenter, 1986).

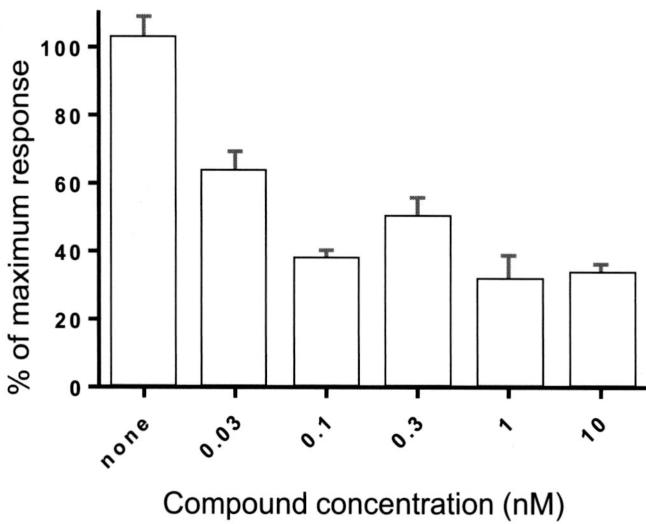

FIGURE 4.4 **Absence of a dose/concentration-response effect.** The figure shows a histogram of an actual data set for a receptor antagonist. This has been redrawn to remove the original detail but shows that the effects of a compound are inconsistent. At 0.03 nM of the antagonist, the system shows a 40% reduction in response. At 0.1 nM, there is a 60% decrease in response. However, increasing the antagonist concentration produced a mixed response. At 0.3 nM, the antagonist is less effective (50% reduction) while at concentrations of 1 and 10 nM, the effect is approximately the same as that seen with 0.1 nM. The mean data point at 0.3 nM may thus be an aberration. In any event, there is no robust response and the data set cannot be used to derive an accurate IC_{50} value for the effect of the antagonist. This figure may be contrasted with the agonist DRC shown in Fig. 4.2, when the effect of the agonist is truly dose/concentration dependent.

4.6.2 Reference Standards

If a known reference standard is not used in a study or fails to show a response consistent with that reported in the literature, there should be cause for concern that the experiment has not been validated (Chapter 2). While most accurately represented in the form of a DRC, data on a reference standard more typically is presented in the form of a single dose/concentration, the latter of which is or should be well-established literature values for potency/efficacy of the reference compound. When competitive reversible antagonists are used to define an agonist-mediated response, their activities present as rightward shifts in the dose/concentration response curve to yield an IC_{50} or Ki value while a Schild plot can be used to derive a pA_2 value (Kenakin, 2014).

Antagonism is optimally defined by whether the interaction is with the agonist binding site or allosteric site and by determining whether the antagonism is competitive, noncompetitive or uncompetitive with the nature of the binding (noncovalent versus covalent) determining whether the antagonism is irreversible or reversible (surmountable) (Page et al., 2006).

4.6.3 Figures and Tables

The decision how and whether to present data in the form of a figure or table depends on the complexity of the data and the number of comparisons. In a table the data can be represented with an absolute numerical value and error, for example, 4.727 ± 0.775 nM, whereas data in a figure requires a reader to make their own approximation of the error, for example,

4.5–4.9 ± 1.0 nM, which can confound initial impressions of the robustness of the data and consequently any differences based on this baseline. Nonetheless, visual presentation of results is highly recommended as it can simplify complex information and facilitate the identification of trends and patterns in the data.

The process of preparing figures has been considerably enabled in recent years by a number of widely available software programs, including those from Microsoft, for example, Excel and PowerPoint. These allow the facile and rapid generation of figures including graphs and histograms—some in 3D format—from a common data spreadsheet that can be modified in real time to select which version most clearly represents the data in conveying its import to the reader. While computer programs provide multiple color options that are visually appealing, care should be taken in selecting colors (Chapter 3). While the argument has been made for the use of dot plots as a more transparent and visually more informative substitute for histograms (Drummond and Vowler, 2011), these are rarely used.

4.6.3.1 *Figure and Table Legends*

An important mandate for both figures and tables is that they should be self-contained and comprehensible without the need to constantly refer back to the methods and results sections of an article to find key information like concentrations and doses, dosing route and regimens, etc. (Mullane et al., 2015). Thus the title and legend should inform the reader as to the content of a table or figure and allow them to fully comprehend what the data presentation reflects in terms of the experiments conducted. Legends should contain sufficient detail to allow the reader to understand what was done and how. This follows from the figure/table title that should describe the content of the figure/table, for example, "The effect of single versus repeat dosing of Compound X on human breast cancer xenografts." The legend would then provide additional top level detail on the actual experimental protocol (e.g., rats were treated with Compound X for 1 and 7 days before tumor growth was measured), the dose/concentrations of compound used that are specifically related to the data in the table or figure that should be clearly stated along with the route and frequency of administration (10 mg/kg po, bid), together with a key to any abbreviations used, indications of estimates of variability (means ± SD or SEM, confidence limits, error bars on individual data points), the outcomes of the statistical analysis and the test used with relevant *p*-values. This is a critical point for authors to appreciate since many experienced reviewers initially judge the merits of a manuscript by immediately reviewing the tables and figures with the text forming the basis of a secondary, more comprehensive evaluation of the results and conclusions where the author can guide the reviewer in his/her appreciation of the data. As already noted, if the figures and tables lack data and/or evidence of being reproduced, (e.g., are absent Confidence Intervals, SD or SEM) and/or a robust DR relationship, or are missing necessary controls, evidence of statistical analysis or a reference compound, this will immediately raise concerns, irrespective of how well the manuscript itself has been written. Increasingly, authors fail to recognize this and spend minimal time in crafting clear, concise, and complete figure/table legends. Instead they anticipate that the reviewer or reader will rely on the text for clarification that can add considerably to the complexity of the peer review process. While full methodological detail must be present in the Methods section, having the specific methodological detail in the legend avoids the reader having to continually refer back to the methods section, making the paper easier to read.

In a table, spaces that contain no data, a dash, or the abbreviations "NT" (not tested) or "ND" (not determined) will raise questions as does the inclusion of data that has been culled from other publications, either that previously published by the author or by another author often without any consideration as to whether the experimental conditions in the published study were at all relevant for comparison to the study being reported. The use of historical data sets to compare with data generated in the studies presented in a manuscript under review is inappropriate. Unless the data are derived from the same experiments under the same conditions, comparisons are invalid and undermine the conclusions of a study. This does not however preclude a section in the discussion section comparing the results obtained in the current study with those reported in other publications. Finally, there are instances where reference compounds are used (e.g., when a compound is used as a tool to probe a mechanism based on previous findings regarding its potency/efficacy and selectivity) but the justification for the concentration/dose is not given or where no detail is given anywhere in the manuscript on the concentration/doses used both of which are unacceptable.

4.6.3.2 Histograms Versus Graphs

The way in which a raw data set is graphically presented markedly impacts its ability to convey to the reader comparisons of different data sets that allow a full understanding of what was found in a study (Frankel and DePace, 2012; Tufte, 2006). More recently, for reasons that are not entirely clear, but may relate to the increased use of single compound concentrations/doses, there is a trend to use histograms rather than a graph to present a continuous data set, for example, a time course or a DR relationship. While histograms are normally used to represent a discontinuous (discrete) series of data, for example, percentage of individuals from a population with different blood types (Fig. 4.5), they can also be used to represent a continuous distribution by assembling common data conditions, for example, controls, manipulation A, manipulation B, etc., also into discrete "bins." As shown in Fig. 4.6, this can produce unwieldy representations that tend to obscure the changes observed with a graph providing a better format to convey the experimental changes. Fig. 4.3 illustrates that presentation of DR data using meaned values and a single average curve fit that will underestimate the true slope. Fig. 4.7 shows that the way in which the y-axis is scaled can grossly exaggerate the size (and even the presence) of a difference between groups.

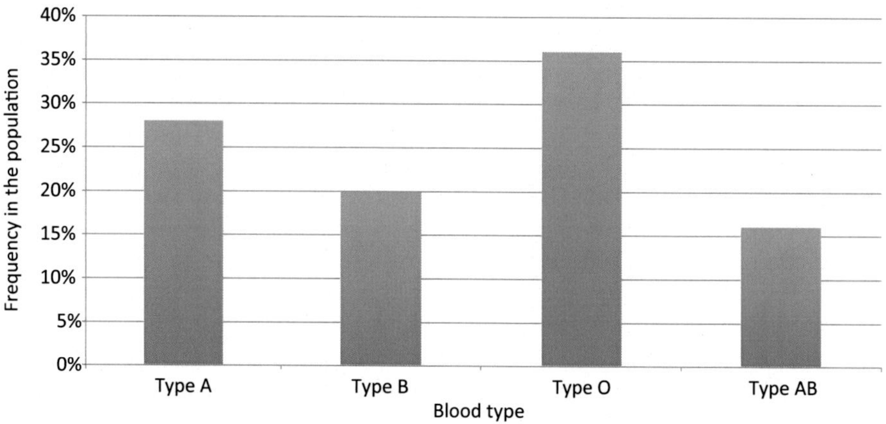

FIGURE 4.5 **Histogram bins of blood types.** This figure shows the utility of a histogram in presenting a discontinuous variable, the human blood type. Each column shows an independent variable.

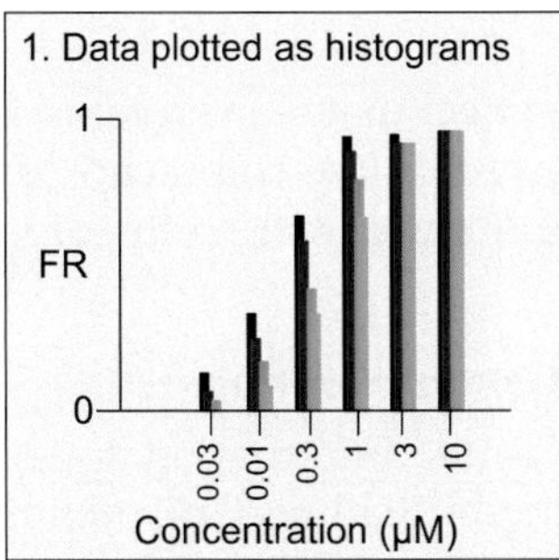

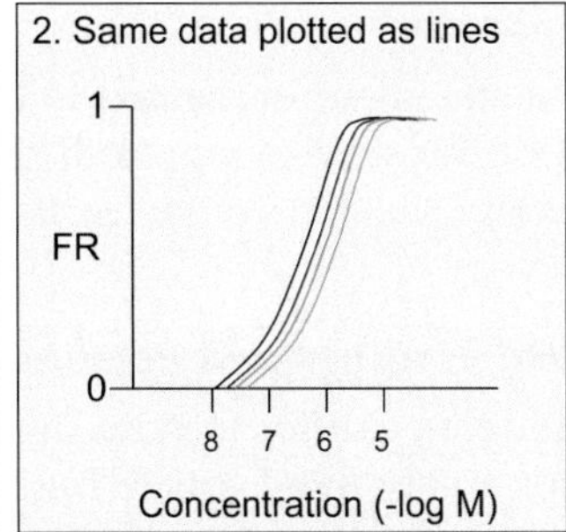

FIGURE 4.6 **Bar graphs versus line graphs.** Panel 1 shows four concentration response curves plotted as continuous histograms. Panel 2 shows the same data plotted as line graphs (with the individual data points not visible). The latter is much easier to interpret—the clarity is dependent in the way data are plotted. FR, Fractional response.

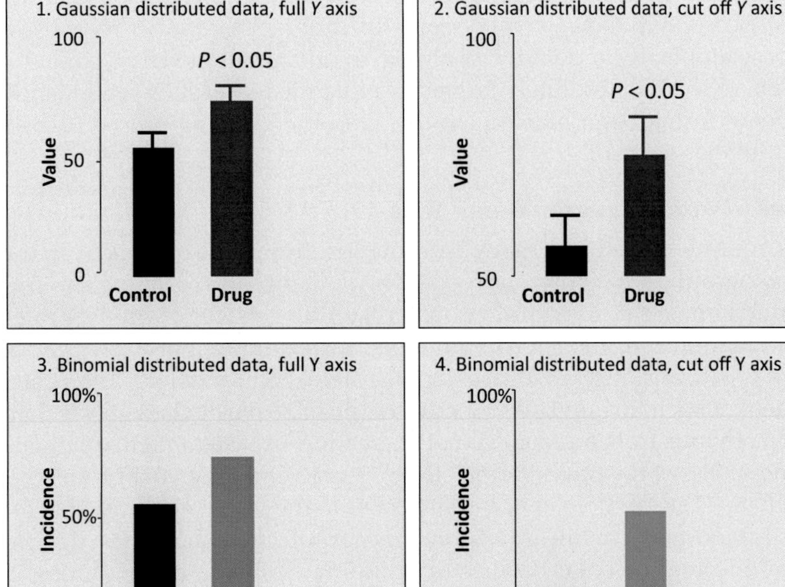

FIGURE 4.7 **Data presentation can influence the way it is perceived.** In Panels 1 and 3 on the left, data that are Gaussian and binomial distributed are presented correctly, using a full *y*-axis. On the right, in panels 2 and 4, the effect of drug appears much greater owing to the manipulated *y*-axis that does not originate at zero. The manipulation is most insidious with the binomial data since the difference appears huge in panel 4, yet the data here are not significantly different. This may allow the author to discuss the data as if there is a real difference without necessarily being challenged in peer review.

4.6.3.3 *Data Not Shown*

The inclusion of a statement in the text to the effect that something of interest did or did not happen as part of the studies reported followed by the comment, "data not shown" is anecdotal and lacks any substantive value. If the finding was worthy of mention it requires presentation.

4.6.3.4 *Unpublished Findings and Personal Communications*

While journal guidelines differ in how they handle information apparently pertinent to the subject matter of a submitted article but unpublished and/or preliminary data that is relevant to the study can be cited as a personal communication from another scientist or as unpublished findings from the authors.

4.6.3.5 *Semiquantitative Scoring—++ + + or 0–5*

Data can also be reported using empirical (+) or empirically numerical (0–5) scales where the occurrence of an event and its intensity/severity is expressed using a scale, either as part of a series of observational paradigms involving animal behaviors or in examining histological slides. The data points are qualitative being expressed as +, ++, +++ or by a number (0–5, 0–10) based on a previously described, often subjective, scale. The latter can be combined to yield an overall score that allows comparisons for different treatment responses within an experiment, for example, for a putative analgesic in an animal model of pain, 0 = no effect; 5 = maximal pain relief. It is absolutely imperative that the use of such semiquantitative scoring methods be performed in a blinded manner, and the blinding procedure be clearly described in the Methods section.

4.6.3.6 *Images—Photomicrographs and Gels*

The images routinely used to convey information from an experiment frequently take the form of photographs—the latter derived from a microscope or animals or from the processing of scanned images derived from polyacrylamide gels, for example, protein staining or radioactive tagging, cardinal coordinate blots, for example, northern (RNA), southern (DNA), eastern (posttranslational), or western (protein) immunoblots. The postexperimentation adjustment of such images to enhance or emphasize particular aspects that might not be immediately obvious from a casual visual inspection—to clean them up or enhance "the clarity and conciseness of the presentation" (Marcus and Oransky, 2012) is an accepted practice provided that—(1) the whole image is subjected to the same manipulation, for example, that the image is modified in a linear fashion, for example, by enhancing brightness, adjusting contrast, or altering the color balance or cropping, resizing, or combining images;and (2) it is made clear that the images have been manipulated (Blatt and Martin, 2013; Marcus and Oransky 2012; Neill, 2006; Newman, 2013; Rossner, 2006; Rossner and Yamada, 2004).

Blot quantification (e.g., westerns) requires normalization to housekeeping proteins, such as actin or GAPDH (glyceraldehyde 3-phosphate dehydrogenase) on the assumption that these levels are quite stable and constant across most cell types. However, such proteins may not always be appropriate controls, depending on the setting and questions being asked. For example, if studying effects on a mitochondrial protein, GAPDH is a cytoplasmic protein and a mitochondrial house-keeping protein, such as VDAC1 (voltage-dependent anion channel 1) may be more relevant, necessitating the consideration of relevant controls for the proteins under study. However, the normalization process is complex, and in most papers, is not explained in terms of the why and how of what was done. Data can be expressed as "fold control" with no indication of how the selected housekeeping protein was used in the normalization process while there is rarely a separate protein control for loading—the amount of sample pipetted onto the gel.

The duration of blot exposure is a balance between ensuring that weak bands can be detected while not allowing strong bands to saturate and obscure or run into weaker bands increasing variability. Measuring the gradation of blot intensity quantitatively is a challenging task. While available software (ImageJ, Image Studio, etc.) allows investigators to select the part of the blot to be "quantitated," this may include greater or lesser amounts of "background" at the discretion of the author. The readout can be average "intensity" or maximum "intensity" (the darkest part). Even when each band is pristine in geometry it is hard to judge where the blot starts and ends. When analysis is not blinded and the investigator is naïve or mendacious, it is possible to generate quantitative readout to fit any requirement, when the intensity of the blot is quantitated.

Image quantitation is not always obvious with either a trend to report western blots in the form of a "representative data sample," for example, an n value of 1, or to report very small SEM values (sometimes as low as 0.01% of the mean rather than the biological variability seen in independent reproductions which rarely is as low as even 1% of the mean) with naïve declarations by the authors that the blot data are the averages of single samples run several times (e.g., "in triplicate"), replicating rather than reproducing the data for an n value of 1, not 3. A subsequent power analysis will then show that $n = 3$ will allow detection of significance at $P < 0.05$ using a t test leading to psuedoreplication (Hulbert, 1984), an inevitable consequence of the pseudovariance (which is, in truth, close to zero variance, as one would expect from a single set of samples run 3 times). The t test thus can become meaningless with $n = 3$ replicates.

Ideally, reporting of western plot images mandates randomization and blinding of the lanes and fully transparent reporting of the quantitation process.

4.6.3.6.1 INAPPROPRIATE IMAGE MANIPULATION

While it has always been possible to manipulate images to misrepresent/"improve" the original data, this process has become easier with the advent of digital imaging and the facile ability of an investigator to alter an image using software like Adobe Photoshop (Neill, 2006). This allows: deleting a band from a gel blot scan, the author having decided that it is irrelevant; adding or duplicating a band because there was an error in conducting the actual experiment; selectively adjusting/enhancing the intensity of a single band; using historical loading control data; adjusting the contrast to drop the gel blot background which may also delete faint bands deemed by the author to be of no interest; splicing lanes together without inserting lines or a gap in the image to indicate that this has been done; obscuring lanes by pasting other images over them; pseudocoloring of micrographs; altering brightness or contrast in fluorescence micrographs; altering a microscope field by combining field images; and altering image resolution (Abbott, 2013; Blatt and Martin, 2013; Neill, 2006; Rossner and Yamada, 2004).

In a metaanalysis that reported that of some 20,000 papers published in 40 scientific journals over the period 1995–2014, 4% had problematic figures, of which 2% were suggestive of

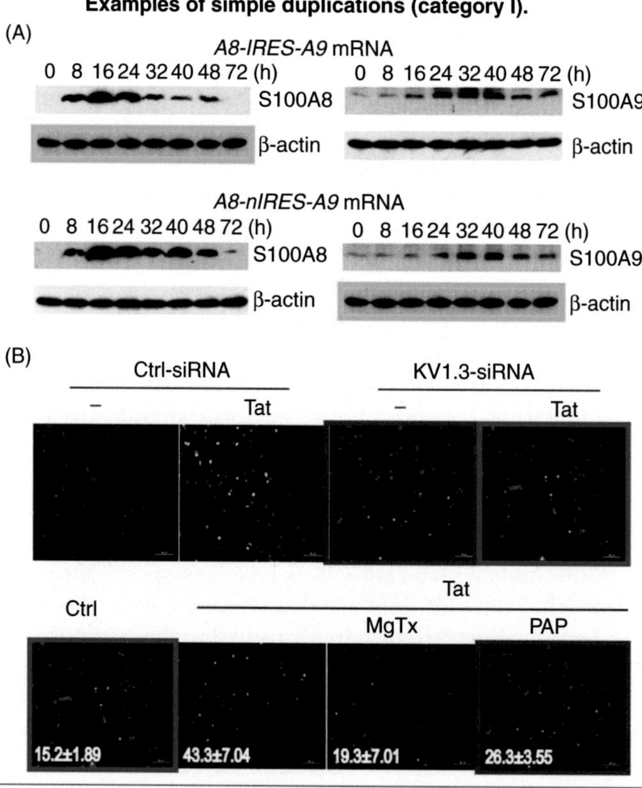

FIGURE 4.8 **Image manipulation—simple duplication.** (A) The beta-actin control panel in the top left is identical to the panel in the bottom right *(green boxes)*, although each panel represents a different experimental condition. This figure appeared in Zou et al. (2013) and was corrected in Zou et al. (2015). (B) Panels shown derived from two different figures within the same paper Liu et al. (2013) corrected in PLoS One Staff (2014). Two of the top panels appear identical to two of the bottom panels, but they represent different experimental conditions *(red and blue boxes)*. All duplications might have been caused by honest errors during assembly of the figures. *Source: Part A: Reproduced with permission from the publisher; Part B: From Bik, E.M., et al., 2016. The prevalence of inappropriate image duplication in biomedical research publications. mBio 7, e00809, figure reproduced under the Creative Commons [CC BY] license.*

deliberate manipulation. Bik et al. (2016) classified inappropriate image duplication into three types : Category 1 (Fig. 4.8), *simple duplication* with figures containing two or more identical panels, for example, βactin loading controls, either in the same figure or between different figures in the same paper, that were designated as representing different experimental conditions; Category II (Fig. 4.9), *duplication with repositioning* where blot images had a clear area of overlap, and one image was shifted, rotated, or reversed with respect to the other and; Category III (Fig. 4.10), *duplication with alteration* that included altered images with complete or partial duplicated lanes, bands, or cell groups. This sometimes involved rotation or reversal with respect to one other, within the same image panel or between panels or figures.

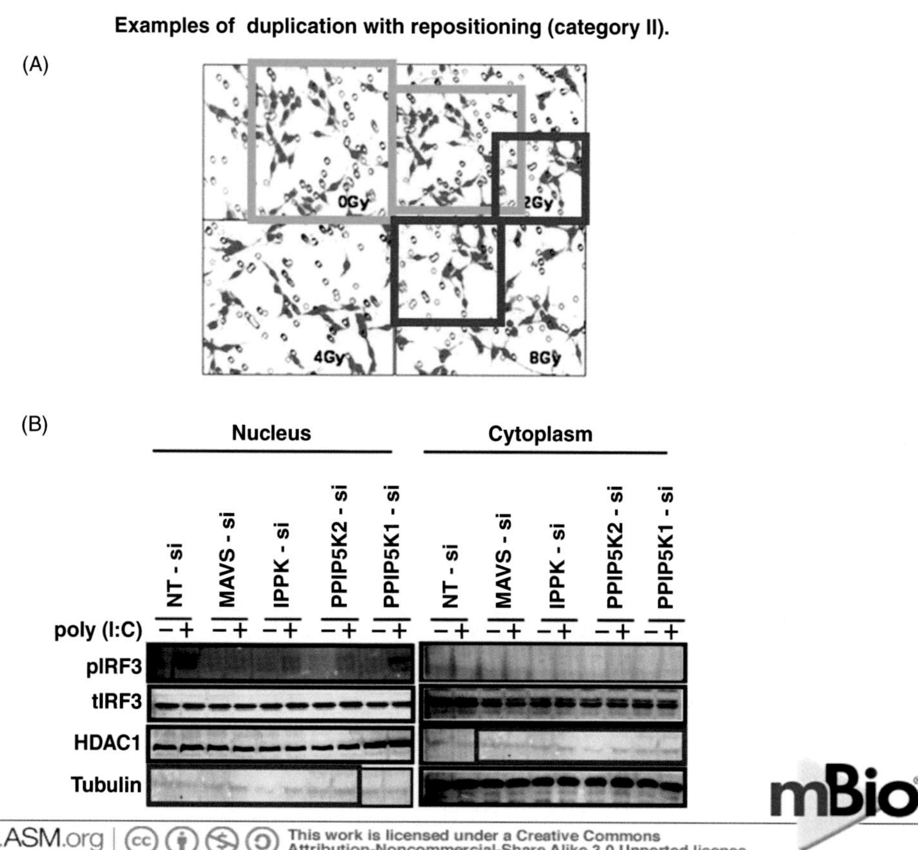

FIGURE 4.9 **Image manipulation—duplication with repositioning.** (A) Although the panels represent four different experimental conditions, three of the four panels appear to show a region of overlap (green and blue boxes), suggesting that these photographs were actually obtained from the same specimen. These panels originally appeared in Liu et al. (2014) and were corrected in reference Liu et al. (2015). (B) Western blot panels that purportedly depict different proteins and cellular fractions, but the blots appear very similar, albeit shifted by two lanes *(red boxes)*. Panels originally appeared in Pulloor et al. (2014), and were corrected by PLoS Pathogens Staff (2014). *Source: Figures in both panels were reproduced under the Creative Commons [CC BY] license, From Bik, E.M., et al., 2016. The prevalence of inappropriate image duplication in biomedical research publications. mBio 7, e00809.*

Duplication in this category also included figures with evidence of: (1) *stamping* where a defined area was duplicated multiple times within the same image; (2) *patching* where part of an image was obscured by an area with a different background; and (3) fluorescence-activated cell sorting (FACS) images sharing conserved regions and other regions in which some data points had been added or removed. A previous screen of 120 articles in three cancer journals by Oksvold (2016) found that 30 of these (25%) showed data duplication. On contacting the authors of these articles, only one case was clarified with supporting data being supplied. When Oksvold contacted the editorial offices of the journals included in this study none responded.

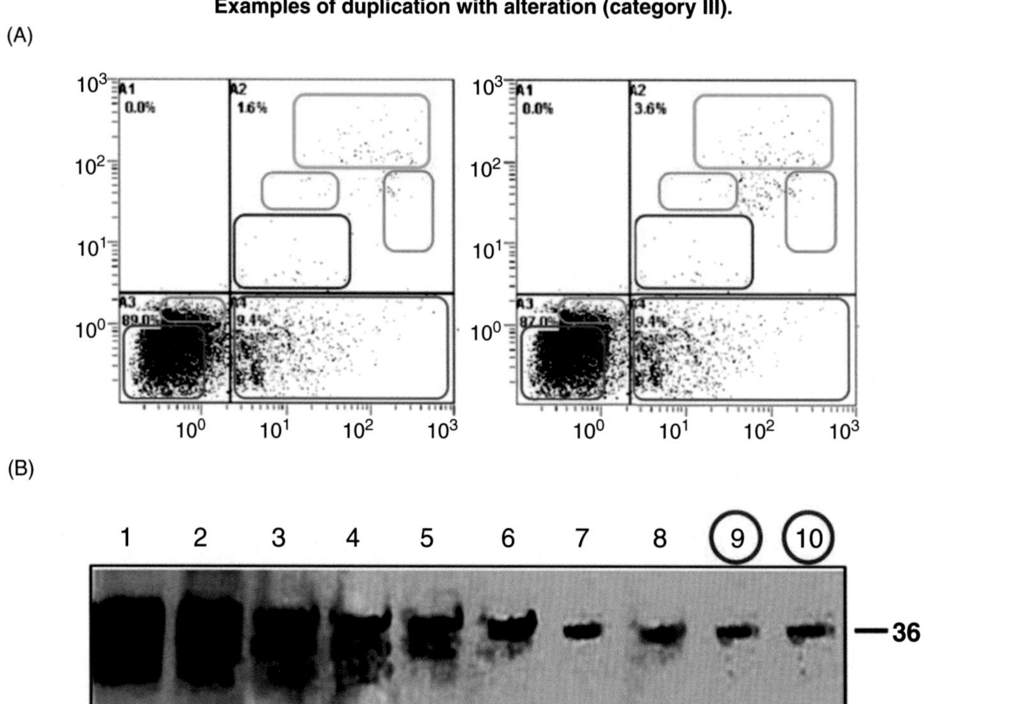

FIGURE 4.10 **Image manipulation—duplication with alteration.** The (A) and (B) FACS panels represent different experimental conditions and show different percentages of cell subsets, but regions of identity *(colored boxes)* between the panels suggest that the images have been altered. [This illustration originally appeared in Xu et al. (2012); the article was retracted in Xu et al. (2015)]. (B) displays Western blotting results for 10 different protein fractions isolated from a density gradient. The figure appears to show a single blot, but the last two lanes *(red circles)* appear to contain an identical band. Exposure was altered to bring out details in Friedman-Levi et al. (2013) the figure was corrected in Friedman-Levi et al. (2015). *Source: Reproduced with permission from the publisher. From Bik, E.M., et al., 2016. The prevalence of inappropriate image duplication in biomedical research publications. mBio 7, e00809, figure reproduced under the Creative Commons [CC BY] license.*

With the trend to report image data in the form of "representative data," for example, an n value of 1, and the nonstop detection of rampant image manipulation, multiple retractions in the areas of stem cell (Kennedy, 2006; Nature, 2014), cardiovascular (Oransky, 2011; Shervick, 2015), and cancer (Zimmer, 2012) research have occurred. As a result there has been a major effort to improve guidelines for the use of gel blot images in the biomedical literature (Newman, 2013). This includes ensuring the inclusion of real time, for example, simultaneous, rather than historical controls; quantitation of bands with replication to give an n value of *at least* 3, the latter of which are normalized to reference protein or housekeeping genes within the individual experiments; controls for antibody specificity (Chapter 2) in cardinal coordinate immunoblots; defined quantitative statistical analysis with conventional means and estimates of error and p-values; the description of all modifications in digital images submitted for peer review along with the provision of raw image data and associated instrument settings to the journal on submission of a paper; and the use of forensic detection tools from the NIH's Office of Research Integrity (https://ori.hhs.gov/advanced-forensic-actions) and forensic droplets (Newman, 2013) that can derive a specific history of image manipulation. The problem of fraudulent image manipulation is of sufficient magnitude that Mike Rossner, a former managing editor of the *Journal of Cell Biology* has started a company to screen for data manipulation in submitted manuscripts (Couzin-Frankel, 2016).

4.6.3.8 Images—Heat Maps

Output from omics based disciplines where changes in multiple cellular pools of biological molecules, for example, the genome, transcriptome, proteome, metabolome, interactome, miRNAome, etc., lead to the creation of extremely large data sets. The simultaneous generation of many hundreds to thousands of individual data points are presented as heat maps (Deng et al., 2014; Eisen et al., 1998; Tarca et al., 2006; Wilkinson and Friendly, 2009). These are complex, multicolored figures produced using statistical algorithms to arrange genes, proteins, etc., according to similarities in their expression pattern. Their display in the form of a heat map provides a means to simultaneously illustrate their clustering and associated expression in a pseudo-quantitative manner to show differences in analyte expression as differences in the size or, more often, intensity in color of individual biological molecules. Thus heat maps can be generated to compare and contrast expression patterns for different experimental conditions, for example, differences in metabolomic profile between a smoker and nonsmoker (Spitale et al., 2013; Fig. 4.11.)

Since heat maps do not depict variations in the quantitated values, the reproducibility of discrete components of the data is usually unknown since the color depicted represents the average of the usual three determinations. Consequently, only overall trends and patterns can be gleaned, and interest in select expression profiles requires follow up with more robust quantitative methods.

While the reproducibility and manipulation of heat maps do not yet appear to date to have been a significant issue in the literature, certain of the aspects involved in their generation are similar to those that have plagued gel and micrographic images. Since the scope and complexity of the data involved in creating a heat map necessarily involves replication/iteration/integration experiments that occur over time periods from days to months, these often result in "nonbiological experimental variation" also known as batch effects that require the use of specialized software (Johnson et al., 2007) to allow data sets to be combined. Such software and the use of Adobe Illustrator in the context of heat map data presentation may eventually reach similar levels of concern as those for gel imaging.

4.6.4 Supplemental Materials

As discussed, many high profile, high impact journals had space considerations due to the large number of papers they received for publication and those that they published leading to

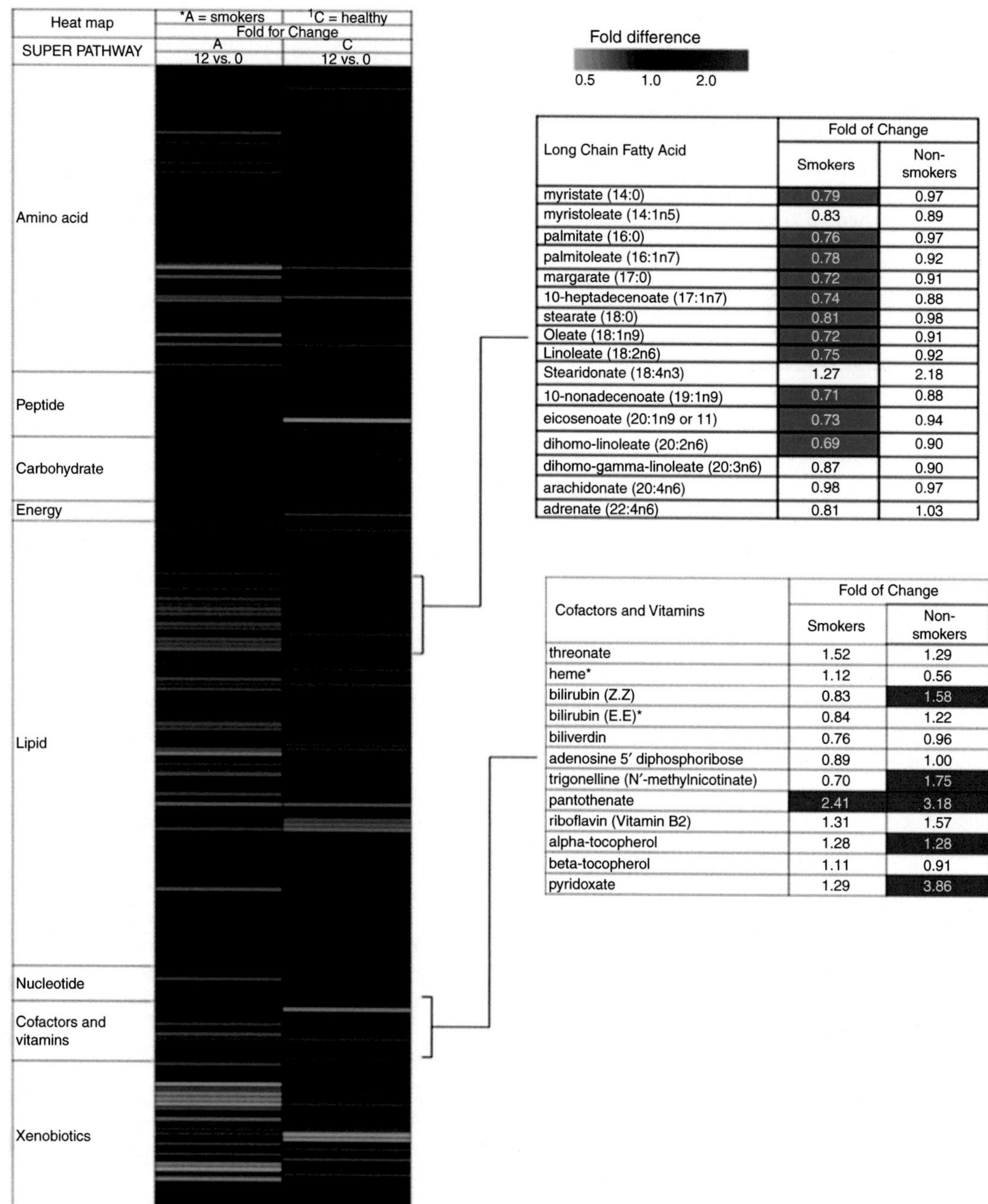

Long Chain Fatty Acid	Fold of Change	
	Smokers	Non-smokers
myristate (14:0)	0.79	0.97
myristoleate (14:1n5)	0.83	0.89
palmitate (16:0)	0.76	0.97
palmitoleate (16:1n7)	0.78	0.92
margarate (17:0)	0.72	0.91
10-heptadecenoate (17:1n7)	0.74	0.88
stearate (18:0)	0.81	0.98
Oleate (18:1n9)	0.72	0.91
Linoleate (18:2n6)	0.75	0.92
Stearidonate (18:4n3)	1.27	2.18
10-nonadecenoate (19:1n9)	0.71	0.88
eicosenoate (20:1n9 or 11)	0.73	0.94
dihomo-linoleate (20:2n6)	0.69	0.90
dihomo-gamma-linoleate (20:3n6)	0.87	0.90
arachidonate (20:4n6)	0.98	0.97
adrenate (22:4n6)	0.81	1.03

Cofactors and Vitamins	Fold of Change	
	Smokers	Non-smokers
threonate	1.52	1.29
heme*	1.12	0.56
bilirubin (Z.Z)	0.83	1.58
bilirubin (E.E)*	0.84	1.22
biliverdin	0.76	0.96
adenosine 5′ diphosphoribose	0.89	1.00
trigonelline (N′-methylnicotinate)	0.70	1.75
pantothenate	2.41	3.18
riboflavin (Vitamin B2)	1.31	1.57
alpha-tocopherol	1.28	1.28
beta-tocopherol	1.11	0.91
pyridoxate	1.29	3.86

FIGURE 4.11 **Heat maps demonstrating fold change of metabolites in smokers and nonsmokers.** The heat maps show that smokers had significantly decreased levels of (long chain fatty acids) but not the cofactors and vitamins listed later, after 12 weeks of study supplementation. *Green band,* decrease; *red,* increase. *Source: From Spitale, R.C., Cheng, M.Y., Chun, K.A., Gorell, E.S., Munoz, CA1, Dale, G., Kern, D.G., et al., 2013. Differential effects of dietary supplements on metabolomic profile of smokers versus non-smokers. Genome Med. 4, 14, reproduced under the Creative Commons [CC BY] license.*

the establishment of limits for the number of printed pages, figures, and tables. This required that materials deemed to be of less interest to their mainstream readership be published as supplemental materials (Collins and Tabak, 2014) As examples, the *Proceedings of the National Academy of Sciences of the United States of America* (PNAS) and the *Journal of Biological Chemistry* (JBC) published 18,799 and 31,198 pages, respectively in 2015 (interestingly down 13% (21,552) and 25% (41,568) from 2012). While many journals have moved to online publication only, thus removing the physical limitations to page content, the virtual pages published online still require resources for their production leaving many journals still restricting page count and/or requiring supplemental materials.

Extent is about the size of the methods section, additional figures and tables and/or video files which typically have their own distinct reference section. *Inclusion* ensures that despite page constraints all the materials required by more specialist readers of a journal are available, usually as an online or archived file. While this streamlines the paper and its take home message without diminishing access to the associated materials thus meeting the needs of the publisher, it also encourages the "verbosity" of some authors who tend to publish anything and everything associated with their studies, failing to be concise in communicating their message. What insights, for instance, can the reader infer from an author-uploaded photograph or video?

A major concern with supplemental materials is what criteria should and are used to relegate data to the supplemental section from the main text and whether this section undergoes any form of peer review (see Chapter 5). Pop and Salzberg (2015) cite two genetics papers where the supplementary materials exceeded by orders of magnitude the page count for the actual article. The paper by Werren et al. (2010) was 6 pages in length and was accompanied by 165 pages of supplemental materials that included 210 references. Another by Meyer et al. (2012) was 5 pages in length with 144 pages of supplemental materials.

Confirming the jaundiced view of peer review of supplemental material, in some instances, material alluded to in the main text as being included in the supplemental materials is often absent either through oversight or intent. In other instances, key data that are essential to validating the materials and methods, the experimental design, etc., are in a supplemental materials section that is not automatically linked to the paper. This makes reading the paper challenging especially when access to these materials is behind a pay wall as a pdf. Pop and Salzberg (2015) reinforce the long-standing concerns as to whether supplemental materials undergo any peer review (or are even read by editors and reviewers) and whether the references included with them are citable.

Finally, there are few, if any guidelines for how supplemental materials are collated and presented, leaving the reader to fend for themselves in interrogating the materials if they have the time and patience to do this. Some journals do not allow supplemental materials because of concerns regarding the arbitrary nature in selecting the data to be relegated to a separate section (Mullane et al., 2015). If the data are necessary to provide a coherent narrative to support the conclusions, what objective criteria are used to place certain data sets and details in a separate section of the submission? None.

Finally, it should be noted that these concerns are distinct from situations involving massive data archives as in the case of GWAS (genome wide association studies), which are included as supplemental materials as part of an open data mandate (Chapter 5.3.2.6).

4.7 DISCUSSION/CONCLUSIONS

The discussion and/or conclusions section in a manuscript follows directly from the reported results and provides the author the opportunity to logically and sequentially recapitulate the key data with their insights and interpretation as to its implications for the hypothesis being tested and the research field in general. The discussion must focus on placing the actual data reported in the paper in the context of the field as a whole and should not merely recapitulate the results. It should also avoid becoming a mini review that contains material that while relevant is tangential to the actual data being discussed. For example, an extensive discussion of the pathways involved in the functional effects of the receptor studied in the paper would be irrelevant unless the findings reported specifically address an aspect of signaling functionality.

Following a brief one or two sentence summary of the initial objectives, the top-level findings should be presented with a brief discussion as to how these reflect/extend/refute/confuse these objectives. This section should then segue into a more detailed elaboration on the actual results, placing these in context of other findings reported in the published literature discussing differences and areas that require resolution remembering that the data reported provide the framework for "testable hypotheses and observed associations, rather than rigorous proof of cause and effect" (Chipperfield et al., 2010). But the bottom line is simple, the author must tell the reader what he or she believes the data means.

If there are limitations to the study in terms of methodology, the statistical analyses and powering (see Chapters 2 and 3) or unexplained outcomes, these should be highlighted by the authors in their discussion. In citing the literature, care should be taken to include any substantive work that contradicts the findings, with an attempt to reconcile differences. Finally, in summarizing the overall conclusions from the study, these should not be generalized or used as an opportunity to speculate and overstate the interpretation of the data, for example, using a data set in a cell line to suggest that a compound is the next best drug to treat cancer or neurodegeneration. Such in vitro studies are overtly reductionist with some authors seemingly unaware of the cell/tissue/animal hierarchy in biomedical research (Kenakin, 2014). Fig. 4.12 highlights the levels of interrogation of receptors/drug targets. Historically, the elucidation of disease mechanisms and compound evaluation took place using animals and animal tissues. With the advent of cloning and expression, human receptors could be transfected into surrogate, often immortalized cell lines, both animal and human. For the data to have validity, compounds and putative disease targets must be evaluated at human receptors from diseased tissues to avoid conclusions being made using data from surrogate cells, where signaling mechanisms may be engaged (or not) that are not present in the diseased human tissue or are artifactually engaged due to overexpression of the target (Kenakin, 2014).

4.7.1 Overstating Research Findings

In developing a hypothesis, there is often a hierarchy of data generation, each step of which moves progressively closer toward the evaluation of a disease mechanism or of a novel compound modulating the mechanism in humans. Initial testing routinely occurs in animal cells transfected with the human target of interest moving into human cell lines similarly transfected with the human target. There are often major issues with the stoichiometric relationship of a transfected target with the cellular signaling repertoire of the host cell and also in the nature

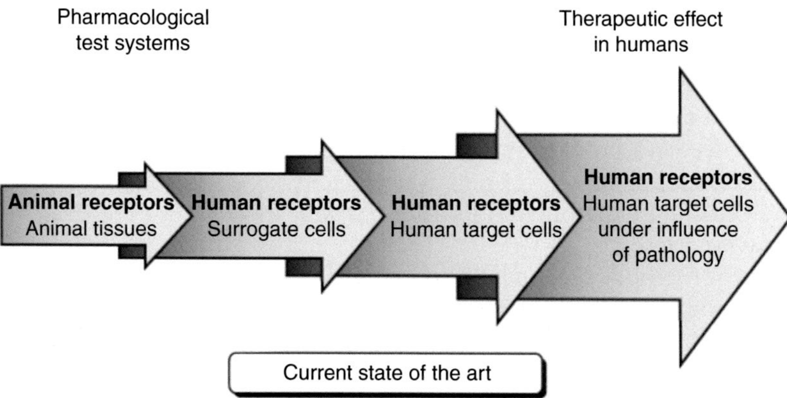

FIGURE 4.12 **Hierarchical evaluation of disease mechanisms and compound evaluation.** Historically research in pharmacology has been depended on studies with animals and animal tissues. With the advent of drug target cloning and expression in the 1980s, human receptors could be abundantly expressed in surrogate cells and in human cell lines. This could often lead to artifacts where pathways not present in a particular type of human cell could be engaged by the expressed target or target/receptor overexpression altered the stoichiometry of molecular interactions in the cell membrane and signaling pathways. Irrespective of a compelling finding that target X unexpectedly engaged signaling pathway A in a surrogate cell line, the validity of this effect needed to be established both in human cells and in diseased human tissue. *Source: Figure from Kenakin, T., 2014. A Pharmacology Primer: Techniques for More Effective and Strategic Drug Discovery, fourth ed. Elsevier Academic Press, San Diego, CA, p. 6, used with permission.*

of the signaling response. This often results in papers that announce that "receptor X interacts with signaling pathway Y" a finding that may be an artifact of the transfected target/host cell combination rather than a reflection of a unique cellular phenotype that reflects the native target/signaling relationship. For this reason, a hypothesis/compound should mandatorily be tested in native as well as transfected cell lines, animal and/or human and human tissues derived from patients before evaluation in appropriate animal models (McGonigle and Ruggeri, 2014; Kenakin, 2014).

As a result of these limitations, describing an interesting new lead compound as a novel treatment for a given disease state is hyperbole in the absence of the compound being sufficiently well characterized to qualify it as a clinical candidate. A compound that interacts with micromolar potency at a known drug target in a transfected cell line is rarely, if ever, a potential drug candidate that is ready to be tested in humans. Similarly, finding that the transfected human μ-opioid receptor forms a heterodimer with the transfected serotonin 5-HT_6 receptor in the immortalized rat adrenal pheochromocytoma PC12 cell line is a long way from implicating this molecular association with the etiology of a human psychiatric disease or for a physiologically-relevant effect of 5-HT on pain modulation.

A final point in the context of baseless hyperbole is that a target is rarely validated as a result of the findings reported in a series of preclinical research papers. Rather, the validation of a putative drug target can only occur when a compound with appropriate potency, target selectivity and drug-like properties is administered to patients with the disease and shows beneficial effects versus a control group. The data generated up this point serves to enhance confidence that the drug target and compounds that interact with it might have value, but this is not validation (Kopec et al., 2005).

4.8 MATERIALS AND METHODS

The methods and materials section should contain a detailed and accurate account of precisely how the data included in the submitted paper was both obtained and analyzed. This section therefore needs to provide sufficient detail to allow another researcher, one "skilled in the art," an individual with appropriate training and experience, to both gain an in-depth understanding of actual experimental procedures and, most importantly, be able to easily repeat the study reported.

Traditionally, the Methods section in the majority of journals appears after the introduction immediately preceding the results, a natural segue to set the stage for the results and conclusions. However, as noted, in some of the more high profile journals, the Methods are relegated to a position after the conclusions and before the references. In this position the section is also often truncated or rendered in a smaller font with much of the actual detail being relegated to the supplementary section, introducing additional hurdles to accessing this critical information. An additional outcome to relegating the methods section to a less prominent position is that it may become inaccurate and incomplete (Collins and Tabak, 2014; Nature 2012). In some instances, this is a reflection of carelessness, in others, a deliberate strategy to prevent other researchers from being able to repeat the study to verify the findings. This can represent a patent protection strategy originating from a biotech where publication of new data is necessary to maintain visibility to maintain investor interest in the activities of the company but where key information (a step in a procedure, the structure of a key compound) is omitted or misrepresented to avoid revealing details of a patent application before its publication. When challenged, authors who subscribe to this approach may not comprehend the ethical aspects of what they considered a standard operating procedure. A figure illustrating the treatment regimen(s) in a study can often more easily convey how a complex study was conducted than a paragraph or two of text. However, assigning codes to the control and treatment groups, for example, Groups I–V and using this designation throughout the remainder of the manuscript, including the figures and tables, can make for difficult reading. If Group I is the control group it should be designated as such and Group II etc. as the experimental manipulations used, for example, Group II = 1 mg/kg ip of Compound X, Group II = 3 mg/kg ip of Compound X, etc.

4.8.1 Materials

This section should begin with the details and a list of the sources of all the materials, reagents and instrumentation used in the study with their origin [e.g., ATCC (American Type Culture Collection), Manassas, VA; Millipore Sigma, St. Louis, MO; Cell Signaling, Danvers, MA; Dr. A.J. Clark, University College London; drug company contact, colleague etc.] or the distributor for a particular geographical locale being provided, for example, for ATCC in France, the current distributor is LGC Standards, S.a.r.l., in Molsheim.

When experimental compounds are used, a published reference should be given to their synthesis and structure which, if novel, should be included as a figure to avoid confusion in interpretation of the chemical name. While exceptions remain, few journals will review a paper without all the structures of the compounds reported being known although there are some "minimum threshold" chemical journals that allow authors to deliberately obfuscate the structure by including a generic chemical formula where the substituents, R1, R2, etc., for the compound of interest are not clearly identified. Similarly, citing a patent where neither the structure nor the full route of synthesis is clearly identified is another form of investigator obfuscation.

For natural products, a topic of increasing interest in light of ongoing efforts to exploit the Chinese *Materia Medica* using 21st century pharmacological approaches to identify new leads/drugs (Bian et al., 2012; Sucher, 2013), the isolation of active ingredients from complex biological mixtures is a high priority. This requires a clear description of the source of the original material, detail on its characterization and the preparation, composition, consistency, and relative ratios of known active ingredients. Thus, in order to facilitate the replication of the isolation of active ingredients, it is mandatory that precise details are provided on the origins, including taxonomy and geographical location, of a plant, venom, soil sample, etc., together with the molecular composition and the proportions of the constituents in various extracts as well as the structure and amounts of any isolated ingredients that are deemed to be active entities.

4.8.1.1 Authentication/Validation of Materials

Authors should indicate whether all the materials used in an experiment were assessed for purity and authenticity. While this not yet a mandatory condition for manuscript submission, in all the likelihood it will become so. With the exception of chemical standards like sodium chloride etc., this includes the chemicals received from a colleague or commercial supplier that have been on the shelf for months or years, cell lines (Almeida et al., 2016; Geraghty et al., 2014; Neimark, 2015), antibodies (Bordeaux et al., 2010; Roncador et al., 2016; Uhlen et al., 2016), RNA interference (RNAi; Fellmann and Lowe, 2014; Kaelin, 2012) including siRNA (http://www.abcam.com/protocols/recommended-checks-and-controls-for-sirna-experiments) and animals, especially transgenics (Jones-Bolin, 2015; NRC, 2011). Detailed background on authentication/validation and experimental guidelines can be found in Chapter 2.2.4.1 and Chapter 5.10. While authentication/validation is often treated as optional, a failure to ensure that the reagents being used are what they are thought to be can have major impact on experimental outcomes (Freedman et al., 2015) and may contribute to discrepancies in results between different laboratories including the failure to replicate studies. As a result, NIH grant applications require descriptions of "methods to ensure the identity and validity of key biological and/or chemical resources used in the proposed studies" (Lauer, 2016).

4.8.1.2 Vehicles, Controls, Doses, and Concentrations

It is also critical to report the vehicles used for the various compounds used in a study (which may differ depending on the solubility of individual compounds) as well as their final concentrations of the solutions used in vitro and to administer compound solutions in vivo. The latter should also indicate the volumes used and the route and frequency of administration. Final doses and concentrations of compounds and any anesthetic used in vivo should be clearly stated in the text.

4.8.1.3 Units and Abbreviations

The units used for concentration and for animal dosing should follow standard IUPHAR guidelines (Neubig et al., 2003) with in vitro concentrations being expressed in molar units (not mg/mL) and in vivo as micrograms (μg) or milligrams per kilograms (mg/kg) with the route and frequency of administration clearly stated, for example, mg/kg/po bid. The form in which the units are expressed (e.g., mg/kg or $mg.kg^{-1}$) varies for individual journals. A failure to routinely qualify the route in the abstract, methods, text, figure legends, and tables limits the information being conveyed and adds an additional hurdle to ease in reading and to reproducing a study.

4.8.2 Methods

The methods should describe in clear and simple detail how the experiments were conducted including full information on the design of the experiment, powering, group numbers and treatments, randomization, blinding, outcomes, data normalization, exclusion criteria, and statistical analysis (Curtis et al., 2015).

4.9 ABSTRACT

While secondary in content to the results and discussion section, the article abstract assumes greater importance as the marque by which interest in an article is generated as it is often the only information that the casual reader will see. For the majority of researchers, and certainly the mainstream scientific media, this will be the main opportunity to judge whether to access and read the actual manuscript, a task which requires swift value judgement especially given the exponential growth in publications that increased from 100,000/year in 1951 (Boissier, 2013) to 2.4 million in 2013 (Plume and van Weijen, 2014). The quality and content of the abstract is thus critical in ensuring that an article is noticed, read, and cited with individual researchers on average reading 270 articles per year (Ware and Mabe, 2015).

In the absence of a defined template there is an expectation that the abstract should begin with a sentence that provides context/background for the studies being reported. This should then be followed by an indication of the biological system used (cell line, tissue, whole animal) and the species from which it was derived (wild or transgenic mouse or rat, primate, transfected cell line, isolated human tissue), and an indication of the major finding(s) of the study, and any additional details on the mechanisms involved. It should conclude with a general consideration of the implications of the study. The conclusion in the abstract should be limited to, and accurately reflect, the actual data contained in the manuscript and should avoid being overly speculative and prone to hyperbole. Readers should avoid citing a paper based on the abstract alone and should read the actual paper to ensure that the conclusions are justified by the data (which are rarely disclosed in detail in the abstract). Abstracts should not contain literature citations.

4.10 INTRODUCTION

The Introduction is the section of the paper that sets the stage for the study being reported. It should provide a topical "fair and scholarly reflection of the literature...[and a]...cogent structuring of hypotheses" (Curtis et al., 2015) that provides appropriate context to the current state of research in the area which the results in the paper would build on—not a mini review—ensuring that seminal and recent (typically published within 2 years of the date of writing) key papers in the area are cited. Studies 5 or 10 years old should not be qualified as being "recent." The authors should also appropriately and objectively acknowledge the contributions of other researchers in the field. This is a collegial, scholarly and eminently logical approach; not acknowledging the contributions of other researchers is likely to antagonize the individuals most likely to be invited to act as reviewers.

The Introduction should clearly indicate the hypothesis being tested, its relevance and why the outcomes of the study should be of interest to the reader. In many instances, the basis for a submitted study follows a path of least resistance logic, for example, "[Natural Product of interest] has shown anticancer effects in vitro but its effects on platelet function have not been studied" with no discussion as to why any association might exist with interest in the relationship being solely dictated by an abundance of the natural product in the researcher's laboratory. "It has not been examined before" is not sufficient justification for a study.

4.11 TITLE PAGE

The title page of a manuscript, in addition to the actual title, contains information that many authors consider routine, especially if they are the senior or corresponding author, the latter being the role usually taken by a laboratory head or PI. The title page includes the title of the article, a list of the authors and their institutional affiliations, the address of the corresponding author(s), a running title, keywords, acknowledgements for funding, and a statement regarding potential conflicts of interest. Most recently, a number of publishers now require that the corresponding author, at a minimum, provide their unique Open Researcher and Contributor ID (ORCID; http://orcid.org) as a persistent identity to avoid confusion with other researchers as personal names are not unique and may reflect the inconsistent use or abbreviation of first-names, differences in the number of initials and can change with life events including marriage, and gender change. The ORCID can also be an aid in preventing fraud.

4.11.1 Title

Many journals have strict requirements for the title, in terms of maximum numbers of characters, use of abbreviations, use of language, tense, style, declamation, revelation of findings, etc. The title of an article should be succinct, providing the potential reader with as complete an insight as possible to the content of the article. If the reader does not understand the title, the paper will usually go unread. For an animal study, the title would normally include the species investigated, an allusion to the methodology use, e.g., "in human xenograft models of acute lymphocytic leukemia," "in transgenic models of Type II diabetes," "in animal models of anxiety," etc. The title may also be mildly provocative to motivate the reader to click through to the abstract and to the full paper, if the journal permits this.

Research has shown that papers with short titles are cited more often than those with lengthy ones (Letchford et al., 2015). Informational titles of acceptable length include: "Potent antiinflammatory effects of the narrow spectrum kinase inhibitor RV1088 on rheumatoid arthritis synovial membrane cells" (To et al., 2015) and "GSK356278, a potent, selective, brain penetrant PDE4 inhibitor that demonstrates anxiolytic and cognition enhancing effects without inducing emesis in preclinical species" (Rutter et al., 2014). These contrast with a recent title, "Treatment with Adenosine Receptor Agonist Ameliorates Pain Induced by Acute and Chronic Inflammation" (Montes et al., 2016), which would have been more informative to the potential reader if the identity (LASSBio-1359) and receptor selectivity (A_{2A}) of the agonist were included in the title along with information on animal models used and the species (Male Swiss mice), for example, "The adenosine A_{2A} Receptor Agonist, LASSBio-1359, ameliorates pain in mouse models of acute and chronic inflammation."

These titles contrast with the epic "The nucleotide sequence of a 3.2 kb segment of mitochondrial maxicircle DNA from *Crithidia fasciculata* containing the gene for cytochrome oxidase subunit III, the N-terminal part of the apocytochrome *b* gene and a possible frameshift gene; further evidence for the use of unusual initiator triplets in trypanosome mitochondria" (Sloof et al., 1987) that is often cited as the longest title yet published in the biomedical literature.

Articles with provocative titles can also pique reader interest and include a runner up in length to the longest title: "What is the future of peer review? Why is there fraud in science? Is plagiarism out of control? Why do scientists do bad things?" Is it all a case of: "All that is necessary for the triumph of evil is that good men do nothing?" (Triggle and Triggle, 2007); the breezy "My Perspective on Time, Managers - and Scientific Fun" (Lipinski, 2013); and the intriguing "Navigation-Related Structural Change In the Hippocampi of Taxi Drivers" (Maguire et al., 2000). These titles all tend to pale however in comparison with "An In-Depth Analysis of a Piece of Shit: Distribution of *Schistosoma mansoni* and Hookworm Eggs in Human Stool" (Krauth et al., 2012), a paper with intentional "shock" value that while inconsistent with traditional mores in scientific publishing, led to more than 30,000 views within 20 days of its publication.

4.11.2 Keywords and Running Title

This section of the title page should include the running title per the journal's instruction to authors and keywords, the number of which will be designated by the journal. Since the latter will be used as search terms used to discover a manuscript, these should convey the maximum information rather than be duplicative, for example, "nicotinic" and "nicotinic receptor" are redundant while "drug discovery" and "neurodegeneration" are terms that are far too broad to form the basis of a useful search term unless qualified in some manner, for example, by a mechanism, a disease, or a compound. If the manuscript describes the properties of a new compound (including its structure), its specific code number and CAS number should be included as keywords as should a human disease state if any of the data reported specifically relates to this aspect. The running title is typically an abbreviation of the full title, for example, for the paper "GSK356278, a potent, selective, brain penetrant PDE4 inhibitor that demonstrates anxiolytic and cognition enhancing effects without inducing emesis in preclinical species" (Rutter et al., 2014), this was "GSK356278, a Brain-Penetrant PDE4 Inhibitor."

Additional author related aspects that involve the title page are affiliations and contact information for the corresponding author. These are usually a relatively straightforward part of the submission except in those instances where authors may choose to use an academic affiliation address to obfuscate their affiliation with a company, an indirect conflict of interest.

4.11.3 Conflict of Interest

Conflicts of Interest (CoI) in scientific publishing reflect situations where the objectivity of a researcher in conducting, interpreting, and reporting a biomedical research study, primarily clinical as well as preclinical, may be viewed as suspect due to the perceived potential for some form of material gain. These can be in the form of financial rewards and/or may be career related. They can be *direct*, in the form of consulting fees or stock awards from a for-profit third party or payment in kind (vacations, entertainment, food, travel, etc.), meeting sponsorship, research support for an individual's laboratory, speaker bureaus; or *indirect*—the latter reflecting equity ownership in a company having interests in the outcomes of biomedical research or the clinical realm, the provision of educational materials including textbooks, product samples, advertising materials, etc., at professional meetings where pharma typically provides support (Goldacre, 2012; Lo and Field, 2009). Because of the potential for a CoI, many journals require a clear statement that a CoI may occur. It is important to note that *a statement of potential conflict does not necessarily confirm a conflict*. Thus a scientist noting that he or she has consulted or consults for Company X, holds stock in Company Y, is a founder of Company Z and/or has received laboratory support from an industry-supported research foundation, or is an employee of Company A (something that should be obvious from the author's affiliation) is a formal "heads-up" rather than an indication that every research activity he or she undertakes is tainted by financial considerations or that every publication from their laboratory is biased by these.

As an example, there has been a longstanding controversy related to conflicts in the membership of the expert Advisory Committees that the FDA uses to review New Drug Applications (NDAs) from industry and recommend to the FDA whether they should be approved or not. Many have criticized this process suggesting that *any* conflict, perceived or real, has the potential to jeopardize both patient safety and the fidelity of basic research (Epstein, 2012; Kaiser, 2009; Young, 2009). As an example, the credentials required for an individual to serve on an FDA Drug Advisory Committee, that provides the agency with "independent advice from outside experts on issues related to human and veterinary drugs, vaccines and other biological products, medical devices, and food" (http://www.fda.gov/AboutFDA/Transparency/Basics/ucm222191.htm), is the same expertize for which a pharma company would use the same individual as a key opinion leader (KoL)/consultant—experience, insight, credibility, and professional standing. To summarily dismiss a physician or researcher scientist with financial connections to a pharmaceutical company automatically as "corrupt shills" (Epstein, 2012) places their personal ethical and professional integrity in the same context as the declining ethics in society in general (Freeza, 2013), while depriving the FDA of the best possible input in making its decisions on which new drugs to approve—hardly the smartest of decisions. This is reflected in the current controversy over the approval of the 2017 nominee for Commisioner of the FDA (Kaplan, 2017). However, despite many examples of the appearance of conflict having no impact on an ethically derived outcome (Cannon, 2016), examples of serious CoIs, however few, add fuel to the CoI debate. One, described as the "largest insider trading scheme reported" (Popper and

Vlasic, 2012) involved a neurologist participating in the clinical trials of a putative biological treatment for Alzheimer's disease who became part of an insider stock trading scheme by revealing information on negative clinical trial results before they were publically available (Keefe, 2014).

There is also a tendency by the community at large to assume that researchers in industry are automatically guilty of a CoI until proven otherwise (Goldacre, 2012; Mecca et al., 2015; Steinbrook et al., 2015) even when the potential for conflict is purely scientific and has absolutely nothing to do with any activities associated with their employer. Like the CoIs mentioned earlier where the perception for conflict far exceeds the actuality, isolated examples are taken as the norm (Ebrahim et al., 2016; Goldacre, 2012).

As a result of the many concerns regarding the potential for fiscally-driven conflicts to cloud the ethical judgment and behavior of scientists and clinicians (Smith et al., 2014), multiple initiatives in establishing transparent guidelines for identifying and dealing with CoI in biomedical research have been proposed (Radd and Appelbaum, 2012; Rockey and Collins, 2010; http://grants.nih.gov/grants/policy/coi/). That said, it can be argued that if by publishing a paper in a high JIF journal leads to rewards, the author is promoted or retains their position, then surely every researcher is conflicted? The key issue is disclosure.

4.12 ACKNOWLEDGMENTS

Many journals include a final section after the conclusions and before the references that provides the opportunity to acknowledge granting bodies and researchers involved with the study but who did not qualify as authors, for example, technical staff involved in IT support, centralized sequencing or high throughput binding operations, colleagues with whom the experimental design and its interpretation were discussed or who provided materials, shared unpublished data or encouragement, and grant awarding bodies. Since the listing of individuals in the acknowledgments section is often taken to imply support of the study and its conclusions, their permission should be obtained before they are included. In some instances, while these individuals have provided support and input, they may not agree with the author's conclusions and ask that they not be acknowledged.

4.13 REFERENCES

References should be organized in the paper in the recommended format specifically designated by the journal. This may involve a restriction in number—50 or so for an original research paper, 100–200 for a review—and will require the use of one of two major conventions within the text. The first is numerical, where each citation is listed in the text in the order that it appears, for example, either as "[1]" or a superscript [1]. The reference section then lists the references in numerical order. The second system designates citations in the text in the form, for example, Jones et al. (2000), with the reference list then arranged in alphabetical order. The actual listing of references can include the complete list of authors or the first author, the first three or the first six (depending on the house style of the journal) followed by "et al.," this can be handled relatively easy by software programs some 30 plus of which exist, https://en.wikipedia.org/wiki/Comparison_of_reference_management_software, including the ubiquitous, EndNote (http://endnote.com), that can automatically reformat a complete manuscript into a different style suitable for submission to another journal.

Some newer journals and also established ones do not follow the conventional year, volume, page range format (e.g., 433–438) for citation and instead use a unique identifier for the article, for example, *Science Translational Medicine, BMJ*. These list the volume number and designate the type of article. For *Science Translational Medicine*, in volume 6, the article designated as 234fs18 is a focus article in issue 234, 234cm3 is a commentary and 234ra55, a research article, with the final number being a specific identifier for the individual article. Others, like the various *PLoS* journals and the *International Journal of Alzheimer's Disease* often use the year of publication as the volume number and provide an identifying number for the article, for example, ID 349249.

Citations for websites that contain information that is not part of the traditional scientific literature, for example, blogs, reports from regulatory and policy bodies (FDA, WHO, etc.), and the media (*Forbes, Wall Street Journal*, etc.), are usually cited using hyperlinks placed within the text or in the reference list. Some journals also require inclusion of the date that the website was accessed.

4.14 THE COVER LETTER

The preparation of a cover letter is the final step in submitting a manuscript. In the age of the Internet, this is often viewed as perfunctory both by the authors and the reviewer but is appreciated. It should be on the letterhead of the submitting institution and provide background to the submission and why it is considered to be of interest to the journal to which it is being submitted. The cover letter is the forum in which to raise any concerns about potential reviewers who the authors feel are conflicted and should conclude with thanks to the Editor for the opportunity to have their work considered. One major faux pas that can occur in the cover letter that was highlighted by Neill (2007) in her excellent article *"How to write a scientific masterpiece"* is to "make sure to address your cover letter to the correct journal" and make sure that the names of any files uploaded are not a legacy from a previous submission. This can avoid giving the journal to which the paper is being submitted "a negative predisposition" based on "The realization that a competitor rejected the paper."

4.15 PRACTICAL ASPECTS OF THE PEER REVIEW PROCESS

The peer review process, the pros and cons and reactions to which are discussed in detail in Chapter 5, begins with the submission of a manuscript to the Editorial Office of the selected journal. In submitting the manuscript, it is important to adhere to the journal's guidelines for the various sections as discussed earlier. In addition to style factors, there are generally guidelines regarding the length of each section and the number of references since, even though journals are increasingly only accessed online, they are still subject to page budgets to cover the administrative and production costs which are independent of the final format.

It is highly recommended that the reader consult the excellent slide presentation by Ushma Neill, an Editor at the *Journal of Clinical Investigation*, "How to get your work published in the best journals" (https://www.google.com/webhp?nord=1#nord=1&q=JCI+How+to+get+your+work+published+in+the+best+journals).

4.15.1 Results From the Peer Review Process (Table 4.2.)

4.15.1.1 Rapid Reject

Immediately on submission an Editor can decide to rapidly reject—a desk reject—a manuscript because: it shows evidence of overt plagiarism; does not meet the standards of the journal in terms of either science or writing; is of limited or no interest in terms of content to the journal readership; or lies outside the scope of interest and/or expertise of the journal. Authors should note that while a manuscript may be perfectly sound, it can be rejected by a journal because the editorial board members lack the necessary expertise (which usually means that the article is out of scope for the journal). Since journals may be members of a publishing consortium, an article that is deemed unsuitable for one journal can, with the permission of the authors, be referred to another journal where the subject matter may be a better fit, a process known as cascade review (see Chapter 5.5.3.3).

Addition reasons for a rapid rejection include that the hypothesis being proposed is non-existent, illogical or has already been examined multiple times, for example, lacks novelty; the methodology used is unsuitable to address the hypothesis; the manuscript contradicts itself (a not unusual failing) or the data is of questionable provenance, does not support the conclusions or has been misinterpreted (a mismatch between data and inference). Lack of novelty should not be a reason for rejection if the paper is defined as a reproducing or refuting findings in previous publications. Examples of mismatch between data and inference are when authors describe a concentration/dose response for a particular effect and there is an "all or none" response where, for example, a concentration of 10 nM has no effect but 30 nM produces a modest effective but increasing the concentration further (100 nM, 300 nM, 1 μM) has no further effect (Fig. 4.2) or when the authors conclude that the absence of an anticipated *increase* in the effect of a compound confirms their hypothesis but they fail to notice, note, or comment on robust concentration-dependent *decreases* because this is not what they were looking for. Such mismatches may not be immediately obvious during the initial editorial triage of a manuscript and may come to light only after more in-depth peer review.

4.15.1.2 Reviewer Selection

Once the Editors have decided to advance a submission to the peer review stage, reviewers are assigned from the journal's database of experts, typically 5 or more from a list of 100s—1000s and from other sources like PubMed or Google Scholar of which 2–3 will accept the review depending on their immediate work load. Authors can request that certain individuals be avoided, usually because of competitive aspects of their research or previous interactions that have been unproductive or suggest reviewers who are experts in the field, but independent of the PI or submitting laboratory. Such requests are not always honored by the Editors for a number of reasons. These may include previous experiences with the author-requested reviewers who often fail to respond to invitations or the naivety of the authors who assume

TABLE 4.2 Review Issues and Responses with thanks to Ushma Neill, Journal of Clinical Investigation, "How to get your published in the best journals" (https://www.google.com/webhp?nord=1#nord=1&q=JCI+How+to+get+your+work+published+in+the+best+journals)

	Editorial/reviewer Issues	Author consideration/Response	Inappropriate response from authors
Rapid reject	• Plagiarism ◦ Sections copied from other manuscripts and detected by software, for example, iThenticate CrossCheck, Google Scholar search ◦ Submission of paper submitted and accepted elsewhere (self-plagiarism) ◦ Pirating a published paper and submitting it with different authors and institute of origin • Does not meet criteria/have a sufficient high score to warrant peer review. • Topic lacks interest or relevance to journal readership. • Hypothesis is poorly formulated, wrong or absent. • Methods and/or results are not intelligible or consistent. • Manuscript is in a format for another journal and is being resubmitted without any attempt to follow journal guidelines, for example, medicinal chemistry versus biological journal. • Manipulation gel/photomicrograph images ◦ Using same image to represent different experimental procedures • Sources of reagents not provided. • Reagents not commercially or collegially available. • Issues with use/authentication/validation of reagents.	• Is the article suitable for peer review? • Was the journal an inappropriate venue for your submission? • Reflect, rewrite, and submit elsewhere recognizing that there may be a high likelihood of the same reviewer being selected at the newly selected journal.	• Editors have failed to properly assess obvious value of submission. • Requesting a new Editor unwise since the Editor in Chief is highly unlikely to override his/her choice of Editors. • Accusations of incompetence and bias in reviewer feedback.

(*Continued*)

TABLE 4.2 Review Issues and Responses (*cont.*)

	Editorial/reviewer Issues	Author consideration/Response	Inappropriate response from authors

Reject	• No independent replication of data sets
	• Findings add nothing to the existing literature
	• Findings conflict with the existing literature due to technical issues
	• Major issues with conclusions/data that does not support conclusions
	• Study is so poorly designed or underpowered—no valid conclusions can be made
	• Additional studies required to clarify loose ends
	• Contradictions in text—inconstancies beteween text/figures/tables
	• Manuscript fails to follow basic journal guidelines
		• Appeal decision?
		  ◦ Wait a day or two before responding to reviewer comments
		  ◦ Is a rebuttal letter in order?
		  ◦ Is debating feedback worth the effort?
		  ◦ Did the Editors/Authors misconstrue aspects of the study and submitted manuscript?
		• Reflect, rewrite, and resubmit
		• Submit elsewhere
			• Accusations of incompetence and bias in reviewer feedback
			  ◦ Guessing who they might be
			  ◦ Arguing via "celebrity endorsements," for example, Nobel laureates thought submission was excellent so what do the reviewers know?
			  ◦ "My last paper was accepted for publication".
			• Author comparing rejected manuscript with another manuscript published in the journal that he/she deems to be far worse than the rejected submission
			• Editors are censoring data for their own reasons
			• Claims that reviewer comments unjustified
			• Requesting a new Editor
			• Resubmitting a revised manuscript without addressing reviewer comments.

Invitation for revision	Detailed review with list of specific issues for the authors to address:
	• Conclusions not supported by data
	• Data missing
	• Manuscript poorly written
	• Poor grammar/usage
	  ◦ Not proof read
	  ◦ Not clear
	• Novel concepts requiring additional data support
		• Thank reviewers for sharing their expertise and time to improve submission
		• Prioritize comments
		• Revise submission addressing *each and every* issue on a point by point basis clarifying areas that may have been misinterpreted by the reviewer or rebutting—in collegial tones—those with which the author disagrees
		• Request time to conduct additional experiments if necessary
		• If additional experiments cannot be done, explain why, and ask Editor(s) for guidance
		• Provide redline revision of original submission indicating changes/additional data
			• Treating reviewer as an adversary rather than colleague
			• Superficial rewrite ignoring reviewer comments
			• Blanket statement "about addressing issues" without specifics
			• Cherry picking points to respond to—ignoring others
			• Submitting a revised manuscript without addressing reviewer comments.
			• Addressing reviewer issues in point by point rebuttal to editor/reviewers rather than adding these key points to the actual manuscript

Acceptance		• Thank you	
Production issues	Proofs with production questions Q1, Q2, clarification of meaning, updating references, etc.	• Formally address all points raised in the pdf text using Adobe notes function	

that since their work is in the same field, a Nobel laureate and/or head of the Crick Research Institute is an appropriate reviewer of their work. Once reviewers accept the assignment, the manuscript is then sent out for electronic review as a pdf file, greatly expediting the turnaround time. Since many reviews may be conducted in the absence of a paper copy, authors can greatly enable the review process (and keep the reviewers happy) by numbering all the pages in the pdf and also the sentences so that the reviewer can more easily refer to sections that are in need of the author's attention without having to manually count the lines. The review process for the majority of journals is conducted in a single-blinded manner with the reviewers knowing the identity of the authors but remaining anonymous themselves. This system has many critics and this is discussed in detail in Chapter 5.3.

4.15.2 Responding to Reviewers and Editors

The review process typically takes a maximum of 2–6 weeks. Given that peer review is usually a *pro bono* activity undertaken by researchers who are busy with their own activities, there can be unavoidable delays. Most journals establish timelines for the various aspects of the review process and use alerting software that sends out regular reminders to the reviewer. A thorough review of a manuscript with detailed written feedback can take easily take the reviewer 2–3 h with many taking 6–8 h especially if the reviewer takes time to reflect on their feedback. This can then be revised to ensure it is understandable, will aid in improving the manuscript and be received by the authors as objective, collegial and professional input worthy of consideration in revising the manuscript. This contrasts to the summarily dismissed output of "pit bull" reviewers (Walbot, 2009).

The balance between the nuanced collegiality that is firmly established in the existing peer review process and the world weary reviewer–whose encounters with a "good" paper are the exception rather than the rule–can often lead to exasperation especially when the author refuses to accept any responsibility for the issues being raised by the reviewer.

4.15.3 Editorial Oversight of Peer Review Process

Reviewer comments are submitted to the responsible Editor with a recommendation as to whether the submission should be rejected, modified, or accepted. This is accompanied by written feedback—anywhere between a paragraph and several pages long. The recommendation together with the accompanying comments and input from other reviewers, are reviewed by the Editor before being forwarded to the authors. These may be edited for clarity or to avoid misunderstanding or offence. Typically, the reviews received are consistent, with the two (or more) reviewers being in general agreement in their recommendations. However, there are instances where there are conflicting opinions and the reviewers are diametrically opposed in their recommendations, one recommending "accept" or "minor revision" while the other indicates "reject" with appropriate comments. At this juncture the Editor, who is responsible for making the final decision, can resolve these discrepancies either by further interaction with the reviewers to resolve the differences or by seeking a "tie breaker" review from a third reviewer when it is clear that one or other of the reviews is out of line or, break the tie themselves. In some instances, a tie breaker can be asked to adjudicate the conflicting reviews by being privy to the previous reviews. In others, an additional reviewer will be invited without forwarding the first set of reviews. Both approaches have merit; the first is faster but the second is unbiased. Once these reviews are collated, the decision, plus the reviews are communicated to the author.

4.15.4 The Review

The reviewer comments received by the authors from the Editor handling their submission generally fall into five categories focusing on:

1. the lack of adherence of the authors to the journal guidelines and publishing norms that is reflected in an inability to write a focused, cohesive, and logical manuscript;
2. whether the experimental design and execution are capable of—and have provided—a valid, rigorous, and complete data set to support or dismiss the hypothesis being tested;
3. whether the conclusions drawn are an accurate interpretation of the data reported;
4. whether there glaring gaps, assumptions, or contradictions; and
5. more recently, whether the requirements concerning design and analysis required by the journal have been followed (Curtis et al., 2015; Mullane et al., 2015).

The review generally begins with a paragraph that provides the top line comments for the decision made regarding the submission. This is then followed by a detailed, point-by-point discussion of specific issues that the reviewers have identified usually referring to a specific, figure, table, and page and lines. Included in these would be issues with the experimental design, the powering and execution of the experiment, data generation and analysis etc., the guidelines for which are outlined in detail in Chapter 2.5 and 2.6 since these aspects represent the core of the manuscript. If major issues are identified here, the remainder of the manuscript is unlikely to receive additional comment as the basic premise of the experimentation is flawed.

The authors should bear in mind that experienced reviewers will, after reading the abstract, gravitate automatically to the data presentation and its analysis, for example, the figures and tables, and consider these before turning their attention to other parts of the manuscript, for example, the methods section to see if the data may have been clearly understood, to the introduction to review the rationale for the study and, finally, to the conclusions to read the author's interpretations of the data.

Some considerations that can be gained from reviewing the figures and tables include: Have the data been independently reproduced and statistically analyzed in an appropriately rigorous manner? Have convincing dose/concentration curves been generated, do they really show a dose/concentration response or instead, a binary "all or none" effect? Have the data been independently reproduced and analyzed to derive a quantitative measure, for example, IC_{50}, EC_{50} values? Is the study exploratory or confirmatory (Chapter 2.3) or have the authors, in their excitement at finding some statistically significant effect in an exploratory study rushed to publish it without performing the necessary prospectively defined and designed confirmatory study? For an in vivo experiment are the doses, route, and frequency noted, is there evidence for the compound being bioavailable and is the half-life compatible with the assay time frame? These points are among the checklists recently developed as prompts for authors and reviewers (Curtis et al., 2015; Mullane et al., 2015) that many biomedical research journals have added to their Instructions/Guidelines for Authors and reflect the FASEB Recommendations "Enhancing Research Reproducibility" https://www.faseb.org/Portals/2/pdfs/opa/2016/FASEB_Enhancing%20Research%20Reproducibility.pdf).

No number of tables or figures, however complicated, colorful, and rich in data that result from the use of multiple and highly sophisticated techniques can rescue a series of experiments that have been poorly conceived, executed, and/or analyzed or that are based on a flawed assumption.

If a preliminary assessment of the data passes muster, the reviewer will then read the remainder of the manuscript identifying and pointing out areas for clarification, for additional information and possibly suggestions for additional experiments that will add to the existing body of data and strengthen the case for the author's conclusion(s). These include inconsistencies in logic, discrepancies between different parts of the manuscript [text information that does not match that in the tables/figures (and vice versa), concentrations/doses that are inconsistent or missing] and conclusions that are not supported by the data. The Editor should consider the recommendations and their implications. Often a reviewer will suggest some experiments they feel would be "nice to see" (Snyder, 2013). Whether these experiments should be done is a matter for the Editor who must consider the time, difficulty, expense, and value, and weigh this up against the value and completeness of the data set already included in the paper.

4.15.5 Author Responses to Reviewer Feedback

4.15.5.1 Rapid Rejection

A rapid rejection is occasionally a starting point for productive correspondence with the editor and/or the journal on the unfairness of the decision and the scientific credentials of the Editors. However, it is not *carte blanche* to modify the original submission and resubmit it as a "revision."

When receiving a rapid reject, the authors should step back and take an objective view of the feedback to avoid wasting additional time, both theirs and that of the journal, and consider whether the fundamental flaws that led to the initial rejection can be corrected before submitting it to another journal.

4.14.4.2 Rapid Acceptance

An Editor also has the option to accept a manuscript in its original form, an event that while extremely rare does occur, especially in the case of short review-type submissions, and can often be counted by most experienced Editors on their fingers.

4.14.4.3 Reject

While the rapid rejection of a manuscript is the result of an editorial decision, the decision to reject after peer review is based on the recommendations of the reviewers. Authors may dispute the decision with the journal asking that a new Editor and/or reviewers be assigned to the review, but this is uncommon (and unwise since it implies that the journal has not acted appropriately, and the author would need good evidence of this, and would need to take the time to ensure that the evidence is presented in a clear and courteous manner). Alternatively, they can submit a revised, de novo version of the rejected manuscript.

4.14.4.4 Request for Revision

When a manuscript is returned with the request "invite to revise" this should be viewed by authors as an opportunity to collegially and proactively address the points raised by the reviewers, in a revised version of the manuscript. Author feedback to reviewer comments should be provided on a point-by-point basis not by a blanket statement or by just submitting a revised manuscript. In fact, many journals require the author's point-by-point responses be accompanied by a redline of the original submission clearly indicating where changes have been made, showing both inserts and deletions. A revision that insufficiently addresses the issues or "cherry picks" those to be dealt with is likely to be rejected (Neill, 2007). In the rare instances that this does not occur and the journal then requests a second series of revisions, the authors should carefully consider and incorporate all the changes requested. In some instances, this may require that additional experiments be conducted. Conversely, when a reviewer requests that the author clarify aspects of a particular data set and this leads to the data set being removed by the author as "not important," this is cause for concern especially when this happens more than once in a submission without the conclusions of the study being revised.

Reviewer feedback should always be viewed as an opportunity to improve a manuscript making it a stronger addition to the literature with the authors and the journal benefiting from the activities of the reviewers. Reviewer comments often raise issues with clarification. When the author responds to the reviewer he/she should also consider whether adding the information shared with the reviewer might also be useful to include in the revised manuscript. If the text may not have been clear to the reviewer it may also not be clear to the intended reader.

Material issues require that the authors must either make the requested changes or present valid reasons as to why they are unable to accept the reviewer comments, for example, the reviewer's feedback is incorrect or unjustified or performing additional experiments is already part of a subsequent study—science being a dynamic discipline—is cost prohibitive for the authors or permission from an IUCAC (Institutional Animal Care and Use Committee) for additional animal studies is unlikely to be forthcoming. Where this requires the citation of additional references in the feedback, it is often a best practice to actually incorporate the selected references in the revised text so that the reader as well as the reviewer can gain better insight into the author's rationale. If the reviewer's comment is justified, the author may need to rewrite a major section of the text to restate or reframe an incorrect, ambiguous, or oblique section of the text in light of the reviewer's inability to understand/agree with the author's approach.

Additional points may represent question of style, writing, and presentation, or semantics that can easily be addressed by the author's changing or clarifying a statement (or a figure) that the reviewer has misunderstood due to a lack of clarity, perceived or real. The inability of the reviewer to clearly understand the content of the original submission can be taken as a given that the reader will have the same issues. In these circumstances, it is generally to the author's advantage to accept the reviewer comments in order to expedite the review process. In the majority of instances this process represents a quasi-, non-confrontational negotiation where both parties can resolve differences via objective accommodation.

4.14.4.5 Language Revision

A frequent issue that is not always addressed satisfactorily in the review process is when an author whose first language is not English receives a request along the lines of having an

"English-speaking colleague or one more familiar with written English usage" read through the revised manuscript to ensure that it is acceptable. Often, whether through a misunderstanding or misplaced pride of the author in their assumed prowess in their second language (often spoken rather than written), the revision will be only marginally improved and will still include major grammatical errors (use of plural instead of singular, inappropriate mixing of past and present tenses, sentences that lack subjects, objects, and verbs), inappropriate word usage due to multiple options in translation, idiosyncratic sentence structure and often amusing neologisms that still render the manuscript unacceptable.

There are many fee-for-service professional proof reading and copy editing services that can aid in improving the English. However, some of these may distort/misrepresent the author's actual intent. While ostensibly charged solely with improving the grammar and style, a copy editor—due to a misplaced confidence in their own expertize or a passing familiarity with the scientific literature—may alter the nuances and intent of the text, introducing problems that are more serious than those they were engaged to solve such that a colleague can be the best solution to resolving this issue.

4.14.4.6 *Conflict in the Review Process*

While it is recommended that the authors *refrain* from any immediate knee jerk responses to reviewer feedback, when this caution is ignored there may be intemperate comments to the journal that are a direct challenge to the reputation of the journal, the impartiality and birthright of the Editors and reviewers and the integrity of all parties involved in the review process that are unlikely to lead to a satisfactory outcome. Similarly, citing celebrity "endorsements" of the paper (A Nobel Laureate thought it was great), arguing that the journal had previously published a paper by the same authors 2, 5, or 10–15 years ago and this one was "just as good" or that the journal had recently published a paper far worse than the one under review does not facilitate or advance the discussion (Neill, 2007).

If the authors become incensed, the review decision should be used as the basis for the rapid submission of the manuscript to a different journal. Authors should bear in mind that the reviewer pool in a specialized research area is often finite. Thus in the absence of some revisions to address the points raised in a proximal revision, submission of the same article in a cosmetically revised version to another journal can result in it being rapidly rejected since it may be reviewed by the same reviewers who had provided the initial feedback who relay the same concerns. At this point the reviewers should reflect on the feedback, consider the input in the spirit in which it was offered and address the issues raised.

4.14.4.7 *Revision Followed by Reject*

If the decision following the submission of a revised manuscript is another rejection, this usually indicates that the authors have not adequately addressed the points raised by the reviewers. The final decision to reject is often rendered by the Editor after reviewing the author's response in the context of the original reviewer comments. In some instances, where the subject matter and/or concerns are complex or where there are nuances in the author responses, the Editor will either request that the original reviewers review the revised manuscript to ensure that they are satisfied with the responses or assign the revised manuscript to a new set of reviewers, the latter often being dictated by the lack of availability of the original reviewers. This may however, lead authors to wrongly conclude that a reviewer has disregarded their time-consuming revision and introduced new criticism which requires editorial oversight to amicably resolve.

4.14.4.8 Revision—Accept

Once a manuscript has been accepted it may still require additional attention from the author to ensure that all sections are present and are in the correct format and that the figures conform to the journal style.

4.14.4.9 Digression From Acceptable Norms

In her slide presentation, "How to get your work published in the best journals" (https://www.google.com/webhp?nord=1#nord=1&q=JCI+How+to+get+your+work+published+in+the+best+journals), Neill provides author feedback that she has received "You can rest assured that I will not consider JCI for future manuscripts from my group as long as incompetent individuals will continue to run JCI and I will discourage anybody from subjecting themselves to the harassment we had to suffer," a not unusual author sentiment directed toward editors and reviewers.

4.15.6 Production

The next steps in the publication process involve the online publication of the accepted manuscript as a pdf of the final edited article. This is followed several weeks later by the availability of the typeset proofs (sometimes referred to as galleys) of the article which need to be proofread to address production questions (typos, updating of references, checking author affiliations, points for clarification etc.) and to identify any errors that were missed or have been introduced during the typesetting process. It is important to ensure that the punctuation is consistent with clarity in meaning and that the Production Editor, often a third party contracted by the publisher, has not dispensed with the Chicago Manual of Style (2010) in its entirety. In one instance, MW was unable to reconcile the proofs of an article with what he had originally written. It turned out that the production editor had, for unknown reasons, removed ALL the punctuation except the periods from the submitted text, introducing sentence structures and run-ons that were meaningless.

Depending on the journal policy, the author may also be able to include a very brief note "added in proof" of important "breaking news" reflecting articles that have appeared since the manuscript was accepted, the inclusion of which further enhances the value of the final article to the reader and the journal.

4.15.7 Invited Articles

Some submitted manuscripts—for example, reviews and commentaries and those for a special issue of a journal on a selected topic that are invited from the authors by an *ad hoc* Guest Editor, are accepted for peer review following agreement with the journal and are generally handled slightly differently from regular articles. This involves additional leeway in the review process to ensure that revisions are expeditiously performed and tends to involve a more interactive dialog than that which occurs with a regular submission. Generally, unless the submission detracts egregiously from what was anticipated from the author(s) when the invitation was extended, there is an implicit assumption that the article will be accepted after peer review and appropriate revision such that the role of both Editors and reviewers is more proactive, often with an active email dialog between the authors and responsible editor to ensure that the level, tone, and content of the article fulfills the original intent of the invitation.

For a Commentary/Viewpoint where an author has been expressly asked to provide personal and/or provocative views on the advances in, and future direction of a field, the Editor may, in this spirit, once ensuring that the article is well written, sufficiently didactic for newcomers to the area and scientifically logical, choose to overlook negative reviewer feedback that argues against the view(s) presented by the author because they do not conform with the current *status quo* of research in the area addressed. Such feedback compromises the content of an article and its intrinsic and its consequent value to its target audience.

Ultimately a journal will have its agenda set by its publisher, scientific society and/or Editor-in-Chief and the way it manages peer review will be tailored to meet the needs of the journal. Meeting the needs of authors and reviewers will be part of the rubric for journals with a middle range JIF (3–8). When the JIF is greater than 8, a journal may find it has far more papers submitted than it can publish, and may adopt a less author-friendly reviewer-friendly stance.

References

Aad, G., Abbott, B., Abdallah, J., Abdinov, O., Aben, R., Abolin, M., et al., 2015. Combined measurement of the Higgs Boson mass in *Pp* collisions at $s\sqrt{}=7$ and 8 TeV with the ATLAS and CMS experiments. Phys. Rev. Lett. 114, 191803.

Abbott, A., 2013. Image search triggers Italian police probe. Nature 504, 18.

Alberts, B., Kirschner, M.W., Tilghman, S., Varmus, H., 2014. Rescuing US biomedical research from its systemic flaws. Proc. Natl. Acad. Sci. USA 111, 5573–5777.

Almeida, J.L., Cole, K.D., Plant, A.L., 2016. Standards for cell line authentication and beyond. PLoS Biol. 14, e1002476.

Bavdekar, S.B., 2012. Authorship issues. Lung India 29, 76–80.

Bhattacharjee, Y., 2011. Saudi Universities offer cash in exchange for academic prestige. Science 334, 1344–1345.

Bian, Z., Chen, S., Cheng, C., Wang, J., Xiao, H., Qin, H., 2012. Developing new drugs from annals of Chinese medicine. Acta Pharm. Sinica. B 2, 1–7.

Bik, E.M., Casadavell, A., Fang, F.C., 2016. The prevalence of inappropriate image duplication in biomedical research publications. mBio 7, e00809.

Blatt, M., Martin, C., 2013. Manipulation and misconduct in the handling of image data. Plant Physiol. 163, 3–4.

Boissier, M.-C., 2013. Benchmarking biomedical publications worldwide. Rheumatol. 52, 1545–1546.

Bordeaux, J., Welsh, A.W., Agarwal, S., Killiam, E., Baquero, M.T., Hanna, J.A., et al., 2010. Antibody validation. Biotechniques 48, 197–209.

Bratton, B., 2013. We need to talk about TED. Guardian, Available from: http://www.theguardian.com/commentisfree/2013/dec/30/we-need-to-talk-about-ted.

Bulmiller, E., 2010. We Have Met the Enemy and He is PowerPoint. New York Times, Available from: http://www.nytimes.com/2010/04/27/world/27powerpoint.html.

Bylund, D.B., Toews, M.L., 2014. Quantitative versus qualitative data; the numerical dimensions. Biochem. Pharmacol. 87, 25–39.

Cannon, M.F., 2016. How to minimize conflicts of interest in medical research. Forbes. Available from: http://www.forbes.com/sites/michaelcannon/2016/04/04/how-to-minimize-conflicts-of-interest-in-medical-research/#3941f9ef588f.

Carpenter, J., 1986. A method for preparing and presenting dose-response curves. J. Pharmacol. Methods 15, 283–303.

Chicago Manual of Style, 2010. Sixteenth ed. University of Chicago Press Staff. Univ Chicago Press, Chicago.

Chipperfield, L., Citrome, L., Clark, J., David, F.S., Enck, R., Evangelista, M., et al., 2010. Authors' submission toolkit: a practical guide to getting your research published. Curr. Med. Res. Opin. 26, 1967–1982.

Collins, F.S., Tabak, L.A., 2014. Policy: NIH plans to enhance reproducibility. Nature 505, 612–613.

Couzin-Frankel, J., 2016. Bringing image manipulation to light. Science. DOI: 10.1126/science.caredit.a1600143.

Curtis, M.J., Walker, M.J.A., 1986. The mechanism of action of the optical enantiomers of verapamil against ischaemia-induced arrhythmias in the conscious rat. Br. J. Pharmacol. 89, 137–147.

Curtis, M.J., Bond, R.A., Spina, D., Ahluwalia, A., Alexander, S.P.A., Giembycz, M.A., et al., 2015. Experimental design and analysis and their reporting: new guidance for publication in BJP. Br. J. Pharmacol. 172, 2671–2674.

Dance, A., 2012. Authorship: who's on first? Nature 489, 591–593.

Deng, W., Wang, Y., Liu, Z., Cheng, H., Xue, Y., 2014. HemI: a toolkit for illustrating heatmaps. PLoS One 9, e111988.

Drummond, G.B., Vowler, S.L., 2011. Show the data, don't conceal them. Br. J. Pharmacol. 163, 208–210.

Ebrahim, S., Bance, S., Athale, A., Malachowski, C., Ioannidis, J.P.A., 2016. Meta-analyses with industry involvement are massively published and report no caveats for antidepressants. J. Clin. Epidemiol. 70, 155–163.

Eisen, M., 2014. I confess, I wrote the Arsenic DNA paper to expose flaws in peer-review at subscription based journals.it is NOT junk blog. Available from: http://www.michaeleisen.org/blog/?p=1439.

Eisen, M.B., Spellman, P.T., Brown, P.O., Botstein, D., 1998. Cluster analysis and display of genome-wide expression patterns. Proc. Natl. Acad. Sci. USA 95, 14863–14868.

Epstein, R., 2012. Unshackle the FDA from rules that kill innovation. Forbes. Available from: http://www.forbes.com/sites/richardepstein/2012/03/15/unshackle-the-fda-from-rules-that-kill-innovation/.

Fellmann, C., Lowe, S.W., 2014. Stable RNA interference rules for silencing. Nat. Cell. Biol. 16, 10–18.

Flaherty, S.K., 2013. Ghost- and guest-authored pharmaceutical industry–sponsored studies: abuse of academic integrity, the peer review system, and public trust. Ann. Pharmacother. 47, 1081–1083.

Frankel, F.C., DePace, A.H., 2012. Visual strategies: A Practical Guide to Graphics for Scientists and Engineers. Yale University Press, New Haven, CT.

Freedman, L.P., Cockburn, I.M., Simcoe, T.S., 2015. The economics of reproducibility in preclinical research. PLoS Biol. 13, e1002165.

Freeza, B., 2013. We Are gathered here today to witness the burial of the protestant work ethic. Forbes. Available from: http://www.forbes.com/sites/billfrezza/2013/07/16/we-are-gathered-here-today-to-witness-the-burial-of-the-protestant-work-ethic/.

Friedman-Levi, Y., Mizrahi, M., Frid, K., Binyamin, O., Gabizon, R., 2013. PrP(ST), a soluble, protease resistant and truncated PrP form features in the pathogenesis of a genetic prion disease. PLoS One 8, e69583.

Friedman-Levi, Y., Mizrahi, M., Frid, K., Binyamin, O., Gabizon, R., 2015. Correction: PrPST, a soluble, protease resistant and truncated PrP form features in the pathogenesis of a genetic prion disease. PLoS One 10, e0133911.

Frommer, F., 2012. How PowerPoint Makes You Stupid: The Faulty Causality, Sloppy Logic, Decontextualized Data, and Seductive Showmanship That Have Taken Over Our Thinking. New Press, New York.

Garwood, A., 2014. How luxury brands are damaging science. Lab Times. 1-2014, 16–21.

Geraghty, R.J., Capes-Davis, A., Davis, J.M., Downward, J., Freshney, R.I., Knezevic, I., et al., 2014. Guidelines for the use of cell lines in biomedical research. Br. J. Cancer 111, 1021–1046.

Glasziou, P., Altman, D.G., Bossuyt, P., Boutron, I., Clarke, M., Julious, S., et al., 2014. Research: increasing value, reducing waste 5 reducing waste from incomplete or unusable reports of biomedical research. Lancet 383, 267–276.

Goldacre, B., 2012. Bad Pharma: How Drug Companies Mislead Doctors and Harm Patients. Fourth Estate, London, pp 287–299.

Goldacre, B., 2016. Make journals report clinical trials properly. Nature 530 (7).

Goodyear, D., 2016. The stress test. Rivalries, intrigue, and fraud in the world of stem-cell research. New Yorker. Available from: http://www.newyorker.com/magazine/2016/02/29/the-stem-cell-scandal.

Grassley, C., 2010. Ghostwriting in Medical Literature. Minority Staff Report, 111th Congress, United States Senate Committee on Finance. Washington, DC. Available from: http://www.grassley.senate.gov/about/upload/Senator-Grassley-Report.pdf.

Hoogenboom, B., Manske, R.C., 2012. How to write a scientific article. Int. J. Sports Phys. Ther. 7, 512–517.

Hulbert, S.H., 1984. Pseudoreplication and the design of ecological field experiments. Ecolog. Monogr. 54, 187–211.

ICMJE (International Committee of Medical Journal Editors), 2010. Uniform requirements for manuscripts submitted to biomedical journals: writing and editing for biomedical publication. J. Pharmacol. Pharmacotherap. 1, 42–58.

Ioannidis, J.P.A., 2014. How to make more published research true. PLoS Med. 11, e1001747.

Johnson, W.E., Li, C., Rabinovic, A., 2007. Adjusting batch effects in microarray expression data using empirical Bayes methods. Biostatistics 8, 118–127.

Jones-Bolin, S., 2015. Guidelines for the care and use of laboratory animals in biomedical research. Curr. Protoc. Pharmacol. 59, A.4B.1–A.4B.9.

Kaelin, Jr., W.G., 2012. Use and abuse of RNAi to study mammalian gene function. Science 337, 421–422.

Kaiser, J., 2009. Senate probe of research psychiatrists. Science 325, 30.

Kaplan, S., 2017. Scott Gottlieb preps for FDA's top post with a resume that cuts both ways. STATnews.com, Available from: https://www.statnews.com/2017/04/05/gottlieb-fda-profile/.

Keefe, P.R., 2014. The empire of edge. New Yorker. Available from: http://www.newyorker.com/magazine/2014/10/13/empire-edge.

Kenakin, T., 2014. A Pharmacology Primer: Techniques for More Effective and Strategic Drug Discovery, fourth ed. Elsevier Academic Press, San Diego CA.

Kenakin, T., 2016. Pharmacology in Drug Discovery and Development, Second edn. Elsevier Academic Press, London, pp. 7–14.

Kennedy, D., 2006. Editorial retraction. Science 311, 355.

Kopec, K., Bozyczko-Coyne, D.B., Williams, M., 2005. Commentary: target identification and validation in drug discovery: the role of proteomics. Biochem. Pharmacol. 69, 1133–1139.

Krauth, S.J., Coulibaly, J.T., Knopp, S., Traoré, M., N'Goran, E.K., Utzinger, J., 2012. An in-depth analysis of a piece of shit: distribution of *Schistosoma mansoni* and Hookworm eggs in human stool. PLoS Negl. Trop. Dis. 6, e1969.

Lauer, M., 2016. Authentication of key biological and/or chemical resources in NIH grant applications. Open Mike. NIH Extramural Nexus. Available from: https://nexus.od.nih.gov/all/2016/01/29/authentication-of-key-biological-andor-chemical-resources-in-nih-grant-applications/.

Lee, C.J., Sugimoto, C.R., Zhang, G., Cronin, B., 2013. Bias in peer review. J. Am. Soc. Info. Sci. Technol. 64, 2–17.

Letchford, A., Moat, H.S., Preis, T., 2015. 2015 The advantage of short paper titles. R. Soc. Open Sci. 2, 150266.

Leung, W., Shaffer, C.D., Reed, L.K., Smith, S.T., Barshop, W., Dirkes, W., et al., 2015. *Drosophila* Muller F elements maintain a distinct set of genomic properties over 40 million years of evolution. G3 (Bethesda) 5, 719–740.

Lipinski, C.A., 2013. My perspective on time. managers-and scientific fun. Ann. Rep. Med. Chem. 48, 15–22.

Liu, J., Xu, P., Collins, C., Liu, H., Zhang, J., Keblesh, J.P., Xiong, H., 2013. HIV-1 Tat protein increases microglial outward K^+ current and resultant neurotoxic activity. PLoS One 8, e64904.

Liu, Y., Liu, Y., Sun, C., Gan, L., Zhang, L., Mao, A., Du, Y., Zhou, R., Zhang, H., 2014. Carbon ion radiation inhibits glioma and endothelial cell migration induced by secreted VEGF. PLoS One 9, e98448.

Liu, Y., Liu, Y., Sun, C., Gan, L., Zhang, L., Mao, A., Du, Y., Zhou, R., Zhang, H., 2015. Correction: carbon ion radiation inhibits glioma and endothelial cell migration induced by secreted VEGF. PLoS One 10, e0135508.

Liumbruno, G.M., Velati, C., Pasqualetti, P., Franchini, M., 2013. How to write a scientific manuscript for publication. Blood Transfus. 11, 217–226, 2013.

Lo, B., Field, M.J., 2009. Conflict of Interest in Medical Research, Education and Practice. Institute of Medicine/National Academies Press, Washington DC.

Mack, C., 2013. How to write a good scientific paper: authorship. J. Micro/Nanolith. 12, 010101.

Maguire, E.A., Gadian, D.G., Johnsrude, S., Good, C.D., Ashburner, J., Frackowiak, R.S.J., et al., 2000. Navigation-related structural change in the hippocampi of taxi drivers. Proc. Natl. Acad. Sci. USA 97, 4398–4403.

Marcus, A., Oransky, I., 2012. Can we trust western blots? Lab Times 2-2012, 41. Available from. http://www.labtimes.org/labtimes/issues/lt2012/lt02/lt_2012_02_41_41.pdf.

McGonigle, P., Ruggeri, B., 2014. Animal models of human disease: challenges in enabling translation. Biochem. Pharmacol. 87, 162–171.

McKee, K., 2016. Communication Tools and Strategies for the 21st Century Scientist. Prezi. Available from: https://prezi.com/juqabcmbo56n/communication-tools-and-strategies-for-the-21st-century-scientist/.

Mecca, J.T., Gibson, C., Vincent Giorgini, V., Kelsey, E., Medeiros, K.E., Mumford, M.D., Connelly, S., 2015. Researcher perspectives on conflicts of interest: a qualitative analysis of views from academia. Sci. Eng. Ethics 21, 843–855.

Messerly, M., 2014. Citations for sale. Daily Californian. Available from: http://www.dailycal.org/2014/12/05/citations-sale/.

Meyer, M., Kircher, M., Gansauge, M.T., Li, H., Racimo, F., Mallick, S., et al., 2012. A high-coverage genome sequence from an archaic Denisovan individual. Science 338, 222–226.

Montes, C.C., Hammes, N., da Rocha, M., Montagnoli, T.L., Fraga, C.A.M., Barreiro, E.J., et al., 2016. Treatment with adenosine receptor agonist ameliorates pain induced by acute and chronic inflammation. J. Pharmacol. Exp. Therap. 358, 315–323.

Mullane, K., Enna, S.J., Piette, J., Williams, M., 2015. Guidelines for manuscript submission in the peer-reviewed pharmacological literature. Biochem. Pharmacol. 97, 224–239.

Mullane, K., Williams, M., 2017. Enhancing reproducibility: failures from reproducibility initiatives underline core challenges. Biochem. Pharmacol. 138, 7–18.

NASEM (National Academies of Sciences, Engineering, and Medicine), 2017. Fostering Integrity in Research. The National Academies Press, Washington, DC.

Nature, 2012. Editorial. Must try harder. Nature 483, 509.

Nature, 2014. Editorial. STAP retracted. Nature 511, 5–6.

Neill US, 2006. Stop misbehaving! J. Clin. Invest. 116, 1740–1741.

Neill US, 2007. How to write a scientific masterpiece. J. Clin. Invest. 117, 3599–3602.

Neimark, J., 2015. Line of attack. Science 347, 938–940.

Neubig, R.R., Spedding, M., Kenakin, T., Christopoulos, A., 2003. International Union of Pharmacology Committee on receptor nomenclature and drug classification. XXXVIII. Update on terms and symbols in quantitative pharmacology. Pharmacol. Rev. 55, 597–606.

Newman, A., 2013. The art of detecting data and image manipulation. Elsevier Connect. Editor's Update. Available from: https://www.elsevier.com/editors-update/story/publishing-ethics/the-art-of-detecting-data-and-image-manipulation.

NRC (National Research Council), 2011. Guide for the Care and Use of Laboratory Animals, eighth ed. National Academies Press, Washington DC.

O'Connor, T.R., Holmquist, G.P., 2009. Algorithm for writing a scientific manuscript. Biochem. Mol. Biol. Educ. 37, 344–348.

Oksvold, M.P., 2016. Incidence of data duplications in a randomly selected pool of life science publications. Sci. Eng. Ethics 22, 487.

Oransky, I., 2011. Montreal Heart Institute researcher dismissed following two retractions for image manipulation. Retraction Watch. Available from: http://retractionwatch.com/2011/09/03/montreal-heart-institute-researcher-dismissed-following-two-retractions-for-image-manipulation/.

Page, C.P., Curtis, M.J., Walker, M.J.A., Hoffman, 2006. The general mechanism of drug action. In: Page, C.P., Curtis, M.J., Walker, M.J.A., Hoffman, B.B. (Eds.), Integrated Pharmacology. third ed. Mosby, London, UK, pp. 15–50.

Plavén-Sigray, P., Matheson, G.J., Schiffler, B.C., Thompson, W.H., 2017. The Readability Of Scientific Texts Is Decreasing Over Time. bioRxiv, doi: http://dx.doi.org/10.1101/119370.

PLoS One Staff, 2014. In: Correction: HIV-1 Tat protein increases microglial outward K^+ current and resultant neurotoxic activity. PLoS One 9, e10921.

PLoS Pathogens Staff, 2014. In: Correction: human genome-wide RNAi screen identifies an essential role for inositol pyrophosphates in type-I interferon response. PLoS Pathog. 10, e1004519.

Plume, A., van Weijen, D., 2014. Publish or perish? The rise of the fractional author…. Research Trends. Available from: https://www.researchtrends.com/issue-38-september-2014/publish-or-perish-the-rise-of-the-fractional-author/.

Pop, M., Salzberg, S.L., 2015. Use and mis-use of supplementary material in science publications. BMC Bioinform. 16, 237.

Popper, N., Vlasic, B., 2012. Quiet doctor, lavish insider: a parallel life. New York Times. Available from: http://www.nytimes.com/2012/12/16/business/sidney-gilmans-shift-led-to-insider-trading-case.html?pagewanted=all.

Pulloor, N.K., Nair, S., McCaffrey, K., Kostic, A.D., Bist, P., et al., 2014. Human genome-wide RNAi screen identifies an essential role for inositol pyrophosphates in type-I interferon response. PLoS Pathog. 10, e1003981.

Radd, R., Appelbaum, P.S., 2012. Relationships between medicine and industry: approaches to the problem of conflicts of interest. Ann. Rev. Med. 63, 465–477.

Rang, H.P., 2006. The receptor concept: pharmacology's big idea. Br. J. Phramacol. 137 (Suppl. 1), S9–S16.

Rasko, J., Power, C., 2015. What pushes scientists to lie? The disturbing but familiar story of Haruko Obokata. Guardian. Available from: http://www.theguardian.com/science/2015/feb/18/haruko-obokata-stap-cells-controversy-scientists-lie.

Rockey, S.J., Collins, F.S., 2010. Managing financial conflict of interest in biomedical research. JAMA 303, 2400–2402.

Roncador, G., Engel, P., Maestre, L., Anderson, A.P., Cordell, J.L., Mark, S., Cragg, M.S., et al., 2016. The European antibody network's practical guide to finding and validating suitable antibodies for research. MAbs 8 (1), 27–36.

Ross, J.S., Hill, K.P., Egilman, D.S., Krumholz, H.M., 2008. Guest authorship and ghostwriting in publications related to Rofecoxib: a case study of industry documents from Rofecoxib litigation. JAMA 299, 1800–1812.

Rossner, M., 2006. How to guard against image fraud. The Scientist. Available from: http://www.the-scientist.com/?articles.view/articleNo/23749/title/How-to-Guard-Against-Image-Fraud/.

Rossner, M., Yamada, K.M., 2004. What's in a picture? The temptation of image manipulation. J. Cell. Biol. 166, 11–15.

Rutter, A.R., Poffe, A., Palmina Cavallini, P., Davis, T.G., Schneck, J., Negri, M., 2014. GSK356278, a potent, selective, brain penetrant PDE4 inhibitor that demonstrates anxiolytic and cognition enhancing effects without inducing emesis in preclinical species. J. Pharmacol. Exp. Ther. 350, 153–163.

Schekman, R., 2013. How journals like Nature, Cell and Science are damaging science. Guardian 2013. Available from: http://www.theguardian.com/commentisfree/2013/dec/09/how-journals-nature-science-cell-damage-science.

Shervick, K., 2015. Heart stem cell researchers' lawsuit against Harvard and BWH dismissed. ScienceInsider. Available from: http://www.sciencemag.org/news/2015/07/heart-stem-cell-researchers-lawsuit-against-harvard-and-bwh-dismissed.

Shidham, V.B., Pitman, M.B., Demay, R.M., 2016. How to write an article: preparing a publishable manuscript! CytoJournal 9, 1.

Sismondo, S., 2009. Ghosts in the machine. Publication planning in the medical sciences. Soc. Studies Sci. 39, 171–198.

Sloof, P., van den Burg, J., Voogd, A., Benne, R., 1987. The nucleotide sequence of a 3.2 kb segment of mitochondrial maxicircle DNA from Crithidia fasciculata containing the gene for cytochrome oxidase subunit III, the N-terminal part of the apocytochrome b gene and a possible frameshift gene; further evidence for the use of unusual initiator triplets in trypanosome mitochondria. Nucleic Acids Res. 15, 51–65.

Smith, A., 2015. How PowerPoint is killing critical thought. Guardian, Available from: https://www.theguardian.com/commentisfree/2015/sep/23/powerpoint-thought-students-bullet-points-information.

Smith, R., Gøtzsche, P.C., Groves, T., 2014. Should journals stop publishing research funded by the drug industry? BMJ 348, g171.

Snyder, S.H., 2013. Science interminable: blame Ben? Proc. Natl. Acad. Sci. USA 110, 2428–2429.

Spector, T., 1994. Writing a scientific manuscript. Highlights for success. J. Chem. Edu. 71, 47–50.

Spitale, R.C., Cheng, M.Y., Chun, K.A., Gorell, E.S., Munoz CA1, Dale, G., Kern, D.G., et al., 2013. Differential effects of dietary supplements on metabolomic profile of smokers versus non-smokers. Genome Med. 4, 14.

Steinbrook, R., Kassirer, J.P., Angell, M., 2015. Justifying conflicts of interest in medical journals: a very bad idea. BMJ 350, h2942.

Stern, S., Lemmens, T., 2011. Lega remedies for medical ghostwriting: imposing fraud liability on guest authors of ghostwritten articles. PLoS Med. 8, 1–5.

Sucher, N.J., 2013. The application of Chinese medicine to novel drug discovery. Exp. Opin. Drug Discov. 8, 21–34.

Tarca, A.L., Romero, R., Draghici, S., 2006. Analysis of microarray experiments of gene expression profiling. Am. J. Obstet. Gynecol. 195, 373–438.

The 1000 Genomes Project Consortium, 2010. A map of human genome variation from population-scale sequencing. Nature 467, 1061–1073.

To, W.S., Aungier, S.R., Cartwright, A.J., Ito, K., Midwood, K.S., 2015. Potent anti-inflammatory effects of the narrow spectrum kinase inhibitor RV1088 on rheumatoid arthritis synovial membrane cells. Br. J. Pharmacol. 172, 3805–3816.

Triggle, C.R., Triggle, D.J., 2007. What s the future of peer review? Why is there fraud in science? Is plagiarism out of control? Why do scientists do bad things? Is it all a case of: "All that is necessary for the triumph of evil is that good men do nothing?". Vasc. Health Risk Manag. 3, 39–53.

Truzzi, M., 1978. On the extraordinary: an attempt at clarification. Zetetic Scholar 1, 11–22. Available from: http://rr0.org/time/1/9/7/8/Truzzi_OnTheExtraordinaryAnAttemptAtClarification/index.html.

Tscharntke, T., Hochberg, M.E., Rand, T.A., Resh, V.H., Krauss, J., 2007. Author sequence and credit for contributions in multiauthored publications. PLoS Biol. 5, e18.

Tufte, E.R., 2006. Beautiful Evidence, second ed. Graphics Press, Cheshire CT, pp. 162–185.

Uhlen, M., Bandrowski, A., Carr, S., Edwards, A., Ellenberg, J., Lundberg, E., et al., 2016. A proposal for validation of antibodies. Nat. Methods 13, 823–827.

Wager, E., 2009. Recognition, reward and responsibility: why the authorship of scientific papers matters. Maturitas 62, 109–112.

Walbot, V., 2009. Are we training pit bulls to review our manuscripts? J. Biol. 8, 24.

Ware, M., Mabe, M., 2015. STM Report, fourth ed. Available from: http://www.stm-assoc.org/2015_02_20_STM_Report_2015.pdf.

Werren, J.H., Richards, S., Desjardins, C.A., Niehuis, O., Gadau, J., Colbourne, J.K., et al., 2010. Functional and evolutionary insights from the genomes of three parasitoid Nasonia species. Science 327, 343–348.

Wilkinson, L., Friendly, M., 2009. The history of the cluster heat map. Am. Stat. 63, 179–184.

Wislar, J.S., Flanagin, A., Fontanarosa, P.B., DeAngelis, C.D., 2011. Honorary and ghost authorship in high impact biomedical journals: a cross sectional survey. BMJ 343, d6128.

Xu, X., Liu, T., Zhang, A., Huo, X., Luo, Q., Chen, Z., et al., 2012. Reactive oxygen species-triggered trophoblast apoptosis is initiated by endoplasmic reticulum stress via activation of caspase-12, CHOP, and the JNK pathway in Toxoplasma gondii infection in mice. Infect. Immun. 80, 2121–2132.

Xu, X., Liu, T., Zhang, A., Huo, X., Luo, Q., Chen, Z., et al., 2015. Retraction for Xu et al. reactive oxygen species-triggered trophoblast apoptosis is initiated by endoplasmic reticulum stress via activation of caspase-12, CHOP, and the JNK pathway in Toxoplasma gondii infection in mice. Infect. Immun. 83, 1735.

Yahia, M., 2012. Are Saudi universities buying their way into top charts? House of Wisdom blog, Nature Middle East. Available from: http://blogs.nature.com/houseofwisdom/2012/01/are-saudi-universities-buying-their-way-into-top-charts.html.

Young, S.N., 2009. Bias in the research literature and conflict of interest: an issue for publishers, editors, reviewers and authors, and it is not just about the money. J. Psychiatry Neurosci. 34, 412–417.

Zimmer, C., 2012. After mistakes, scientists try to explain themselves. New York Times 2012. Available from: http://www.nytimes.com/2012/04/17/science/after-retractions-scientists-try-to-explain-themselves.html?r=0.

Zou, X., Sorenson, B.S., Ross, K.F., Herzberg, M.C., 2013. Augmentation of epithelial resistance to invading bacteria by using mRNA transfections. Infect. Immun. 81, 3975–3983.

Zou, X., Sorenson, B.S., Ross, K.F., Herzberg, M.C., 2015. Correction for Zou et al., augmentation of epithelial resistance to invading bacteria by using mRNA transfections. Infect. Immun. 83,1226–1227.

CHAPTER

5

Addressing Reproducibility: Peer Review, Impact Factors, Checklists, Guidelines, and Reproducibility Initiatives

Michael Williams, Kevin Mullane and Michael J. Curtis

OUTLINE

Research in the Biomedical Sciences. http://dx.doi.org/10.1016/B978-0-12-804725-5.00005-7

5.1 INTRODUCTION

Previous chapters in this monograph have focused on the scope, nature and impact of data irreproducibility in biomedical research, and provided guidelines (and some initiatives) for improvements in experimental design, validation, and standardization of experimental reagents, experimental execution (Chapter 2), data analysis (Chapter 3), and data reporting (Chapter 4). These guidelines are intended to improve the quality, relevance, and reproducibility of the data produced to support experimental outcomes and advance scientific knowledge.

Additional aspects for ensuring the relevance and reproducibility of research findings lie in the dissemination of these data in the form of peer reviewed publications. The present chapter focuses on the contentious topics of peer review for both journals and grant applications, disruptive forces in the traditional model of scientific publication, and the incentive systems for publication, one of which is the much derided but universally used *Journal Impact Factor* (JIF). Finally, additional guidelines and checklists for the submission of manuscripts to journals and for grant applications are discussed along with the current status of *Reproducibility Initiatives* (RIs).

5.2 PEER REVIEW

Scientific peer review is a quality control process that is generally considered to be one of the "sacred pillars of the scientific edifice" (Goodstein, 2000) that "signals to the body politic that the world of science and scholarship takes seriously its social responsibilities as a self-regulating, normatively driven community" (Lee et al., 2013). Additionally, peer review is viewed as "an essential and integral part of consensus building ... inherent and necessary to the growth of scientific knowledge" (Kronick, 1990) with the reviewer being the lynchpin about which the whole business of Science is pivoted" (Csiszar, 2016). Such comments are inevitably subjective, and it can logically be argued that the ultimate arbiter is more likely an editor or the chairman of a grant award committee, the individuals who make the final decision with input from the reviewers, and are responsible for ensuring that the peer review process is thorough, fair, and in general terms, appropriate.

Peer review has an important, career-defining role as its decisions "determine who gets published, who gets funded, and who gets promoted" (Souder, 2011) and thus it represents a key element in the economics of biomedical research that dictate the distribution of professional rewards, meritocratic, and otherwise (Ioannidis, 2014).

Analysis of peer review has largely been restricted to the topic of scientific publication, and the majority of the quotes and comments below refer almost exclusively to this arena. Peer review has also been described as "scientifically conservative" and "biased" in terms of the actions of both reviewers and editors (Lee et al., 2013); highly selective (Walker and Rocha da Silva, 2015); corrupt (Dzeng, 2014); a hurdle to innovation (Begley and Ioannidis, 2015; Foster et al., 2015; Horrobin, 1990); potentially a form of censorship (Casadevall and Fang, 2009) that inevitably "lends a spurious authority to reviewers…is unreliable, unfair, and fails to validate or authenticate….[and]…. a human activity...[where]… reviewers, like editors, may be partial, biased, jealous, ignorant, incompetent, malicious, corrupt, or incapacitated by conflicts of interest" (Rennie, 1999). This, is not necessarily a criticism of peer review itself, but rather a criticism of the way it is conducted and managed.

Opinions about peer review are not uniformly hostile. For example, some authors consider peer review as "a learning opportunity" (Taylor and Francis, 2015) and "a hurdle… that encourages authors to be self-critical prior to submission" (Ware, 2011). This has led to the widely held belief that "Anything that isn't peer reviewed...[is]...worthless" (RIN, 2010). Furthermore, while peer review does not guarantee that the decision to accept a research paper or fund a grant proposal is necessarily correct, it is difficult to conceive of any value in an alternative system that publishes or funds any and all papers or research proposals or, worse, makes a decision to publish/fund or not on an arbitrary basis that makes no attempt to consider the merit of the content of the submitted manuscript/grant. Peer review is therefore an obligatory part of biomedical research. The issue under discussion is how it can be effectively accomplished and validated. Hereon the focus is primarily on research publications.

Since the JIF is intimately associated with *traditional* or *classical prepublication peer review* (*CPR*), and will not be dealt with in detail until Section 5.7, it is helpful in the context of the following discussion to note that the JIF was originally developed by the distinguished bibliographer, Eugene Garfield, as a "quantitative tool... for ranking, evaluating, categorizing, and comparing journals" for use by librarians (Garfield, 1955, 2006). It has since been misappropriated as a widely used and contentious metric (Callaway, 2016) to measure to rank the productivity of researchers and grant applicants and, consequently, their work (Kaiser, 2013; Van Noorden, 2013a).

5.2.1 The Evolution of the Peer Review Process

While it is generally agreed that the first formal documentation of a peer-review process was made in Syria by Al Rahwi (854–931) in his book *Ethics of the Physician* (Spier, 2002), the origins of the present *CPR* system are less clear (Kronick, 1984). Some accounts note that the Royal Societies of London and Edinburgh introduced peer review in 1661 (Mulligan et al., 2013; Spier, 2002) and 1731 (Benos et al., 2007; Ware, 2008), respectively, while other accounts (Biagioli, 2002; Siegelman, 1998) document a similar role for the Académie Royale des Sciences of Paris around the same time.

The present format of the traditional prepublication peer review process, the *CPR* (Table 5.1) owes its genesis: (1) as a means to provide accountability for the increased funding of scientific research that occurred at the end of World War II (Biagioli, 2002; Burnham, 1990); (2) to the development of the photocopier to facilitate the sharing of documents (Grivell, 2006) and; (3) to the Internet with the latter facilitating electronic submission, review, revision and publication. This has exponentially accelerated the process, further facilitating anonymity, while increasing transparency. For authors, it is relatively easy to submit a manuscript online and editors find it easier to identify and recruit reviewers with the appropriate skill sets and expertize using a PubMed search for keywords from the submitted manuscript. For reviewers, searchable online access to the relevant literature can greatly facilitate the peer review process not only in terms of crosschecking author statements and references but also in assessing the author's familiarity with the current state of an area of research. Thus the peer review process has become far more streamlined, informed, and markedly faster as the online review process avoids the need for the delays that used to occur with any form of physical mail (including Fedex) and the "clunkiness" and unreliability of facsimile-based communications.

5.2.1.1 Peer review is...

Peer review involves the evaluation of a manuscript submitted to a journal by qualified experts who, with appropriate editorial guidance and oversight, attempt to ensure the competence, significance—biological as well as statistical—and originality in the subject matter of the manuscript and its presentation. There are only two final outcomes of the peer review process: *acceptance*—the submitted manuscript is accepted for publication—or *rejection*. Regardless of the dialog between the author and the journal editor(s) this is the binary endpoint toward which the trajectory of the process is targeted.

There is a widespread misconception that anything published following peer review must, by definition, be correct. Despite such expectations, peer review does not provide, and was never intended to provide, a reliably effective means for the detection of fraud, transparency, and/or reviewer bias (Hames, 2014) nor, to the disappointment of fringe elements in science, can it assess and consequently guarantee whether a submitted article is reproducible (Santori, 2016). Accordingly, Rennie (1999) has noted that peer review should not be equated "with the granting of a stamp of perfection." Rather it assures the intrinsic quality of the submission which can accordingly increase the probability of reproducibility.

TABLE 5.1 Peer Review and Author Feedback—Categories and Definitions

Type	Subcategory	Format	Pros	Cons	Comments
PEER REVIEW					
Classical or Traditional Peer review—*CPR* (Section 5.3.2.1)	*Single Blind*	• Editor invites reviewer(s) who know the identity of author • Author does not know who reviewers are. • Publication- • Online via DOI followed by volume/page citable "print" version	• Reduced concern of retribution from author to reviewer • Encourages frank response	• Lacks transparency • Facilitates ◦ reviewer bias ◦ censorship ◦ vindictive behavior • Conflicts of interest • Abuse of privileged information	• Unfair • Increased likelihood of detecting plagiarism
	Double Blind	• Editor knows identity of author and reviewers • Author and reviewers do not know identity of one another. • Publication—Online via DOI followed by volume/page citable "print" version	• Reduces reviewer bias and Conflicts of interest • Reduced concern of retribution from author to reviewer • Encouragement of frank response.	• Circumvented by reviewer guessing identity of author based on self-citation of references in manuscript, awareness of researcher focus in area, writing style	• Logistically demanding—sanitizing submitted manuscript • Increase turnaround time • Unpopular with reviewers
Open Peer Review—OPR (Section 5.3.2.2)	• Peer reviews published with full attribution to the reviewer • Reviews published anonymously	• Identity of author and reviewer(s) known to one another. • Publication—Online via DOI followed by volume/page citable "print" version	• Transparent • Reduces reviewer bias and Conflicts of interest • Reviews more positive and collegial	• Bias for positive reviews to avoid retribution • Fear of retribution from author to reviewer with negative review • High reviewer refusal rate reviewers "too busy, and lack[ed] sufficient career incentive to review" (Nature, 2006a,b) • Longer times for peer review process	• Logistically complicated

(*Continued*)

TABLE 5.1 Peer Review and Author Feedback—Categories and Definitions (*cont.*)

Type	Subcategory	Format	Pros	Cons	Comments
Postpublication Peer review—PPPR (Section 5.3.2.3)	1. Review by formally invited reviewers, after publication of the un-reviewed article 2. Review by volunteer reviewers*, after publication of the un-reviewed article. 3. Comments on blogs or third party sites, independent of invitation and any formal peer review that may have already occurred on the article.	"Publish first/referee openly later" Model, Following editorial assessment of quality, absence of plagiarism, suitability, and adherence to journal guidelines. Submitted manuscript, published on line with a DOI and formally reviewed. Reviewer reports published on journal website the name of the reviewer Revisions and additional reviews appended as they occur. Publication • Online via DOI • Reviewed article with reviews • Volume/page citable "print" version that can be updated as different versions	• Transparent • Reviews more positive and collegial • *F1000Research* uses a system similar to "traditional" peer review	• Rigor of initial editorial assessment questionable • Expertise of often anonymous volunteer reviewers unknown • Confounded by idiosyncrasies in use of PPPR between different journals/publishers, for example, formal, volunteer, and blog reviewers	Challenges • Low levels of participation in voluntary PPPR results in articles not being reviewed. • Ensuring voluntary reviewer expertise –*Science Open* requires potential reviewers to have five articles in ORCID, *PubMed Commons* at least one article in *PubMed.* However, neither can check that reviewer's previously published work and expertise is relevant to in the article being commented on. • PPPR discussion fragmented occurring on multiple sites, for example, *PubPeer, PubMed Commons, F1000Research, Researchgate*, blogs and *Twitter* (Van Noorden, 2014a).
Open Data Peer Review—ODPR (Section 5.3.2.6)		Peer review of raw data supporting a submitted manuscript either accompanying a submission as supplemental material or in a data database (*Dryad Digital Repository figshare, GigaScience Scientific Data*) to ensure data are in a format standard for the field, understandable, and can be reused (Shaklee and Cousijn, 2015)	• Makes data more discoverable, interpretable and reusable • Peer review involves an assessment "of the technical rigor of the procedures used to generate the data, the reuse value of the data, and the completeness of the data description (http://www.nature.com/sdata/for-referees#criteria) • Facilitates reanalysis of raw data supporting conclusions in a manuscript using "high-quality information ...for the possibility that new nuggets of useful data are lying there, previously unseen" (Longo and Drazen, 2016).	• No clear guidelines for use (Murphy, 2016). • If supplementary material is important why is it not in body of paper? • Lack of "data persistance" (Murphy, 2016). • Enables research parasites (Longo and Drazen, 2016).	Mandated by federal and private research funding bodies, for example, Gates Foundation, Biotechnology and Biological Sciences Research Council (BBSRC), Canadian Institutes of Health Research (CIHR), Cancer Research UK, Howard Hughes Medical Institute (HHMI), the UK Medical Research Council (MRC), the National Science Foundation (NSF), the National Institutes of Health (NIH), and the Wellcome Trust (Hahnel, 2015).

FEEDBACK TO AUTHOR					
Postpublication Commentary—*PPC* (Section 5.3.2.4)	Comments on blogs or third party sites, independent of any formal peer review that may have already occurred on the article	• Uncurated, unstructured • Multiple forums (Van Noorden, 2014a).	In high interest areas, multiple comments result in immediate response—for example, STAP (Stimulus-triggered acquisition of pluripotency; Obokata et al., 2014a,b) papers (Rasko and Power, 2015)	• Anonymous • "vast majority of *PPC* is negative" (Knoepfler, 2015). • response rates only 5%–13%. • "online commenting and blogs have generally contributed little. Of the thousands of papers published every year, only a few attract substantive comments" (Nature, 2010) • Easily gamed	• Associated in researchers' view with Open Access (OA) journals • Issues with actual impact of PPPR/*PPC* (Allison et al., 2016)
Preprint services/ online data repository (Section 5.6.1)	*bioRxiv*—free online archive and distribution service for unpublished preprints in the life sciences *PeerJ Preprints*	Not peer-reviewed, edited, or typeset before being posted online but do undergo basic screening for offensive and/ or nonscientific content and are checked for plagiarism.	Rapid dissemination—generally free Citable	• Ingelfinger rule (Marshall, 1998) precludes being publishable in regular journals • Value and extent of reviewer feedback • Funding agencies and academic committees view of value of unrefereed preprints unknown.	*Nature Precedings* (http://precedings.nature.com) discontinued as the result of "technological advances and the needs of the research community …[that]… have evolved to the extent that the *Nature Precedings* site is unsustainable as it was originally conceived."

Peer review can aid in preventing the publication of work that: (1) has been poorly conceived (often lacking any logical concept of a hypothesis), designed and executed; (2) has flaws in its design or methodologies (no a priori definition of endpoints or statistical testing, no randomization, etc.); (3) is ambiguously, selectively, and/or incorrectly reported; (4) reaches invalid conclusions that are inconsistent with the data reported, and; (5) lacks context to the literature, failing to acknowledge the work of others (Hames, 2014). Thus critical and objective feedback from effective peer review can in theory help in ensuring that the results reported are from properly designed experiments, are correctly interpreted, are neither too preliminary nor too speculative, and avoid impeding innovation or stifling creativity. In this context, journal editors and peer reviewers have an obligation to be especially skeptical and ensure that the burden of proof for "extraordinary claims" (as in the case of the 2014 STAP fiasco (Chapter 1.10.6.4. Normile and Vogel, 2014; Rasko and Power, 2015)) is reflected in the necessary "extraordinary data" (Skeptico, 2008). On this topic, Begley and Ioannidis (2015) further noted that "Reviewers can have no knowledge as to what data investigators have chosen to exclude. Reviewers cannot know whether data was strung together post hoc simply to create the best story. Of necessity, they take the work at face value" and further commented that "The principal responsibility for research findings rests with the investigator and their host institution" with the latter being viewed as providing oversight and penalties for noncompliance with guidelines for research integrity (Begley et al., 2015). Thus, rather than expecting that peer reviewers should function as the primary filter for fraud and plagiarism, it must be the responsibility of the biomedical research community as a whole, researchers, institutions, and granting bodies, to develop practical solutions that can have a positive and tangible impact on addressing reproducibility and detecting fraud.

5.2.1.2 *Surveys on the Perceived Usefulness of Peer Review*

Peer review has been the topic of regular surveys, with 10 covering the period 1999–2015 (Mulligan et al., 2013; Taylor and Francis, 2015). These have repeatedly reported that effective peer review improves both the quality and transparency of the content of a scientific manuscript and its readability, providing a "signal of quality" (Armstrong, 1997).

In a 2007 survey conducted by the UK Publishing Research Consortium (PRC) (Ware, 2008), 64% of respondents were reported as being satisfied with the current system of peer review with only 12% being dissatisfied (Table 5.2). Of the different forms of peer review, 56% of respondents indicated that they preferred double blind with 25% preferring single blind, 13% open, and 5% postpublication review. In terms of the effectiveness of the peer review process, 71% of respondents thought that double blind was the most effective, 52% single blind, 31% postpublication and 25% open. Astute readers will note that the cumulative totals assessing the effectiveness of peer review add up to 187%, a reflection of multiple responses being allowed (Ware, 2008). Subsequent surveys (Mulligan et al., 2013; Taylor and Francis, 2015) reinforced many aspects of the PRC survey (Table 5.2).

5.2.2 Concerns With Peer Review

Few topics in biomedical research, other than research funding and the JIF, the latter of which is a direct output of the peer review process, have galvanized opinion in biomedical researchers to the same degree as peer review with numerous erudite and often emotive statements

TABLE 5.2 Peer Review User Survey Results

Criteria		Ware (2008)	Mulligan et al. (2013)	Taylor and Francis (2015)
	Number of respondents	3,040	4,037	7,438
Current Peer Review System	Satisfied	64%	69%	78%
	Dissatisfied	12%	9%	
	Plays vital role	84%	84%	
	Could be improved	32%	32%	
	Unsustainable	19%	19%	
	Enjoy		86%	
	Improved submission	90%	91%	77%
Preference	Double Blind	56%		8/10
	Single Blind	25%		
	Open	13%		
	Post-publication	5%		
Effectiveness	Double Blind	71%	76%	
	Single Blind	52%	45%	
	Open	27%	20%	
	Post-publication	37%	47%	
Author Data Review				Neutral

that reflect on the innate bias of reviewers and editors, inconsistencies in reviewing standards, fairness and a lack of transparency, relevance, and effectiveness. This has led to the proposal of newer, Internet-based alternatives, all of which, to a major extent, suffer from the same flaws of those in *CPR* while adding new concerns. These are reviewed in detail in Section 5.3.2.

Both grant (Fang and Casadevall, 2009) and publication (Smith, 2006) review have been a subject of seemingly endless and contentious debate with many, if not the majority, of researchers concluding that it is a "fundamentally flawed" process (Jennings, 2006; Smith, 1997, 2006, 2010) with structural weaknesses that include the voluntary nature of the review process and lack of standardization in both the process and criteria (Ferreira et al., 2016). If these concerns are correct, can peer review be "fixed" or should it be abandoned?

It can certainly be argued that attempts should be made to improve it, and that it should not be simply abandoned in favor of a new system. To paraphrase Churchill, as nearly everyone who has written on the topic of peer review has already done, "peer review is the worst means by which to ensure scientific validity, except for all the others" (Rennie, 1999; Triggle and Triggle, 2007). This viewpoint was also echoed in the comments of Moore (2006) who stated, "that it's the least-bad system that can be devised, and that, although it might need tinkering with, its fundamentals should remain intact." Similarly, at presentations to the House of Commons Science and Technology Committee (2011) in the UK, Beddington (2011) noted that "If you posed the question, "Is the peer review process fundamentally flawed?" I would say absolutely not. If you asked, "Are there flaws in the peer review process which can be appropriately drawn to the attention of the community?," the answer is yes," while Goodlee (2011) the Editor of the *BMJ* added "At its best I think we would all agree that it …[peer review]...does improve the quality of scientific reporting and that it can improve, through the pressure of the journal, the quality of the science itself and how it is performed."

Finally, Wilsdon et al. (2015) as an outcome of the seminal UK *Metric Tide* report, made the definitive conclusion that "Peer review is far from perfect, but it is the least worst form of academic governance we have, and should remain the primary basis for assessing research papers, proposals and individuals." Despite such favorable assessments, peer review remains open to improvement and probably will be improved as long as research is undertaken by human beings (Rennie, 1999). Furthermore, it is unlikely that a credible peer review machine or effective peer review software will be invented—witness the iThenticate plagiarism software that while useful in theory, is ineffective without qualified human input - that of editors and reviewers.

5.2.3 Peer Review—Ineffective?

Faulty peer review can lead to the acceptance of papers that should have been rejected (e.g., the STAP papers from Obokata et al., 2014a,b) and conversely reject papers that in hindsight should have been published.

A frequently cited example of the failure of peer review in the biomedical sciences is elaborated in a paper published some 35 years ago by Peters and Ceci (1982). These authors selected 12 articles in the field of psychology that had been previously published by authors from leading psychology departments in the United States, one from each of the 12 most highly regarded American psychology journals that had rejection rates of over 80% and blinded reviewer practices (e.g., the reviewers knew the identity of the authors but not vice versa). Substituting fictitious author names and institutions (removing the cachet and implicit integrity and expertize of the original authors and their institutions), the altered manuscripts were then resubmitted to the same journals that had originally refereed and published the articles 18–32 months previously. Of 38 editors and reviewers, only three (8%) identified the articles as resubmissions (albeit with new, fictitious authors; Ceci and Peters, 2014). As a result, 9 of the 12 articles continued through the peer review process. Astonishingly 8 of the 9 (all of which had previously been accepted) were rejected. The grounds for rejection were in many cases described as "serious methodological flaws." Clearly the referees involved in the peer review took a different viewpoint to work that emanated from comparatively unknown institutions and authors compared with what were deemed more respectable ones, and justified their different attitude on the basis of the intrinsic value of the work itself—an archetypal example of bias in nonblinded peer review.

Regardless of the provenance of the authors' institution, there are other well-known examples of seminal research articles that were initially rejected. These include Krebs' paper on his eponymous cycle (Krebs and Johnson, 1937), Mullis' paper on the polymerase chain reaction (PCR; Mullis and Faloona, 1987) and Warren and Marshall's (1983) work on *H. pylori* and peptic ulcers that appear to have been initially rejected simply because they were unusually innovative—that is, "outside the mainstream"—but nonetheless the work reported garnered Nobel Prizes for all these researchers.

Such outcomes were consistent with Horrobin's (2001) comments that "Peer review is the core system by which the scientific community allots prestige (in terms of oral presentations at major meetings and publication in major journals) and funding is a nonvalidated charade whose processes generate results little better than does chance." This was an advance on similar comments that he had made a decade previously that peer review was adverse to novel ideas, for example, innovation, and instead was focused on publishing science that conformed to current mainstream thought (Horrobin, 1990). Horrobin's jaundiced viewpoint of the value of peer review found support some 25 years later in a study that was focused on the clinical literature (Siler et al., 2015).

In a cohort of 1,008 papers submitted to three "elite' medical journals, the *Lancet*, *BMJ* and *Annals of Internal Medicine* only 62—approximately 6% of the total—were accepted with 722 being editorially (desk) rejected before peer review. Of the 808 manuscripts that were eventually published elsewhere, the 14 that were most highly cited were part of the 946 submissions rejected by the three journals with 12 of the 14 not considered worthy of peer review. Siler et al., concluded that while the peer review system added value it was "ill-suited to recognize and gestate the most impactful ideas and research."

Other scathing, yet erudite, views on the limitations of peer review include those of the controversial *Lancet* editor, Richard Horton (2000), who noted, "The mistake, of course, is to have thought that peer review was any more than just a crude means of discovering the acceptability—not the validity—of a new finding. Editors and scientists alike insist on the pivotal importance of peer review. We portray peer review to the public as a quasisacred process that helps to make science our most objective truth teller. But we know that the system of peer review is biased, unjust, unaccountable, incomplete, easily fixed, often insulting, usually ignorant, occasionally foolish, and frequently wrong."

Regardless of such comments, there is an awareness that there is a home in print for every work. Rennie (1986) who, in writing what has been described as "the greatest sentence ever published in a medical journal" (Smith, 2010), noted that "there seems to be no study too fragmented, no hypothesis too trivial, no literature too biased or too egotistical, no design too warped, no methodology too bungled, no presentation of results too inaccurate, too obscure, and too contradictory, no analysis too self-serving, no argument too circular, no conclusions too trifling or too unjustified, and no grammar and syntax too offensive for a paper to end up in print." This statement is even more relevant with the expansion of the sector—the creation of new journals, many of which are virtual and do not produce a print copy—that make use of the infinite amount of space available online (Peres-Neto, 2016).

Critics have argued that there is little in the way of consistent evidence that peer review is effective (Smith, 2010), with the existing process being viewed as managed and undertaken by untrained persons (what training does anyone get in "peer review?") appointed on the basis that those who play the game can best referee the game (i.e., that poachers make the best gamekeepers). Such a view is supported by studies like that of Peters and Ceci (1982). That said, there are few solid facts about the process, allowing the proliferation of strongly held opinions with the hubris of complacent proponents being matched only by that of the vituperation of their opponents. In this context, lack of proof does not indicate a proof of a lack of standards, or efforts to implement standards.

Another recent viewpoint (Snyder, 2013) is that, due to the facile techniques of molecular biology, reviewers have become unreasonable in requesting manuscript revisions. While these are feasible to do and nice to have, they can often take many months to complete and often add little to the existing content of a manuscript under review while significantly delaying its publication This has led Snyder to suggest that journals "should provide expeditious reviewing and reasonable requests for revision...[and]...reviewers should be trained in this modified approach." In such circumstances, Triggle and Triggle (2017) suggest that "the editor ....assert their ultimate authority and better referee the referees."

Concerns about the peer review process were encapsulated in yet another seminal comment by Smith (1997) who noted that peer review is "slow, expensive, profligate of academic time, highly subjective, prone to bias, easily abused, poor at detecting gross defects, and almost useless for detecting fraud." This is probably no longer true, since there are numerous examples of efforts to make peer review faster, fairer and critically, more rigorous (Section 5.3.2). Nonetheless, nothing really substantive appears to have yet changed in the 20 years since the landmark Smith article while the recent development of guidance and checklist initiatives for publication have not yet had time to be evaluated for effectiveness. Unfortunately the slow progress suggests that the erudite, often sarcastic and entertaining statements made 20 years ago (and since) might be perceived as less about improving the peer review process and more an opportunity to engage in a spirited intellectual discourse reminiscent of an 18th Century London coffee house (Green, 2013).

5.2.4 Peer Review Oversight and Transparency

While peer review is far from irredeemably flawed, it may be considered to lack adequate oversight. Firstly, in regard to the criteria used to recruit reviewers. Having published in an area does not automatically qualify a researcher to have the knowledge, experience, insights, and temperament necessary to be a reviewer. However, initial editorial oversight as to the quality, clarity, and timeliness of a reviewer can provide guidance to their suitability and there are instances where a single substandard review from a reviewer can immediately curtail any additional assignments. Secondly, oversight of peer review can be viewed as an exercise in futility. If every piece of peer review were itself subjected to peer review, then those responsible for peer reviewing the peer review—the editors—may just as well have undertaken the original peer review themselves. In such circumstances, peer review is indeed intrinsically flawed in as much as it is entirely dependent on trust in the skill and integrity of the reviewers, many of whom are unknown to the editors.

5.2.4.1 Reviewers

As independent experts in their field, reviewers perform an invaluable and *pro bono* service to the scientific community. However, verifying independence and expertize can be challenging especially as the system relies upon the selection of individuals who appear to be expert (have been successful in the existing peer review system, have well-funded laboratories and a large list of relevant publications). Their independence is tested only in terms of whether the journal editor perceives a conflict of interest (CoI; Dunn et al., 2016) avoiding inviting colleagues from the same institution as the author to act as reviewers, or reviewers from the biopharma industry who are automatically viewed as conflicted (Stossel, 2015), or if the invited reviewer voluntarily declares a CoI. In such instances the declaration of a CoI or the perception of a CoI may have no relevance if the reviewer behaves in an ethical manner and if the conflict has no relevance in how the individual functions as a peer reviewer. Indeed, it can be argued that if a researcher has the required expertize by working in the same field, they are conflicted by definition. This is a Catch 22 situation that has been a consistent constant issue with FDA Advisory Committees, members of which are selected because they are experts in a field of medicine and have the necessary insight, experience, and experience to advise the agency on new drug approvals (Husten, 2015). These individuals are often called out by the mainstream media and politicians as being conflicted (Kaplan, 2017), and in some instances because of individual ethical lapses they indeed are (Keefe, 2014). It should be noted however, that often the potential for conflict is not the same as an individual actually being conflicted and the vast majority of the FDA's experts perform an invaluable public service. It would also be lunacy to use an "expert" to advise the FDA based on their lack of perceived CoI rather than true expertize. The same situation exists in peer review.

Selecting individuals to perform peer review to date assumes that both they and the authors are ethical and honest. Many reviewers use their access to submitted manuscripts—the content of which is not always in the public domain—to legitimately stay on the cutting edge of science. Others may use it to gain competitive advantage either by gleaning what is confidential information or by rejecting the paper so that that they can be the first to publish. More recently, some authors have used the anonymity and lack of transparency of the Internet to "scam" the peer review process by "reviewing" and recommending their own papers (Callaway, 2015; Ferguson et al., 2014). Such situations may be the result of inadequate Editorial Office oversight that allows reviewers to submit papers using non-institutional email addresses, for example, gmail, Yahoo, or others of a suspect nature (Section 5.3.1) with no recorded institutionally affiliated email. This is a situation that can be potentially rectified by the use of the *ORCID* (Open Researcher and Contributor ID) digital identifier to authenticate the origin of the corresponding author (Meadows, 2015a).

5.2.5 Grant Peer Review

Peer review is also a key component of the grant application review process. However, the assessment of a submitted grant application often places a higher priority on the consideration of issues tangential to the grant proposal per se. Among these is the JIF of the journals in which a grant applicant has recently published; too low a JIF and the publication will be discounted as a measure of the significance of a grant applicant's work. Given that grant peer review is often a more interactive and formalized process that contrasts with the often solitary environment of journal peer review with the application being actively discussed among members of the study section. These are then directly ranked against other applications with "guidance" from the Chair of the Study section that can favor "allegiance-bias favoring grants that align with their own beliefs and interests" (Begley and Ioannidis, 2015); it is thus distinct from *CPR*.

5.2.6 A Clinical Bias in the Critique of Peer Review?

An important caveat regarding peer review is that its discussion has historically been limited to the realm of clinical medicine and has only recently begun to include preclinical research (Rennie, 1986; Rennie et al., 2003; Smith, 1997, 2006, 2010). Consequently, it has been colored by the preeminence of reports on randomized clinical trials (RCTs); by case reports that report subjective evaluations of findings often from single patients; by meta-analyses, the majority of which are part of the Cochrane Collection (Goldacre, 2012; http://www.cochranelibrary.com/cochrane-database-of-systematic-reviews/); by concerns regarding the impact of pharmaceutical companies in influencing what is published (Smith, 2010; Goldacre, 2012, 2015, 2016); by funding that is typically twice that in basic research, a reflection of the cost of clinical trials that individually can range from $1.4 million to $52.9 million (Sertkaya et al., 2016); and by an unusual high preponderance of iconoclasts in the form of editors.

There is a good reason for the level of scrutiny of the clinical literature and this relates to its provenance: clinical research directly affects human lives by directly impact new drug approvals and guiding standards of treatment in health care. Despite the major differences between preclinical and clinical research, the use of human subjects, the critical nature of the outcomes that directly impact new drug approvals and health care practices, the many lessons learnt in the clinical research area regarding peer review are of relevance for systematic application in preclinical research. Additionally, preclinical research is the area from which the majority of drug candidates are derived. Such research is critical for first in human (FIH), Phase I studies that cannot occur until comprehensive preclinical research is complete. Therefore the peer review of the preclinical research used for an IND submission should ensure that the data are reproducible thus removing this as a perceived barrier to effective translation (Dolgos et al., 2016; Plenge, 2016) although these data are not always readily available (Federico et al., 2014).

5.2.7 The Literature on Peer Review

In 2011, some 200 papers were published each year on the topic of peer review (Ware, 2011). By 2015, the number of hits for the term "peer review" was over 27,000 on *PubMed* and 45,000 on *Google Scholar*. The increase reflects the importance of the topic and also an increase in what has been termed "research on research" that involves metaanalyses of the published literature post publication. Such "research" typically involves little in the way of original data generation on the part of the authors, with outcomes that often lack any useful insight, are highly subjective and repetitive, and owing to the absence of the independent/original information (access to readouts from individual instances of peer review) have limited degrees of freedom (Ioannidis, 2016a). An additional problem with research on peer review is that it typically lacks any useful insights on improving the process other than to complicate it further while decreasing transparency although this may be changing with new journals like BMC's *Research Integrity and Peer Review* (Harriman et al., 2016).

5.3 PRACTICAL ASPECTS OF PEER REVIEW

Peer review begins when a manuscript is submitted by the corresponding author to a *bona fide* journal thus establishing a formal dialog between the author, the journal editors and the peer reviewers as to the suitability of the manuscript for publication. Each step of the process varies depending on the journal in terms of the type of peer review (Kriegeskorte, 2012), the time that it takes (Powell, 2016; Voshall, 2012), and the specific publication forum for its dissemination (Schekman, 2013a). This has led to considerable debate on how to improve the peer review process (Section 5.3.4).

As a point of clarification, the designation of what constitutes a journal is evolving as many journals are virtual as they do not exist in physical, printed form but are only published online. Printed or not, a *bona fide* biomedical research journal is one that has an Editor and Editorial Board who are known to the journal readership, has a physical office, originates from an established publisher or a learned society, conducts rigorous peer review, has an established history of publishing on a regular basis, a robust number of articles (that occupy between several hundred to the 31,000 pages per annum published by the *Journal of Biological Chemistry*) and are of sufficient interest to be routinely indexed in publisher or governmental databases, for example, National Library of Medicine NLM); MedLine/PubMed; DOAJ (Directory of Open Access Journals); EMBASE (Excerpta Medica database); OVID; SCIE (Science Citation Index-Expanded); SCOPUS; or Web Of Science and can be routinely searched on Google Scholar, PubMed etc., and contribute to a JIF that is regarded as "acceptable," that is, in the upper 20%–40% of journals in a given research sector. Such journals can be differentiated from predatory journals which are truly virtual to the extent of having no *bona fide* physical presence, no credible peer review and often have no published content (Beall, 2012, 2016; Shen and Björk, 2015).

5.3.1 The Manuscript Submission Process

As noted in Chapter 4.15, once submitted, a manuscript is typically checked for authenticity and assigned a reference number and/or, for example, a DOI (Digital Object identifier). It then undergoes a preliminary editorial assessment of its conformity to the interests, style, and format of the journal. The authenticity of the source of the manuscript can be checked by the Editorial Office to ensure that it is: (1) complete and (2) neither computer-generated nonsense (van Noorden, 2014a) nor fake (Bohannon, 2013), something that while occurring infrequently is highly visible and makes excellent copy for the mainstream media. Authenticity can usually be assessed by a cursory reading of the content and by checking the author's affiliation and email with the originating institute. It is often not clear whether either actually exists (Marcus and Oranksy, 2016; Sorokowski et al., 2017) and establishing this can be a major challenge especially when dealing with universities in certain countries where the use of institutional email addresses may not be the norm. Checking the website of an unknown research institution often leads to a site that is unverifiable. To address this situation, some journals will decline submissions where the corresponding author uses a personal email from servers like gmail.com, aol.com, yandex.ru, 126.com, 163.com, etc. unless there is additional registered information on their institutional affiliation on a publisher's website or via an ORCID affiliation.

Once these fundamental checks are complete, a submitted manuscript can be run through a plagiarism software program like iThenticate's CrossCheck (http://www.ithenticate.com/products/crosscheck) that identifies the textual overlap of the manuscript against CrossCheck's database of published works and Internet sources to generate a Similarity Score (SS). This indicates the percentage of the text in the submitted manuscript that matches text in other published documents or web page. There is no hard and fast number for a suspect

SS although typically, in the authors' experience 40% or greater requires an editor to assess whether the SS results originate from one or two external sources or is simply a cumulative score with 20 or 30 sources. Unless this software is specifically instructed to not do so, it will identify the articles cited in the bibliography as plagiarized, since many will have been cited elsewhere, so this needs to be considered when setting a threshold for triggering the need for human input to check for the possibility of plagiarism. This is easily to do when whole sentences adjacent to one another are highlighted by the CrossCheck software which should lead to rejection with plagiarization being present if 80% of the words in a sentence matched those in a previous published paper (Higgins et al., 2016). A high SS may however merely be a reflection of common scientific words that make a cumulative contribution. While iThenticate's CrossCheck can aid in detecting plagiarism, it is dependent on human oversight, e.g., an experienced reviewer, to interpret the output. Citations, references, quotations, and the title page are usually excluded from the SS, albeit this has to be custom programed and not all journals do this. Interestingly, in an article on plagiarism in the medical literature where iThenticate was used to analyze the abstract, introduction, results and discussion/conclusion sections, the median scores were in the range of 17–32% (Higgins et al., 2016).

A submitted manuscript can be rapidly rejected—a process also known as *desk* or *triage rejection*—by journal editors due to issues that include its lack of concordance with the aims and scope of the journal (scope mismatch), a lack of: comprehensibility; any reasonable justification for the study; inclusion of appropriate controls; transparently inappropriate data processing and statistical analysis; its adherence to the journal guidelines and house style; its completeness and whether the conclusions made are justified by the data (Thrower, 2012).

Assuming that the article is not rapidly rejected, or rejected following the first round of peer review (undertaken by reviewers assigned by the editor), authors may be asked to: reconsider/rewrite some of the text to improve the clarity of the presentation; make figures or tables clearer; reformat tables into figures, and vice versa; perhaps reanalyze some data and/or generate additional data (Snyder, 2013); and refine/redefine the logic of the initial hypothesis based on the data and/or the experimental design and execution. While this enters "P-hacking territory", the intent is to use the data, assuming it has not been been biased in its selection to adjust the hypothesis rather than the opposite. Any and all of these revisions can improve the original submission, an outcome that the majority of researchers perceive as being a positive aspect of peer review (Hames, 2014; Mulligan et al., 2013; Taylor and Francis, 2015; Ware, 2008, 2011; Wilson, 2012).

In predatory journals or those with a low JIF, peer review is often perfunctory or nonexistent. Conversely journals with a high JIF often encourage reviewers to proactively seek reasons to reject a paper because the journal "received more papers" than they can publish with many Editors expecting a 70%–80% or greater reject rate. This can lead to the paper being submitted elsewhere (and reviewed by the same reviewers with the same outcomes) or to the generation of lengthy and self-serving letters of rebuttal. MW recalls one paper where an author provided 21 single spaced pages rebutting 37 distinct points raised by three reviewers following review of the second version of the submitted manuscript. Rather than collegially and productively addressing the specific issues raised, the author chastised the reviewers as being "so called experts" who "trumped up concerns," were "misinformed" and "nonsensical" in their comments, lacked any "breadth of expertize" such that their feedback was "bordering on ignorance." The paper was rejected.

5.3.2 Formal Peer Review: Historical, Innovative, and Controversial

Peer review can loosely be grouped into: (1) traditional or *CPR* which includes single-blinded and double-blinded review; (2) *open peer review* (OPR) and; (3) *postpublication peer review* (PPPR) (Table 5.1). There are a number of other forums, formal, quasiformal and informal, where readers of the scientific literature can provide comments or "rate an article in a manner similar to" *Amazon.com; rottentomatoes.com; TripAdvisor.com* etc., (Swoger, 2014; Van Noorden, 2014b) although the value of such informal 5-star rating systems has been questioned since the criteria are unclear and the comments are skewed and rated for popularity rather than quality (Ware, 2011). They also represent a self-selecting section of the research community who like others to be aware of their viewpoints. The latter will be discussed to the extent possible under the rubric of *Post Publication Commentary* (*PPC*; Table 5.1).

5.3.2.1 Classical Prepublication Peer Review *(Table 5.1)*

CPR includes single-blinded (sometimes misleadingly called "open" as the referee knows the identity of the author) and double-blinded review *CPR*, both of which function to provide feedback to journal editors on the suitability of a manuscript for publication and to authors in terms of collegially constructive feedback that can be used to improve the quality of a manuscript.

5.3.2.1.1 TRADITIONAL OR "SINGLE-BLINDED" PEER REVIEW

Traditional or "single-blinded" peer review is the predominant form of *CPR* (Table 5.1) that occurs when the editors and reviewers of a journal are aware of the authorship of a submitted manuscript but the identity of the reviewer is not formally disclosed to the author even when the author requests the journal to use specific reviewers. This of course does not preclude the author guessing the identity of a reviewer, or prevent the reviewer from deliberately or inadvertently disclosing their identity by the nature of their comments or by insisting that their name be included in the feedback to the author. In blinding reviewer identity, single blinded *CPR* is intended to allow frank and uncensored comment without fear of retribution (Triggle and Triggle, 2007). This contrasts with the situation in OPR where a lukewarm or negative review offers scope for retribution whereby the author may subsequently give a low score for a subsequent grant application submitted by the reviewer or retaliate when reviewing a paper where the reviewer is an author. *CPR* also has the advantage of allowing the reviewer to place the submitted manuscript in the context of other publications from the same authors. In some instances, this has usefully led to the identification of self-plagiarism (Bonnell et al., 2012) when an author submits essentially the same manuscript to two different journals (Garner et al., 2013) or instances where authors misappropriate text from Wikipedia and data from the published work of others and present it as an original contribution (PLoS ONE Editors, 2016). A recent review (Higgins et al., 2016) of plagiarism in the medical literature that focused on 66 articles identified China as the major source of plagiarized articles (23) with the United States of America coming in as a distant second (10).

From the author perspective, single-blinded *CPR* lacks transparency thus providing an element of secrecy that facilitates both editor and reviewer bias, subconscious or overt, perceptions of censorship and conflicts where personal histories can impact the process. To avoid these possibilities, an author can request the exclusion of a particular reviewer(s) when submitting a manuscript for peer review. However, journals do not automatically acquiesce to this, and there are no agreed rules of engagement.

Additional shortcomings of single-blinded *CPR* include: the facilitation of vindictiveness via "pit bull like-behaviors" (Walbot, 2009) where reviewers having a bad day can experience the catharsis of providing a "devastatingly negative review of a manuscript representing years of work ..... on a difficult, unsolved question" without any fear of consequences. Conflicts of Interest (CoI) and reviewer bias (Section 5.4) are part of a wider issue that can evolve into the abuse of access to the "insider information" contained in the submitted manuscript. The reviewer, a competitor in the same area of research, can reject or delay publication of a manuscript while using its content to gain a competitive advantage in advancing their own research goals or to establish priority. Some have even misappropriated entire papers and published them as their own (Dansinger, 2017; Laine, 2017).

From the reviewer perspective, single-blinded *CPR* can eliminate bias, encourage a diplomatic yet frank response and allows a focus on the quality of the submission by eliminating concerns that would occur if his or her identity were known.

5.3.2.1.2 DOUBLE BLIND *CPR*

Double blind *CPR* reflects the situation where neither the author nor the reviewer is aware of the identity of one another but the editor is aware of both. This is perceived to be a fairer process than single blinded *CPR* as it precludes reviewer bias and, to a degree, CoIs. Despite its intent, double blind *CPR* can be easily circumvented by the reviewer guessing the identity of the author(s) based on author self-citation in the references (Lee et al., 2013; Snodgrass, 2007) and by experts in a given research area being well aware of one another's work, experimental approach, and writing style (Emerson et al., 2010; Nature, 2008). Double blind *CPR* can also be a very cumbersome process placing extra burdens on the logistics of the editorial process in sanitizing submitted papers (removing author names, institutions, etc.) markedly reducing the enthusiasm in reviewing a submitted manuscript while at the same time having the potential to further increase the time for turnaround of the review. If double blinding is extended to the editor as well as reviewers, it requires the journal having access to expert and experienced administrators to manage the submission process. This situation may also undermine the role of the editor.

5.3.2.2 Open Peer Review (OPR)

Open Peer review (OPR) is a form of *CPR* where the identity of the author of a submitted manuscript and the reviewer(s) are known to one another and where the reviewers' names—and sometimes the actual review and author responses—are published as part of the final accepted version of a paper. This allows the reader of a published manuscript to understand and follow the process of the review and its sequential iterations, avoiding any element of secrecy by making reviewers more accountable while improving the transparency of the review process (Amsen, 2014a).

Advocates of OPR view it as effectively addressing long standing issues of transparency, fairness, and accountability in the peer review process by making reviewers more honest and accountable, and also providing recognition to reviewers for their work (Mulligan et al., 2013; Ware, 2011). Additional perceived upsides are that the reviewers may be more collegial in their feedback and indeed there is anecdotal evidence that open reviews are somewhat more positive than traditional single blinded *CPR*. Its downsides however include: the inevitable concerns of repercussion from authors receiving less than favorable reviews from their peers (especially younger reviewers) with the subject authors being potentially involved in grant review and tenure/promotion recommendations of the reviewers (Triggle and Triggle, 2007); a higher refusal rate by invitees to act as reviewers (van Rooyen et al., 1999) and longer times for the peer review process (Lee et al., 2013). Furthermore, while OPR may be conceptually viewed as an improvement over single or double blinded *CPR*, its practical implementation and acceptance have been questioned (Nature, 2006a; Ware, 2011). Various OPR initiatives conducted by *Nature* concluded (Nature, 2006a,b) that *most* reviewers were "too busy, and lack[ed] sufficient career incentive, to venture onto a venue such as *Nature*'s website and engage with public, critical assessments of their peers' work." The qualification "most" in this assessment is disturbing; if most researchers decline to engage with OPR presumably due to workload issues and/or a lack of motivation it can be presumed that these are mostly busy, experienced, and successful researchers, leaving the less busy, unsuccessful researchers to carry the burden of OPR, not an ideal outcome.

The OPR process results in published reviews being appended to the accepted paper with full attribution to the reviewer or being published anonymously (Table 5.1). The *EMBO Journal* reported the anonymous OPR process was effective (Ware, 2011). The final verdict on the practical feasibility and utility of OPR remains a work in progress.

5.3.2.3 Postpublication Peer Review *(PPPR)*

PPPR is a more recent iteration of the peer review process (Table 5.1) where submissions are rapidly published online after a cursory administrative review—what has been termed as an "access review"—by the editorial office, for example, a check for completeness, assignment of a DOI (Digital Object identifier), and then reviewed. Peer review, depending on the journal, can be conducted by formally inviting reviewers, by seeking volunteer reviewers or via open review in a blog-like format which is more reflective of a *PPC* (Section 5.3.2.4) than PPPR per se.

PPPR has its origins in the decade or more of success of *F1000Prime Reports/F1000 Faculty Reviews*, a PPPR directory of top articles in biology and medicine that is selected from the already peer reviewed and published literature by the Faculty of 1000 (F1000), a group of distinguished experts that now numbers in excess of 11,000 (Tracz, 2015). The subsequent F1000 reports are then published on the journal site together with the name and affiliation of the reviewer. This "experts pick" approach is not by any means an alternative to traditional *CPR*, since the work had already been subjected to prior peer review by the original journal in which it was published. Other journals, for example, the *Nature Reviews* series, *Nature, Science* etc., also carry "experts pick"-type reports of articles of interest although these are usually written by journal staffers or freelance science journalists rather than a dedicated faculty.

The success of the F1000 approach led to the founding of the journal, *F1000Research* (Hunter, 2012) that uses a formal, open PPPR process. Like *CPR*, PPPR entails a formal social contract (Blatt, 2015) between the author and a journal editor where the author agrees to the concept of his or her work being judged by his/her peers and the editor assumes the responsibility to identify appropriate reviewers who in addition to being known to the editor—if not the author—are specifically invited to review articles because of their experience and proven expertize and whose feedback to the editor is neither anonymous nor lacking in transparency.

The editor is then responsible for moderating the review process to provide critical and constructive feedback to the reader to facilitate evaluation of the published article.

Thus, following an in-house editorial assessment of the quality and suitability of a submitted manuscript, it is generally published within 7 days of submission and then formally reviewed, in a "publish first/referee openly later" approach (Ware, 2011). The reviewer reports are published on the journal site together with the name and affiliation of the reviewer with revisions and additional reviews being appended as these occur. However, the PPPR process can appear subjective (Eyre-Walker and Stoletzki, 2013), confusing, amorphous, and highly fragmented (Bastain, 2014) with much of the intended benefit (especially when reviewers self-select) being lost to posterity as substantive feedback is often lost among trivial, uninformed and irrelevant comments. There is also a long standing cultural aspect where unsolicited, unpeer-reviewed responses to peer-reviewed work are viewed as impertinent by their recipient. It has also been noted that "there is no incentive to peer review in the current…[PPPR]…system" (Leek, 2016).

The precise meaning of "published" in the context of PPPR is also confusing to the reader (and perhaps the author) as it is used to designate different events—the uploading of an unreviewed article and the subsequent formal publication of an article along with its associated peer reviews. Thus the PPPR can involve three or more iterations of the manuscript: (1) the manuscript upload for PPPR as a DOI; (2) the reviewed article together with the reviews, solicited or otherwise; and (3) the revised article with modifications reflecting the peer review feedback. An example of this process is a paper on Open Access (OA) publishing by Tennant et al. (2016) in *F1000Resarch*. Version 1 was initially published as an upload on April 11, 2016 with a revised Version 2 being published on June 9, 2016. This consisted of a 52-page pdf that included highlights of the amendments made to Version 1 and was updated again on August 25, 2016 "with reservations." Of the 52 pages, 22 comprised the actual revised article followed by approximately 18 pages covering the OPR discussion followed by another 12 pages of discussion.

While OA publishing is topical and obviously of considerable interest and debate, many PPPR papers fail to attract discussion anywhere near that of the Tennant et al. example as a cursory evaluation of articles in *PLoS ONE* or *eLife* will attest. At what point the Tennant et al. (2016) paper will reach a final version is unclear as are the implications of the perceived need for multistage peer review (Pöschl, 2012) and what has been termed "the aftercare of research publications" (Bastain, 2014). Will a research article undergoing PPPR ever be available in a finished format or will science in the 21st century evolve toward papers that are infinite in their iterations and their need for "aftercare"?

Eyre-Walker and Stoletzki (2013) in a study that assessed scientific merit and impact of two datasets of biomedical papers published in 2005 by "a panel of experts," 716 funded by the Wellcome Trust and 5,811 from the F1000 database, both PPPR in nature, concluded that (1) "scientists are poor at judging scientific merit and the likely impact of a paper, and that their judgment is strongly influenced by the journal in which the paper is published" and (2) that "the number of citations a paper accumulates is a poor measure of merit and …that the Impact Factor, of the journal in which a paper is published, may be the best measure of scientific merit currently available."

Other journals that also provide PPPR in publishing original research findings include *PLoS One* and *eLife* both of which *also* use *CPR*, *PeerJ*, and *ProcPoS* (*Proceedings of Peerage of Science*), *eLife*, offers "rapid...constructive and fair review .... by "scientist editors" that "in the interests of openness and transparency.... [leads to publication of]... the most substantive parts of the decision letter after review and the associated author responses, subject to author agreement." One online PPPR vehicle, the *Winnower* (https://thewinnower.com/about) combines OA and OPR with a high level of social interaction and reader engagement (Kennison, 2016) and also features cover stories behind research that has had a major impact on science along the lines of *Citation Classics* (the Grain) and also behind research that was retracted (the Chaff).

The usefulness and degree of rigor of the initial editorial assessment for PPPR, at least for *PLoS*, has proven to be of questionable quality in two high profile cases where papers were retracted. Both originated from China. The first was on hospital acquired infection (Lai et al., 2015) and contained a number of instances of outright plagiarism together with writing that was substandard and included the actual misspelling of the title (PLoS ONE Editors, 2016). The second, a contentious paper on creationism (Liu et al., 2016) was retracted after a series of comments on Twitter and threatened resignations from *PLoS* editors that led to *PLoS* being described as an "absolute joke of a journal," having "peer-review standards... [that]... were too low" (Cressey, 2016) and "too many editors, and not enough editorial oversight" (Coyne, 2016).

5.3.2.4 Postpublication Commentary (PPC)

Another form of postpublication feedback is *PPC* (Table 5.1) which is frequently confused with PPPR proper. *PPC* has its origins in the historically established reader response to published scientific research which took the form of editorials, letters to the editor, etc. that while encouraging debate were often untimely with months passing between the published article and the appearance of the response. In 2012, the *PubPeer* initiative built on this "to accelerate the exchange of ideas and scientific progress" (Blatt, 2015; Stoye, 2015). *PPC* websites/blogs exist that can be personal, institutional, learned society or publisher hosted and are associated with a variety of journals—*Nature, Science, Scientific American, The Scientist, Chronicles of Higher Education, Times Higher Education, Royal Society of Chemistry,* and *PLoS*—as well as those that involve specialized biomedical research/drug discovery sites. A partial list of these include: *In the Pipeline* (http://blogs.sciencemag.org/pipeline/archives), *retractionwatch.com*, the *NIH Director's Blog* (directorsblog.nih.gov) *scholarlykitchen* (http://scholarlykitchen.sspnet.org), *sciencebasedmedicine* (http://sciencebasedmedicine.org), *PubPeer* (http://www.pubpeer.com); *bioRxiv* (http://biorixiv.org); *Research Gate's Open Review* (https://www.researchgate.net/publicliterature.OpenReviewInfo.html/); *Tree of Life* (http://phylogenomics.blogspot.com/); *RRResearch* (http://rrresearch.fieldofscience.com/); and the NIH's *PubMed Commons* (http://www.ncbi.nlm.nih.gov/pubmedcommons/). *PubMed Commons* has been described as "a virtual water cooler" (Collins, 2014) that is intended as a forum for scientists to "share opinions and information about scientific publications in *PubMed*." While these sites collectively have the potential to "lower the barrier to creating and spreading conversations" by becoming the equivalent of "on line journal clubs," they have provided a multitude of vehicles for an individual scientist to potentially post every smidgen of their scientific life—including raw data and publications (Faulkes, 2016)—before, during and after they have written them—while opening themselves to feedback, much of which has not been well received (Blatt, 2015; Faulkes, 2014).

A major issue with *PPC* is the ability of a reviewer to remain anonymous allowing he or she "to cross the line to engage in nonconstructive criticism much in the same way that typical anonymous prepublication *CPR* allows referees to cross the line, leading to a certain amount of destructive vindictiveness. It is much worse with *PPC* such that "the vast majority of *PPC* is negative and sometimes intensely so" (Knoepfler, 2015). Whenever peer review is anonymized, unless it is carefully moderated by experienced editors, it can degenerate to the level of one star reviews on *Amazon, Open Table, Trip Advisor,* etc., with no means by which to separate substantive and informed commentary from the irresponsible, malicious and trivial (Bastain, 2014), or to guard against innuendo and hearsay (Faulkes, 2014), which while not being unique to *PPC* can be particularly problematic.

The evidence from journal-hosted *PPC* commenting systems indicates that members of the research community have been reluctant to volunteer comment on the published work of other researchers (Nature, 2013; Neylon and Wu, 2009; RIN, 2010; Ware, 2011) with response rates of only 5%–13%. Nature (2010) noted that in their experience "online commenting and blogs have generally contributed little. Of the thousands of papers published every year, only a few attract substantive comments." This may reflect the 90-9-1 rule of social media (http://www.90-9-1.com/) where 90% of people observe, 9% make minor contributions and 1% are responsible for the majority of original media content (Neylon and Wu, 2009). Ware (2011) has also noted that there is little scope or feasibility for *PPC* to attract sufficient adherents for it to be representative of informed opinion arguing that if a "researcher ... currently reviews 1–2 papers and reads 20–30 papers a month, a move to postpublication review ...would mean having to read the 20–30 papers with the same degree of critical attention currently given to the 1–2 that were reviewed, a substantial increase in workload." Authors of *PPCs* therefore tend to represent a self-selecting, proactive group (the vocal 1%?) that enjoys participating, and also have the ability to prioritize the time to participate in the dynamic discourse of the blog format. They move from being passive consumers of the scientific literature to being active critics often with "primal and voyeuristic" tendencies (Blatt, 2015). Inevitably, there is also a major scope for gaming of the review system via *PPC*.

The *PPC* process may be of little interest and have little perceived value to experienced researchers versed in the traditions and civility of *CPR*. This is likely to skew the overall tenor of the *PPC* process such that instantaneous responses are delivered without appropriate forethought or context by individuals lacking the necessary expertize, experience, or insight in the cognate field of study. As such, *PPC* becomes a crowdsourcing approach that represents a spectrum from the "wisdom of crowds" to the "tyranny of the majority" all orchestrated by "faceless judges" (Couzin-Frankel, 2013) in between which lies an expert, experienced, and insightful Internet-powered form of proactive peer review. While authors would no doubt find insightful commentary to be of value, the overall provenance of *PPC* feedback and lack of curation means the "faceless judges" will prevail, often precluding any possibility of a collegial exchange of ideas.

Indeed, an obvious question is, if the author of a *PPC* has a legitimate concern, why would he or she not contact the author directly to discuss the topic of concern and resolve it (Allison et al., 2016) collegially rather than resorting to the public airing of personal disputes (Whittaker, 1979), regardless of how satisfying this may be to the originating author and amusing to the reader? The answer is quite clear; because if a journal or blog site provides a *PPC* forum, people will use it. There is also the possibility that researchers whose data has been criticized—potentially affecting its integrity and their reputation and career progress—will initiate legal proceedings against anonymous members of the "crowd" (Blatt, 2015; Faulkes, 2014;

Lowe, 2014). In this context, in December 2016, the Michigan Court of Appeals in considering a lawsuit against *PubPeer* to reveal the identity of a *PPC* commentator who asked "Are the identities of anonymous scientists who comment on other scientists' research online protected by the First Amendment?" (https://www.aclu.org/sites/default/files/field_document/20161206_c326667_125_326667.opn_.pdf) - ruled in favor of *PubPeer* (Cushing, 2016).

To conclude, while a minority of researchers have expressed the view that *PPC* (and PPPR; Teixeira da Silva and Dobránszki, 2015) can be a very useful supplement/addition to *CPR* they do not perceive it to be a logical or effective replacement for the latter. At the same time it is hard to see how *PPC* (or unsolicited PPPR) can be used as a supplement to a paper since this means that not only the paper and its authors would be subject to the whims and caprices of the noisy minority of commentators, but the original reviewers would be as well. If a reviewer asks for revisions and eventually accepts a paper and it is then torn to shreds via *PPC* or PPPR, it makes them look foolish or incompetent. Moreover, a reviewer would have no avenue of redress unless he/she waives anonymity or unless his/her anonymity could be preserved within the *PPC* or PPPR environment. However, being the object of trolling is probably not what the original reviewer is likely to have imagined when he/she signed up for peer review. Indeed, it is difficult to comprehend why anyone would agree to act as a reviewer in such an environment making it difficult to see the rational justification for *PPC* or PPPR in this context. Only if all peer review and *PPC* and PPPR were stripped of anonymity, and the process properly curated and moderated would it be remotely worth considering. Even then, its actual impact on reproducibility is likely to be small; if reviewers presently cannot identify sources of potential error or fraud leading to later failures in reproducibility, why are they likely to do better using OPR and PPPR with the addition of *PPC*?

Like the long held assumption that the majority of papers are rarely cited (Bauerlien et al., 2010; Meho, 2007; Remler, 2014) that led to the comment that "Most papers sit in a wasteland of silence, attracting no attention whatsoever" (Davis, 2011), it is probable that the paucity or absence of *PPC* is a similar reflection of a research article that lacks peer interest. This could be because it is obviously correct, obviously incorrect, or just utterly dull. While it has often been noted that "90% of academic papers are never cited" (Remler, 2014), the actual metrics supporting this statement are apocryphal, controversial, and not easily verified (Nelson 2015; Remler, 2014). Clearly the statement is unhelpful given that there is a very long tail of low JIF journals whose metrics distort such calculations.

Some *PPC* sites like *PubMed Commons* and *Open Review* do not allow anonymous or pseudonymous comments (Stoye, 2015), an emerging trend that is reflected in the *Open Evaluation Initiative* (OEI; Kriegeskorte, 2012) and the proposed *Peer Reviewers' Openness Initiatives* (PROI; Morey et al., 2016) that both recognize that removing anonymity is a key hurdle to making PPPR, in any of its myriad forms, a viable and collegial replacement for *CPR* (Bastain, 2014; Neylon and Wu, 2009). This is of especial concern as *PPC* is an obvious information source for altmetrics (Shema et al., 2014), the latter a vehicle for crowd-sourced peer review that has been argued to have the potential to replace the JIF (Priem et al., 2010) if an effective, inclusive and ongoing rating system that cannot be gamed can be established (Eisen, 2013; Kriegeskorte, 2012; Kyriazis, 2013). The Finnish peer review site, *Peerage of Science* (https://www.peerageofscience.org), generates subjective quality indices from reviewers that involve peer review of the peer review.

PubMed Commons has the potential to provide a more focused and structured format for providing open user comments on any of the 25 million plus indexed research articles on PubMed that would be similar to other online review formats (*TripAdvisor*, *Open Table*, *Rottentomatoes Amazon* etc.). Ideally, this would facilitate informed debate together with the documentation of corrections and retractions (Knoepfler, 2015) along the lines of the Tennant et al. (2016) paper. Issues identified with the previously mentioned PROI proposal (Morey et al., 2016) include a lack of reviewer access to the data supporting the submitted manuscript, with many journals not mandating data deposition and where editors and researchers remain unconvinced about the merits of data sharing (Bishop, 2016).

It remains to be seen whether *PubMed Commons* will, over time, become the forum for the centralization of *PPC* that avoids trivial and vindictive feedback (that itself would need to be extensively curated and edited) and the major time commitment involved in accessing multiple sites including those not associated with the original article. This has the potential for comments being more readily available to add to the intended discussion (Van Noorden, 2014b).

While the value in *PPC* as an *adjunct* to either *CPR* or PPPR remains debatable, on its own it appears to have dubious merit especially when it is argued as a means to avoid "the perverse incentive for authors to keep important insights and additional data until a subsequent publication" (Bastain, 2014) suggesting that articles with an open-ended peer review process will come to represent the eighth terrace of Dante's Inferno for the author.

5.3.2.5 PPPR in Practice

Since many individuals involved in the debate on peer review anticipate that PPPR will become an effective, if not superior, *replacement* for *CPR*, it is of interest to understand how well it actually works. Examples (Knoepfler, 2015; Stoye, 2015) of the success of PPPR are the retractions of the high profile STAP (Stimulus-triggered acquisition of pluripotency) papers from Obokata et al. (2014a,b) and the claim that the bacterium, GFAJ-1 could, in contrast to all other known forms of life, use arsenic as a nutrient in place of phosphorus (Wolfe-Simon et al., 2011). As a result of the "profound evolutionary and geochemical importance" of this latter finding, the original paper was accompanied by eight technical comments (Schiermeier, 2012) and was subsequently refuted (Erb et al., 2012; Reaves et al., 2012).

The furor in social media on both the STAP and arsenic findings not only showed the criticality of social media to those papers falling into the category of extraordinary claims—aka "too good to be true" (Blatt, 2015; Hayden, 2012). They also questioned the value of the original *CPR* process as having any effective standards. *CPR* should have caught these fraudulent papers. However, rather than being viewed exclusively as a paean to the virtues of PPPR or merely a further indictment of the shortcomings of *CPR*, it is important to note that these papers were published in the high JIF journals, *Nature* and *Science*, respectively that are: (1) unique in their quest for "impact" and "importance" in their publications and thus more risk oriented in the peer review process; and (2) extremely widely read. Thus the success of the PPPR/*PPC* process in the two instances cited may be unique to high impact journals and unlikely to be replicated in the single digit JIF literature. Seeking other, less mundane examples of the value of PPPR, is challenging beyond the examples cited by Knoepfler (2015).

Finally, the debate around the value and practicality of PPPR/OPR has been complicated by the fact that it is often confused or associated with Open Access Publishing (OAP; Section 5.6). Concerns with an inability to obtain a satisfactory response after providing PPPR/*PPC* feedback have been noted by Allison et al. (2016). In noting statistical errors in peer-reviewed articles that were substantial or invalidating, they found that their efforts to correct these issues were onerous resulting in "ineffective e-mails among authors, editors, and unidentified journal representatives, often without any public statement added to the article. Some journals that acknowledged mistakes required a substantial fee to publish our letter: we were asked to spend our research dollars on correcting other people's errors." Accordingly, they identified six problems in PPPR/*PPC* feedback that need to be "fixed" to make it more effective:

- Editors are often unable or reluctant to take speedy and appropriate action.
- Where to send expressions of concern is unclear.
- Journals that acknowledged invalidating errors were reluctant to issue retractions.
- Journals charge authors of PPPR/*PPC* feedback ($1,716–$2,100) to correct others' mistakes.
- No standard mechanism exists to request raw data.
- Informal expressions of concern are overlooked.

They recommended that "journals, publishers and scientific societies should standardize, streamline and publicize these processes" while also noting that "robust science needs robust corrections."

5.3.2.6 *Open Data Peer Review (ODPR)*

A second source of confusion for OPR is its association with ODPR, where the peer review process is extended beyond the actual data collated and summarized in a submitted manuscript to either supplemental material that accompanies the submitted manuscript (Ware, 2011) or the actual raw data (Grootveld and van Egmond, 2012). While ODPR can, in principle, be a useful addition to the *CPR* paradigm, it brings with it a new set of as yet unresolved issues. For reviewers, it represents an additional set of data, in some instances in the form of supplemental material, arranged in an inconsistent and sometimes incomprehensible manner that requires considerable attention, adding to the time required to review a paper. Also, there is no clear understanding as to precisely what data the peer review is expected to evaluate (Murphy, 2016).

Guidelines developed by Shaklee and Cousijn (2015) for ODPR involve the assessment of whether the protocols for data generation are available, are adequate, well documented and presented in a format that is standard for a given research field, are reusable and whether the authors adequately explain the utility of the data to the research community. With this guidance, from a practical viewpoint, a reviewer would need to be able to access and understand the data and decide if it makes sense. Beyond this it is unclear what is done with the information and how the reviewers would appropriately query the authors.

Nonetheless, in concept ODPR makes it possible, to varying degrees for the reviewers and other scientists, if they have the time and inclination, to reanalyze the actual data used to prepare the manuscript reexamining "high-quality information ...for the possibility that new nuggets of useful data are lying there, previously unseen" (Longo and Drazen, 2016). This would certainly prevent the "one out of six, best story scenario" described by Begley (2012) that contributed to an inability to reproduce published findings in preclinical cancer research (Begley and Ellis, 2012).

Several organizations are involved in making raw data available and include: *Dryad Digital Repository* (datadryad.org); *figshare* (figshare.com); and *GigaScience* (gigascience.

biomedcentral.com) as well as *Scientific Data*, an open-access, peer-reviewed journal that is part of the Springer Nature Group (http://www.nature.com/sdata/) that is focused on "descriptions of scientifically valuable datasets. Our primary article-type, the Data Descriptor, is designed to make your data more discoverable, interpretable and reusable." Other databases provide access to published articles (*Pub Med Central* (PMC), *Web of Science*, *Google Scholar*), and some also provide the ability to add comments (*Zotero, Mendeley, Pub Med Commons*).

Open data policies, as in the case of OA, are being increasingly mandated by both federal and private research funding bodies with 34 of these requiring, and 16 encouraging, data archiving as of July 2015. These include the Gates Foundation, the Biotechnology, and Biological Sciences Research Council (BBSRC), the Canadian Institutes of Health Research (CIHR), Cancer Research UK, the Howard Hughes Medical Institute (HHMI), the UK Medical Research Council (MRC), the National Science Foundation (NSF), the National Institutes of Health (NIH), and the Wellcome Trust (Hahnel, 2015).

For data in a specialized data journal, for example, *Scientific Data*, peer review involves an assessment "of the technical rigor of the procedures used to generate the data, the reuse value of the data, and the completeness of the data description ...(http://www.nature.com/sdata/for-referees#criteria) .... with full public release of the research data through a trusted community data repository ... before publication of the Data Descriptor." In this publication, authors are expected to "convince referees that ...[the]...data are worthy of wider use in the scientific community." It is unclear precisely what this means and the reviewer response is unknown.

Concerns have been raised regarding the curation of raw data in repositories and the effects of fluid content, that is, "dataset persistence" (Murphy, 2016) where published datasets are not fixed entities but are "highly dynamic, regularly growing...duplicated, split, combined, corrected, reprocessed or otherwise altered during use" (Mayernik et al., 2015). Thus there are major differences between the data that is made as "permanently available as possible on the Internet" and the data uploaded by an investigator on a website that is modifiable by the author, and is present without any commitment to digital archiving (no item of record, no DOI). An additional wrinkle in the open data discussion, already mentioned, reflects concerns that the availability and reuse of data will lead to "a new class of research person ..."research parasites"... people who had nothing to do with the design and execution of the study but use another group's data for their own ends, possibly stealing from the research productivity planned by the data gatherers, or even use the data to try to disprove what the original investigators had posited" (Drazen, 2016; Longo and Drazen, 2016). Yet others believe that to maximize data generated is an obligation and promotes transparency, which is why key funding bodies make it a requirement.

A complication, albeit altruistically usefully intended, of the Open Data research model has been highlighted by Humphries (2016) following the release of a "landmark" data set of brain images from the Allen Institute of Brain Research (Shen, 2016). These data were described (Humphries, 2016) as "the first complete set of neural activity recordings released *before* publication..[that is] ....available to anyone, for free" to find something novel that could be the basis of a new publication. Humphries further noted that these data resulting from "a billionaire software designer's philanthropy" has the potential to disrupt the current model of biomedical research where the incentives are to raise money (grants) and increase output (publications) and citations to one where the "quality and rigor of science can be prioritized over ...output and money as measures of "success" providing "deep insights ...[into]

neuroscience" that could not have occurred outside a privately funded institute. This issue is discussed further in Chapter 6.

5.3.2.7 *Supplemental Materials*

The submission of supplemental material together with a submitted manuscript reflects a practice initiated by high profile journals to save print space and avoid diluting "the message" with mundane information like methods. This mandated that authors place essential methodological and validatory information in a separate section of supplementary material, leaving space in the paper itself only for the "message" (Collins and Tabak, 2014). These supplementary materials were often assumed by reviewers to be merely informative and thus exempt from formal review. When published, they were often in a much smaller font than the main text of the article (and often abbreviated) or only available online, in both instances to save space. The separation of the conclusions from the evidence, with the latter in a silo of uncertain provenance that was not peer reviewed—and often not peer reviewable—was a trend that added to the challenges of reproducibility that only now is being addressed and reversed. The meaning of supplementary material is rarely defined: what do journals expect reviewers to do with it? And why is it not in the paper itself? Additional discussion on this topic can be found in Chapter 4.6.4.

5.3.3 Reducing Entropy in Peer Review

Rejection of a manuscript by triage or after peer review is a normal outcome of the peer review process. While often perceived as a negative event, it represents a considerable investment of time and effort not only on the part of the author but also the reviewer. That said, a reviewer who finds that a paper is unfit to be published in a respectable journal can hardly claim his/her time has been wasted if the outcome is that the paper is rejected. Few researchers take on a peer review task with the mindset that it is worthwhile only if the outcome is acceptance. Nevertheless when a paper is rejected, it inevitably ends up in a journal with a lower JIF whose reviewers do accept it. Whether this is a reasonable outcome is a different issue. Once an author receives a rejection he or she can appeal it by discussion with the editor (although most journals do not entertain this), resubmit a revised manuscript that addresses the issues raised in the review process (this is actually a good strategy although it is not encouraged by journals as they are aware that resubmitted papers tend to be accepted almost by a process of attrition) or submit it to another journal and go through another round of peer review. The latter is the most common outcome, although not usually the best since it often means that serious flaws are disregarded by authors who hope that the reviewers from the second journal may not notice them; after all, if the problems identified by the first journal can be remedied, why not remedy them and resubmit? When the manuscript is rejected because it is unsuitable because of a disconnect with the scope of interest of a particular journal, it can be revised to conform with the requirements of another journal and resubmitted. For example, a paper that focuses on the characterization of a new drug candidate is inappropriate for a journal that has as its focus experimental methodologies, but it may be possible to change the emphasis of the paper and resubmit it. However, even when submitting to a different journal after the paper has been unequivocally rejected by the first, it is always prudent (and honorable/honest) for the author to address as many as possible of the issues identified in the initial review. Indeed, research communities are small and, members of the original group of reviewers may be asked by the editor of the second journal to review the manuscript that they had previously rejected. If no changes have been made the paper is likely to be rejected again by irritated referees who are likely to be developing an increasingly disfavorable attitude towards the author, adding a further loss of time and opportunity to the publication process.

5.3.3.1 *Transferable or Cascade Peer Review*

A number of initiatives to streamline the prepublication peer review process have been focused on introducing standardized review procedures, transferable peer review reports, and the new phenomenon of trickle down, or trickle over "cascade" peer review. This involves higher JIF journals that triage rejections on the basis of priority and pass the paper on to a lower JIF (often OA) secondary journal owned by the same publisher, thus avoiding the endless cycle of reviews to improve the efficiency of the process. To avoid an endless round of new submissions and reviews, rejected manuscripts together with their accompanying reviews can be submitted to a another journal in the same publishing house or to a publisher consortium (Economist, 2013a; Nature Neuroscience, 2008; Theissen, 2015; Warne, 2014) formalizing a long standing informal process in effect at several publishers to facilitate the publication of scientifically sound manuscripts. One interpublisher consortium is the *Neuroscience Peer Review Consortium* (http://nprc.incf.org) which was established in 2008 (Nature Neuroscience, 2008) and involves some 50 journals in the neuroscience area from a variety of different publishers that include the *American Physiological Society, the American Psychological Association, BioMed Central, Cold Spring Harbor Press, Elsevier, Hindawi, Imperial College Press, Informa Healthcare, IOS Press, Karger, Nature, Sage, the Society for Neuroscience, Springer, Wiley*, and *Wolters Kluwer*. Another cross-publisher consortium was formed in 2013 between *eLife*, *PLoS*, and the European Molecular Biology Organization (EMBO; Harold, 2013). If an author agrees, a manuscript rejected by one of the consortium members can be resubmitted to another journal in the consortium together with the reviewer's comments to date. The latter are usually anonymous although reviewers have the choice of identifying themselves. A key issue in regard to cascade peer review is the intent of finding a suitable venue for a scientifically sound manuscript, not to identify the lowest common denominator "peer reviewed" journal to propagate "bad science".

5.3.3.2 *Presubmission Peer Review*

Another iteration in the peer review process is prepeer review where peer review is conducted independently of any publisher via third party entities like *Axios Review, Enago, Endaz, Editage, EditEon, Peerage of Science*, and *Rubriq*. Some of these companies forward papers submitted to their services to prevetted and approved reviewers for peer review before they are submitted to a journal (Economist, 2013b; Harold, 2013; Yandell, 2015). This early feedback is intended to provide a reality check to authors regarding the publishability of their work and also to ensure that it will pass the basic standards of peer review. These organizations have different business models; some are free to authors, deriving their income from publishers, etc., while others charge authors a fee ($250–$850) for handling each manuscript (Yandell, 2015). Some of the companies incentivize their reviewers by paying for their reviews. The end product of the third party review system (the content of which varies with the vendor) includes written peer reviews of the submitted manuscript with specific recommendations being made for its improvement, various content rating systems that are intended to provide a quantitative assessment of the quality of research in the manuscript and its novelty, impact and interest, and "peer review validation seals" similar in intent to the CrossMark identification service (http://www.crossref.org/crossmark/) that is used by many journals and verifies to the reader that they are accessing the most recent and reliable version of a paper. An unofficial variation on presubmission peer review occurs when authors submit an early version of a manuscript to a journal to obtain reviewer feedback. Then they use these to position and improve the manuscript to either resubmit or send elsewhere, i.e., free peer review to test the acceptability of a hypothesis and findings.

Peer review approval seals are used by AAAS' *PRE* (Peer Review Evaluation: http://www.pre-val.org) for the establishment of best practices and the support and validation of quality peer review and by *Rubriq* (https://blog.rubriq.com/2014/12/05/rubriq-adds-sound-research-stamps/) in the form of a "Sound Research Stamp" that certifies that a manuscript, at least in the view of *Rubriq's* independent peer reviewers, is fundamentally sound and suitable for publication. Finally, a submitted manuscript may also undergo a plagiarism screen (e.g., CrossCheck) and the service may also provide recommendations for a suitable journal to which the authors can submit the paper together with the prereview reviews once it is revised.

5.3.4 How's It All Working?

Informal comments suggest that the transferable peer review is a viable option for authors, at least in terms of an in-house publishing cascade, where publishers, because of the structured relationship with their editors and journals, can easily and efficiently facilitate the transfer. There is very little public information available however, on how portable peer review is working within the transferable peer review consortia.

Similarly, there is minimal information on how third party prereviews are managed by journals and the level of agreement between journal editors and reviewers as to the quality of third party prepeer reviews and whether the papers themselves meet the established standards of formal peer review at the journal (Jubb, 2016). The NPG in commenting on a fast track peer review experiment with *Rubriq* were quoted as saying that "the quality of reviews received was similar between the private company and NPG's own efforts" (Cressey, 2015a) which may not be saying very much given the STAP fiasco. However, it was not made clear how quality was objectively assessed. There are also concerns about editorial autonomy (Jubb, 2016) and, by extrapolation, reviewer autonomy. With author-initiated and purchased prepeer reviews in hand, does the journal reviewer in addition to providing their own, independent and perhaps lukewarm or negative assessment of a manuscript then have to justify their review to the Editor in the context of glowing reviews from unknown paid reviewers from a third-party source?

Given the lack of standardization in peer review (within and between journals, in terms of guidance to referees and in terms of verification that reviewer comments and author responses are appropriate) and the large element of chance in the peer review process (Peters and Ceci, 1982; Rothwell and Martyn, 2000; Smith, 2010), it would be surprising if third party prepeer reviews were accepted as fit for purpose. Indeed the likely view is that it would be preferable if prepeer reviews were omitted. Addressing the question of the extent of concordance between prepeer reviews and subsequent prepublication peer reviews at a journal remains to be determined. However, this is an extremely nuanced issue and perhaps less relevant than it may first appear, since it is certainly the case that the degree of concordance between reviewers in *CPR* journals is not well researched and more likely to be variable than it is consistent.

Another controversial facet of the evolving prepeer review process is a pilot project in fast track submission that involved a collaboration between *Rubriq* and NPG's *Scientific Reports* journal. The decision to offer expedited decisions on submissions within 3 weeks of submission for a fee of $750 (Cressey, 2015b) led to major concerns in the research community regarding the establishment of a two-tiered submission system that allowed those able to pay "effectively jump[ing] the queue to publication". A collective letter of resignation from 150 editors of *Scientific Reports* (http://www.peerreviewneutrality.org) expressing concern at the ethical issues of the fast track submission trial was withdrawn when the trial was ended and NPG committed to analyze and discuss the feedback with all stakeholders (Jackson, 2015).

5.3.4.1 *Technology as a Facilitator of Effective Peer Review?*

In reading through the often confusing menu of emerging approaches intended to improve peer review, none of which currently appears to represent a viable alternative to the *CPR* process, one might be struck on the one hand by their naivety and on the other by the opportunity to create new profit centers to take advantage of the situation. Both *CPR* and PPPR depend on "desk decisions" (triage) as an initial filter to eliminate substandard contributions, in the former by experienced Editors and in the latter by "unbiased" technical staff. Clearly both have limitations while the *PLoS* experience with the papers *Metagenomic Human Repiratory [sic] Air* (Lai et al., 2015) and *Hand Coordination* (Liu et al., 2016) does not place the initial steps of the PPPR process in a positive light. Similarly, the rigor of the "desk decisions" at *Scientific Reports* has been ridiculed (Lowe, 2016) based on a paper (Samie et al., 2016) described as "deliriously incompetent fraud" where the chemistry section was inconsistent and erroneous and where the biological data consisted of numerous instances of blatant image duplication. This led Lowe to ask the question "Who looked at this stuff?" and note that the PPPR model "doesn't work if you don't review the damned papers." The answer to the first question is "nobody" with any attempt to pretend there is some sort of triage being ludicrous. Sadly, there seems to be a trend where, because of the long history of finding fault with *CPR*, any new approach, for example, PPPR, is automatically assumed to be superior despite any convincing evidence.

5.4 BIAS IN PEER REVIEW

Bias in peer review is another topic of major concern to authors and has, to date, been generally focused on *CPR* review. It appears to be less of a concern with PPPR, a reflection of the naïve perception that PPPR, since it is not *CPR*, is an improvement on the status quo of *CPR*. It would be anticipated that PPPR (and *PPC*) is likely to be awash with peer review bias especially when reviewers self-select, all of whom are likely to have their own agenda.

5.4.1 Publication Bias

The most frequent cited bias in the publication process is that of publication bias (which actually means acceptance bias) whereby papers with positive results are favored to the exclusion of those containing negative findings (Curtis and Abernethy 2015; Dirnagl and

Lauritzeb, 2010; Fanelli, 2012; Ioannidis, 2005; Ioannidis and Trikalinos, 2005; Song et al., 2013; ter Riet et al., 2012). Many journals are now making greater efforts to publish negative studies instead of taking the viewpoint that an inability to confirm a previously published finding is of limited interest to readers and irrelevant to the issue of reproducibility. While the *Journal of Irreproducible Results (The Science Humor Magazine)* has been an amusing source of information on reproducibility issues for some 60 years, mainstream journals are now more actively encouraging the submission of negative findings. *PLoS One* has created a collection of papers termed *Positively Negative* that focuses on negative, null, and inconclusive results (PLoS, 2015). More recently Elsevier (van Hilten, 2015) and BioMed Central have launched journals devoted to the topic of negative findings—*New Negatives in Plant Science* and the *Journal of Negative Results in BioMedicine*, respectively.

5.4.2 Confirmational Bias

This is another form of publication bias that reflects a bias against manuscripts that describe results that are inconsistent with the current *status quo* in a given area of research and which the reviewer then reflexively chooses to dismiss or manuscripts that replicate already published studies. Leslie (2016) has captured this bias as part of the "human social life" aspect of research "deference to the charismatic, herding toward majority opinion, punishment for deviance, and intense discomfort with admitting to error" which is similar to *conservatism*—a bias against innovative research (Horrobin, 1990; Mahoney, 1977; Nickerson, 1998) and *cognitive cronyism* (Travis and Collins, 1991).

5.4.3 Prestige or Institutional Bias

This relates to the "class structure" in science where "those rich in prestige disproportionately accumulate limited resources" the latter including grants, awards, etc. The concept of prestige bias in science—which obviously extends far beyond peer review—was originally discussed by Merton in 1968 and is widely known as the Matthew effect based on the biblical verse, "For unto every one that hath shall be given, and he shall have abundance: but from him that hath not shall be taken even that which he hath" (Matthew 25:29). A more colloquial variation of the Matthew effect is that "the rich get richer and the poor get poorer," a perfect metaphor for societal norms in 21st century society where the "most aggressive, take-all, calculating managers ...... protect their research fiefdoms" and become the academic celebrities (Shea, 2014) and "science heroes of the 21st century" (Ioannidis, 2016b). In the publication arena, this biases peer review as explained later.

Scientists rich in personal prestige, "Academe's '1 Percent' " (Benderly, 2014) are held in high esteem by their peers and the mainstream media, being perceived to receive preferential treatment via gender pejorative "old boy networks," a long standing bias in terms of *social status* where the status quo differentiates those who operate within a "network" and those who are left outside (Lee et al., 2013). This can also be viewed in terms of an *affiliation bias* that involves formal (same institution, same grant review panels, same societies) and informal (personal relationships, affiliation groups) networks.

Prestige/institutional/social status/bias can be a two-edged sword. In addition to favoring those with prestige, social status etc., it disadvantages those who are excluded from these various groups and can also operate in reverse as noted by Oswald (2008) in an erudite study of peer review bias. He noted that a journal in the field of economics, the *Journal of Political Economy* discriminated against authors from its host institution in Chicago, an example of what was termed as *negative favoritism* and perhaps of familiarity breeding contempt, a phenomenon that is supported by anecdotal evidence in a number of other journals, many in the field of biomedical research.

Thus a manuscript received from a researcher from a top research university may, de facto, be assumed to be in good shape from the start of the review process and thus receives a less rigorous review while a manuscript from an unknown institution/author is subject to extra stringency including a desk reject. Naïve referees assume that because the submission emanates from a famous author it is certain to be more worthy, more substantial and more likely to be correct than if exactly the same paper had emanated from a less well known author.

5.4.3.1 *Printing Drugs?*

An unfortunate example of this type of institutional bias occurred with the paper, "*Drug-Printer: print any drug instantly*" (Chen, 2014), from an author who was affiliated with MIT and also with academic institutions in Taiwan. This paper was accepted for publication and published on line in *Drug Discovery Today* (http://dx.doi.org/10.1016/j.drudis.2014.03.027) in early 2014. The paper described the hypothetical use of a 3D printer and "optical tweezers" to make individual compounds by printing them. The abstract to the paper which can be found in various sources on the Internet reads, "In drug discovery, de novo potent leads need to be synthesized for bioassay experiments in a very short time. Here, a protocol using Drug-Printer to print out any compound in just one step is proposed. The de novo compound could be designed by cloud computing big data. The computing systems could then search the optimal synthesis condition for each bond–bond interaction from databases. The compound would then be fabricated by many tiny reactors in one step. This type of fast, precise, without byproduct, reagent-sparing, environmentally friendly, small-volume, large-variety, nanofabrication technique will totally subvert the current view on the manufactured object and lead to a huge revolution in pharmaceutical companies in the very near future." This paper contained no data and was essentially an exercise in wishful thinking as evidenced by the use of the words "could" and "would" in the abstract. This paper was withdrawn before it was formally published by the author in response to almost universal derision on various chemistry blog sites (http://chemjobber.blogspot.ch/2014/04/i-cannot-believe-this-was-published.html; http://quintus.mickel.ch/2014/04/19/print-your-own/#comment-151625).

It is highly probable that if the author of this paper had not been affiliated with MIT, it would have been rapidly rejected. Nonetheless, it should be noted however that the concept of making compounds or drugs on demand, while reduced to the level of naïve opportunism in this particular example [and in a recent Dilbert comic strip (http://dilbert.com/strip/2017-04-14)] is not, per se fantasy as evidenced by a recent report of a DARPA-funded research project that described the on-demand, continuous-flow production of four commonly used drugs, diazepam, diphenhydramine, fluoxetine, and lidocaine (Adamo et al., 2016), albeit without the use of a hypothetical printer. While the practical utility of this pilot on-demand system, a prelude to producing drugs in remote and inaccessible locales like third world countries, battlefields, and colonies on Mars, given real world situations and regulatory guidelines remains to be determined, it is a major research achievement that will no doubt in time be reduced to practice.

5.4.3.2 *Planck's Principle and Scientific Funerals*

Another facet of prestige bias is reflected in what has been termed "Planck's Principle." In his autobiography, the German theoretical physicist and Nobel laureate, Max Planck, commented that "A new scientific truth does not triumph by convincing its opponents and making them see the light, but rather because its opponents eventually die, and a new generation grows up that is familiar with it." (Planck, 1950). While this sentiment has been the subject of debate, a recent paper by Azoulay et al. (2015) entitled *Does Science Advance One Funeral at a Time?* tested Planck's Principle by examining the impact of the premature death (or superstar extinction) of 452 "elite" or "superstar" life scientists in academia between 1975 and 2003 on the "vitality" of the research area in which they worked by measuring publication rates and funding flows. The criteria for inclusion in this US centric study were one of the following metrics of "superstardom" (see also Chapter 6.6.3): (1) highly funded scientists; (2) highly cited scientists; (3) top patenters; (4) members of the National Academy of Sciences or of the Institute of Medicine; (5) NIH MERIT awardees; (6) Howard Hughes Medical Investigators; and (7) early career prize winners. Overall, the results of this analysis indicated that newcomers to a field with different perspectives were reluctant to challenge leadership within an area of research while the superstar is still alive and that a number of barriers that are reflective of intellectual, social, and resource aspects constrain entry to the field even after the passing of the scientific superstar, questioning Planck's Principle.

5.4.4 Gender Bias

This reflects the situation where women are viewed as being less favorably treated than men during the peer review process whether this is for grants or publications (Bornmann et al., 2007). Evidence for the existence of gender bias is the result of a bibilometric analysis of citations (Larivière et al., 2013) that provided "much-needed empirical evidence of the inequality that is still all too pervasive in science." This study analyzed over 5 million research papers with some 27 million authorships and reported that articles with females in dominant author positions—sole, first- or last-authorship—were less cited than those with males in the same author positions and that publication portfolios from female researchers were more domestic than those of males depriving them of the "extra citations that international collaborations accrue." These conclusions appear questionable. The first statistic does not control for subject area. It may be the case that some publication areas are populated by predominantly one gender and the citation densities consequently may not be the same. The second statistic is an outcome statistic that could be the result of numerous factors including gender representation in different subject areas. Clearly only a study that matches males and females according to age, experience, affiliation, and subject area (i.e., that controls for other sources of variance) would be able to show whether or not women are treated less well than men by peer review. This is supported by Lee et al. (2013) who cite several studies that suggest claims of gender bias in peer review "are no longer valid" including studies by Ceci and Williams (2011) and Marsh et al. (2011). Like most biases in peer review, it is difficult to comprehend why reviewers of either gender would discriminate against authors on the basis of their gender especially when an author name in the form, "A.N. Other," does not typically denote the gender of the author. However, a widely published sexist review from a reviewer at *PLos ONE* (Else, 2015) has revived interest in the topic, where it should be noted that the subject matter of the article that led to the sexist review itself was on the topic of gender bias in academia.

5.4.5 Nationality/Language Bias

Nationality bias is another form of affiliation bias where the geographical location of researchers can sometimes influence the perspective of a reviewer. As an example, some researchers in Europe and Asia are convinced that major US journals discriminate against them on the basis of their country of origin (Oswald, 2008). Likewise some investigators from the United States are under the impression that the *British Journal of Pharmacology* only publishes papers originating from the United Kingdom. Nationality bias however can often be confused with *linguistic bias*, where the grammar, syntax, style, and clarity of a submitted manuscript can be a determining factor in its peer review outcome (Lee et al., 2013). Recent trends in expanding international research networks (Adams, 2012; NASEM, 2017), the changing demographics in the national origin of researchers in western laboratories (Olefksy, 2007; Van Noorden, 2012), and the almost universal use of English as the *lingua franca* (sic) of science accentuate this issue making it inevitable that the language of many submitted articles may be suboptimal, perhaps sufficiently so as to bias the reviewer. This is compounded by an exponential increase in researchers who are unable, *independent of their nationality or native language*, to write a coherent paragraph as a result of not being taught English grammar (Toplansky, 2016) or basic writing skills (McGuire, 2014).

Despite these factors, a recent case study (Walker et al., 2015) while failing to confirm bias in regard to gender, language, or author prestige/institution did find "weak" evidence of regional bias. The authors concluded that their study shows "no evidence that review results are influenced by the personal characteristics of reviewers, and only weak evidence for social bias due to interactions between author and reviewer characteristics. These findings do not rule out generalized bias against authors with specific characteristics or forms of bias not considered in the study". The issue is however far more nuanced. Pattern recognition (bias) is often used to help speed up the peer review process. A paper from certain parts of the world, for which it is evident from a cursory read of the abstract has neither been spell checked nor proofed by a native English speaker, and for which the quality of the graphics may be equally poor, will soon be deemed not worthy of detailed consideration. Unless the journal has an educational remit (the majority do not), then precious time and resources are unlikely to be expended in helping an author to improve their manuscript.

5.4.5.1 Issues with Readability

The issue of the actual readability of a manuscript is another aspect of the peer review process that is becoming increasingly challenging as traditional aspects of academic writing deteriorate (Ludbrook, 2007; Taylor and Francis, 2015). Indeed, a recent analysis of some 700,000 scientific abstracts by Plavén-Sigray et al., (2017) concluded that the readability of science has decreased, further exacerbating issues with reproducibility. Requirements include style and grammatical usage that together are conducive both to the critical thinking and analysis that are necessary to develop a logical and cogent argument to present the rationale and conclusions of a paper and to its communication. Increasingly this is being replaced by the dubious, and often opaque, writing skills of a generation of scientists that is more familiar with the vernacular of "techspeak," for example, Facebook and Twitter, and whose writing skills are compromised by software autocorrect functions that create "a sense of laziness in students, inhibiting their ability to work through a challenge and stay focused on a task" (Bronowicki, 2014).

When poor syntax and idiosyncratic grammar is evident it can lead to rapid rejection based on comprehensibility rather than scientific merit (Thrower, 2012). While the majority of biomedical journals are published in English, many authors and referees do not have English as their first language. While this can be rectified during the peer review process by the use of editorial services, when a paper is written by a nonnative speaker, and the peer review is undertaken and managed by a colleague with a similar background, the likelihood that an accepted article is published with suboptimal syntax can almost be guaranteed.

5.4.6 Improving Peer Review? Better Education and Training

The following comment made by ORCID's Meadows summarizes the key issues in improving peer review. "In the meantime, the one thing I believe would most improve the submission and review process is better education and training. At present, this is virtually nonexistent, at least in any consistent or comprehensive way. Individual PIs and professors may teach their students, many publishers and societies offer in-person or online training, and organizations like Sense About Science also provide support—but there are still way too many reviewers and authors who have never received any formal training at all. And it's starting to show—especially with the emergence of new players, such as China, which is set to overtake the United States shortly in terms of article authorship" (Michael, 2015).

5.5 GRANT PEER REVIEW

Despite the more formalized nature of grant review, in terms of the numbers of boxes that are required to be ticked by reviewers, and the number of sometimes quite esoteric and nuanced questions about novelty and impact that a reviewer is required to address, it still reflects many of the same issues associated with journal peer review. The two systems are also highly interdependent since the assessment of the productivity and success of an existing grant that forms the basis for a new application are typically measured in terms of publications originating from the grant—peer review begets peer review which begets grant applications. The same personnel are often involved (academics are both the applicants and the majority of the peer reviewers). In contrast to journal peer review, that for grants has a well-defined and well-known customer—the grant awarding agency—and a structured system and process—at least within a given funding agency—that has as its defined end point the disbursement of the available funding by allocation based on criteria and specifications that the funding body is entirely free to define according to its own requirements.

An important point to again bear in mind in assessing the scope and relevance of published analyses of grant review is that this literature tends to be biased with a greater focus on clinical than basic research with the perception (Sackett, 2004) that granting bodies have been viewed as being hijacked by basic biomedical scientists who "have erected research policies that place greater value in servicing their own personal curiosities than in serving sick people."

While the majority of available evidence-based insights on the grant peer review process tend to be NIH-centric given the size of its budget (460,000 grants totaling $200 billion over the 9 years from 2002 to 2011; Nicholson and Ioannidis, 2012) and the global reach of its grant activities. A survey (Schroter et al., 2010) of 28 funding organizations in 19 counties (Australia, Belgium, Canada, the Czech Republic, Denmark, Eire, Estonia, France, Germany, Hungary, Italy, the Netherlands, New Zealand, Norway, Poland, Sweden, Switzerland, and the United Kingdom) noted that peer review of grant applications was challenging with a growing workload and fewer quality reviews being available in a timely manner, a lack of academic and practical support and lack of clear guidance from funders. The latter in turn advocated the development of uniform requirements for the format and peer review of proposals. Some observers have also noted a shortage of willing reviewers.

5.5.1 Issues with Grant Peer Review

Like journal peer review, grant peer review has many detractors as evidenced by the provocatively titled paper "*NIH Peer Review Reform - Change We Need, or Lipstick on a Pig?*" from Fang and Casadevall (2009) that documented the outcomes from the 2007 NIH "Enhancing Peer Review" initiative designed to reduce the "enormous administrative burden" (Daniels, 2015; Fang and Casadevall, 2009; Germain, 2015; Schroter et al., 2010) where researchers spend a good deal of their time writing grant applications instead of actually conducting research. This is frequently an exercise in futility given current success rates of approximately 10% in the United States (Fang et al., 2016) and the European Union (Anonymous Academic, 2014) reflecting a growing imbalance between the number of grant applicants and the available resources to fund them (Alberts et al., 2014; Germain 2015; Newman, 2015). The ERC (European Research Council) reportedly awarded 300 starting grants to new researchers in 2013 against the backdrop of 100,000 PhD degrees being awarded in 2010 (Anonymous Academic, 2014).

The grant peer review process has been variously described as an imprecise, statistically weak, and biased process (Johnson, 2008) that "encourages nonconformity if not mediocrity" (Nicholson and Ioannidis, 2012). It is also perceived as antithetical to innovative research (Danthi et al., 2014; Ioannidis and Nicholson, 2012; Mervis, 2014), can be skewed against applicants from small institutions (Murray et al., 2016) and can discourage "bright individuals" from seeking a career in research (Germain, 2015). In the context of discouragement, Woodgett (2015) noted in the context of an article, the theme of which was "The worst piece of peer review I've ever received" that the NIH peer review process tends to be egregiously poor with an overt personal bias reflecting "efforts to shape the world in the reviewer's image." In this context, Flier (2017) has noted that a researcher applying for a grant is placed in " a position where data is filtered through the lens of the stated hypothesis, in a way that promotes expectation of a particular result, and biases against, or promotes rejection of, contradictory evidence. It is easy to see how this construct puts great pressure on a scientist to avoid falsifying the hypothesis upon which their grant was funded, even when evidence suggests this is the most rational approach."

Anonymous peer review management of grant proposals is in many ways more brutal and devastating than journal peer review. It is less well managed and the reviewers appear to have undue influence in that their recommendations seem to hold sway without much in the way of moderation by an "editor." This is partly because the "editor" is a committee chair who often has no time to read the proposal or indeed the detail of the peer review before the decision date. Additionally, once the proposal has been rejected the author may have nowhere else to go to obtain the necessary funding. While a research paper has numerous alternative venues for its publication (Peres-Neto, 2016), the same is not true for a grant proposal.

5.5.2 Grant Peer Review Outcome Metrics

A controversial study based on a cohort of nearly 1500 NIH-funded cardiovascular clinical grants found no association between the percentile ranking that resulted from the grant review process and scientific impact (Danthi et al., 2014; Lauer et al., 2015) with projects that received the lowest priority scores having as many publications and citations as those with the highest scores, This suggested that the peer review ranking was "close to random" (Mervis, 2014), a finding supported in a subsequent study that analyzed data on 102,740 research project grants funded by the NIH in the 28 years from 1980 through 2008 (Fang et al., 2016). The perceived randomness of the grant peer review process led to the suggestion (Jacobs, 2013) that the 24,000 reviewers cited as being involved in NIH grant review by Mervis (2014) could "all be replaced by a random number generator."

In contrast, Li and Agha (2015) by tracking publication, citation, and patenting activity from over 130,000 NIH R01 grants from 1980 to 2008 found that better peer-review scores were consistently associated with better research outcomes. This viewpoint has however been questioned by Lindner and Nakamura (2015) who noted that "retrospective analyses of the correlation between percentile scores from peer review and bibliometric indices of publications resulting from funded grant applications are not valid tests of the predictive validity of peer review at the NIH." It has also been argued that a lottery may be just as effective as the current grant peer review system and indeed Fang et al. (2016) have reported that the NIH has considered a modified lottery system while the New Zealand Health Research Council has already instituted a lottery system as part of its Explorer Grants program.

5.5.3 Bias in Grant Peer Review

Bias in the grant peer review process is thought to involve major components of the "old boy networks," a Matthew effect that reflects age bias (Robertson, 2015). Thus the number of Principal Investigators (PIs) receiving NIH RO1 grants at age 66 years or above in 2014 was nearly 5 times that of PIs aged 35 years or younger, and has been argued as a reflection of a trend for older Americans to put off retirement and remain active in the workforce (Rockey, 2015). Concerns regarding this outcome have been compounded by a proposal for an "emeritus award" [that] would "permit a senior investigator to form a partnership with a junior faculty member in order to hand off his or her line of research inquiry in an efficient and cost-effective way," that has been viewed as "an entitlement for senior investigators."(Kaiser, 2015a). The average age of first-time RO1-funded PIs has remained at 42 years for several years (Barr, 2015) and has led to a "pipeline of researchers in serial postdoctoral positions" (Alberts et al., 2015a) that reflects a decades long imbalance in supply and demand in the current biomedical research system (Alberts et al., 2014; Goodstein, 1995; Triggle and Miller, 2002). This system of insiders and outsiders, reflecting tenured faculty and the like and newly minted PhD graduates, has been further discussed in the context of the concept of the "dualization" of labor markets by Alfonso (2013). In this study comparisons were made between postdoctoral positions and membership in criminal drug gangs where a dual labor market creates outsiders who are "ready to forgo wages and employment security in exchange for the prospect of uncertain security, prestige, freedom, and reasonably high salaries that tenured positions entail."

Additional biases involve both prestige/institutional and national where the processes for prioritizing EU consortia grants, for example, *Horizon 2020*, are viewed as "reversed Robin Hood" schemes with most of the funding going to "well off countries" with researchers in Eastern Europe having "next to no chance of getting an ERC grant." Of 300 startup grants in 2013, 222 were to researchers in France, Germany, Israel, the Netherlands, Switzerland, and the United Kingdom (Anonymous Academic, 2014).

In toto, these findings question the effectiveness of the grant peer review activities and their cost both in time and money, with the NIH spending $110 million annually to administratively support these activities. This has led to discussions on the topics that include: verifying the "scientific premise"; the "validate-ability" of grant applications; the independent validation of key findings and transparency in data reporting (Wadman, 2013); the utility of guidelines for federal grant applications and awards (Collins and Tabak, 2014; Schmidt, 2014; Wadman, 2013) and the value of "big science" projects on researcher autonomy (Eisen 2017).

5.6 OPEN ACCESS PUBLISHING (OAP)

The present system of scientific publishing is in a markedly disruptive phase of its existence with numerous models for increasing transparency and improving access to its peer reviewed product (Triggle and Triggle, 2017). Some of these initiatives have been short lived, and in retrospect, poorly thought through and executed.

One that has become firmly entrenched is OAP which had its origins in the *Budapest Open Access Initiative* (Amsen, 2014b; Suber, 2012; BOAI; http://www.budapestopenaccessinitiative.org/read;) that stated—

> "By "open access" to this literature, we mean its free availability on the public Internet, permitting any users to read, download, copy, distribute, print, search, or link to the full texts of these articles, crawl them for indexing, pass them as data to software, or use them for any other lawful purpose, without financial, legal, or technical barriers other than those inseparable from gaining access to the Internet itself. The only constraint on reproduction and distribution, and the only role for copyright in this domain, should be to give authors control over the integrity of their work and the right to be properly acknowledged and cited."

The BOAI was followed in 2003 by the *Bethesda Statement on Open Access Publishing* and the *Berlin Declaration on Open Access to Knowledge in the Sciences and Humanities* that collectively defined OA contributions in the following terms (Suber, 2012). "For a work to be OA, the copyright holder must consent in advance to let users 'copy, use, distribute, transmit, and display the work publicly and to make and distribute derivative works, in any digital medium for any responsible purpose, subject to proper attribution of authorship.'"

It has been estimated that approximately 75% of scientific articles are inaccessible to readers without a formal subscription to a journal or journal bundle (Bergstrom et al., 2014), both of which are usually beyond the financial means of an individual thus requiring some form of institutional affiliation or repository access (Helland, 2011). This situation has been interpreted (Tennant et al., 2016) as a violation of Article 27 of the United Nations Declaration of Human Rights where access to scientific knowledge and information is regarded as a fundamental aspect of global human equality, for example, "Everyone has the right to freely participate in the cultural life of the community, to enjoy the arts and to share in scientific advancement and its benefits." The invocation of this altruistic UN Article indicates OA as an inevitable endpoint in scientific publishing (Triggle and Triggle, 2017) and also to comparisons of OA with access to online music sources. If one equates the established scientific publishers with recorded music conglomerates, one might predict "the development of the 'Napster moment' to force the established industry to change" (Tennant et al., 2016). This view is somewhat simplistic and fails to acknowledge the long standing and increasing proactive efforts of the established scientific publishers—conventional publishers in the OA vernacular—to adopt and improve on the useful elements of the OA model. These efforts are in marked contrast to the denial, disorganization, hubris, and incompetence of the global recording industry when faced with the inevitability of the development of websites like *Napster, iTunes, Gnutella, Spotify, Google Play, Pandora*, etc. (Goodman, 2010; Witt, 2015). An additional concern however, is how the online music access model (and associated e-book models) represented by *iTunes, Amazon Prime*, etc., will factor into future iterations of OA publishing.

In 2007, STM, the global trade association for academic and professional publishers published the *Brussels Declaration on STM Publishing* (http://www.stm-assoc.org/public-affairs/resources/brussels-declaration/) that emphasized the role of conventional publishers in organizing, managing, and financially supporting peer review processes and creating "rights-protected archives that preserve scholarship in perpetuity."

The BOAI publication model became reality in the early 2000s via publishers like *BioMed Central* (2000) and *PLoS* (2003) and were enabled by the Internet (Tracz, 2015).

OA publishing currently exists in two forms:

- *Gold OA*, where articles are freely accessible at the moment of publication and where publication is made possible by the author or institution paying an article processing charge (APC).
- *Green OA* where an author can self-archive a version of the peer-reviewed article that can be posted to an online repository like *PubMed Central, arXiv* and *bioRxiv* or an author or laboratory sponsored website that is accessible to the public. For a period (normally a year), the paper is embargoed to all but the universities and Industry that pay for access. After a year the work is freely available.

"Work that is not open access, or that is available only for a price, is called *toll access* (TA)." Suber (2012).

An OA article, whether it is in a dedicated OA journal, in an online repository or in a traditional journal, is provided free of charge to all readers to read, unlike the situation in the traditional "pay-to-read" (TA) scientific publishing model where a paid subscription is required or where access to an individual article is provided on a pay-per-view basis. It should be emphasized that there is a key distinction between access to reading an article versus access to the original raw data, the latter of which is becoming available in Open Data repositories (Section 5.3.2.6).

5.6.1 Preprint Servers

Another venue for OA publication, *PPC* and informal PPPR are preprint servers. This forum for publication has been in use in the physical sciences for over 20 years as *arXiv* (pronounced "archive"). *arXiv* is a public repository of electronic preprints, or e-prints, of scientific articles that are not peer reviewed. It contains over 1 million e-prints in Physics, Mathematics, Computer Science, Quantitative Biology, Quantitative Finance, and Statistics (https://arxiv.org). These e-prints can subsequently be submitted to peer reviewed journals for publication but many are only available on the preprint server especially as many journals currently view preprint publication as a formal publication per se.

bioRxiv ("bio-archive") is a free online archive and distribution service for unpublished preprints in the life sciences operated by the nonprofit, Cold Spring Harbor Laboratory (CSHL). Posting preprints to *bioRxiv* allows impatient authors to avoid the slow pace of traditional publishing (Callaway and Powell, 2016; Voshall, 2012) and "make their findings immediately available to the scientific community and receive feedback on draft manuscripts before they are submitted to journals" (http://biorxiv.org/about-biorxiv). E-prints are not peer-reviewed, edited, or typeset before being posted online but do undergo basic screening for offensive and/or nonscientific content and are checked for plagiarism. CHSL does not endorse the methods, assumptions, conclusions, or scientific quality of a *bioRxiv* e-print. Once posted on *bioRxiv*, articles are citable and can be updated but cannot be removed.

As of April, 2017 *bioRxiv* contained approximately 3,000 e-prints and is viewed with concern by biological researchers because of the possibility of being scooped and missing out on credit for new ideas (Callaway and Powell, 2016). Supporters of preprint servers anticipate that preprints/e-prints will only become accepted only "if the life-sciences community develops a consensus that preprint publication establishes a priority for any discovery" and that quality is maintained in this nonpeer-reviewed forum. Additional concerns include: (1) how funding agencies and academic committees view the value of unrefereed preprints as part of their funding/hiring/promotion decisions and; (2) the fact that articles previously posted to preprint servers may not be accepted by peer-reviewed journals. The latter reflects the Ingelfinger rule (Marshall, 1998) that was introduced by the *New England Journal of Medicine* in 1969 to prevent researchers from talking to the media about their research findings before these findings were peer reviewed. While it was widely adopted in biomedical research publishing this appears to be changing as preprints can now be directly transferred from *bioRxiv* to not for profit journals via a service termed B2J (bioRxiv to Journals) that is partnered with some 50 plus journals including many of the *PLoS* journals, *eLife, Journal of Biological Chemistry, Journal of Pharmacology and Experimental Therapeutics* and the *Proceedings of the National Academy of Sciences, U.S.A.* all of which apparently accept preprints for peer review.

Other preprint forums include *PeerJ Preprints* (Hoyt and Binfield, 2013) and the now defunct *Nature Precedings* (http://precedings.nature.com) that published some 5,000 papers from its inception in 2007 through 2012 (Walker and Rocha da Silva, 2015) and was discontinued as the result of "technological advances and the needs of the research community ... [that]... have evolved to the extent that the *Nature Precedings* site is unsustainable as it was originally conceived." (http://precedings.nature.com).

5.6.2 Financing OA Journals

Of the approximately 8,000 OA journals registered with the Directory of Open Access Journals (DOAJ) in 2011, some 30% charged the author for publication via various iterations of a publication fee, the APC (Kozak and Hartley, 2013), as do OA and hybrid journals published by mainstream scientific publishers. The APC varies greatly depending on the publisher and is often substantial, but estimates reported in the literature vary widely and are impossible to substantiate owing to the anecdotal nature of their reporting with contradictions from one report to another. Solomon and Björk (2012) reported an APC range from \$8–\$3,900 with an average of \$900; Van Noorden (2013a) reported a range of \$3,500–\$4,000 per article and noted that *Nature* had reported internal costs of \$20,000–\$30,000 per paper—a major outlier. Tennant et al. (2016) cite a range of \$6.50–\$10 for other journals. The APC may be specific to an article or may be free if the author has purchased a "lifetime" subscription to an OA journal.

The remaining 70% of OA journals are termed, rather misleadingly, "no-fee" because the cost of publication is paid directly or indirectly by sources other than the author or reader, namely via collective agreements with consortia of institutions (universities, government agencies, laboratories, research centers, libraries, hospitals, museums, learned societies, foundations). As an example, *eLife* is supported by grants from the Wellcome Trust, the Max Planck Society, and the Howard Hughes Institute (Van Noorden, 2013a). Additionally, publication costs may be offset by revenue from non-OA publications from the same publisher (*PLoS Biology* and *PLoS Medicine* receive subsidies from *PLoS ONE* (Van Noorden, 2013a)). Some OA publishers defray costs via advertising, society membership dues, endowments, reprint sales, or a print or premium edition (Suber, 2006). Since the OA publishing model has the potential to create new sources of revenue (by diverting income from the established scientific publishers) this makes it attractive to serial entrepreneurial publishers.

Nonetheless, Tennant et al. (2016) have argued that OA can facilitate innovation by providing researchers, especially those in low- and middle-income developing countries, full access to the scientific literature. They further argue that OA is key to advancing the societal case for "citizen science initiatives" that allow anyone to be actively engaged with research thus removing the "ivory tower" perception of research and accordingly improve transparency and literacy in science that is key to encouraging continued public funding of biomedical research (Alberts et al., 2014).

5.6.3 Piracy

An additional facet of the traditional publishing/OA publishing debate is the support of OA by individuals who hold the view that copyright to all forms of literary, musical, or artistic work should be abolished as it creates monopolies that are detrimental to a societal free market. In this context, Article 27 of the United Nations Declaration of Human Rights, also states that "everyone has the right to the protection of the moral and material interests resulting from any scientific, literary or artistic production of which he is the author" seemingly representing the yang to the yin of another basic tenet of the OA movement also enshrined in Article 27, "the right to freely participate in the cultural life of the community."

Those who advocate the violation of existing copyright laws via the Internet by republishing published work for free play a cat and mouse game with federal authorities and publishers affected by the violation by frequently changing website domains and servers (Schiermeier, 2015). As a result, OA has its own versions of the original peer–to-peer file online file sharing service, *Napster*, in the form of *Sci-Hub* and *Library Genesis (LibGen)*. *Sci-Hub* is an online repository hosted in Russia that is intended "to remove all barriers in the way of science." As of April, 2017 contains over 62 million scientific papers (Greshake, 2017) of dubious provenance that are available to readers for free (Rosenwald, 2016). In a lawsuit in New York an injunction was issued suspending the *Sci-Hub* web domain where it was ruled that "simply making copyrighted content available for free via a foreign website, disserves the public interest" (Schiermeier, 2015).

5.6.4 OAP in the Context of Predatory, Virtual, and Trash Journals

For better or worse, many scientists consider all OA journals to be of lower quality than the traditional journals that are published by established scientific publishers and learned societies (Allen, 2016; Gross and Ryan, 2015) even when these journals originate from the same sources. Furthermore, these are deemed to be associated with substandard or nonexistent peer review (Wicherts, 2016) and a low rejection rate (Lin and Zhan, 2014), a viewpoint that is reinforced by the unfortunate if erroneous association of OAP with what have been termed predatory, virtual and "trash" OA journals (Beall, 2012; Bohannon, 2015; Lin and Zhan, 2014; Shen and Björk, 2015) that often claim JIFs, authentic online indexing status (Tennant et al., 2016) and traditional CN and ISSN numbers (Lin and Zhan, 2014). While this view appears to be gradually changing (NPG, 2015; Tennant et al., 2016), this is against a backdrop of an exponential rise in the number of predatory OA journals from 225 in 2013 to 693 in 2015 (Shen and Björk, 2015).

Predatory publishing has distinct author and publisher demographics (Tennant et al., 2016; Xia et al., 2015) with authors often being inexperienced early-career researchers from developing countries including Asia and Africa while the publishers are overwhelmingly represented by a presence in India. The existence of predatory journals has been justified as a means to aid scientists in their efforts to gain promotion and tenure by publishing research articles that are unacceptable in the mainstream biomedical literature—another facet of the publish or perish mentality—that creates a hierarchy of journal quality. This conforms with the "infinite amount of space online" concept of Peres-Neto (2016), who has argued that a submitted manuscript that is ethically produced—a "minimum publishable unit" (MPU)—will eventually find a journal that will accept it. This is a trend in publication described as "where rather than if" that has been enabled by new publication technologies, forums, and journals. The assumption that any paper that reaches a state of finalization, and can be submitted for peer review, will be accepted somewhere may explain the recent explosive growth in predatory journals (Shen and Björk, 2015). While the exponential growth in the number of MPUs creates opportunities for new publication venues, it also adds to the phenomenon of attention decay in the reader (Parolo et al., 2015). While the perception that there is an infinite amount of space online to publish papers may be theoretically correct, even in the absence of *bona fide* peer review, these papers still require formatting and copyediting which requires qualification of the concept of the term, infinite.

Some 72% of predatory journals charge an APC which is relatively modest—an average of \$178—leading to mean revenue streams per journal of \$8,399 per annum with a range of \$1,900–\$143,535 depending on the APC and the number of papers (Xia, 2015). Many of these papers, once published, are 1–2 pages in length and often lack references (Lin and Zhan, 2014). An interesting outcome of the predatory journal debacle is an increase in the peer reviewed literature on the topic (Beall, 2012; Lin and Zhan, 2014; Shen and Björk, 2015; Xia, 2015; Xia et al., 2015).

5.6.5 Your Name Here—Pay to Publish and Paid to Publish

An additional element in the predatory journal debate is a trend for authors to actually purchase authorship of papers, specifically in China (Hvistendahl, 2013) with some 27 Chinese companies offering authorship slots on legitimate or plagiarized papers to create a $150 MM business (Economist, 2013b). Schekman (2013b) has cited a memo "published in April by the Chinese Academy of Sciences showing that cash payoffs are being offered simply for getting into prominent journals. A paper in *Science* is worth about $33,000, which goes into a scientist's pocket, not his or her lab."

The rise in "fabrication....scholarly corruption and.... plagiarism" in Chinese research (Xie et al., 2014) has been ascribed to a misguided, rigidly hierarchical, and politicized research culture that is obsessed with quantity over quality and where "doing good research is not as important as schmoozing with powerful bureaucrats and their favourite experts" (Economist, 2016). This functions via a "top down system of administration" for funding with "warped incentives" that base "personal academic performance...solely on an author's SCI (Science Citation Index) papers." This is coupled with a "lack of severe punishment in the evaluation system; excessive pursuit of personal profits; and a lack of scientific ethics" (Lin, 2013). None of these factors is necessarily unique to the Chinese system of biomedical research as the serial offenders from other countries (Chapter 1) attest and may be part of the growing pains of a biomedical research system that has grown exponentially in a relatively short period of time (Xie et al., 2014). This has however been accompanied by rampant plagiarism with "a third of more than 6,000... researchers at six leading institutions ...[in China]... admitting to plagiarism, falsification, or fabrication" (Economist, 2013b) and with "opportunists ready to take advantage of the situation with padded CVs, fraudulent and plagiarized articles, bogus medicines, and medical procedures carried out without clinical evidence" (Nature, 2012a).

The continuing growth of Chinese biomedical research in the global research community (Lin, 2013; Van Noorden, 2016) and the recognition of the need to resolve these ethical problems is the context of initiatives like the Thousand Talent Program's "Recruitment Program of Global Experts" to recruit western-trained Chinese expatriates—termed "sea turtles"—back to China (Xie et al., 2014) is intended to rapidly narrow the gap in the quality and transparency of research in China and put it on a par with other technologically advanced nations. Examples of this trend are the awarding of seven Howard Hughes Medical Institute grants to Chinese expatriates returning to China (Wines, 2012) and a 44% improvement over the period 2012–15 "in a *Nature* index that rates the production of high-quality research papers from each country" (Economist, 2016). During the same time period, the index for the United States of America fell by 8%. China's leader, Xi Jinping has recently advocated "the development of a system in which science policy is created by scientists, rather than at the whim of officials" (Nature, 2016a).

While the same standards of rigor used in the rest of the world are routinely applied by the mainstream journals to submissions from China—and from India, Korea, Egypt, Tunisia, Turkey, Iran, Iraq, Saudi Arabia, etc.—effecting changes in the issues discussed earlier will take considerable time due to the career/financial incentives to fabricate and plagiarize and the relative absence of consequences for such behavior. Unfortunately, these same incentives remain an issue in the West given a similar absence of consequences (Alberts et al., 2014; Begley and Ioannidis, 2015).

5.6.6 Citations for Sale

Another variation on the "paid to publish" theme is the "highly-cited researcher program" that originated from the King Abdulaziz University (KAU) in Saudi Arabia. This offered researchers with high citation credential positions as adjunct faculty at KAU with a $72,000 a year contract that requires these individuals to add KAU as an affiliation on papers that are sent to journals included in Thomson's *Web of Science* database (Bhattacharjee, 2011; Messerly, 2014; Yahia, 2012). This program, also described as "Citations for Sale" (Eisen, 2014) led to KAU being rated as number 7 in the 2014 global math rankings list published by US News and World Report, outranking MIT even though the KAU math department was only 2 years old. This was ascribed by Pachter (2014) "to the fact that KAU employs (as adjunct faculty) *more than a quarter* of the highly cited mathematicians at Thomson Reuters."

5.7 THE JOURNAL IMPACT FACTOR (JIF)

While functional neuroimaging studies in a group of researchers showed a positive correlation between increased reward signal in the *nucleus accumbens* and the anticipation of a publication with a high JIF (Paulus et al., 2015), the JIF (or IF) is generally equal only to peer review as a major cause of dissatisfaction with scientific publishing among biomedical researchers globally. The JIF is part of the so called "publish or perish culture" that despite continuing to be the cultural norm in academia and, to lesser extent, the biopharmaceutical industry, "has outlived its usefulness" (Vanclay, 2012) and has been vilified in terms of its distortion of incentives, tendency to corrupt science (Colquhoun, 2011; Ioannidis, 2014) and consequently to slow or destroy innovation (Foster et al., 2015; Horrobin, 1990). It tends to be less of an issue in the industrial research area due to the fact that researchers are required to work more closely together with shared and highly measurable and transparent goals and incentives to produce information that is proprietary, where the timing of publications is for the corporate rather than individual good, and where there are tangible consequences for scientific malfeasance.

The publish or perish culture places a high premium on citations of peer reviewed publications as a quantitative proxy of individual scientific productivity which many researchers feel is an inappropriate measure of the quality of the science or individual achievement. In his obituary of Fredrick Sanger, the twice honored (1958 and 1980) Nobel laureate in Chemistry, Brenner (2014) commented that "Sanger would not survive today's world of science. With continuous reporting and appraisals, some committee would note that he published little of import between insulin in 1952 and his first paper on RNA sequencing in 1967 with another long gap until DNA sequencing in 1977. He would be labelled as unproductive." Similar comments have been made by Peter Higgs, the joint Nobel laureate in Physics for 2013 for the identification of the Higgs boson, who noted that a similar breakthrough could not be achieved in today's academic culture, because of the expectations on academics to collaborate and "keep churning out papers" (Aitkenhead, 2013).

Nonetheless, whether in academia or, to a lesser extent in industry, individual career impact and progress have inevitably been evaluated on the basis of the number of publications in peer-reviewed journals with "high" JIFs, an associative rather than direct relationship that is more a measure of the cache of a journal than the intrinsic contributions of the researcher.

As noted in the introduction to this Chapter, the JIF concept was originally proposed by the doyen of bibliographic metrics, Eugene Garfield of the Institute for Scientific Information (ISI; Garfield, 1955), as a "quantitative tool... for ranking, evaluating, categorizing, and comparing journals" based on the number of citations that a journal receives. It was never intended to evaluate individual scientists, the scientific importance of their work or their standing in the research community (Wang et al., 2013; Alberts, 2013) but rather as "a tool to help librarians identify journals to purchase, not as a measure of the scientific quality of research in an article" (DORA, 2012).

The JIF, the data for which are gathered and sold by Thomson Reuters (since 2016, Clarivate Analytics), is a numerical term derived from the ratio of the number of citations for a journal in a given year to the articles published in that journal for the previous 2 years divided by the total number of articles published in the same 2-year period (Garfield, 2006). Journals with high JIFs that include *Cell* (JIF 2015 = 28.71), *Nature* (JIF 2015 = 38.14), and *Science* (JIF 2015 = 34.66) were termed "luxury" or "glamor" journals by the Nobel laureate, Schekman (2013a) in a high profile, widely read opinion piece in the *Guardian*. In his view, luxury journals represent "the epitome of quality, publishing only the best research…artificially restricting the number of papers they accept." In doing so they have established a "tyranny of exclusivity" based on their JIF, "a deeply flawed measure, pursuing which has become an end in itself—and is as damaging to science as the bonus culture..[on Wall St.]…. is to banking" with the actual science being secondary to the metrics—the tail wagging the dog (Pulverer, 2015). Accordingly, Schekman viewed the JIF as driving risk-taking in science to "encourage the cutting of corners, and contribute[s] to the escalating number of papers that are retracted as flawed or fraudulent." Instead, he advocates publication in open-access web journals like *e-Life* that reflect a scientist's ideal, "free…have no expensive subscriptions to promote….can accept all papers that meet quality standards…[and]… are edited by working scientists, who can assess the worth of papers without regard for citations… publishing world-class science every week." Not incidentally, Schekman is the editor of *e-Life*. His views have been preceded/reflected in comments from many other biomedical researchers (Alberts, 2013; Colquhoun, 2011; Misteli, 2013; Rossner et al., 2007).

The luxury journals have been differentiated by *Nature* (2012b) from "minimum threshold journals" that have single digit JIFs like *PLoS ONE* (JIF 2015 = 4.41) and *Scientific Reports* (JIF 2015 = 5.228) where "the misleading extension of the JIF's significance to three decimal places" (Pulverer, 2015) is used to rank journals. Is a JIF 5.076 any different from a JIF of 4.948 to anyone else than the Editors of the journal whose JIF is marginally above 5? To many researchers, as well as grant and tenure review committees, journals with JIFs below 5 are often discounted, being viewed as having zero value in the career evaluation process (Alberts, 2013; Fersht, 2009). Of relevance to the JIF debate is that for many journals a small number of papers account for the majority of their citations. For instance, in 2005, *Nature* had estimated that 89% of their citations came from 25% of the papers published (Nature, 2005). Thus the methods used to derive the JIF are thought to contribute to it being an arbitrary metric that is secretive (Macdonald and Kam, 2007), lacks transparency (DORA, 2013), and reproducibility (quantixed, 2015, 2016), and is readily open to manipulation by editors and authors (Arnold and Fowler, 2011; Macdonald and Kam, 2007) via cronyism (Pulverer, 2015). Major issues with the JIF noted by Seglen (1997) are that the citation rates of the articles determine the JIF and not vice versa and is that it "conceals the difference in article citation rates as articles in the most cited half of articles in a journal are cited 10 times as often as the least cited half."

5.7.1 Debating the Value of the JIF—DORA

The use of the JIF as an indirect measure for individual scientific impact has been questioned by many scientists both in print (Alberts, 2013; Chandrashekhar and Narula, 2015; Misteli, 2013; Rossner et al., 2007; Seglen, 1997) and in online blogs that are far too numerous to mention. The discontinuation of author metrics based on the JIF has long been advocated (Misteli, 2013) with the Schekman opinion piece (Schekman, 2013a) noting that "funding and ..[academic]... appointment panels often use place of publication as a proxy for quality of science." This viewpoint parallels that published in the *San Francisco Declaration on Research Assessment* (DORA, 2012), which "recognizes the need to improve the ways in which the outputs of scientific research are evaluated" and has its goal to stop the use of JIFs as the sole measure in assessing the value of a scientist's work. In the 6 months after its publication, the original 82 support organizations for DORA had grown to 623 with the number of individual signatories growing from 155 to 12,713, a measure of the level of dissatisfaction with the JIF.

5.7.2 Alternatives to the JIF

Numerous alternatives to the JIF have emerged and are generally variations on the basic theme of article citation many of which involve relatively complex assumptions and mathematical derivations that can be difficult to calculate and interpret (Bollen et al., 2009), the result of which is that they appear to be less transparent and potentially more arbitrary/subjective than the JIF they are intended to replace. Accordingly, the reader is directed to the original publications for more detail than that provided later. Among these alternatives in approximate order of their appearance are:

5.7.2.1 The h-Factor or h-Index

The h-factor or *h-index* (Hirsch, 2005) which has been described as a "fuzzy, Sugeno-like integral," is an author-level, self-calculable metric that is intended to provide an estimate of the importance, significance, and broad impact of the cumulative research contributions of a researcher rather than their productivity. Thus the *h*-index is based on citations to the actual articles not the journals in which they are published capturing the number of published papers cited in other papers at least *h* times, An "*h*-index" of 25 means that the individual researcher has published 25 papers each of which has been cited at least 25 times. This is generally considered as a more accurate measure of individual scientific achievement and as a result is not skewed by either one or two well-cited and influential papers or a large number of papers that are poorly cited—if at all. However, by default, older, more established researchers, especially those deemed to be key opinion leaders often get invited to submit reviews to high JIF journals that are read and cited more frequently thus enhancing their *h*-index. Another concern with the *h*-index is its currency. Is it a current statement of the impact of a researcher or a reflection of historical contributions? The influence of the *h*-index is reflected in an increase in its downloads, from 262 in November 2005 to 110,000 plus in April, 2009 (Fersht, 2009). Hirsch (2007) has also presented evidence that the *h* index is better than the total citation count, citations per paper, and total paper count in predicting future scientific achievement. The *h*-index has also been adapted as the basis of Google's "Scholar Metrics" (Anderson, 2012a).

5.7.2.2 The Eigenvector Score

The Eigenvector Score (Bergstrom et al., 2008; Fersht, 2009) is a measure of the total importance of a journal to the scientific community and is based on the *Google PageRank,* an algorithm used by Google Search to rank websites in its search engine in a manner that correlates with human concepts of importance. The Eigenvector Score recapitulates a "random walk through the scientific literature.. [that creates]....a citation network" using an algorithm (http://www.eigenfactor.org) corresponding to a model where readers follow citation chains as they move from a citation in one paper in to a paper in another journal. It can be used to differentiate "small journals in the lowest tiers of the publishing hierarchy...[from]...large, important journals, such as *Nature* or *PNAS.*"

5.7.2.3 The SCImago Journal Rank

The SCImago Journal Rank (SJR; González-Pereira et al., 2010; Guerrero-Bote and Moya-Anegon, 2012) is a "measure of scientific influence of scholarly journals that accounts for both the number of citations received by a journal and the importance or prestige of the journals where such citations come from .. [being]...a variant of the eigenvector centrality measure used in network theory... [that].. ranks journals by their 'average prestige per article'" (http://www.journalmetrics.com/sjr.php). The SJR uses the *Elsevier Scopus* journal database which contains greater than 20,000 journals versus the 13,000 in the Thomson/Clarivate *Web of Science* database (http://guides.lib.uw.edu/c.php?g=99232&p=642081). The SJR for a journal is often different from the JIF and may reflect the differentiation of popularity from prestige in the JIF factor with "popular journals cited frequently by journals of low prestige have high Impact Factors and low SJRs, whereas journals that are prestigious may be cited less but by more prestigious journals, giving them high SJRs but lower Impact Factor" (Butler, 2008). The SJR2 indicator is a more recent size-independent indicator of the influence of a journal that "takes into account not only the prestige of the citing scientific journal but also its closeness to the cited journal using the cosine of the angle between the vectors of the two journals' cocitation profiles." The interested reader is directed to Guerrero-Bote and Moya-Anegon (2012) for additional clarification and insight on this potentially interesting metric.

5.7.2.4 The Author IF (AIF)

The Author IF (AIF) (Pan and Fortunato, 2014) is a dynamic iteration of the JIF that measures "individual scholarly impact, where instead of the papers published in a journal one considers the papers of an author. The AIF of an author A in year *t* is the average number of citations given by papers published in year *t* to papers published by A in a period of Δt years before year *t*. Due to its intrinsic dynamic character, AIF is able to capture trends and variations of the impact of the scientific output of scholars in time." The dependence of the AIF on the JIF questions its utility.

5.7.2.5 Relative Citation Ratio (RCR)

Relative citation ratio (RCR) (Hutchins et al., 2015, 2016) is a newer metric developed by the NIH to measure the impact of a scientific paper based on the use of "the cocitation network of each article to field- and time-normalize by calculating the expected citation rate from the aggregate citation behavior of a topically linked cohort. An average citation rate [ACR] is computed for the network, benchmarked to peer performance, and used as the RCR denominator" to create a unique citation network that is relevant to the article. This "allows articles to be assessed on the basis of their relevance in their own field, and highly influential articles will be recognized even if they are published in an obscure journal" (Bloudoff-Indelicato, 2015). Using a set of 88,835 articles published between 2003 and 2010 by NIH grant awardees who "occupy relatively stable positions of influence across all disciplines" and a web tool, *iCite* (https://icite.od.nih.gov), the authors concluded that "the values generated by this method strongly correlate with the opinions of subject matter experts in biomedical research."

Where precisely *opinions* fit into a transparent and objective metric is unclear and while some researchers have described the results obtained using the RCR (Hutchins et al., 2015) as "stunning," others have criticized its complexity, lack of transparency and subjectivity.

5.7.2.6 Citation Distributions

Citation distributions (Larivière et al., 2016) represent an impact measure that is based on findings for the distributions of 11 journals and indicates that: (1) many citations are skewed with an average of 70% of citable papers being below the JIF and; (2) 15%–25% of the articles in a journal account for 50% of its citations. These findings indicated that the JIF is not a reliable predictor of the citation performance of individual papers.

Based on these findings, Larivière et al. made the following recommendations: (1) "We encourage journal editors and publishers that advertise or display JIFs to publish their own distributions using the earlier method, ideally alongside statements of support for the view that JIFs have little value in the assessment of individuals or individual pieces of work"; (2)"We encourage publishers to make their citation lists open via Crossref, so that citation data can be scrutinized and analyzed openly; (3)"We encourage all researchers to get an ORCID_iD, a digital identifier that provides unambiguous links to published papers and facilitates the consideration of a broader range of outputs in research assessment." These recommendations add additional qualifications to the use of JIF that are discussed further in Section 5.7.

5.7.2.7 Altmetrics

Altmetrics is an article level metric described as a form of crowd sourcing peer review that is intended to assesses the impact of a paper at multiple levels (Bornmann, 2014; Priem et al., 2010). These categories include: *viewed* (HTML views, downloads, YouTube): *discussed* (scholarly and scientific blogs; Shema et al., 2014): *journal comments* (*impactstory, Twitter, Facebook, Reddit, Pinterest, linkedin, Wikipedia*); *saved* (*Mendeley, BibSonomy, CiteULike,* and other social bookmarks); *cited* (in the scientific literature, on *Wikipedia, Sina Weibo,* policy documents) and *recommended* (Editorials, *F1000Prime, Publons, PubPeer,* etc.). A metric missing from this list which probably should be equally mandatory is *retracted* (*retraction watch*). Among the publishers currently using *Altmetrics* are *PLoS, BioMed Central, Nature,* and *Elsevier,* the latter of which has 250 journals using this tool (http://blog.sciencedirect.com/topics/alternative-metrics) and which acquired its own altmetrics provider, Plum Analytics (http://plumanalytics.com) in early 2017.

The utility of bibliometric tools like *Altmetrics* has become increasingly important in measuring research output. This is the result of changes in funding policy by the US National Science Foundation (NSF), which in 2012 stated that it would focus on "products" rather than "publications" with the statement that "Acceptable products must be citable and accessible including but not limited to publications, data sets, software, patents, and copyrights" (Piwowar, 2013).

An example of the use of *Altmetrics* is shown in Fig. 5.1. from a paper by Hiyama et al. (2012) in *Science Reports* chosen at random on the basis of a *Nature* press release (http://www.nature.com/press_releases/article-metrics.html). In addition to indicating 304,681 page views and the total citations for the paper on *Web of Science, CrossRef* and *Scopus*, the *Altmetric* also shows the amount of online attention in terms of an *Altmetric Score* (AS) calculated by an automated logarithm that is based on two main sources of online attention, social media and mainstream news media and is weighted. The AS is calculated on the assumption that "the

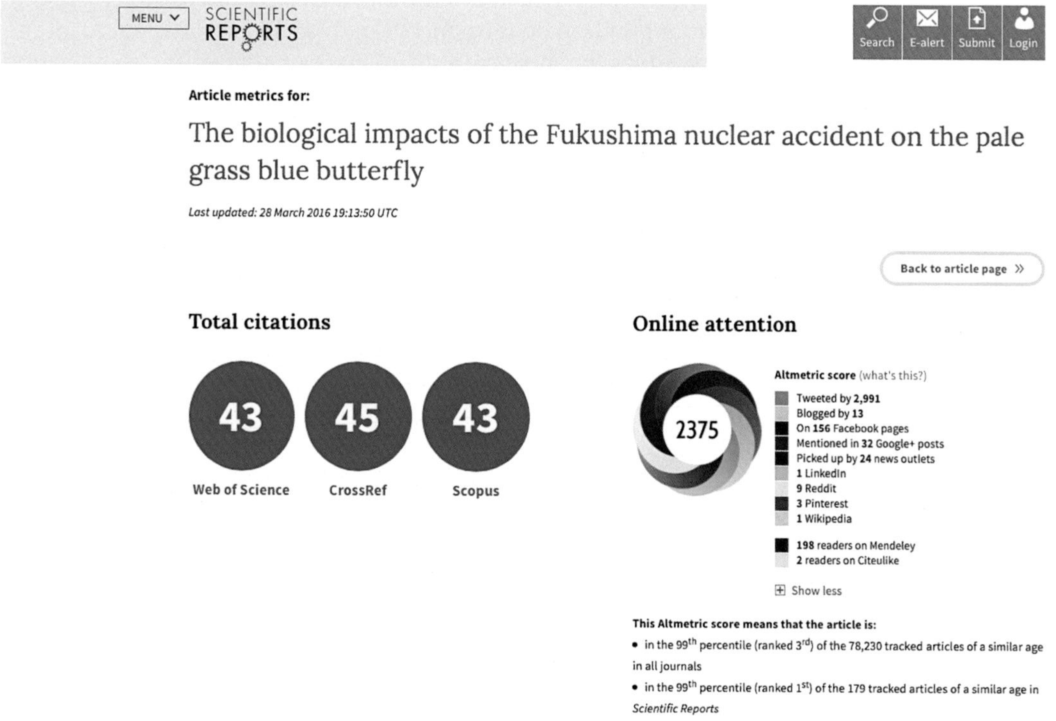

FIGURE 5.1 **Altmetric Scoring Template: Scientific Reports.**

average newspaper story is more likely to bring attention to the research output than the average tweet" (https://help.altmetric.com/support/solutions/articles/6000060969-how-is-the-altmetric-score-calculated-). Thus a news outlet is given a default weight of 8, a mention on *Twitter* or an *F1000* highlight, 1, and a *Facebook* or *YouTube* mention, 0.25. The "average newspaper" mentioned further differentiates *The New York Times* from other news media (although the online attention impact of the *NYT* needs to be tempered in light of its subscriber paywall—TA—which limits access. In comparison, an equally serious newspaper, the *Guardian* is OA and additionally has far greater credibility and insight than the *NYT* when it comes to the topic of biomedical research), while the impact of *Twitter* posts are modified on the reach, promiscuity, and bias of the author.

As examples, on their website, *Altmetrics* notes that a "Tweet from a publisher journal account will count for less than a tweet from a researcher who is unconnected to the paper. ..[while]...if a hugely influential figure were to tweet about a research output, this could contribute 1.1 to the score, which would then be rounded up to 2." These caveats are unfortunately infinite, subjective, and obtuse and raise legitimate questions regarding the transparency and subjectivity of the *Altmetrics* weighting system.

Would celebrities like Katy Perry, Cher, Bono, or Piers Morgan qualify as "hugely influential figures?" Or should such individuals by default be scientists who few, beyond some unnamed individual at *Altmetrics,* are aware of? Is an uninformed tweet from a Nobel laureate in physics on a potentially important finding on the genetics of drug abuse sufficient to inflate the generic tweet value? And why does the intrinsic logic of mathematics suffer at the hands of the automated AS algorithm with the illogical rounding *up* a value of 1.1 to 2 rather than rounding it *down* to 1.0? Overall the AS appears more a black box than a transparent vehicle with many of the metrics being overtly subjective and thus the antithesis of what supporters of *Altmetrics* presumably assume is its objective.

Returning to the paper by Hiyama et al. (2012), at the time of sampling (June, 2016), it had an AS of 2,375 as represented by the colored threads in the AS circle logo, had 198 readers on Mendeley, had been tweeted 2,991 times, blogged by 13 individuals, mentioned on 156 Facebook pages, and 32 Google+ posts and picked up by 24 news outlets, all interesting metrics regarding the attention an article has received but saying little about its intrinsic quality or scientific relevance.

This latter point has been addressed along with a number of other issues related to 21st century biomedical publishing by Colquhoun and Plested (2014) in a typically scathing yet thought provoking article that *eLife* declined to publish entitled *Scientists don't count: why you should ignore Altmetrics and other bibliometric nightmares.* In this article, the authors argue that *Altmetics* and other forms of bibliometrics are "surrogate measures....ambiguous to the point of dishonesty" and note that "Scientific works get tweeted about mostly because they have titles that contain buzzwords, not because they represent great science." Since only 1 in 40 researchers used *Twitter* in 2012 (Priem et al., 2010), the degree of contribution from *bona fide* experts to this forum may be questionable.

5.7.2.8 The **PQRST** *Index*

The PQRST (*productivity, quality, reproducibility, sharing, translational potential*) (Begley and Ioannidis, 2015; Ioannidis and Khoury, 2014) index is yet another alternative metric to the JIF that has the potential to identify "reward quality... reproducibility.. [and relevance]... rather than volume and quantity." It is part of an ongoing initiative by Ioannidis and others (Alberts et al., 2014; Ioannidis, 2014; Schekman, 2013a,b) to highlight the less altruistic motivations of researchers, namely the dysfunctional reward system that is a key component of the reproducibility paradigm and part of the perverse incentives discussed in Chapter 6.

5.7.2.9 CiteScore

CiteScore (CS; Zijlstra and McCullough, 2016) is a journal level citation metric score that was introduced in December, 2016 by Elsevier. CS differs from the JIF in the following ways: (1) CS is based on the approximately 22,000 sources in the Scopus database, double the 11,000 sources in the Clarivate Analytics Web of Science database; (2) CS is based on yearly citations that use the previous 3-year citation window as the denominator rather than the 2-year window used for the JIF with the 3-year window being sufficiently long to capture the citation peak in the majority of disciplines (Lancho-Barrantes et al., 2010); (3) CS includes all published documents in the Scopus database including "front materials" that include editorials, letters, news, notes, conference papers, "experts pick"-type reports of published articles of interest, erratum, etc., in addition to the "citable items"—the articles and reviews—traditionally used in deriving the JIF; (4) CS is calculated on a monthly basis (as CS Tracker) rather than annually like the JIF; (5) CS is arguably more transparent than the JIF since it does not use "secret algorithms or hidden details" and provides the exact figures used to determine the score; and (6) unlike the JIF, CS is an accessible metric that a researcher can calculate on their own (Bergstrom and West, 2016) using the Scopus database (http://supportcontent.elsevier.com/RightNow%20Next%20Gen/Scopus/Files/5221_CiteScore_FAQ.pdf).

CS has been variously described as "a simple and easy to understand metric" (Waltman, 2016), a "flawed…game changer" (Davis, 2016) and a "flashy new rival" to the JIF (Van Noorden, 2016b). A major concern with CS is its ranking of the citation performance of high Impact Factor journals versus that reported in the JIF. Thus, while "pure" research journals like *Cell* (2015 JIF = 28.71; CS = 23.62) and *Pharmacological Reviews* (2015 JIF = 17.10; CS = 18.92) show a congruence between the two citation scores, other high impact journals that contain a high proportion of "front material", like *Science* (2015 JIF = 34.66; CS =13.12), *NEJM* (2015 JIF = 59.56; CS =12.5) and the *Lancet* (2015 JIF = 44.00; CS = 7.72) and those in the NPG family, like *Nature* (2015 JIF = 38.14; CS = 14.38) and *Nature Reviews Drug Discovery* (2015 JIF = 47.12; CS = 9.17) show a marked difference in their CSI score relative to the JIF (Bergstrom and West, 2016). Unlike the Clarivate JIF, CitScore like all Scopus metrics is freely accessible at www.journalmetrics.com.

5.7.2.10 The Leiden Manifesto

The Leiden Manifesto (Hicks et al., 2015), like DORA, is a consensus document that seeks to address the distortions in the assessment of researcher performance and productivity as these pertain to recognition, funding, and career advancement that are the result of the "abuse" of metrics like the JIF and the *h*-index and publication in high impact journals. To these familiar publish or perish issues, the Leiden Manifesto also adds the "obsession" of universities with their global ranking, another topic addressed in Chapter 6.

In order to encourage a more balanced judgment of research metrics by evaluators and researchers, the authors in "offer[ing] a distillation of best practice in metrics-based assessment" list 10 principles and in noting that "simplicity is a virtue in an indicator because it enhances transparency" and conclude that "research metrics can provide crucial information that would be difficult to gather or understand by means of individual expertize. But this quantitative information must not be allowed to morph from an instrument into the goal."

The 10 principles of the Leiden Manifesto are:

1. quantitative evaluation should support qualitative, expert assessment;
2. measure performance against the research missions of the institution, group, or researcher;
3. protect excellence in locally relevant research;
4. keep data collection and analytical processes open, transparent, and simple;
5. allow those evaluated to verify data and analysis;
6. account for variation by field in publication and citation practices;
7. base assessment of individual researchers on a qualitative judgment of their portfolio;
8. avoid misplaced concreteness and false precision;
9. recognize the systemic effects of assessment and indicators; and
10. scrutinize indicators regularly and update them.

Additional details on each of these top-level points is present in the original Hicks et al. (2015) article and on the Leiden Manifesto website (http://www.leidenmanifesto.org).

5.7.2.11 *Metric Tide*

The *Metric Tide* report on the Role of Metrics in Research Assessment and Management (Wilsdon et al., 2015) which interfaces with the Research Excellence Framework (REF) metric to assess the quality of research in UK higher education institutions (http://www.ref.ac.uk/about/) has also developed 20 recommendations for the responsible use of metrics that covers the following topics: supporting the effective leadership; governance and management of research cultures; improving the data infrastructure that supports research information management; increasing the usefulness of existing data and information sources; using metrics in the next REF iteration; and coordinating activity and building evidence.

5.7.3 Citations, Research Metrics, and Facebook Likes

Despite the NSF mandate on using "products" versus "publications" as a measure of productivity (Piwowar, 2013), many biomedical researchers will find it challenging to consider that many of the proposed metrics-based sources to assess the total impact of a research article are in any way equivalent to a *bona fide* citation. Many of the newer paradigms used to judge the intrinsic "value" of a paper successfully recapitulate many of the perceived flaws of the JIF (if not in fact adding new ones). They also appear to be limited due to the subjective nature of the weighting metrics that are woefully lacking in logic and transparency and thus preclude them being considered as serious alternatives. In this context, Faulkes (2016) has characterized current trends in scientific publishing as "mission creep" with the mantra of "publish or perish" morphing into "be visible, or vanish."

5.8 ADDRESSING THE FUTURE OF PEER REVIEW AND THE JIF

As evidenced by the previous discussions, the literature on scientific peer review is bewildering in its volume, scope and intent with endless debates on many conceptually, a priori improvements (e.g., anything that is not *CPR* has to be good) that fail to live up to expectations when introduced into use (Smith, 2010). Numerous articles on the topic that date back over 20–30 years while unrelentingly critical, dismissive, and vituperative have made the topic of peer review like complaining about the weather, slow Internet speed, income taxes, and airport security lines, a pet peeve unique to those in the biomedical research community with a need to find cathartic relief from various frustrations without resorting to actual: (1) physical violence or (2) constructively addressing the issues.

Despite the negative views of peer review, the *Metric Tide* (Wilsdon et al., 2015), an independent review of the increasingly important role of metrics in research assessment and management conducted by the Higher Education Funding Council for England (HEFCE) has, as previously mentioned, definitively concluded that "Peer review, despite its flaws and limitations, continues to command widespread support across disciplines. Metrics should support, not supplant, expert judgement. Peer review is far from perfect, but it is the least worst form of academic governance we have, and should remain the primary basis for assessing research papers, proposals, and individuals."

5.8.1 Stakeholders

While much of polemic surrounding peer review has focused on the shortcomings of the existing *CPR* system with reviewers receiving the brunt of the complaints, other perceived culprits include journal editors and publishers. For grant peer review, the major shortcomings are seen as the result of less federal support than the grant bodies request with the solution—unequivocally obvious to all those involved—being the increased provision of sufficient taxpayer monies to fund all requests. There are however other stakeholders—researchers, authors, as well as the funding bodies—who tend to behave as long suffering martyrs whose treatment justifies behaviors that only add to the dysfunctional aspects of peer review.

In discussing stakeholder contributions, many of the roles and consequent viewpoints are necessarily interchangeable and can lead to inconsistencies in attitude and behaviors that tend to be hypocritical and, at times, overtly schizophrenic. For example, the reviewer who views his/her feedback to an author as being the invaluable, altruistic, indispensable and, at times, infallible product of their expertize, experience, and objectivity rapidly takes umbrage when functioning as an author on receiving similarly intended feedback that is in turn perceived as arrogant, biased, and ignorant.

5.8.1.1 Researchers

Researchers are the scientists actually involved in the planning of experiments, who generate and analyze the data and write the papers at which point they transition to being authors. As researchers, these individuals are prone to an "experimenter effect" where a researcher investigating a phenomenon is more likely to confirm the phenomenon if they need to believe in it (Wilson, 2016). While conducting research they also function interchangeably as peer reviewers, as readers of the peer reviewed literature and less often as editors.

5.8.1.2 Authors

Authors play an obviously indispensable role in providing the raw material for the peer review process and are the individuals who bear the brunt of what is perceived as dysfunctional peer review feedback from reviewers. The vast majority of authors appreciate that a manuscript submitted to a journal should be formatted into four main sections: introduction, methods and materials, results and discussion; be written in a logical, transparent, and consistent style that is grammatically and stylistically correct; and follows the house style of the journal (Chapter 4). This typically involves an iterative process that follows on from the actual experimentation and requires that a draft of the manuscript be written, discussed by the authors and then revised, reviewed, and subjected to an objective process of quality control that ensures it is comprehensible both in terms of composition and the logical progression from hypothesis to logical conclusions, and is complete.

More recently however, there has been an increasing trend for authors to be less rigorous and objective in preparing a manuscript for submission, demonstrating a distinct absence of pride in their work. As a result, the Editorial Office of a journal tends to receive what in essence is a working draft of the article, putting an extra burden on the journal to ensure that the manuscript is complete (not missing tables, figures, sections, etc.) and is reasonably in line with the guidelines to authors. This involves assessing whether the article is written intelligibly and is organized in a logical manner before even judging whether the science is of sufficient quality and substance to merit submitting it to reviewers.

With these substandard submissions, editors can either perform a perfunctory rapid reject or provide detailed, collegial feedback indicating to the author where improvements need to be made, with the editor in essence acting as a copyeditor. If an editor fails to do this effectively, either as the result of workload or oversight, the reviewers selected may decline the invitation to review the manuscript seeing it as a waste of their time, or will have to assume the role neglected by the oversight or indifference of the editor. Over time, this can sour the enthusiasm of the reviewer to commit their personal time and experience pro bono to aid in improving and maintaining the quality of the manuscripts that are accepted in the journal. Like the editors, they are not copyeditors.

While authors also use the peer review process as a valid and collegial means to improve their submission—the reviewer usually being far more objective than the author(s) in reading a manuscript and representing an additional perspective in data interpretation—some "high maintenance" authors devote minimal attention for preparing the manuscript, leaving this task to junior members of their group—graduate students, postdocs, and visiting researchers—assuming that any shortcomings will be taken care of by the reviewer (and the journal) leaving them to do the work that the senior author failed to take the responsibility for thus abusing the peer review process. When changes are requested, be they scientific, grammatical, and/or technical/ methodological, or questions are asked for clarification and insight, the more extreme of these individuals metaphorically scream about the incompetence and lack of expertize of the reviewers. Others cry censorship when poorly revised articles that consciously and/or consistently ignore feedback are still found wanting and further question whether the editors or reviewer are conflicted, have a vendetta against the author or are just not up to the job. Such hostile feedback provides yet another reason for reviewers to avoid the peer review process.

5.8.1.2.1 AUTHOR BIASES

Other less emotional biases evidenced by authors have been identified in the *Cochrane Handbook for Systematic Reviews of Interventions* (Sterne et al., 2011). While primarily focused on the clinical literature many are applicable to basic research and include: *Time lag bias*—the rapid or delayed publication of results depending on their nature and direction (positive versus negative outcomes); *multiple publication bias*—publishing multiple papers from a single study—slicing and dicing the data and releasing it in dribs and drabs, which is distinct from the self-plagiaristic practice of publishing essential the same paper in multiple venues; *selective reporting bias*—publishing some data from a study and ignoring any other data that fails to support the preconceived outcome (Begley, 2012); *location bias*—selecting journals based on their indexing profile and ease of access—which includes releasing peer reviewed data to the media before the formal publication and thus violating the Ingelfinger rule; *citation bias*—the citation and/or noncitation of research findings that support or dispute results depending on their nature and direction; and *language bias* which has already been covered (Section 5.4.5) and is gradually disappearing as English in its many forms becomes the accepted *lingua franca* of scientific publishing. Additionally, there are many language editing services that can be used to improve the grammar, syntax, and clarity of manuscripts some of which are offered by publishers for accepted manuscripts. A major drawback of these services is that they can often go beyond editing of the grammar and begin to change the meaning intended by the authors, adding to the problem rather than remedying it.

5.8.1.2.2 CITATION AND CITATIONAL BIAS

In rounding out potential issues with author bias, as a result of ease of access to the scientific literature provided by the Internet, many authors will develop the context of the hypothesis being tested in their submission by conducting an online literature search that typically uses the *Google Scholar* search engine, which, as of 2014, provided access to approximately 100 million scholarly articles (Khabsa and Giles, 2014).

The pages of "hits" presented by an Internet search engine can be skewed by the use of subjective search terms that can select for articles that support a particular point of view—a confirmatory bias on the part of the author—and either ignore any articles presenting alternate findings or relegate these deep in the numerous pages that result from the initial search. The author's evaluation can also be biased by not reading anything beyond the abstract of an article which is not always faithful to its content due to either space limitations or the tendency of the author(s) to highlight or hype particular findings and ignore others.

Researchers also tend to assume that the order of presentation of results from a search engine represents the importance of the paper. Rather it may reflect the cumulative "preranking" as reflected in Google's description of the role of *Google Scholar*—"Google Scholar aims to rank documents the way researchers do, weighing the full text of each document, where it was published, who it was written by, as well as how often and how recently it has been cited in other scholarly literature" (https://scholar.google.com/intl/en-US/scholar/about.html).

While the precise meaning regarding "weighing the full text of each document" is open to interpretation, the remainder of this statement raises concerns about additional biases in the ranking and presentation of "hits": "where it was published" is a reflection of prestige or institutional bias; "who it was written by" is clearly an affiliation bias; while "how often and how recently it has been cited in other scholarly literature" relates to the general perception *that certain journals are over-cited* because of their fame" (Oswald, 2008), which by default reinforces author priorities in selecting a publishing venue. As noted by Hutchins et al. (2016), "a widely known paper is more likely to be referenced than an obscure one of equal relevance" and add that "each time a paper is cited, it is a priori more likely to be cited again," a phenomenon termed "preferential attachment" (Wang et al., 2013) that is a variation on the Matthew effect. In the case of the *Google Scholar* PageRank algorithm outlined on the Google website, the rank order of the citations can be a function of preferential attachment which, because of the tendency of researchers to only check the first page or two of a search, leads to the papers in these pages to be repeatedly cited, thus further enhancing their future citation and distorting their metrics. This can also provide unscrupulous researchers with the opportunity to game the search process by citing their own articles via repetitively clicking on a paper ensuring that it will appear more frequently in the search results. Indeed, there have been instances where a seminal paper in a given research area is buried deep (e.g., on page 13) of a *Google Scholar* search where newcomers to an area would be unlikely to find it. Indeed, it is sometimes bewildering that a seminal, well known paper in a given research area cannot be easily found via the use of web search engines thus depriving newcomers to the area of a complete picture of the key research conducted to date.

Other real world aspects of the citation/JIF process that can skew the outcomes include: *self-citation* where authors, reviewers and editors can inflate the JIF by citing their own publications either at the personal level, with reviewers coercively "suggesting" to the author to add citations from the reviewer's publiactions and/or editors ensuring that their reviewers encourage authors to preferentially cite articles published in the journal (Lăzăroiu, 2013); *preponderance* which reflects the number of researchers working in one area of biomedical research versus another, for example, colon cancer versus Huntington's disease, the former of which, from a multidisciplinary perspective, makes a disproportionate contribution to citations, and; *herd behavior* where researchers limit their reading to certain journals, specialist, or otherwise, and tend to only cite the papers from these journals. This can result in papers of limited interest and value being better cited than excellent and valuable papers on the same or related topics that are published in venues that have minimal interest to the reader. For instance, a potential new drug target or lead compound published in a *Nature* journal, *Cell* or *Science* will garner far more attention than one published in a traditional pharmacology or medicinal chemistry journal. Additional iterations on the "value" theme are *topicality* and *access*. In the former, researchers by choice (or inertia enabled via the ease of use of *Google Scholar* or *PubMed* database searches) limit their scope of interest in the literature to the present and 1–2 years prior, making it relatively easy to be; (1) current and (2) oblivious of the seminal literature. The *access* category reflects an increasing trend where basic researchers have limited access to journals that are outside the budget of their institution (or discretionary grant funds) or to relatively obscure or specialized journals. By default, this leaves them having to depend on abstracts and reviews and/or read multidisciplinary or OA journals, further increasing the unintentional subjectivity of their citations.

This of course assumes that the integrity of the citer requires that he/she has actually read the paper rather than judging its merit for citation based on reading the title, the abstract, a secondary "expert opinion"/commentary/blog source or mainstream media coverage that often distorts the meaning and context. In toto, these various biases, even with PPPR, markedly skew existing citation metrics and distort the intrinsic value of the research on which they are based.

Another realm of citation manipulation involves coercive citations and cross-citing journal cartels (Martin, 2016).

5.8.1.3 *Reviewers*

Despite the fact that peer reviewers are the focus of much of the criticism related to the peer review process and receive little in the way of reinforcement (Ioannidis, 2014), they remain highly motivated in performing peer review from an altruistic standpoint (Ware, 2008). Some 90% of scientists (Taylor and Francis, 2015) view their role as reviewers as a means of playing their part as a member of the scientific community to reciprocate the benefit to reviewers when they are authors, as well as helping improve the quality of submissions and science in general. In a 2012 study, van Dalen and Henkens (2012) noted that "Population scientists find publishing in ...refereed journals and being cited by other scholars the most rewarding element in their job" but also noted that "Writing referee reports...rank[s] among the least appreciated elements of their work," conclusions that are no doubt equally applicable to scientists in other branches of the biological sciences. Also, like those who advocate and actively participate in Internet-based discussions as part of the PPPR/*PPC* approach, the individuals involved in *CPR* usually self-select in their willingness to participate in the existing process.

A peer reviewer can add immeasurably to the quality of an article by providing diligent, experienced, and collegial feedback in the form of a written review that is fair, balanced, and constructive, civil in its tone yet clearly communicates issues identified without personalizing shortcomings or antagonizing the author. In some instances, this approach falls short of its intent with the author taking umbrage at any and every point discussed misconstruing the feedback while in other, albeit infrequent but nonetheless welcome instances, an author will acknowledge reviewer feedback as being helpful.

In providing written feedback, peer reviewers often use a historically collegial approach that contrasts markedly with that of the pit-bull (Walbot, 2009) or vigilante (Blatt, 2015). This typically involves linguistic nuances that have the potential to be misunderstood or even ignored by new generations of authors. As examples, a reviewer noting that an article is of "limited interest" is more an indication that there is actually "no interest." Similarly, "archival" is a subtle indication that "no-one will probably read this," while "confusing" is a euphemism for "incomprehensible." In all instances this requires that the reviewer, while remaining collegial, needs to be more direct in his/her feedback to avoid confusion. In this same context, a rapid reject email often informs the author that his or her submission has been rejected because of a generic "low priority score," a phrase that while no doubt true, more probably reflects "mediocre, seriously flawed, illogical, and/or incomprehensible." However, to require a reviewer to write a detailed response on the shortcomings in a thoroughly substandard submission as a prelude for engaging in an extended discussion/ debate/ negotiation/analysis with the author is rarely a productive use of what is their pro bono time.

Nonetheless, there is an exceptionally fine balance between the nuanced and deferential collegiality that represents the cornerstone of the established peer review process and the more direct response of the often jaded or time constrained reviewer whose encounters with submissions worthy of peer review have become the exception rather than the rule. And even those submissions that are rapidly rejected require time to assess. The reviewer, who is sometimes perceived as participating in the peer review process for the sole purpose of enjoying the personal catharsis in providing a " devastatingly negative review of a manuscript representing years of work ..... on a difficult, unsolved question" (Walbot, 2009), strives to provide constructive and collegial feedback that will aid the author in improving a manuscript. Nonetheless, as communication in the 21st century favors spontaneity, speed, and brevity, in the area of peer review it may result in the substitution of collegial insight and rigor with a more abrasive approach that is ephemeral and counterproductive.

5.8.1.3.1 REVIEWER INCENTIVES

Despite much of the feedback received being negative, peer reviewers tend, from their alter ego perspective as authors, to be familiar with the frustrations that authors have with the peer review process. Absent overt personal attacks, they are thus used to working with authors to produce a publishable final product acceptable to both parties.

Despite such altruism, there is an increasing dearth of competent peer reviewers with the demand for reviewers exceeding their supply. Fox and Petchey (2010) have identified this as an example of the *Tragedy of the commons* (Hardin, 1968) where "individuals have every incentive to exploit the "reviewer commons" by submitting manuscripts, but little or no incentive to contribute reviews," representing with the emergence of a group who regularly decline invitations to review, a reflection of their "enlightened self interest" or selfishness (Perry et al., 2012), and apparent immunity to appeals of altruism and professional ethics (Fox and Petchey, 2010).

Despite the uniformly positive surveys on the motivation of peer reviewers, participants tend to be self-selecting in that the constituency interviewed is already a part of the system (Willis, 2016) while those exploiting the system are absent. Many newly graduated scientists in the 21st century do not find peer reviewing of interest or of sufficient priority to include in their professional activities while many of those currently engaged in the process are retiring, resulting in a reduction in the available pool of expertize, experience, and training for the process that is not being replaced.

The net result is that it has become increasingly difficult to find individuals with the appropriate competencies to be part of the peer reviewer pool. To spend 5–10 h a week (or more) on reviewing activities, either journals or grants, is 5–10 h that may be needed to conduct experiments, to write yet another grant application, find a new job or revise a manuscript in light of reviewer feedback, all of which must rate as higher personal priorities than peer review. This is again reflective of escalating issues with the "hypercompetitive" (Alberts et al., 2014; Begley and Ioannidis, 2015) culture in current biomedical research. Added to this is the large number of well-trained research scientists who are still seeking full time employment which includes those who have been unable to assume viable tenure track positions in academia many years after graduation (Benderly, 2016) as well as the many researchers let go from the biopharmaceutical industry as companies continue to consolidate, merge and outsource.

Thus, the time spent in peer reviewing while personally satisfying and at times enlightening must be balanced with other, more urgent priorities. Since peer reviewing receives little in the way of tangible recognition that can be readily accessed and assessed by colleagues and those making decisions for tenure, promotion, or grant prioritization, it is unfortunately an activity that adds little to job security or career advancement.

While there is currently no equivalent of the JIF that can be used as a metric to reflect the time spent in reviewing papers, in his conceptual system of research currencies, Ioannidis

(2014) in commenting "That.. [peer review].. is so little valued and rewarded is not calculated to encourage its benefits and minimize its harms" has proposed that individuals involved in peer review receive a currency credit equivalent to that of publishing a paper—*a publication unit*—making it a far from insignificant part of any scientific reward system. Various other suggestions have been made to recognize and quantify peer reviewer metrics (Meadows, 2015b,c). A major challenge however, given the general anonymity of the peer review process is how to ensure that any credit is appropriately earned and given. One scheme put forward by Fox and Petchey (2010) envisaged a scheme that privatizes the peer-review system, where authors can "pay" for their submissions using a virtual "currency", *PubCreds* that can be earnt by performing reviews. To enable schemes, such as this, there is a need to identify mechanisms to measure peer review as a *bona fide* research activity consistent with it being a cornerstone of publication. One such platform is *Publons* (Van Noorden, 2014c; https://publons.com Economist, 2017), which allows peer reviewers "to get credit for their contributions (without breaking reviewer anonymity) in a format they can include in promotion and funding applications." *Publons* allocates unique Digital Object Identifiers (DOI) to peer reviews allowing them to be cited and discovered proving "an incentive for reviewers to provide high quality reviews, improving both the underlying science and commentary." Another means to facilitate this process is *ORCID's Peer Review* initiative developed in partnership with *F1000* (Meadows, 2015a).

ORCID (Open Researcher and Contributor ID) is an open, nonprofit, community-driven registry of unique researcher identifiers that provides a transparent method of linking research activities and outputs to these identifiers. As conceived, *ORCID*'s *Peer Review* is intended to allow reviewers to capture their peer review output by providing their *ORCID iD* when accepting a peer review assignment thus gaining recognition for the activity while maintaining anonymity.

Another metric for the recognition of peer review activities, both time and expertize, *the R-index*, has been proposed by Cantor and Gero (2015). This builds on existing metrics by adding the annual list of reviewers for a journal, the number of papers reviewed by each reviewer and the total number of words in an article which is then multiplied by the square root of the JIF and is then is weighted by editorial feedback on individual revisions, using an excellence score from poor to exceptionally good quality. The utility of the R-index system is ultimately dependent on the JIF which makes its transparency and quality somewhat suspect. The interested reader is referred to the original article for further detail.

5.8.1.4 Editors

The role of editors in scientific publishing is to ensure that their journal is consistent in its standards of publication that ensure it consistently attracts authors, readers, and reviewers by publishing quality articles that are of relevant interest. Generally, the editors of a journal are well known to their colleagues in their cognate field of research and thus provide a degree of confidence that a submitted manuscript will be reviewed with appropriate expertize and experience with editors acting as "a neutral bridge" between the two parties. This contrasts with many of the unknown or quasi-familiar journals that have a title that is a variation on that of a well-established journal, for example, the *British Journal of Biochemical Pharmacology*, an amalgam of the titles of *Biochemical Pharmacology* and the *British Journal of Pharmacology*, or the *Journal of Nature and Science*, that originate from publishers outside the mainstream and who send out daily or weekly e-mail requests for authors to submit papers.

Among the specific roles of journal Editors that are recommended by the CSE (Council of Science Editors; http://www.councilscienceeditors.org/resource-library/editorial-policies/white-paper-on-publication-ethics/2-1-editor-roles-and-responsibilities/) are: treating authors with fairness, courtesy, objectivity, honesty, and transparency; providing guidelines to authors for the preparation and submission of manuscripts that include authorship criteria, policies on conflicts of interest; clearly communication of all editorial policies and standards; protection of the confidentiality of submitted manuscripts; establishment and maintenance of an effective system for rapid, informed, and transparent peer review that allows for expeditious editorial decisions to be communicated to authors in a clear, unambiguous, and constructive manner.

While the quality, sustainability, and reader recognition for the accomplishments of a journal is a reflection of how well the editors and reviewers work together, both as a team and as individuals committed to high standards, the reality of editing a journal is also an economic one, reflecting a business, such that unless a certain number of quality manuscripts are submitted, reviewed, accepted, and published on a consistent basis to generate a sustainable and relevant JIF, the journal will not be successful. This can then lead to an overhaul of the editorial board and/or what some may view as a relaxation of editorial standards. Thus if 90% of submitted manuscripts are rejected, then a new level of 85% may reflect an appropriate compromise to maintain a flow of publishable material.

5.8.1.5 Publishers—Oligarchs, Con-Artists, Altruists, and the Naïve

Added to all the concerns related to peer review and the role of the JIF, the world of scientific publishing is itself in a major state of disruptive flux (Schmitt, 2015; Tracz, 2015) where traditional scientific publishing houses, the oligopoly (Larivière et al., 2015) that includes Reed-Elsevier, Springer Nature, Taylor and Francis, Wolters Kluwer, and Wiley-Blackwell, often play catch up in adopting the more useful of the many ideas emerging from the new initiatives to address the shortcomings of the peer review process.

As of 2014, some 28,000 peer-reviewed journals were published in English with an additional 6,500 published in other languages (Ware and Mabe, 2015). Of these, 10,000 were OA (Schmitt, 2015). These are part of a Scientific, Technical, and Medical (STM) information publishing market that generates income of $25.2 billion annually (Ware and Mabe, 2015).

The profitability of STM publishing based on content that originates from taxpayer-funded research is another major issue that can prevent reader access to content in journals that is not OA and is often behind paywalls that require a journal subscription for access (Triggle and Triggle, 2017). In the United States, there has been a long standing legislative initiative that began in 2006 with the Federal Research Public Access Act (FRPAA) that was intended to mandate free online public access to read articles in the scientific literature that resulted from taxpayer-funded research. A strengthened version of this act, the Fair Access to Science and Technology Research Act of 2015 (FASTR) moved into full US Senate consideration in July, 2016 with a 41% chance of being enacted (https://www.govtrack.us/congress/bills/114/s779).

While OA initiatives (McKenna, 2015; Tracz, 2015) readily solve the issue of article access, they have in turn created new, alternate profit centers based on researcher payment for publication from their grants in order to disseminate research findings. The sustained viability of this approach remains to be determined, especially as research funding, specifically that in the biomedical arena, is contracting as are employment opportunities in academia, government, and the biopharmaceutical industry (Alberts et al., 2014).

5.9 DEEMPHASIZING THE JIF?

In July 2016, the American Society of Microbiology (ASM) in noting that "the relentless pursuit of high-IF publications has been detrimental for science" (Casadevall et al., 2016) made the following announcement: "The editors-in-chief of ASM journals and ASM leadership have decided to no longer advertise the Impact Factors of ASM journals on the journals' websites. This decision was made in order to avoid contributing to a distorted value system that inappropriately emphasizes high IFs. High-IF journals limit the number of accepted articles to create a perception of exclusivity, and individuals receive disproportionate rewards for articles in high IF journals, while science as a whole suffers from a distorted values system and delayed communication of research" (https://www.asm.org/index.php/asm-newsroom2/press-releases/94299-asm-media-advisory-asm-no-longer-supports-impact-factors-for-its-journals), the first formal foray "to purge the conversation of the Impact Factor and ....make it so tacky that people will be embarrassed just to mention it." (Callaway, 2016b). While other journals are being encouraged to follow the lead of the ASM, finding an appropriate replacement for the JIF will be challenging. In the meantime, grant review bodies and tenure/hiring committees will need to continue their dependence on this flawed metric whether they agree with it or not.

5.10 GUIDELINES AND CHECKLISTS

When a manuscript is submitted for publication to a *bona fide* journal, it is assumed that the data, both its quality and quantity, and the experimental detail provided by authors are accurate, complete, and follow universal guidelines in terms of transparency and experimental validation that were part of undergraduate and graduate training programs in research and put into practice via mentoring from more experienced researchers. Accordingly, editors and reviewers have traditionally tended to give authors the benefit of the doubt when issues arise with incomplete data sets, erroneous data collection and analysis, and unusual or incomprehensible conclusions. Thus this process tends to implicitly assume that the core practices and the ethics of editors and reviewers can be automatically projected onto authors who, unknown to the editors and reviewers, may not been properly trained or mentored and who fail to appreciate the consequences, if any, of sloppiness, misrepresentation, or fraud in their submissions.

While scientists as a group are probably no more inclined toward fraud and misrepresentation than any other social group (with the notable exception of politicians), individuals with such tendencies do exist in the ranks of research scientists and run the gamut from arrogant and devious individuals whose personal world view considers their own research efforts to be immune from mainstream mores and those who are natural candidates to publish in "trash" journals. At the opposite end of this spectrum are the scientists who make genuine mistakes involving faulty and biased oversights that lack logic. The present honor system of the review process is unlikely to either deter or identify those who are intent on fraud or misrepresentation as these will "game the system" irrespective of what processes are in place. However, those authors who gravitate toward the "genuine mistake" category can, along with editors and reviewers benefit significantly from the use of guidelines and checklists.

5.10.1 Guidelines for Biomedical Research

Among the 350 or so sets of guidelines for reporting health care research (www.equator-network.org) are guidelines those for the design and reporting of biomedical research primarily for in vivo animal research (Table 5.3). These include: *ARRIVE* (*Animal Research Reporting In Vivo Experiments*: Kilkenny et al., 2010, 2014; McGrath et al., 2010; Sena et al., 2014); *CAMARADES* (*Collaborative Approach to Meta-Analysis and Review of Animal Data from Experimental Studies*; Ritskes-Hotinga et al., 2014; Sena et al., 2007; http://www.dcn.ed.ac.uk/camarades/);

TABLE 5.3 Biomedical Research Guidelines for Animal Studies

ARRIVE	*Animal Research Reporting In Vivo Experiments*	Design and analysis checklist for studies using laboratory animals	McGrath et al., 2010 Kilkenny et al., 2010
CAMARADES	*Collaborative Approach to Meta-Analysis and Review of Animal Data from Experimental Studies*	Meta-analysis methodology to improve translational research	Sena et al., 2007 Ritskes-Hotinga et al., 2014 http://www.dcn.ed.ac.uk/camarades/
CONSAERT	*Consolidated Standards of Animal Experiment Reporting*	Transparent template for accounting for and reporting all animal studies as a formalized standardized structural report as an online appendix for reviewers and readers	Drucker, 2016
DEPART	*Design and Execution of Protocols for Animal Research and Treatment*	Recommendations for a priori rigor in design and execution of experiments to reduce risk of bias including: objective of study; choice of outcome measures; animal and control selection and sample size	Smith et al., 2016
MIBBI	*Minimum Information for Biological and Biomedical Investigations*	Recommendations for minimum reporting of experimental methods involving 21 discipline/technology-based checklists.	Taylor et al., 2008
MIQE	*Minimum Information for Publication of Quantitative Real-Time PCR Experiments*	Guidelines for the conduct and reporting of qPCR experiments	Bustin et al., 2009
NIHPG	*NIH Principles and Guidelines for Reporting Preclinical Research*	Recommendations for data analysis, reporting, sharing, and publication of refutations	Landis et al., 2012 Moher et al., 2015 http://www.nih.gov/science/reproducibility/index.htm
RIPOSTE	*Reducing Irreproducibility in laboratory STudiEs*	Primer on experimental design and analysis	Masca et al., 2015
GSPC/ SYRCLE	*Systematic Review Centre for Laboratory Animal Experimentation/Gold Standard Publication Checklist*	Methodology for cross study analysis and meta-analyses of preclinical research findings GSPC (gold standard for publication checklist for reducing animal use)	Hooijmans et al., 2010, 2011, 2014. Karp et al., 2015
TOP	*Transparency and Openness Promotion Guidelines*	Center for Open Science guidelines for journal standards	Nosek et al., 2015

CONSORT (*Consolidated Standards of Reporting Trials*; Altman et al., 2001; Moher et al., 2001; Schulz et al., 2010); *DEPART* (*Design and Execution of Protocols for Animal Research and Treatment*; Smith et al., 2016) *MIBBI* (*Minimum Information for Biological and Biomedical Investigations*; Taylor et al., 2008); *MIQE* (*Minimum Information for Publication of Quantitative Real-Time PCR Experiments*; Bustin et al., 2009); *NIHPG* (*NIH Principles and Guidelines for Reporting Pre-clinical Research*; Landis et al., 2012; Moher et al., 2015; Nature 2016b; https://www.nih.gov/research-training/rigor-reproducibility/principles-guidelines-reporting-preclinical-research; *RIPOSTE* (*Reducing Irreproducibility in laboratory STudiEs*; Masca et al., 2015); *GSPC/ SYRCLE* (*Gold Standard Publication Checklist*; Hooijmans et al., 2010, 2011); *Systematic Review Centre for Laboratory Animal Experimentation* risk of bias (RoB) guidelines (Hooijmans et al., 2014; Karp et al., 2015), the latter of which are derived from the Cochrane Library/Collection metaanalysis databases (Higgins et al., 2011); and *TOP* (*Transparency and Openness Promotion Guidelines*; https://osf.io/2cz65/?_ga=1.12207156.188575582.1463347000, Nosek et al., 2015). These guidelines tend to cover many of the same topics with differing degrees of detail and focus.

In vivo experiments represent one of the key and often last steps in the preclinical evaluation of a disease mechanism/new compound entity (NCE) before either is assessed for relevance and/or safety and efficacy in clinical trials. This assessment involves either a proof of concept that a mechanism identified preclinically is relevant to the targeted human disease phenotype and resultant pathophysiology or to ensure that NCE administration to human subjects and patients is a relevant translational extrapolation of the efficacy and safety observed in animal models. To a major extent therefore, these guidelines have borrowed extensively from existing guidelines for the conduct of clinical research, for example, CONSORT, and attempt to predict benchmark randomized clinical trial (RCT) criteria (Table 5.4; Hartung, 2013; van Luijk et al., 2014) and are intended to bring the standards for preclinical research closer to those used in the clinical research setting (Landis et al., 2012; Masca et al., 2015; Schulz et al., 2010).

These guidelines provide a detailed series of recommendations in the planning, execution (including time parameters), and reporting of animal research (Henderson et al., 2013) and include documentation on food and water availability, the type of food, housing temperature, humidity and light/dark cycles, caretaker/investigator gender, and details on compound/drug administration including dose, frequency, and route (McGrath et al., 2010) and, where appropriate, the use of anesthesia and/or analgesics. While these conditions often appear to be a formality, they have a major impact on experimental outcomes and can often explain discrepancies in results between different laboratories including the failure to reproduce studies (Van Bavel et al., 2016).

Additionally, the various guidelines in Table 5.3 are intended to ensure responsible and ethical animal use as encompassed in the 3Rs protocols (Russell and Burch, 1959): *Replacement*—the use of nonanimal subjects and research into development and validation of nonanimal research and testing models; *Reduction*—when replacement is not feasible, minimizing the number of animals used by improved research design, appropriate powering and statistical methods and the use of databases of previous studies where available; and *Refinement*—improving experimental procedures, housing, and husbandry to minimize risks to animal welfare.

In several logical and well-intended initiatives, these ethical requirements are being used by governmental agencies to reduce animal usage in biomedical research, as exemplified by EU Directive 2010/63/EU, which has an ultimate goal of replacing all animal research procedures (http://eurlex.europa.eu/LexUriServ/LexUriServ.do?uri=OJ:L:2010:276:0033:0079:en:pdf).

TABLE 5.4 Comparison of Animal Efficacy Studies and RCTs (Randomized Clinical Trials)

	Animal efficacy/disease mechanism studies	RCTs
Objective	• Identifying disease mechanism(s) • Identifying, characterizing, and evaluating new chemical entities (NCEs) in animal models for efficacy and safety. • Estimating human dosing via animal ADME and allometrics (Caldwell et al., 2004; Fan and de Lannoy, 2014)	• Demonstrating clinical efficacy and safety
Disease manifestation	• Model(s) ◦ Relationship to human disease equivocal ◦ ADME species dependent and often not predictive of human	• Disease naturally present albeit with different chronology, intensity, and duration
Compound/ Drug Intervention Animal models	• Introduced under homogenous, controlled conditions to relatively homogenous animal subjects • Within experimental sets, animals usually treatment naïve and homogenous in gender, age, and genetic background. • Between experimental sets—different species and strains, genetically heterogeneous with a variety of metabolic pathways and drug metabolites, leading to variation in efficacy and toxicity. • Acute animal models for chronic human disease phenomena. • Monofactorial animal disease models versus multifactorial disaese causality in humans • Dosing regimens of uncertain relevance to normal human subjects and patient	• Heterogenous with age, lifestyle, gender, comorbidities previous or concomitant interventions (drug treatment)
Sample size	• Small sample size (n = 3–9) with a priori experimental design not always documenting sample size calculations, for example, powering	• Sample size n = 200 +
Internal validity	• low	• High—due to randomization and blinding
External validity	• Low (extrapolation between species)	• High—extrapolation in one species—human
Research Staffing	• Small teams of researchers—one researcher (jack of all trades) often responsible (with technical support) for deciding—without blinding—on experimental hypothesis, design, execution, outcome assessment, data analysis, interpretation, and reporting. Reporting guidelines evolving, for example, *ARRIVE*, GPSC, NINDS	Large team of experts with individual experience in defined areas with well-established and validated protocols and clearly defined guidelines for methodological quality and reporting—CONSORT guidelines (Altman et al., 2001; Moher et al., 2001; Schulz et al., 2010)
Study type	• Different types of studies, for example, exploratory/hypothesis generating, confirmatory, replication	• Exploratory—dose ranging studies for safety and target occupancy

TABLE 5.4 Comparison of Animal Efficacy Studies and RCTs (Randomized Clinical Trials) (*cont.*)

	Animal efficacy/disease mechanism studies	RCTs
Registration/ Reporting	• No preregistration database for animal studies • Studies often not reported or selectively reported	• Trials registered at clincaltrials. gov. with anticipation of publication • Mandatory publication with registration approval (rarely followed)
Sample availability	• Postmortem data—all tissues available immediately following sacrifice of animals	• Human tissue samples restricted, for example, blood, CSF, urine • Postmortem tissues rarely fresh, for example, brain banking involves preservatives
Outcomes	• Surrogate—Investigator inferred outcomes—frequently biased • No verbal diagnostics	• Patient relevant/disease relevant outcomes—Clinical investigator can communicate with patient to ascertain actual responses

Modified from Hartung, 2013; van Luijk, et al., 2014.

These initiatives can lead to unintended consequences where concerns in reducing the numbers of animals in an experiment leads to experiments that are woefully underpowered to an extent that renders their outcome meaningless such that it becomes unethical to even conduct the work (McGrath and Lilley, 2015).

5.10.1.1 *The* ARRIVE *Guidelines*

The ARRIVE Guidelines (Kilkenny et al., 2010, 2014; McGrath et al., 2010; Sena et al., 2014) comprise a 20-item checklist that covers what authors should report in the *introduction, methods, results, and discussion* sections of a manuscript that is being submitted for peer review.

5.10.1.2 *The* DEPART *Guidelines*

The *DEPART* Guidelines (Smith et al., 2016) address issues with the effective use of the *ARRIVE* Guidelines that may result from researchers consulting the *ARRIVE/GSPC/Syrcle/ NIHPG* guidelines after experiments have been completed, which are being prepared for publication, completion of experiments, at the time of publication.

5.10.1.3 *The* MIBBI *Guidelines*

The MIBBI Guidelines (Taylor et al., 2008) are an extensive, coordinated collation of 21 checklists: CIMR (Core Information for Metabolomics Reporting); MIACA (Minimum Information About a Cellular Assay); MIAME (Minimum Information About a Microarray Experiment) which extends to subgroup checklists for nutrigenomics (/Nutr), toxicogenomics (/Tox), environmental biology (/Env), and phytology (/Plant); MIAPE (Minimum Information About a Proteomics Experiment); MIARE (Minimum Information About an RNA interference Experiment); MIFlowCyt (Minimum Information for a Flow Cytometry Experiment); MiGen (Minimum Information about a Genotyping experiment); MIGS/MIMS (Minimum Information about a Genome Sequence/ Minimum Information About a Metagenomic Sequence/Sample); MIMIx

(Minimum Information required for reporting a Molecular Interaction experiment); MIMPP (Minimum Information for Mouse Phenotyping Procedures); MINI (Minimum Information about a Neuroscience Investigation); MIQAS (Minimum Information for QTLs and Association Studies); MIqPCR (Minimum Information about a Quantitative Polymerase Chain Reaction) which has been supplemented by the *MIQE* guidelines (Bustin et al., 2009); MIRIAM (Minimum Information Requested In the Annotation of biochemical Models; MISFISHIE (Minimum Information Specification For In Situ Hybridization and Immunohistochemistry Experiments); and STRENDA (Standards for Reporting Enzymology Data). The *MIBBI* guidelines/checklists are in two parts: a *portal* that provides summary information on each of the minimum information checklists, and what the authors have termed a *foundry* that allows members of the *MIBBI* community to develop a "suite of self-consistent, clearly bounded, orthogonal, integrable, checklist modules" that can be used to facilitate the development of checklists that are specific for particular research areas using the collective expertize already established within the *MIBBI* guidelines and thus avoiding the need for other researchers to reinvent the proverbial wheel.

5.10.1.4 *The* NIHPG *Guidelines*

The NIHPG Guidelines (https://www.nih.gov/research-training/rigor-reproducibility/principles-guidelines-reporting-preclinical-research) comprise a core set of reporting standards based on those in Landis et al. (2102) that include: *Rigorous statistical analysis; Transparency in reporting* to provide details on standards (to include the ARRRIVE guidelines), replicates, statistics, randomization, blinding, powering, and data-handling that include inclusion and exclusion criteria; *Data and material sharing; Consideration of refutations* and *establishing best practice guidelines* for image based data and the description of biological materials that include antibodies, cell lines and animals (Moher et al., 2015).

5.10.1.5 *The* RIPOSTE *Guidelines*

The RIPOSTE Guidelines (Masca et al., 2015) are focused on reducing irreproducibility in preclinical research by encouraging early discussion of study design and analysis in a multidisciplinary team setting that includes statisticians. Additionally, these Guidelines advocate measures to assess real time progress via "early and regular discussions" rather than the traditional vacuum that occurs between a series of experiments being initiated and their conclusion. In this regard, they follow the evolution of a series of studies similar to the way in which a clinical trial is monitored. The elements of these recommended discussions are provided in a framework that includes:

1. research aims and objectives, specific outcomes and hypotheses; including aims and objectives; outcomes, interventions, and predictors of interest;
2. study planning including logistics; materials and techniques;
3. study design that includes: units of measurement; randomization; blinding; groups, treatments, and other predictors of interest; other potential biases, confounders and sources of variability and sample size considerations;
4. planned analysis including: data assessment and preparation; quality control criteria; data verification, normalization/correction, and outliers; and statistical methods that include interim analyses and replication/validation; and
5. reporting results.

5.10.1.6 *The* GSPC/SYRCLE *Guidelines*

The GSPC guidelines (Hooijmans et al., 2010, 2011) are focused on strategies used in selecting the type of animal model/study used for a given disease state, building on the lamentably few, but critically important, systematic reviews of the preclinical literature like the landmark assessments in the stroke area by MacLeod and colleagues (Bath et al., 2009; Lapchak et al., 2013; Macleod et al., 2008, 2009; Sena et al., 2010).

5.10.1.7 *The* TOP *Guidelines*

The TOP Guidelines (Nosek et al., 2015) include standards for "greater transparency and openness" that address: citation, replication, design, research materials, data sharing, analytic methods, and the preregistration of study design and data analysis to distinguish between exploratory, hypothesis generation studies and those intended to confirm hypotheses. These are documented at the *Open Science Framework* website (https://osf.io/9f6gx/) in either exquisite or excruciating detail depending on the perspective of the reader. As of November 2016, there were 757 journal and 64 organizational signatories to the *TOP Guidelines* (Munafò et al., 2017).

5.10.1.8 *Collated Guidelines*

Key elements from the *ARRIVE*/GSPC/*NIHPG Guidelines* have been collated Table 5.5. as a "best practices" approach for experimental planning and are extended in Table 5.6 as Issues With Experimentation.

5.10.2 Checklists-Additional Considerations

5.10.2.1 *Good Institutional Practice*

Good Institutional Practice (GIP; Begley et al., 2015) guidelines that are intended to encourage research institutions that receive overheads for grants ensure that their employees adhere to a minimum research standard that include the following points which have been slightly modified by the current authors:

- *Routine discussion* of research methods with meetings that dissect manuscripts in preparation where methods and data as well as conclusions would be debated in a manner similar to discussing and critiquing a competitor's paper in a journal club.
- *Reporting systems* that enable colleagues, graduate students, and postdocs to discuss concerns about sloppy science.
- *Training and standards* that involve compulsory institutional training to ensure a common understanding of rigorous experimental design, research standards, and objective data evaluation that should be supported, audited, and acknowledged by the Institution.
- Appropriate *incentive and evaluation systems* to deter noncompliance with guidelines, poor mentoring, and scientific sloppiness. Faculty members with poor records should face loss of laboratory space and trainees, decreased funding, and potential demotion while faculty members who excel as mentors and careful experimentalists should be rewarded.
- *Enforcement*. Institutions should investigate egregious lapses and record them in a routine, transparent way. Departments of research integrity or other centers of excellence should be funded, staffed, and given enough authority to prevent, detect,

TABLE 5.5 *ARRIVE*/GSPC/*NIHPG*/NIH Guidelines—"Best Practices" for Experimental Planning, Reporting and Value Creation

Topic	Scope of activity	Value creation/concerns

Study Intent	• Elaboration of hypothesis under experimental evaluation
	• Context of hypothesis replicating based on relevant literature
	• Selection and intent of validated test methods/assays/models
		• Is the intent of the study relevant and of interest to other researchers?
		  ◦ Is "hypothesis" low priority? In fact is there a hypothesis?
		  ◦ Are all key outcomes defined?
		  ◦ Have the authors demonstrated sufficient knowledge and currency with the area?

Study Design	• Experimental or confirmatory
	• Null hypothesis based
	• Number of groups—vehicle and treatment controls + reference standard(s) + experimental groups with concentration/dose response.
	• Equal group sizes
	• *Written* study plan including detail on materials, sources, solution preparation, cell authentication etc.
	• Experimental readout clearly defined—for example, target engagement/residence time, biochemical/hormonal signal, expression phenotype, animal phenotype/behavior
	• Ethical approval for animal usage
	• Protocols for:
	  ◦ *Randomization* to avoid subjective bias incorporated into experimental design.
	  ◦ *Blinding*—experimenter unaware of which assay well or animals are in which group
	  ◦ *Powering*—appropriate sample size to control effect size including power calculation and prospective selection of statistical analyses http://www.gpower.hhu.de
	  ◦ Prospective (a priori) criteria for data inclusion/exclusion and primary and secondary outcomes
	    – Statistical analysis and defined endpoints
	• Control of physiological/environmental variables, for example, housing and body temperature, humidity, cage size and cleanliness, feeding and lighting cycles, microbiome variations, compound effects on blood pressure, experimenter gender etc.
	• Experimental endpoint(s)
	• Reproducibility criteria—avoiding reporting of single observations and pseudoreplication including group sizes of $n = 3–5$ or per powering analysis and an *independent* n value of greater than 3 for a collated data set
		• Are the design, methods and analyses appropriate
		  ◦ Are they reported in sufficient detail to assess replicability?
		  ◦ Are key assays not duly described, material sources not given to provide authors a competitive advantage?
		• Have potential conflicts of interest been acknowledged?
		• Have elements of investigator bias been adequately addressed?

Experimental detail	• Methodologies • Number of assay wells/animals per group • Animal source, strain, weight, gender genetic background • If human tissues used—source, institutional approval and documentation of anonymous case histories of both experimental subjects and matched controls. • Anesthesia (if used). • Pharmacokinetic data used for dosing of reference standard + experimental. • Compound administration—volume of administration, route, frequency (acute versus chronic), target coverage, duration of injection, time of day • Documented detail and validation of all reagents—sources, batch number, salts, purity, activity, selectivity • Authentication of cell lines • Data independently replicated • Specific section on *Design and statistical analysis*	• Is execution efficient in terms of critical path, cost, and ethical use of animals according to *ARRIVE*/GSPC guidelines?
Data analysis	• Data collection and analysis randomize • Post hoc rationale for inclusion/ exclusion of data • Primary endpoint • Secondary outcomes • Documentation of missing or excluded data with reason • Use of a single p-value avoiding multiple levels of significance	• All data sets should be decided a priori—not *ad hoc* or *post hoc* • Provide actual p-value, for example, $P = 0.39$ • After analysis of variance, *post hoc* tests only valid if F shows no variance inhomogeneity and achieves the necessary level of statistical • Significance, for example, $P < 0.05$.
Results/ Outcomes	• Conclusions to experiment • Unexpected outcomes • Next steps	
Data sources	• Data file location • Notebook—page number/electronic file • Timely witnessing of data by "one skilled in the art"	• Witnessing key to validating experiment and protecting Intellectual Property (IP) (see Chapters 2 and 4)
	• Interpretation of data in the context of study objectives and hypothesis • Study limitations • Publication of all data—not just representative sets or cherry picked to support hypothesis • Inferences • Generalizability and predictiveness • Adverse events—compound or procedure related	• Are the findings appropriately discussed in the context of other published studies? • Do the data support the conclusions? • Is preliminary in vitro data extrapolated to inappropriate descriptions of compound potential in the treatment of human disease?

Modified from Hooijmans, et al., 2010; Hooijmans, et al., 2011; Hooijmans et al., 2014; Karp et al., 2015; Kilkenny et al., 2010; Kilkenny, et al., 2014; Landis et al., 2102; McGrath et al., 2010; Moher et al., 2015; Moher et al., 2016; Sena, et al., 2014.

TABLE 5.6 Issues with experimentation

Topic	Issues	Minimum requirements
1. Experimental design and conduct	• Null hypothesis approach - experiment designed to support a preconceived difference rather than refute a hypothesis • HARKing - hypothesizing after results are known - reanalyzing data and creating different subgroups until "an intriguing relationship is found" (Motulsky, 2014). Different from a hypothesis generating experiment. • Unrecognized/unaccounted variables • Methodology not validated or inadequately documented	• Experiment designed on the premise that there is no difference between control and treated groups • Ethical approval for study from institutional IACUC (Institutional Animal Care and Use Committee) • Endpoints and experimental groups clearly defined before start of experiment.
2. Experimental powering	• The smaller the n value (number of independent replicates within an experiment) the greater the risk the result is invalid, irrespective of the resultant *p*-value—*t* value determined by group size and variability in data. • Many in vitro results are reported as n=3 as this is the minimum number required to plug into a statistical software program, not because it has been determined a priori that this is the correct number to appropriately power a given experiment • Frequently considerable ambiguity and obfuscation as to what constitutes an "n" value for data reporting. • Underpowered/unblinded experiments leading to experimental noise that is often interpreted as a positive finding	• Use of power analysis based on exploratory statistical experiments to determine variability • If power analysis not feasible group size should be an a priori minimum of n = 3–5 • Reproducibility—execution of an experiment on three (preferably more) separate occasions—using new cell/tissue preparations (different sources, passages, etc.), a naïve animal cohort new solutions, etc.
3. Randomization	• Fewer than 30% of published preclinical studies report randomization and less than 5% report blinding.	• All studies, both in vitro and in vivo, should be randomized with regard to factors that include order of treatment, order in which reagents are added, assay position on a plate, assignment of animals to test and control groups, housing of animals, etc (see Chapter 2).

4. Blinding and bias	• Data biased?
	• Were experiments blinded to the investigators to avoid bias and the possibility of placebo-like effects?
	• Data manipulated
	• No dose/concentration response for compounds used to interrogate systems
	• Single, often supramaximal, and consequently-non-selective, dose/concentration effects of receptor and signaling pathway agonists, antagonists/blockers, for example, nanomolar selective compound used at micromolar concentrations (see Chapters 2 and 4).
	• Data fabricated
	• Incentives for premature publication
	  ◦ Academic tenure
	  ◦ Grant application/renewal
	  ◦ Biotech press release/fund raising/ partnering activities
		• Ensure that the data is not biased by experimental approach, for example, using dorsal root ganglion cells—"choosing cells based on ease of dissection, somatic size, or viability in culture all may lead to a more tractable experimental design, but risk introducing a significant sampling bias that would be absent in a purely random sample."
		• The experimenter running an experiment and involved in data collection and analysis should not be aware of the treatment protocol, due to the obvious risk of introducing bias, not matter how unintentional. A review of in vivo studies found that failure to blind the experimenter increased the odds of finding a statistically significant effect by 3.2-fold.

5. All data reported	• "Data not shown"—unacceptable • Evidence of antibody selectivity compromised by incomplete data sets. • Independent replication of the study to determine if the outcome is the same is essential to validating the results. • Number of animals described in methods and those reported in results are different. • Obfuscation as to whether "*n*" value represents number of experiments, number of replicates or histological sections within a single experiment, or combinations thereof • Selection of data for inclusion into the manuscript that supports the desired story (selective outcome reporting)	• Criteria to exclude any data must be defined *before* the study commences, including dealing with outliers and experiments in which controls did not perform as required to validate the assay • These criteria must be applied consistently • Predefined inclusion/exclusion criteria with documentation of why samples/animals were excluded • All data must be reported.
6. All experiments repeated	• Independent replication of the study to determine if the outcome is similar or the same is essential to validating and interpreting the results. • Many Western blots and RNA interference studies often performed once. • Mechanism/pathway-based system effects usually only conducted in a single cell line focused on one pathway member, for example, NFκB using a single compound at one concentration	• Reproducing versus replicating experimental results • Independent replication of an experiment • Reproducing experiment with same cells, same solutions is not an independent observation. • Assess several structurally and functionally different pathway blockers at concentrations consistent with reported selectivity. • If feasible derive quantitative data by running a concentration-response curve.

7. Inclusion of positive and negative controls as part of study design	• Controls not shown, new compounds not directly compared with reference standards or drugs (where these exist)
	• No positive control(s) for:
	  ◦ Vehicle(s) and/or solvents
	  ◦ target antagonist or blocker on its own—data only shown in combination with activator/agonist
	  ◦ scrambled oligonucleotide sequences
	  ◦ "house-keeping" proteins to quantify protein levels being co-localized with the protein of interest (e.g., cytoplasmic versus mitochondrial) to reference standards
	  ◦ Surgical manipulations
	• No negative control(s) for:
	  ◦ Compounds that should not block a given response/phenotype given their pharmacology.
		• No historical controls—all controls must be run in all experiments at same time as experimental cohorts
		• In microtiter plates use several controls in different positions to account for edge—and time-dependent effects (drift)
		• Sham operated animals
		• Contralateral injections
		• Include negative as well as positive controls, for example, propranolol, a nonselective β-adrenoceptor antagonist should block the effects of norepinephrine on tissue responses. Neither the opioid receptor antagonist, naloxone nor the nicotinic receptor antagonist, methyllycaconitine should have any pharmacological effect on a directly assessed norepinephrine-evoked response.

(*Continued*)

TABLE 5.6 Issues with experimentation (*cont.*)

Topic	Issues	Minimum requirements
8. Validation of methods and reagents	• Validation of test methods/assays/reagents • Avoidance of hypothesis testing at the same time as assay/methods validation • Data showing that antibodies have been validated as selective for antigen with high specificity, affinity and avidity AND available to other investigators • RNAi-mediated gene suppression "is somewhat unpredictable and often not as efficient as desired", and "not all designed sequences are target-specific." • "Off-target" effects that involve sequence homology of the RNAi with nontargeted transcripts • Kinase inhibitors—given the nature of the ATP binding site, very few are selective for their kinase target but are less so when used at high concentrations (see Chapter 2)	• Validation of reagents especially antibodies • Authentication of cell lines and transgenic animals
9. Pharmacokinetic assessment of compound bioavailability (Fan and de Lannoy, 2014)	• Many compounds characterized in vitro as interesting fail to show any effect in vivo—half life/bioavailability incompatible with animal model or route of administration	• Assess bioavailability, including duration systemic exposure of compound exceeds target levels (e.g., 3 × EC_{50}/ IC_{50}) • For CNS compounds establish PK profile in the brain • Residence time (Copeland, 2016)
10. Conclusions challenged using alternative approaches	• Principal findings should be corroborated using an alternative approach attempting to *refute* the generated hypothesis, for example ◦ Relationship of a gene to a disease can involve both deletion and overexpression of the gene with opposite outcomes ◦ Role of a signaling pathway in altering phenotype or mediating effects of compound confirmed using pharmacologic inhibitors, RNAi, and other approaches with attendant effects on pathway components or activities ◦ Different source of cells to corroborate results, for example, HUVECs from ATCC versus XX vs. primary cells, versus microvessel endothelial cells	Two particular examples are provided to highlight a strategy that can be adopted, and applied in various settings • Corroborate RNAi findings with another technique ◦ Dominant-negative mutants, pharmacological agents, ◦ Overexpressing the RNAi target to produce changes opposite to the RNAi, ◦ Use of other methods to inhibit gene function (e.g., CRISPR, TALENs) ◦ Rescue of cellular phenotype using si/shRNA-resistant mRNA for gene of interest • Use of kinase inhibitors to implicate a specific pathway in a response should entail: ◦ Corroboration of findings with two different "specific" kinase inhibitors, each with a different selectivity profile ◦ Approaches such as mutation of the gatekeeper site in p38 MAPK or knock-in mice containing a mutant JNK that alter sensitivity to inhibitors to determine parallel effects on the response

11. Statistics (see Chapters 2 and 3)	• *"three types of lies – lies, damn lies and statistics"*
	• *"If your experiment needs statistics, you ought to have done a better experiment"* (E. Rutherford
	• *"Bloody Obvious Test"* (Kitchen, 1987)
	• Lack of reproducibility of many studies attributable to inappropriate use of data analysis tools and statistical packages.
	• Standard errors more frequently used since they are smaller than standard deviations (making the data visually more attractive) but they do not transparently represent variability.
	• Post hoc studies to interpret and power data?
		• "Significant trend", "approaching significance", "almost significant," and similar are meaningless and have no place in the research literature.
		• One tailed requires a priori prediction of direction of effect
		• Author(s) must decide before conducting experiments what *p*-value denotes significance
		• A larger sample size will improve confidence in the outcome but may reduce
		• Show variability of datasets
		• Report effect sizes (difference), confidence intervals, data exclusions, manipulations
		• Avoiding use of statistical hypothesis testing
		• Do not use SEM, but the standard deviation (SD) or if a regression, the 95% confidence intervals
		• Application of a Bayesian framework—probability as the plausibility rather than potential frequency of an outcome.
		• "Preregistered replication" via Open Science Framework etc.
		  ◦ Two exploratory studies to gather findings
		  ◦ Registration and conduct of replication studies
		• In each figure or table, clearly state whether sample size was selected in advance
		• If *P*-hacking is used, describe conclusions as "preliminary"

(*Continued*)

TABLE 5.6 Issues with experimentation (*cont.*)

Topic	Issues	Minimum requirements

12. Western blots (WBs) (Bik et al., 2016)	• Western blots cropped to only show a single lane of a gel (see Chapter 4).
	• Images manipulated using:
	  ◦ cloning and healing tools in Abode Photoshop
	  ◦ change in brightness and/or contrast
		• WBs should be quantified and statistically analyzed.
		• Primary and secondary antibodies must be described in the Materials and Methods section. If the antibodies are not of commercial origin, their characterization must be described
		• Overall quality of WB. Bands should be well-marked.
		• Images gathered at different times or locations should not be combined into a single image unless indicated
		• Do not submit a WB:
		  ◦ Where experimental data is emphasized (e.g., by changing bias controls) versus controls
		  ◦ With fuzzy or smearing bands.
		  ◦ With over-loaded or over-exposed bands that cannot be quantified.
		  ◦ Where the samples to compare have been loaded on more than one gel
		    - Without the proper loading controls
		    - With the same loading controls in several different WBs—this is fraudulent
		• WBs
		  ◦ If gels are spliced - clearly delineate the point of splicing and avoid overextending quantitative interpretation across splices
		  ◦ Should not be overexposed.
		  ◦ Be made using proteins extracted under the same conditions as the analyzed proteins, for example, if a nuclear protein is analyzed, the loading control should be made with a nuclear protein and not a cytoplasmic protein
		• Image of the original, unprocessed WB image must be deposited with the publisher

13. Grammar, writing style, manuscript structure, and integration	• Authors lack expertize in English grammar • Authors use colloquialisms unsuitable for the scientific literature • Text and tables have not been proof read by authors • Tables and figures have not been properly integrated into the text. • Figure and table legends do not provide sufficient detail for reader	• Consult colleague with greater familiarity with English grammar and usage • Read carefully through the text using Spellcheck software • Ensure figure and table legends are standalone and do not require repeated reference to text

Collated from Bustin et al., 2009; Curtis et al., 2015; Hooijmans et al., 2010, 2011, 2014; Karp et al., 2015; Kilkenny et al., 2010. McGrath et al., 2010; Mullane et al., 2015; Landis et al., 2012; Moher et al., 2015, 2016; Marino, 2014; Motulsky, 2014; Ritskes-Hotinga et al., 2014; Schulz et al., 2010; Taylor et al., 2008.

investigate, and penalize poor-quality research. They should also be charged with promoting an institutional culture that nurtures robustness.

These guidelines also echo the seminal Editorial from Nature (2014b) entitled *"The Greater Good"* that noted that "universities have a key role [in] ensuring a greater degree of researcher training and of lab stewardship in the quality of outputs than is often happening … [and that]….funding agencies... [must]...make clear their intentions in promoting rigorous lab standards..[with]… a concomitant pressure on universities and institutes to demonstrate quality assurance of lab practices and culture." Added to these should be the quality assurance (QA) control systems highlighted by Baker (2016). QA guidelines for biomedical research, covered in additional detail in Chapter 2 while not required by the NIH, involve securing raw data, maintaining laboratory equipment and documenting the care of experimental materials and are often seen by researchers as a constraint to their research activities.

5.10.2.2 The Lancet Guidelines (Table 5.7)

The Lancet Guidelines (Moher et al., 2016) derive from a series of articles published in the *Lancet* in 2014 that addressed the interrelated topics of increasing value and decreasing waste in biomedical research. Prompted by a 2009 report (Chalmers and Glasziou, 2009) that was primarily focused on clinical research activities that estimated that 85% of research funding was wasted, these articles (Chalmers et al., 2014; Chan et al., 2014; Glasziou et al., 2014; Ioannidis et al., 2014; Macleod et al., 2014; Salman et al., 2014), highlighted five dimensions for reducing waste: research priorities; research design, conduct and analysis; research regulation and management; accessibility; and reporting (Moher et al., 2016), some of which are addressed in Chapter 2. To reemphasize, these dimensions were primary related to clinical studies such that basic biomedical researchers tended to disagree "with the concept and believed that waste was less important in their specialty (e.g., "… to state that 85% of research funding is wasted is an insult to current research efforts"; "There is no … waste in pure, basic science") (Moher et al., 2016). Viewpoints on aspects of the *Lancet* Guidelines from basic and clinical researchers and potential solutions, along with the roles of research funders, institutions and journals/editors are outlined in Table 5.7. One notable point of agreement between basic and clinical researchers is placing planned research in context via a systematic review. The *Lancet* group notes that "initiation of research without a systematic review of already known evidence is unethical, unscientific, and wasteful, particularly when the research involves people or animals" (Moher et al., 2016). Basic researchers argue that there is "no such thing as all available evidence" while clinical researchers note that it is "expensive and time consuming to do full systematic reviews and most researchers aren't good at it" (Table 5.7). It would appear that the *Lancet* guidelines are in need of additional real world refinement.

5.10.3 Preregistration of Preclinical Research Studies

The suggested preregistration of preclinical research studies (Jansen of Lorkeers et al., 2014; Kimmelman and Anderson, 2012; Moher et al., 2016; Nosek et al., 2015; Wagenmakers et al., 2012) is conceptually intended to: (1) avoid publication bias (by not publishing negative results); (2) ensure that new studies are placed in the context of previous work; and (3) distinguish hypothesis-generating exploratory studies from those focused on hypothesis testing/confirmation with the ultimate goal of minimizing false discoveries, avoiding missing serendipitous findings, and reducing animal usage. Among the information proposed to be

TABLE 5.7 The *Lancet* Guidelines—Recommendations for the Barriers to Reduce Waste in Research: Researcher Comments and Facilitators to Increasing Research Value

Recommendation	Basic research issues	Clinical research issues	Facilitator remediation
Perform a systematic review of all available evidence before planning a study	• "....primary barrier vast amount of information to be surveyed." • "no such thing as all available evidence. What constitutes evidence for a particular study is integral part of the conceptualization of the study. Different people have legitimately different methods in using evidence. Too much evidence, some of which is just bad data, can be paralyzing and prevent innovation"	• "..expensive and time consuming to do full systematic reviews and most researchers aren't good at it"	Funders to make a systematic review a condition for grant submission; funders and journals to collaborate on developing educational instruments for research in context; institutions to provide methodological and logistical support to researchers to do systematic reviews
Systematically register study protocol at inception	• "will add extra work and a collection of information that will not correspond to the actual experiment"	• "...little knowledge of how and when to register"	Develop an appropriate register for basic scientists; develop researcher guides for use of the WHO's International Clinical Trials Registry Platform, PROSPERO, and other relevant repositories
Make the full protocol publicly available	• "would make new break-throughs impossible for smaller groups, even though the idea was theirs"	• "...takes time and innovative ideas might be hard to publish once in the public domain"	Develop an appropriate repository for basic scientists; provide specific funding and logistical support to researchers to make these documents and data available; funders, institutions, and editors to reward researchers making the protocol, analysis plan, results, and raw data publicly available
Make the analysis plan publicly available	• "these questions are not for basic research but for applied clinical research"	• "love to do this, but usually time to complete the analysis plan is too short"	Develop appropriate repository for basic scientists; provide specific funding and logistical support to researchers to make these documents and data available; funders, institutions, and editors to reward researchers making the protocol, analysis plan, results, and raw data publicly available

TABLE 5.7 The *Lancet* Guidelines—Recommendations for the Barriers to Reduce Waste in Research: Researcher Comments and Facilitators to Increasing Research Value (*cont.*)

Recommendation	Basic research issues	Clinical research issues	Facilitator remediation
Systematically make results publicly available	• "...time waste, need lot of time to write negative experiments"	• "...negative results are less likely to have enthusiasm for publication"	Develop appropriate repository for basic scientists; provide specific funding and logistical support to researchers to make these documents and data available; funders, institutions, and editors to reward researchers making the protocol, analysis plan, results, and raw data publicly available
Make raw data publicly available	• "...scarcity of suitable repositories and little funding to establish these"	• would create many issues such as confidentiality, which would need to be redacted and would waste time; probably reluctance to give access to such data exists because others could use them for their own purposes; and mass sharing of data could lead to inappropriate use, as the context of data collection and the objective of the study are necessary to understand their meaning"	Develop appropriate repository for basic scientists; provide specific funding and logistical support to researchers to make these data available; funders, institutions, and editors to reward researchers making the protocol, analysis plan, results, and raw data publicly available

Modified from Table 3 in Moher et al., 2016.

required for preregistration are: (1) the hypothesis being tested and the context in which the hypothesis has been developed; (2) a description of the intended study including steps made to avoid bias; (3) the power calculations that dictate the number of animals to be included in the study; (4) the reference standards to validate the study; (5) the primary and, if relevant, secondary outcomes from the study; (6) the intended statistical analysis including *post-hoc* tests; (7) the prospective criteria used to exclude data from analysis; and so on (see Chapter 2).

The preregistration concept has been formalized by the journal *Cortex* in the form of *Registered Reports* (Chambers, 2013). These divide the peer review process into two distinct stages: the first, a description of the proposed study in an initial manuscript that includes "a slimline introduction, hypotheses, experimental procedures, analysis pipeline, a statistical power analysis...to a minimum power level of 90%, and pilot data where applicable" that is preregistered and peer reviewed before substantive data are collected. This process is intended to result in the study either being rejected or "accepted in principle" with the former guaranteed to stifle innovation by focusing on the adoption of hypotheses consistent with favored concepts deemed important, e.g., amyloid and Alzheimer's disease where it has been viewed that "if you were not working in the amyloid field you were not working on Alzheimer's disease" (Buee quoted in Whalen, 2013) the latter decision being based on the importance of the hypothesis under investigation and the validity of the method, "not whether results are novel or statistically significant." The latter, in addition to potentially preventing *P*-hacking (Motulsky, 2014), data exclusion, and selective reporting (Begley, 2012), guarantees that the results of the actual study will be published, even if the findings are negative, provided that the authors adhere precisely to their registered protocol. Once experimentation is completed, authors are then expected to resubmit their full manuscript for rereview, publicly sharing the raw data and, pending quality checks and what the reviewers view as a "sensible interpretation of the findings," the manuscript is then published. While commendable, the outcome depends on the expertize and diligence of those conducting the peer review.

The *Registered Report* concept has also become an integral part of the reproducibility initiative, *Reproducibility Project: Cancer Biology* (Section 5.13). The content of the *Registered Report*s which are published in *eLife* as part of the *RP:CB* is, based on a limited random evaluation of published reports, comprehensive and impressive in detail (Sharma et al., 2016) and certainly represents a conceptually major step forward in avoiding obstacles to reproducibility *if:* (1) the conditions outlined in the *Registered Report* are adhered to by the authors and are appropriately reflective of those in the original peer reviewed publication: (2) this practice becomes more widespread (reaching beyond organized initiatives like *RP:CB*): and (3) the published outcomes prove to be unambiguously definitive, transparent, and robust which to date has not been the case (Section 5.12).

To incentivize study preregistration, the *COS* is offering faux "prizes" in the amount of $1,000 to scientists who preregister their studies and agree with "an outside reviewer exactly what hypothesis they are seeking to test before they run their experiment" (Matthews, 2016). Another initiative is the *preclinicaltrials.eu* site (http://www.preclinicaltrials.eu/index.html) from the regenerative cardiovascular medicine research group at the University Medical Center Utrecht in the Netherlands which is based on the recommendations of the Jansen of Lorkeers et al. (2014) paper.

While studies that are preregistered are deemed by those advocating this approach as having "a substantially higher truth value than regular studies," others dismiss them as setting a new gold standard for reporting research due to concerns that preregistration will denigrate "the vast majority of great research and allow a number of serious constraints to be placed on it" (Scott, 2013). If a lack of preregistration does not prevent publication in a high JIF journal then what is the incentive for researchers to engage?

5.10.3.1 Preregistration of Preclinical Research in the Context of Translational Medicine

An additional positive facet of the preregistration initiative is ensuring that preclinical experiments on NCEs that enter clinical trials are publicly reported in order to provide access to the data that justifies initiation of a clinical trial (Federico et al., 2014; Hartung, 2013). Given that human trials often fail to reproduce the effectiveness of a therapeutic demonstrated in preclinical studies (Kimmelman and Anderson, 2012) or there are unexpected side effects as in the cases of the CD28 monoclonal antibody, TGN1412 (Hünig, 2012) and the fatty acid amide hydrolase inhibitor, BIA 10-2474 (Callaway and Butler, 2016), both of which had extremely serious and unexpected side effects in Phase I clinical studies, access to peer reviewed preclinical data may have provided some context for these outcomes.

In a metaanalysis of 2,462 efficacy studies in animals, Federico et al. (2014) found that the number of efficacy studies reported for a new drug candidate was a function of how far the NCE had advanced in clinical development, the further along—the greater the number of studies reported—with the "vast majority" of animal efficacy studies being published after initial clinical trial reports. This may be reflective of straightforward commercial concerns or an unwillingness of drug companies to share their strategic approach to drug discovery and thus lose what is deemed as a competitive edge. Unfortunately, this may disguise possible discrepancies between positive preclinical findings and failures in clinical development, precluding meaningful interrogation of why clinical trials fail and which preclinical approaches and models may be (1) informative and (2) intrinsically flawed.

5.10.3.2 *ClinicalTrials.gov. as a Precedent for Preclinical Preregistration?*

However well intended the preregistration of preclinical research studies might be, the potential success of this proposed initiative needs to be objectively assessed in the context of the similar-in-intent NIH website for clinical trials, *ClinicalTrials.gov* (https://clinicaltrials.gov/ct2/about-site/link-to), "a registry of clinical trials information for both federally and privately funded trials conducted under investigational new drug (IND) applications to test the effectiveness of experimental drugs for serious or life-threatening diseases or conditions."

The entry of information on clinical trials on *ClinicalTrials.gov*—which was established in 2000 (Ross et al., 2009)—resulted in the preregistration of new drug trials to allow the public to identify those in which they might qualify as participants. Mandatory elements for registration include the title of the trial, identification of the drug candidate, the disease state for which it is/was being studied, the trial design, and the intervention studied. These elements were expanded via the Food and Drug Administration Amendments Act of 2007 (FDAAA), as a result of multiple and continuing concerns regarding the registration, accuracy, topicality, and transparency of the actual registered trials and their dissemination, especially in those instances where industry sponsored clinical trials were found to have been selectively reported, manipulated or suppressed (Goldacre, 2012, 2015, 2016; Wynia and Borden, 2009). The validity of such concerns was further exemplified in a seminal metaanalysis conducted by the German regulatory authority, IQWiG (Institut für Qualität und Wirtschaftlichkeit im Gesundheitswesen—Institute for Quality and Efficiency in Health Care) of the selective serotonin reuptake inhibitor (SSRI) antidepressant, reboextine (Eyding et al., 2010). While approved in the European Union "for the acute treatment of depressive illness or major depression," reboextine failed to gain approval in the United States. In analyzing both published and unpublished clinical studies (the latter being available via IQWiG's status as a regulatory authority), IQWiG concluded that reboextine "to be an ineffective and potentially harmful antidepressant...[with published evidence]...affected by publication bias, underlining the urgent need for mandatory publication of trial data."

Regulations based on the FDAAA and its subsequent iterations required clinical researchers in the United States, both in academia and industry, to register clinical trials with the start date, and to report primary and secondary outcomes on *ClinicalTrials.gov* within 2 years of trial completion. Penalties for noncompliance with the FDAAA include withholding NIH grant funding and civil monetary penalties of up to $10,000 a day and appear to have been rarely—if ever—enforced—reflecting a continuing inability or unwillingness of the US federal government to enforce its own laws (Anderson, 2016). The FDAAA mandated a time frame of 12 months for completed clinical trial results to appear in an open repository, for example, *ClinicalTrials.gov,* irrespective of whether these results were formally published. This recommendation led the editors of many mainstream clinical journals (De Angelis et al., 2004) to establish guidelines that stated that only those clinical trials registered with *ClinicalTrials.gov* could be published in their journals to further ensure greater transparency and accountability in the results thus allowing all interested parties to track clinical trials from initiation through to completion (Anderson, 2016).

5.10.3.3 Assessing Compliance with ClinicalTrials.gov Preregistration

Despite the ethical and legal mandates behind the *ClinicalTrials.gov* initiative, multiple metaanalyses have consistently concluded that compliance is a major problem. In one analysis (Lee et al., 2008), 50% of trials supporting registration of FDA-approved drugs were found to be unpublished more than 5 years after their approval with an analysis of 7,515 randomized controlled trials (RCTs) registered on *ClinicalTrials.gov* coming to the same conclusion (Ross et al., 2009). A separate metaanalysis that used the 15 primary registries in the World Health Organization International Clinical Trials Registry Platform (ICTRP) of 447 clinical trials for postherpetic neuralgia, diabetic neuropathy, and fibromyalgia also reported that the results of 50% of the registered trials had not been published (Munch et al., 2014). Another metaanalysis (Anderson et al., 2015) reported that only 13% of the 13,327 RCTs registered with *ClinicalTrials.gov* that were completed or terminated between January 1, 2008 and August 31, 2012 were reported within the FDAAA mandated 12-month period with an additional 25% (for a total of 38%) being reported within 60 months, the reporting period for the study.

The metaanalysis of Anderson et al. (2015) also compared reporting compliance between RCTs conducted by the pharmaceutical industry and in federal and academic studies funded by the NIH. Using a random, albeit somewhat very small, sample of 205 of the 13,327 RCTs covered, this analysis noted that 80% of the industry-funded trials were either reported within 60 months or had legal reasons for delay. In contrast, only 50% of the NIH-funded studies reported within 60 months or had legal reasons for delay. An additional metaanalysis of 4,347 RCTs conducted at 51 academic medical centers in the United States with primary completion dates between October 2007 and September 2010—again registered on *ClinicalTrials.gov*—reported (Chen et al., 2016) that the results of 36% of the trials were published within 24 months of completion with an additional 26% being disseminated after the initial 24-month period. However, 30% of the trials were not published. This led the authors to conclude that "Despite the ethical mandate and expressed values and mission of academic institutions, there is poor performance and noticeable variation in the dissemination of clinical trial results across leading academic medical centers" reflecting the "drift toward irrelevance and obsolescence" of *ClinicalTrials.gov* (Anderson, 2016).

This lack of compliance despite the obvious ethical (e.g., researchers not upholding their part of the informed consent agreement with trial subjects; Economist, 2015) and the potential financial consequences in the form of fines by the FDA has been argued as being due to both a lack of motivation on the part of investigators with *ClinicalTrials.gov* being a difficult website to navigate and, in academia, a lack of the necessary resources necessary for reporting. A more probable contributing cause is the lack of any tangible consequences that are reflected in the compliance mandated by the FDAAA being almost universally flouted. Given that the stakes in the clinical arena are infinitely higher (human health) than in preclinical research (Ioannidis, 2016b), it is either naïve or foolish to think that what has not worked in the transparent reporting of results in the clinic will work for preregistration of studies in preclinical research, irrespective of the intent.

5.10.3.4 Other

Additional systematic reviews of the animal models used in multiple sclerosis, cancer pain, spinal cord injury, Parkinson's disease, and Alzheimer's disease research are part of the growing *CAMARADES* (Collaborative Approach to Meta-Analysis and Review of Animal Data from Experimental Studies; http://www.dcn.ed.ac.uk/camarades//) database (van Luijk et al., 2014).

From a translational perspective, systematic reviews of animal studies are intended to aid in these becoming more congruent with the gold standard RCTs conducted in humans (Table 5.6; Henderson et al., 2013; Muhlhausler et al., 2013) with a much greater emphasis on optimizing experimental design to improve the resultant data and its analysis by reducing unintentional bias to improve study quality, validity, and replicability (Curtis et al., 2015; Green, 2015; Marino, 2014; Motulsky, 2014; Mullane et al., 2015; van Luijk et al., 2014). Aspects of the existing guidelines are collated in Tables 5.5 and 5.6 and may represent a useful basis on which to build formal checklists for authors, editors, and reviewers (Kilkenny et al., 2014; Masca et al., 2015; Nosek et al., 2015; Taylor et al., 2008). Unsurprisingly, many of the concepts developed in relation to in vivo animal-focused guidelines are equally applicable to in vitro studies as discussed in detail by both Taylor et al. (2008) and Masca et al. (2015).

Despite the criticality of the recommendations in the seminal *ARRIVE* Guidelines and related guidelines, compliance has been infrequent, let alone routinely mandated/enforced, either by journals (Avey et al., 2016; Baker et al., 2014; Smith et al., 2016) or grant bodies (*NIHPG*; https://www.nih.gov/research-training/rigor-reproducibility/principles-guidelines-reporting-preclinical-research).

5.11 CHECKLISTS

Embedded in several of the Guidelines discussed are elements of a formal checklist. While checklists have generally been disparaged in research and viewed as unnecessary for individuals with graduate degrees, they have proven to be a highly effective standard operating procedure by which to improve accuracy, efficiency, and safety in both airplane cockpits (Degani and Weiner, 1993) and surgical operating rooms (Gawande, 2009). However, these successful checklists tend to be mandatory rather than mere guidelines and are enforced in the one instance by the airlines, the FAA and IATA, and, on the other, by hospitals, the federal government, and the breed of lawyers known as ambulance chasers. They have, unlike preclinical reporting /publication checklists, real consequences.

The first formal checklist related to research reproducibility was Begley's Six Rules that were published as the "six red flags" checklist in *Nature* (Begley, 2013) which asked the following key questions:

1. Were experiments performed blinded?
2. Were basic experiments repeated?
3. Were all the results presented?
4. Were there positive and negative controls?
5. Were reagents validated?
6. Were statistical tests appropriate?

The long awaited NIH guidelines for preclinical research (https://www.nih.gov/research-training/rigor-reproducibility/principles-guidelines-reporting-preclinical-research; Collins and Tabak, 2014; Moher et al., 2015) while providing greater granularity than that in the "six red flags," was little different in identifying key concerns. However, this represented a critical first step in the development of more universal guidelines (Wadman, 2013) while covering the basic elements required to plan, execute, analyze, and report experiments. Unfortunately, the lack of clarity and the complexity of subsequent guidelines have tended to overwhelm researchers and lack sufficient definition to provide a useful model for practical submission checklists.

In 2015, the *Nature* Publishing Group and the *Science* family of journals instituted formal and, in some instances, detailed checklists as part of the submission process for their various journals (McNutt, 2014; Nature, 2014a,b). Like an increasing number of basic (https://www.faseb.org/Portals/2/pdfs/opa/2016/FASEB_Enhancing%20Research%20Reproducibility.pdf) and clinic journals (Moher et al., 2016), mainstream pharmacological journals including *Biochemical Pharmacology* (BCP, Mullane et al., 2015) and the *British Journal of Pharmacology* (BJP, Curtis et al., 2015) also created checklists that build on aspects not only of the *Nature*, *NIHPG*, and *MIBBI guidelines* (Table 5.3.) but also those in the *ARRIVE* and *GSPC/SYRCLE* guidelines (Table 5.5). The contents of many of the checklists are reflective of one another, an indication that common themes are involved across the various biomedical research disciplines and that there is value in collective wisdom rather than reinventing the wheel.

However, a "core curriculum" of requirements that may be tailored for specific disciplines (i.e., requirements for informed consent may be deleted for animal research, and requirements for standardization of anesthesia may be deleted for cell culture, etc.) is required. Presently there is a growing redundancy with a tendency for each and every journal or learned society to generate its own guidance/checklists from scratch (https://www.faseb.org/Portals/2/pdfs/opa/2016/FASEB_Enhancing%20Research%20Reproducibility.pdf). Additionally, too many of the guidelines are "best practice" rather than mandatory.

The editors/authors of this monograph were intimately involved in the development of the BPC and BJP checklists (Table 5.8), which, along with their coauthors, they viewed as requirements to proactively improve the peer review process to increase the

TABLE 5.8 Representative Biomedical Publication Checklists

BCP	*Biochemical Pharmacology Guidelines*	Manuscript construction, experimental design and analysis check list	Mullane et al., 2015
BJP	British Journal of Pharmacology Guidance	Experimental design and analysis check specifications	Curtis et al., 2015
NIH Rigor and Reproducibility	*NIH Principles and Guidelines for Reporting Preclinical Research*	Recommendations for data analysis, reporting, sharing, and publication of refutations	Landis et al., 2012 Moher et al., 2015 http://www.nih.gov/science/reproducibility/index.htm
MIBBI	*Minimum Information for Biological and Biomedical Investigations*	Recommendations for minimum reporting of experimental methods	Taylor et al., 2008
NPG Check List	Nature Publishing Group Checklist	Methodology and analysis check list: required for all life science submissions	http://www.nature.com/authors/policies/checklist.pdf

likelihood of reproducibility—not just passive guidelines. The intent of these checklists was threefold:

1. to ensure that a manuscript submitted for publication was complete containing all the required detail and depth on group size determination, independent reproduction of experiments, and appropriate data quantitation, with sufficient clarity to ensure that the manuscript can be objectively peer reviewed;
2. to provide similar guidance to both editors and reviewers as the submission went through the peer review process; and
3. more ambitiously, to provide a template for authors in the planning, execution, and analysis of their research activities to ensure that their data would support the preparation of a manuscript that was suitable for peer review (Table 5.8).

The content of the BCP checklist (Table 5.9; Mullane et al., 2015) comprises 36 questions that were culled from a list of over 60 that require a binary yes/no answer from authors submitting manuscripts with the results of all but 4 of the questions (which deal with specific formatting issues) being used by the editors and reviewers of the journal as part of the peer review process. While there were initial concerns that responding to 36 questions would deter authors from submitting papers, this has not been an issue other than in those instances where the authors compulsively check all the "yes" boxes or answer the question wrongly.

The current BCP guidelines mandate a minimum $n = 3$ for reproduction (Table 5.8, question 21), taking the position that while $n = 5$ or greater provides a more robust outcome to avoid false positives (Curtis et al., 2015; Motulsky, 2014; Chapters 2 and 3), this represents a cultural change that will take time to occur.

TABLE 5.9 The *Biochemical Pharmacology* (BCP) Checklist (Mullane et al., 2015; https://www.elsevier.com/wps/find/journaldescription.cws_home/525454?generatepdf=true)

Table 1. Scientific Submission Checklist

Please answer the following questions with "Yes", "No", or "Not applicable".

Formatting - The submission will automatically be rejected if these first four questions are not marked "yes"

1. As *Biochemical Pharmacology* does NOT publish supplemental data with the exception of audio or video files, are all necessary data included in the body of the manuscript?
2. Are all tables and figures numbered and appropriately titled with descriptive legends that permit stand-alone interpretation?
3. Are all data shown in the figures and tables also shown in the text of the Results section and discussed in the Conclusions?
4. Does the e-mail address for the corresponding author indicate an affiliation with a research- based institution or has the author provided a separate statement written in English on institutional letterhead and signed by an official responsible for research activities for the institute verifying the affiliation listed by the corresponding author, along with the official's institutional e-mail contact information?

INTRODUCTION

5. Is there a clear statement with background describing the hypothesis being tested by this study?
6. Are the primary endpoints clearly described?

MATERIALS AND METHODS

7. The sources of all study materials are clearly indicated and is it acknowledged that reports containing experiments conducted with chemical mixtures, plant or animal extracts are not considered for publication unless the chemical structures and precise concentrations of all substances are reported?
8. Is (are) the chemical structure(s) of any new compound(s) presented as a figure in the manuscript or referenced in the peer-reviewed literature?
9. Are the source(s) and passage number of cell lines indicated?
10. Were cell lines authenticated by you or the vendor?
11. If used, has the selectivity of antibodies and/or interference RNA been validated and is/are their source clearly indicated?
12. If used, has the species, strain, sex, weight and source of the animals been provided?
13. Is a statement included in the text indicating compliance with regulations on the ethical treatment of animals including the identification of the institutional committee that approved the experiments?
14. Is the rationale provided for the selection of concentrations, doses, route and frequency of compound administration?
15. Are quantified results (e.g., IC_{50} and/or EC_{50} values) of concentration- and dose-response experiments included in the manuscript?
16. If used, is the method of anesthesia described?
17. Are all group sizes approximately the same?
18. Were the criteria used for excluding any data from analysis determined prospectively and clearly stated?
19. Was the investigator responsible for data analysis blinded to which samples/animals represent control and treatment groups?
20. Is the exact sample size (n) for each experimental group/condition clearly indicated in the text and/or in the tables and figures?

TABLE 5.9 The *Biochemical Pharmacology* (BCP) Checklist (Mullane et al., 2015; https://www.elsevier.com/wps/find/journaldescription.cws_home/525454?generatepdf=true) (*cont.*)

21. Are the reported data displayed as the mean +/- standard deviation (SD) of three or more independent experimental replications?
22. Is the number of replicates used to generate an individual data point in each of the *independent* experiments clearly indicated and is it equal to or greater than 3?
23. Were the statistical tests used to analyze the primary endpoints predetermined as part of the experimental design?
24. Is the threshold for statistical significance (p-value) clearly indicated?
25. Were the data normalized?
26. Were *post-hoc* tests used to assess the statistical significance among means?
27. Were human tissues or fluids used in this study?

RESULTS

28. If western blots are shown, are the following included: i)appropriate loading controls for each western blot, ii)replication data, iii) quantification, and iv) the results of a statistical analysis?
29. If PCR and RT-PCR are included, were *MIQE* guidelines followed?
30. Was a reference standard (positive or negative controls) included in the study to validate the experiment?

DISCUSSION

31. Are all the findings considered within the context of the hypothesis presented in the Introduction
32. Are the primary conclusions and their implications clearly stated?
33. Are any secondary endpoints reported and are these sufficiently powered for appropriate statistical analysis?
34. Are the limitations of the current study or alternative interpretations of the findings clearly stated?

CONFLICT OF INTEREST/FINANCIAL SUPPORT

35. Indicate by checking the box at right that a conflict of interest statement is included in the manuscript
36. Indicate by checking the box at right that all organizations providing funding for this work are listed in Acknowledgements.

5.12 REPRODUCIBILITY INITIATIVES (RIs)

The issues with reproducibility outlined in detail in Chapter 1 have led to the creation of a number of reproducibility initiatives (RIs) from various stakeholders all of whom are seeking to provide workable models to improve reproducibility in the biomedical sciences. These include:

5.12.1 Science Exchange

Science Exchange (https://www.scienceexchange.com) is a venture capital-funded initiative (Anderson, 2012b) that provides researchers with the ability to contract out their experiments to "the world's best service providers" with *Science Exchange* overseeing the process of compliance, contracts, and payments. In collaboration with the open access publisher, *PLoS One*, Elsevier's research collaboration platform, *Mendeley* and the data depository, *figshare*, *Science Exchange* has developed a *Reproducibility Initiative (RI)* platform to provide independent validation of experimental results on a case by case basis (Fulmer, 2012) and broader therapeutic area/disease related initiatives that include: the *RP:CB* (Errington et al., 2014) initial results from which were published in early 2017 (Nosek and Errington, 2017) with additional reports being published in *eLife* in the 2017–18 timeframe; *Reproducibility Project: Prostate Cancer (RP/PC)* that like the *RP:CB* involves the *Center for Open Science (COS)*, "a non-profit technology company providing free and open services to increase inclusivity and transparency of research. COS supports shifting incentives and practices to align more closely with scientific values" (https://cos.io) in collaboration with the Prostate Cancer Foundation and the Movember Foundation; and a project to reanalyze four key papers in the HIV area under the auspices of the Gates Foundation-funded *International Initiative for Impact Evaluation* (3ie) in collaboration with the University of Nebraska Center for Collaboration on Research Design and Analysis.

With a stellar Scientific Advisory Board that included many of the key opinion leaders leading the debate in reproducibility in biomedical research, the stated purpose of the *RI* was to "identify and reward high quality reproducible research via independent validation of key experimental results" and in doing so "benefit stakeholders from across the research spectrum, including research scientists, drug companies, publishers, funders, and patient groups, all of whom agree that independent confirmation of results improves science and speeds discovery" (http://blogs.plos.org/everyone/2012/08/14/plos-one-launches-reproducibility-initiative/).

5.12.2 NIH Initiative on Enhancing Research Reproducibility and Transparency

The *NIH Initiative on Enhancing Research Reproducibility and Transparency* (also known as *Rigor and Reproducibility the NIH Reproducibility initiative* (Schmidt, 2014; Thayer et al., 2014; Wadman, 2013) (https://www.nih.gov/research-training/rigor-reproducibility/updated-application-instructions-enhance-rigor-reproducibility) is focused on: deficiencies in reporting experimental methodology; bias in randomization and blinding; the numbers of subjects to detect meaningful differences between the treatment and control groups; and problems in data handling and analysis all of which can lead to false positive results. As of October, 2016 this NIH initiative is primarily focused on training grants with a single Funding Opportunity Announcement (FOA) to develop "novel, reliable, and cost effective" Tools for Cell Line Identification (SBIR [R43/R44]).

5.12.3 Reproducibility Project:Psychology Data (RP:PD)

Reproducibility Project:Psychology Data (RP:PD) sponsored by the COS involved a consortium of 270 scientists, the Open Science Collaboration (OSC) that was focused on reproducing the results of 100 published studies in psychology (OSC, 2012), the results of which were published in 2015 (OSC, 2015) but not without considerable controversy (Anderson et al., 2016; Gilbert et al., 2016; Reuell, 2016; Mullane and Williams, 2017).

5.12.4 Maintaining Motorcycles—the Cure Huntington's Disease Initiative (CHDI) Foundation Initiative

The CHDI, a not for profit disease advocacy group, proposed formalizing data replication as an integral part of their research funding activities (Munafò et al., 2014). Thus, it was proposed that the publication of selected studies, described as those that appear to represent a significant breakthrough in the basic science related to the mission of CHDI, would be delayed to allow an independent laboratory—either academic or a CRO (Contact Research Organization)—to reproduce the study (or not) with the joint publication of the original and reproduced studies. This was stipulated in the grant funding agreement with the initiative no doubt being intended to help in avoiding a repetition like the failed translation of numerous promising preclinical drug candidates to the clinic in the ALS area (Chapter 1; Perrin, 2014).

The requirement for formalized, independent replication of research findings before publication as part of the funding agreement was noted by the authors (Munafò et al., 2014) as "counter to the strongly held notion of academic freedom" thus reflecting a "substantial cultural change" that was beneficial to all parties. As of October, 2016 the status of this initiative remained unclear with the CHDI establishment of an Independent Statistical Standing Committee (ISSC; http://chdifoundation.org/independent-statistical-standing-committee/) to provide independent, unbiased and objective evaluation and expert advice for all aspects of experimental design and statistics being the focus of CHDI's activities (Munafò et al., 2017).

5.12.5 Prepublication Independent Replication (PPIR)

A similar strategy to the CHDI proposal is embodied in the PPIR (Schweinsberg et al., 2016), a collaborative crowdsourcing project that proposes a "nonadversarial replication process" to improve reproducibility by having original findings in psychology reproduced in qualified, independent replicator laboratories selected by the authors before they are published. In a pilot study, *the Pipeline Project*, 10 original studies focused on moral judgement effects from a single laboratory were repeated with an overall replication rate of 60%. The authors concluded that their findings hold "enormous potential for building connections with colleagues and increasing the robustness and reliability of scientific knowledge, whether in psychology or in other disciplines" and also the potential "to create an online market place for laboratories interested in replicating each other's work" (http://retractionwatch.com/2016/03/31/what-if-we-tried-to-replicate-papers-before-theyre-published/).

5.13 RIs IN PRACTICE: CONCERNS AND OUTCOMES

5.13.1 Tactical Aspects of RIs: Funding

A major issue in the planning and successful execution of RIs is funding. The original grant from the Arnold Foundation to support the *RP:CB*, originally noted as being $1.3 million and now as $1.6 million, was intended to validate "50 landmark cancer studies" (Errington et al., 2014) with an average of $26,000–$32,000 for each experiment. This proved to be an underestimate due to "time-consuming peer reviews… and costly experiments involving animals" with $40,000 an experiment now being the new estimate (Maher, 2015). Even this may prove to be inadequate given the estimated $0.5–2 million that is used in the biopharmaceutical industry to reproduce published research findings (Freedman et al., 2015) which, rather than being indicative of a bloated bureaucracy with high overheads, is the cost of addressing the issue of reproducibility properly.

The challenges involved in executing *RB:CB* led to the reduction of the 50 targeted studies to a final 29 (Baker and Dolgin, 2017) due to issues that included a lack of access to the materials used in the original study—either because of legal issues or because they no longer existed or had been corrupted/contaminated—and a lack of information on how the original experiment(s) had been conducted. The latter reflected poor—or, in many instances, no—record keeping, leading to an inability to access to the raw data (whether or not it still existed Engber (2016)), to understand precisely what was done to conduct the experiment and contribute to the manuscript content, and/or the unavailability of a key individual who had actually done the experiments. The latter was either due to their physical absence or to faulty memory raising the question of whether the original experiments could actually be repeated—let alone the findings reproduced (Engber, 2016). Anderson (2012b) has noted that the limited funding for the RI will help reinforce the concerns initially raised by Ioannidis (2005) on the use of smaller studies that lead to false outcomes and further noted that "The Reproducibility Initiative takes the inherent barriers to negative, replication, or reanalysis studies and adds further barriers—the cost to do more tests, the potential loss of confidentiality by sharing data with unknown scientists at other institutions, and another cost of publication." This does not bode well for the future activities of the RI.

Ensuring adequate funding rather than operating on a shoestring budget like *RP:CB*, led Pusztai et al. (2013) to propose the adoption of a "three-pronged funding scheme" comprising grants in the areas of innovative research, replication, and product development. *Innovative research grants* would be similar to the current NIH RO-1-type research grants; *replication grants* (more correctly *reproducibility* grants) would be intended to confirm and validate newly published findings that are deemed sufficiently important to warrant this step; and *product development grants* would "be reserved for putting independently verified results into practical use," for example, as a prelude to translational studies (Kimmelman et al., 2014). Pusztai et al., acknowledge that the funds required for the proposed replication and product development grants will need to be reallocated from discovery-oriented research or found via "novel mechanisms," for example, an investor-based "validation research fund," arguing that this cost will be more than offset by reducing the costs of drug discovery by decreasing the attrition rate in translation—an outcome that is highly questionable.

A somewhat similar approach has been suggested by Rosenblatt (2016) in terms of incentive-based approaches to funding academia-industry collaborations in drug discovery. His proposal is for the industrial partner, for example, a biopharmaceutical company or early stage biotech, to provide funding to the academic institution that is the source/owner of the technology of interest to the industrial partner to support the replication of the study. If the study confirms the original study, the host institution can proceed with its activities in executing a collaboration. If the study cannot be replicated, Rosenblatt proposes that the funding is returned, in whole or part, to the sponsor, providing a penalty—a consequence that is missing from so many of the initiatives in reproducibility—to incentivize the institution and its investigator. In turn, research institutions who are "early adopters" of this approach may be viewed as preferred partners to biopharmaceutical companies.

5.13.2 Tactical Aspects of RIs: Status and Outcomes

The status, outcomes, and comments on known RIs as of April, 2017 were as follows:

5.13.2.1 Antiparasitic Peptides and Replication/Validation

The results of the first RI involving the Science Exchange consortium involved an author-requested validation of a 2011 study (Lynn et al., 2011) showing that a peptide, bovine myeloid antimicrobial peptide 28 (BMAP-28) and its isomers had antileishmanial activity in the micromolar concentration range. The initial replication study (Iorns et al., 2014; Van Noorden, 2014d) reported that while BMAP-28 and its isomers had antiparasitic activity, this occurred at concentrations some 10-fold greater than those originally reported. This was found to be the result of the original data being generated using amidated BMAP-28, the form present in cows with leishmaniasis while the initial replication attempt used unamidated peptides. This was because the peptide form had not been clearly specified in the original paper (Iorns et al., 2014; Lynn et al., 2011; Van Noorden, 2014d). Further replication of the initial studies using amidated BMAP peptides found that the activity of the latter while somewhat closer to that originally reported was lower than observed in the original study. Promastigote viability in the presence of L-, D-, and RI-peptides at 2 μM was 94%, 36%, and 66%, respectively in the reproduction study versus 57%, 6%, and 18% in the original study (Iorns et al., 2014). The overall conclusion from this "Replication Attempt" was that the protocol in the original paper failed to accurately document the use of amidated peptides since it was assumed that this was common knowledge.

While the cost of this study was only $2,000 (Van Noorden, 2014d), its only concrete outcome appeared to be the decision to require the publication of a *Registered Report* outlining the protocol before a *Replication Attempt* was initiated (Errington et al., 2014), adding to the peer review burden.

5.13.2.2 Reproducibility Project: Cancer Biology (RP:CB)

As noted, a year after its formal initiation (Errington et al., 2014), the goals of the *RP:CB* were scaled back from replicating the original 50 "high impact" cancer biology articles selected to 29 due to various legal, funding, and peer review issues (Kaiser, 2015b; Maher, 2015). This was accompanied by criticism that the initiative was "time-consuming, out-of-touch with the realities of basic science and unlikely to produce interpretable results" with the comment "It's a naïveté that by simply embracing this ethic, which sounds eminently reasonable that one can clean out the Augean stables of science" (Maher, 2015). Furthermore Begley, a member of the *RP:CB* advisory board stepped down due to concerns that some of the studies selected in the original 50 were seriously flawed, for example, lacking in appropriate controls, such that trying to reproduce these was viewed as a "complete waste of time" (Kaiser, 2015a). The initial results from the first 5 of the 29 papers still targeted for reproduction by the RP:CB were published in early 2017 (Nosek and Errington, 2017) with outcomes that were described as "muddy" (Baker and Dolgin, 2017). Two of the 5 studies were "substantially reproduced", two yielded "uninterpretable results" while the fifth could not be reproduced despite having been independently reproduced by "at least ten laboratories".

5.13.2.3 Reproducibility Project: Psychology Data (RP:PD)

This project involved the attempted replication of findings from 100 published studies in psychology that were selected by the OSC. Of these studies, only 39 were reported as being successfully replicated (OSC, 2015) despite being previewed for methodological fidelity, using materials provided by the original authors and being powered to detect the original effect sizes. This outcome further reinforced the concerns originally raised by Prinz et al. (2011), Begley and Ellis (2012), and the Economist (2013c) regarding a reproducibility crisis in biomedical research (Jarvis and Williams, 2016) and also questioned the intrinsic value of the RI approach. The conclusions from *RP:PD* have been roundly criticized (Gilbert et al., 2016; Reuell, 2016) due to numerous issues with the design of the studies that introduced: statistical errors; low fidelity methodologies including populations that differed from those used in the original studies; a low powered design that was found to have underestimated the replicability of studies known to have a high replication rate; and "an idiosyncratic, arbitrary list of sampling rules" that led to outcomes that could have occurred by chance suggesting that the *RP:PD* was a failure. Other reasons included disconnect in the subjects used in the original and reproducibility studies (Gilbert et al., 2016; Mullane and Williams, 2017) and an overestimation of effect sizes or evidence against the null hypothesis (Etz and Vandekerckhove, 2016).

In rebutting the Gilbert et al., critique, the OSC (Anderson et al., 2016) concluded that both the "optimistic and pessimistic conclusions about reproducibility are possible, and neither are yet warranted."

5.13.3 Publishing RIs

In addition to the efforts of *Science*, *PLoS*, and *eLife* to publish reports documenting the reproducibility efforts, *F1000 Research* has added an open access forum, *Preclinical Reproducibility and Robustness Channel* that is intended as a platform for the open and transparent publication of confirmatory and nonconfirmatory studies in biomedical research (Alberts and Kamb, 2016) setting the stage for publication of the more typical, informal self-correcting manuscripts that more often than not are difficult to publish.

5.14 REPRODUCIBILITY INITIATIVES—PROGRESS TO DATE?

Reporting of research findings has historically been based on an informal, "trust-me" culture (Begley and Ioannidis, 2015) that can no longer be taken at face value due to result of changes in funding, training, mentoring, peer oversight, and culture, in the biomedical research enterprise. These have distorted incentives, enabled hubris, and lowered standards (Mullane and Williams, 2015) while creating a hypercompetitive environment due to the imbalance between an ever-increasing number of graduate-level scientists competing for finite or diminishing research resources and positions (Alberts et al., 2014). This is thought to reduce creative thinking and risk taking thus reducing the likelihood of innovation (Alberts et al., 2015b) and has limited career options for new graduates (Benderly, 2016; Ioannidis, 2014).

As a result of these changes, the time tested process of independently replicating a research finding—the process of self-correction—that proved so successful in the timely identification of the recent STAP fraud (Goodyear, 2016; Rasko and Power, 2015) is in the process of being enthusiastically replaced by various formal processes for reproducibility described under the rubric of RIs. As discussed earlier, while these are logical in intent and have the potential to create significant value to their respective research communities, their track record to date is disappointing—if not embarrassing.

The antiparasitic peptide RI (Iorns et al., 2014; Van Noorden, 2014d) provided equivocal results (Goodman et al., 2016) in an original study (Lynn et al., 2011) the significance of which was unclear, while the results from *RP:PD* (OSC, 2015) were uniformly disappointing, both negatively reinforcing concerns regarding the credibility of research findings in psychology and generating controversy as to the ability of those reproducing to adhere to the conditions of the original study, (Gilbert et al., 2016; Reuell, 2016) instead of behaving like chefs who change the ingredients of an established recipe and wonder why the soufflé fails to rise. The *RP:CB* has proven difficult to get underway with initial results that echo the shortcomings of the *RP:PD* (Baker and Dolgin, 2017), and in many respects resembles a classic example of a well-intentioned approach that is naïve and uninformed in intent that creates bureaucratic constraints and organizational entropy, is woefully underfunded, and results in the waste of researcher time in complying with its needs while casting aspersions on researcher ethics/integrity when the outcome is a failure to reproduce a study (Baker and Dolgin, 2017). Despite the logic of formalizing the process of independent reproduction via the RI platform, many investigators do not view these as a viable alternative to scientific self-replication (Engber, 2016; Kaiser, 2015a). Thus, a researcher whose work was unilaterally selected to be part of the RP:CB initiative despite it already have been reproduced (Kaiser, 2015a) and another whose original work was independently reproduced by "at least 10 other laboratories" before the RP:CB announced their inability to reproduce the original finding (Baker and Dolgin, 2017) tend to view the RP:CB as a "waste of time" (Kaiser, 2015a) with the potential to damage the reputation of a researcher including his/her ability to obtain funding (Baker and Dolgin, 2017). Similarly, other researchers note that that the experimental technologies used in 21st century biomedical research are "numerous, complex, sophisticated, and nuanced" (Kraus, 2014) requiring "a craft to running experiments" (Engber, 2016) with "cutting-edge science...[being]... unbelievably difficult to reproduce" (Wadman, 2013). This has raised concerns about "using scientists without deep expertize to reproduce decades of complicated, nuanced experiments." (Kaiser, 2015a) with third-party efforts in attempting to reproduce original research potentially not being executed with enough time, funding, and resources or with a nuanced understanding of the methodologies involved in the original study (Bissell, 2013).

Demand for researcher-initiated reproduction attempts similar to that attempted in the antiparasitic peptide RI—the initial intent of the *Science Exchange*—has, perhaps unsurprisingly, been limited such that rather than being an "awareness-building experiment" (Engber, 2016), it has added to concerns regarding outcomes of the various RI initiatives, which have become reminiscent of the GIGO (garbage in, garbage out) metric in computer science with the selection of studies for replication being viewed as arbitrary (Kaiser, 2015b) and the outcomes questionable. Thus, rather than being part of a solution to addressing and resolving reproducibility issues they have become part of the problem—only adding to the concerns. The bottom line is that RIs, because of their current structure and the way in which they are conducted, are one-time events that are temporally disconnected from the original study—often by years making them to all intents and purposes irrelevant.

5.14.1 Two Out of Three—Not So Bad—or Worse?

Conceptually, the data in an original paper can be reproduced. Assuming that the attempt to reproduce is conducted properly, the ideal outcome will be considered as binary—the data in the reproducibility study either confirms, to a major degree, the original study or it does not. In the first instance, the world of science has garnered two positive studies. In the second, there is a single "positive" study that has not been reproduced. If the topic of the original study remains of interest, those involved in the RI are then left with the need to perform another reproducibility study to hopefully reach some form of resolution especially as "a single replication cannot make a definitive statement about the original finding" (Nosek and Errington, 2017). Based on the slow pace of execution of the *RP:CB*, this could stretch any definitive outcome out to a decade or more after the original finding making it of questionable value, However, with its limited funding the *RP:CB* has taken the position of "hop[ing that] someone else will follow up on it" (Harris, 2017b), making a second formal *Replication Attempt* unlikely. When the outcome from the reproducibility study is ambiguous, this merely adds to the confusion. In the real world of scientific self-correction, a seminal example of which was the resolution of the STAP fraud (Goodyear, 2016; Normile and Vogel, 2014; Rasko and Power, 2015), 133 independent studies were conducted in 7 laboratories that were unable to reproduce the original findings, within 3 months of the original publication. This led to the retraction of the original STAP studies within 6 months (Chapter 1.10.6.4.). In comparison, the timelines of the RP:CB appear to lack any real-time practical utility in research making the RI approach more an academic exercise than a vital and timely initiative to aid in resolving reproducibility issues and maintain research momentum.

5.15 IMPROVING REPRODUCIBILITY OUTCOMES—WHERE ARE WE?

In the present chapter, various initiatives have been documented and discussed to address key concerns in biomedical research that are related to improve the reproducibility of research findings by improving their quality. These include:

- improving the peer review process via the use of a variety of pre- and postpublication review processes, the value—if any—of which remains to be determined;
- addressing the perverse misuse of the JIF, a "quantitative tool... for ranking, evaluating, categorizing, and comparing journals" (Garfield, 2006), as a key measure for committees, grant, and promotion, to assess the productivity and impact of an individual scientist;
- using submission guidelines and checklists to prepare a manuscript for submission; and
- formalizing the scientific self-correction process in reproducing research findings in the form of various RIs.

Similar initiatives emerged from a symposium held in April, 2015 under the auspices of the UK Academy of Medical Sciences, the Biotechnology and Biological Sciences Research Council (BBSRC), the Medical Research Council (MRC), and the Wellcome Trust to explore how to improve and optimize the reproducibility of biomedical research (Academy of Medical Science, 2015). These included:

- Greater openness and transparency—in methods and data, including the publication of negative results.
- Reporting guidelines to help deliver publications that contain the right sort of information to allow other researchers to reproduce results.
- *Postpublication peer review* to encourage continued appraisal of previous research, which may in turn help improve future research.
- Preregistration of protocols and plans for analysis to counteract some of the practices that undermine reproducibility in certain fields, such as the post-hoc cherry-picking of data and analyses for publication.

The UK Academy of Medical Sciences also added an additional measure—the better use of input and advice from other experts, via collaboration on projects.

Other initiatives in the reproducibility arena include the establishment of Ioannidis' Meta-Research Innovation Center (METRICS) at Stanford University to improve scientific publication standards (Economist, 2014; Harris, 2017a; Newby, 2014); the Global Biological Standards Institute (GBSI, 2015) that is focused on creating "consensus-based standards to ....enhance reproducibility and drive innovations" as well as the evolving Good Institutional Practice (GIP) guidelines discussed earlier (Begley et al., 2015).

The proliferation of these many, well-intended guidelines represents a tremendous effort and commitment of resources, time, and funds, in policing unsound ethical and experimental practices to ensure the quality of published research in terms of its ability to be reproduced. Their implementation and oversight is often conducted by self-qualified custodians of reproducibility, whose scientific qualifications and lack of practical experience at the bench may only serve to exacerbate the problem. This contrasts with training and mentoring researchers from the beginning of their research careers to do the right things right the first time.

While initiatives to improve peer review and use checklists and replication initiatives as potential solutions to reproducibility issues are the subject of an ever-expanding literature, it is necessary to question the point of prioritizing these seemingly logical efforts—important as they certainly can be—when basic safeguards to ensure experimental integrity are routinely ignored. As discussed in Chapter 2, these include better training and oversight in best practices for experimental design, and execution and analysis and the validation and the standardization of experimental reagents that appear, based on progress to date, to be of little to no consequence to researchers in their quest to advance science, their own careers or both. No amount of effort to perfect the peer review process—if that is even possible—and the impact of the JIF on peer review is of any consequence if researchers make minimal if any effort to remove known issues that constrain the outcomes before even conducting their experiments. Improvements in training and mentoring in best practices in experimental design, execution, and analysis, will aid in avoiding the persistence of behaviors that are detrimental to reproducibility, while encouraging responsible research conduct.

References

Academy of Medical Science, 2015. Reproducibility and reliability of biomedical research: improving research practice. Available from: http://www.acmedsci.ac.uk/policy/policy-projects/reproducibility-and-reliability-of-biomedical-research/.

Adamo, A., Beingessner, R.L., Behnam, M., Chen, J., Jamison, T.F., Jensen, K.F., et al., 2016. On-demand continuous-flow production of pharmaceuticals in a compact, reconfigurable system. Science 352, 61–67.

Adams, J., 2012. Collaborations: the rise of research networks. Nature 490, 335–336.

Aitkenhead, D., 2013. Peter Higgs: I wouldn't be productive enough for today's academic system. Guardian. Available from: https://www.theguardian.com/science/2013/dec/06/peter-higgs-boson-academic-system.

Alberts, B., 2013. Impact factor distortions. Science 340, 787.

Alberts, B., Kamb, A., 2016. Publishing confirming and non-confirming data [version 1; referees: not peer reviewed]. F1000Res. 5, 135.

Alberts, B., Kirschner, M.W., Tilghman, S., Varmus, H., 2014. Rescuing US biomedical research from its systemic flaws. Proc. Natl. Acad. Sci. USA 111, 5773–5777.

Alberts, B., Cicerone, R.J., Fienberg, S.E., Kamb, A., McNutt, M., Nerem, R.M., et al., 2015a. Self-correction in science at work. Science 348, 1420–1422.

Alberts, B., Kirschner, M.W., Tilghman, S., Varmus, H., 2015b. Opinion: addressing systemic problems in the biomedical research enterprise. Proc. Natl. Acad. Sci. USA 112, 1912–1913.

Alfonso, A., 2013. How academia resembles a drug gang. Maximising the impact of academic research LSE Blog. Available from: http://blogs.lse.ac.uk/impactofsocialsciences/2013/12/11/how-academia-resembles-a-drug-gang/.

Allen, M., 2016. Is frontiers in trouble? Neuroconscience. Available from: https://neuroconscience.com.

Allison, D.B., Brown, A.W., George, B.J., Kaiser, K.A., 2016. Reproducibility: a tragedy of errors. Nature 530, 27–29.

Altman, D.G., Schulz, K.F., Moher, D., Egger, M., Davidoff, F., et al., 2001. The revised CONSORT statement for reporting randomized trials: explanation and elaboration. Ann. Intern. Med. 134, 663–694.

Amsen, E., 2014a. What is open peer review? F1000Res. Available from: http://blog.f1000research.com/2014/05/21/what-is-open-peer-review/.

Amsen, E., 2014b. What is open access? F1000Res. Available from: http://blog.f1000research.com/2014/04/23/what-is-open-access/.

Anderson, K., 2012. Google's new "scholar metrics" have potential, but also prove problematic. The scholarly kitchen. Available from: https://scholarlykitchen.sspnet.org/2012/04/24/googles-new-scholar-metrics-have-potential-but-also-prove-problematic/.

Anderson, K., 2012. The reproducibility initiative—solving a problem, or just another attempt to draw on research funds? The scholarly kitchen. Available from: http://scholarlykitchen.sspnet.org/2012/08/16/the-reproducibility-initiative-solving-a-problem-or-just-another-attempt-to-draw-on-research-funds/.

Anderson, K., 2016. Why is clinicaltrials.gov still struggling? The scholarly kitchen. Available from: https://scholarlykitchen.sspnet.org/2016/03/15/why-is-clinicaltrials-gov-still-struggling/.

Anderson, M.L., Chiswell, K., Peterson, E.D., Tasneem, A., Topping, J., Califf, R.M., 2015. Compliance with results reporting at clinicaltrials.gov. N. Engl. J. Med. 372, 1031–1039.

Anderson, C.J., Bahník, S., Barnett-Cowan, M., Bosco, F.A., Chandler, J., Chartier, C.R., et al., 2016. Response to comment on "estimating the reproducibility of psychological science". Science 351, 1037c.

Anonymous Academic, 2014. European research funding: it's like Robin Hood in reverse. Guardian. Available from: http://www.theguardian.com/higher-education-network/2014/nov/07/european-research-funding-horizon-2020.

Armstrong, J.S., 1997. Peer review for journals: evidence on quality control, fairness, and innovation. Sci. Eng. Ethics 3, 63–84.

Arnold, D.N., Fowler, K.K., 2011. Nefarious numbers. Notices Amer. Math. Soc. 58, 434–437.

Avey, M.T., Moher, D., Sullivan, K.J., Fergusson, D., Griffin, G., et al., 2016. The devil is in the details: incomplete reporting in preclinical animal research. PLoS One. 11, e0166733.

Azoulay, P., Fons-Rosen, C., Graff Zivin, J.S., 2015. Does science advance one funeral at a time? NBER Working Paper No. 21788. Available from: http://www.econ.upf.edu/~fonsrosen/images/planck_complete_12-02-2015.pdf.

Baker, M., 2016. How quality control could save your science. Nature 529, 456–568.

Baker, M., Dolgin, E., 2017. Reproducibility project yields muddy results. Nature 541, 269–270.

Baker, D., Lidster, K., Sottomayor, A., Amor, S., 2014. Two years later: journals are not yet enforcing the ARRIVE guidelines on reporting standards for pre-clinical animal studies. PLoS Biol. 12, e1001756.

Barr, R., 2015. R01 teams and grantee age trends in grant funding. Inside NIA Blog. Available from: https://www.nia.nih.gov/research/blog/2015/04/r01-teams-and-grantee-age-trends-grant-funding.

Bastain, H., 2014. A stronger post-publication culture is needed for better science. PLoS Med. 11, e1001772.

Bath, P.M., Gray, L.J., Bath, A.J., Buchan, A., Miyata, T., Green, A.R., 2009. Effects of NXY-059 in experimental stroke: an individual animal meta-analysis. Br. J. Pharmacol. 157, 1157–1171.

Bauerlien, M., Gad-el-Hak, M., Grody, W., McKelvey, B., Trimble, S.W., 2010. We must stop the avalanche of low-quality research. Chron. Higher Edu. Available from: http://chronicle.com/article/We-Must-Stop-the-Avalanche-of/65890/.

Beall, J., 2012. Predatory publishers are corrupting open access. Nature 489, 179.

Beall, J., 2016. Beall's list of predatory publishers 2016. Scholary Open Access. Available from: http://scholarlyoa.com/2016/01/05/bealls-list-of-predatory-publishers-2016/.

Beddington, J., 2011. Evidence given to the UK House of Commons Science and Technology inquiry into peer review. HC856, Q 294 p.15. Available from: http://www.publications.parliament.uk/pa/cm201012/cmselect/cmsctech/856/856.pdf.

Begley, S., 2012. In cancer science, many "discoveries" don't hold up. Reuters. Available from: http://www.reuters.com/article/2012/03/28/us-science-cancer-idUSBRE82R12P20120328.

Begley, C.G., 2013. Reproducibility: six red flags for suspect work. Nature 497, 433–434.

Begley, C.G., Ellis, L.M., 2012. Raise standards for preclinical cancer research. Nature 483, 531–533.

Begley, C.G., Ioannidis, J.P.A., 2015. Reproducibility in science: improving the standard for basic and preclinical research. Cir. Res. 116, 116–126.

Begley, C.G., Buchan, A.M., Dirnagl, U., 2015. Institutions must do their part for reproducibility. Nature 525, 25–27.

Benderly, B.L., 2014. Academe's '1 Percent' Science. Available from: http://www.sciencemag.org/careers/2014/04/academe-s-1-percent.

Benderly, B.L., 2016. Postdoc mysteries. Science. Available from: http://www.sciencemag.org/careers/2016/06/postdoc-mysteries.

Benos, D.J., Bashari, E., Chaves, J.M., Gaggar, A., Kapoor, N., LaFrance, M., et al., 2007. The ups and downs of peer review. Adv. Physiol. Edu. 31, 145–152.

Bergstrom, T.C., Courant, P.N., McAfee, R.P., Williams, M.A., 2014. Evaluating big deal journal bundles. Proc. Natl. Acad. Sci. USA 111, 9425–9430.

Bergstrom, C.T., West, J., 2016. Comparing Impact Factor and Scopus CiteScore. EigenFactor.org. Available from: http://eigenfactor.org/projects/posts/citescore.php.

Bergstrom, C.T., West, J.D., Wiseman, M.A., 2008. The Eigenfactor metrics. J. Neurosci. 28, 11433–11434.

Bhattacharjee, Y., 2011. Saudi Universities offer cash in exchange for academic prestige. Science 334, 1344–1345.

Biagioli, M., 2002. From book censorship to academic peer review. Emergences 12, 11–45.

Bik, E.M., Casadevall, A., Fang, F.C., 2016. The prevalence of inappropriate image duplication in biomedical research publications. mBio 7, e00809–e816.

Bishop, D.V.M., 2016. Open research practices: unintended consequences and suggestions for averting them. (Commentary on the peer reviewers' openness initiative). R. Soc. Open Sci. 3, 160109.

Bissell, M., 2013. Reproducibility: the risks of the replication drive. Nature 503, 333–334.

Blatt, M.R., 2015. Vigilante science. Plant Physiol. 169, 907–909.

Bloudoff-Indelicato, M., 2015. NIH metric that assesses article impact stirs debate. Nature. Available from: http://www.nature.com/news/nih-metric-that-assesses-article-impact-stirs-debate-1.18734.

Bohannon, J., 2013. Who's afraid of peer review? Science 342, 60–65.

Bohannon, J., 2015. How to hijack a journal. Science 350, 903–905.

Bollen, J., Van de Sompel, H., Hagberg, A., Chute, R., 2009. A principal component analysis of 39 scientific impact measures. PLoS One 4, e6022.

Bonnell, D.A., Buriak, J.M., Hafner, J.H., Hammond, P.T., Hersam, M.C., Javey, A., et al., 2012. Recycling is not always good: the dangers of self-plagiarism. ACS Nano 6, 1–4.

Bornmann, L., 2014. Do altmetrics point to the broader impact of research? An overview of benefits and disadvantages of altmetrics. J. Informat. 8, 895–903.

Bornmann, L., Mutz, R., Daniel, H.-D., 2007. Gender differences in grant peer review: a meta-analysis. J. Infometr. 1, 226–238.

Brenner, S., 2014. Frederick Sanger (1918-2013). Science 343, 262.

Bronowicki, K.A., 2014. Technology's adverse effects on students' writing: an emphasis on formal writing is needed in an Academic Curriculum College at Brockport: State University of New York, Digital Commons @Brockport. Education and Human Development Master's Theses. Paper 392. Available from: http://digitalcommons.brockport.edu/cgi/viewcontent.cgi?article=1399&context=ehd_theses.

Burnham, J.C., 1990. The evolution of editorial peer review. JAMA 263, 1323–1329.

Bustin, S.A., Benes, V., Garson, J.A., Hellemans, J., Huggett, J., Kubista, M., et al., 2009. The MIQE guidelines: *m*inimum *i*nformation for publication of *q*uantitative real-time PCR experiments. Clin. Chem. 55, 611–622.

Butler, D., 2008. Free journal-ranking tool enters citation market. Nature 451, 6.

Caldwell, G.W., Masucci, J.A., Yan, Z., Hageman, W., 2004. Allometric scaling of pharmacokinetic parameters in drug discovery: Can human CL, Vss and t1/2 be predicted from in vivo rat data? Eur. J. Drug Metabol. Pharmacokinet. 29, 133–143.

Callaway, E., 2015. Faked peer reviews prompt 64 retractions. Nature. Available from: http://www.nature.com/news/faked-peer-reviews-prompt-64-retractions-1.18202.

Callaway, E., 2016. Beat it, impact factor! Publishing elite turns against controversial metric. Nature 535, 210–211.

Callaway, E., Butler, D., 2016. Researchers question design of fatal French clinical trial. Nature. Available from: http://www.nature.com/news/researchers-question-design-of-fatal-french-clinical-trial-1.19221.

Callaway, E., Powell, K., 2016. Biologists urged to hug a preprint. Nature 530, 265.

Cantor, M., Gero, S., 2015. The missing metric: quantifying contributions of reviewers. R. Soc. Open Sci. 2, 140540.

Casadevall, A., Fang, F.C., 2009. Is peer review censorship? Infect. Immun. 77, 1273–1274.

Casadevall, A., Bertuzzi, S., Buchmeier, M.J., Davis, R.J., Drake, H., Fang, F.C., et al., 2016. ASM journals eliminate impact factor information from journal websites. mSphere 1, e00184-16.

Ceci, S.J., Peters, D.P., 2014. The Peters & Ceci study of journal publications. The Winnower. Available from: https://thewinnower.com/discussions/7-the-peters-ceci-study-of-journal-publications.

Ceci, S.J., Williams, W.M., 2011. Understanding current causes of women's underrepresentation in science. Proc. Natl. Acad. Sci. USA 108, 3157–3162.

Chalmers, I., Glasziou, P., 2009. Avoidable waste in the production and reporting of research evidence Lancet 374, 86–89.

Chalmers, I., Bracken, M.B., Djulbegovic, B., Garattini, S., Grant, J., Gülmezoglu, A.M., et al., 2014. Research: increasing value, reducing waste 1. How to increase value and reduce waste when research priorities are set. Lancet 383, 156–165.

Chambers, C.D., 2013. Registered reports: a new publishing initiative at Cortex. Cortex 49, 609–610.

Chan, A.W., Song, F., Vickers, A., Jefferson, T., Dickersin, K., Gøtzsche, P.C., et al., 2014. Research: increasing value, reducing waste 4. Increasing value and reducing waste: addressing inaccessible research. Lancet 383, 257–266.

Chandrashekhar, Y., Narula, J., 2015. Challenges for research publications: what is journal quality and how to measure it? J. Am. Coll. Cardiol. 65, 1702–1705.

Chen, C.Y.-C., 2014. Withdrawn: DrugPrinter: print any drug instantly. Drug Discov. Today. Available from: http://dx.doi.org/10.1016/j.drudis.2014.03.027, https://dl.dropboxusercontent.com/u/54354531/Drug%20printer/drugprinter.pdf.

Chen, R., Desai, N.R., Ross, J.S., Zhang, W., Chau, K.H., Wayda, B., et al., 2016. Publication and reporting of clinical trial results: cross sectional analysis across academic medical centers. BMJ 352, i637.

Collins, F., 2014. PubMed commons: catalyzing scientist-to-scientist interactions. NIH Director's Blog. Available from: http://directorsblog.nih.gov/2014/08/05/pubmed-commons-catalyzing-scientist-to-scientist-interactions/.

Collins, F.S., Tabak, L.A., 2014. Policy: NIH plans to enhance reproducibility. Nature 505, 612–613.

Colquhoun, D., 2011. Publish-or-perish: peer review and the corruption of science. Guardian. Available from: http://www.theguardian.com/science/2011/sep/05/publish-perish-peer-review-science.

Colquhoun, D., Plested, A., 2014. Scientists don't count: why you should ignore altmetrics and other bibliometric nightmares. DC's Improbable Sci. Available from: http://www.dcscience.net/2014/01/16/why-you-should-ignore-altmetrics-and-other-bibliometric-nightmares/.

Copeland, R.A., 2016. The drug–target residence time model: a 10-year retrospective. Nature Rev, Drug, Discov. 15, 87–95.

Couzin-Frankel, J., 2013. The Web's faceless judges. Science 341, 606–608.

Coyne, J., 2016. PLos ONE publishes paper giving credit to God for designing the human hand. Why evolution is true blog. Available from: https://whyevolutionistrue.wordpress.com/2016/03/03/plos-one-publishes-paper-giving-credit-to-god-for-designing-the-human-hand/.

Cressey, D., 2015a. Thumbs down for the freemium model? Researchers reject Nature's fast track peer review experiment. Available from: http://scholarlykitchen.sspnet.org/2015/05/05/so-much-for-the-freemium-model-researchers-reject-natures-fast-track-peer-review-experiment/.

Cressey, D., 2015b. Concern raised over payment for fast-track peer review. Nature. Available from: http://www.nature.com/news/concern-raised-over-payment-for-fast-track-peer-review-1.17204.

Cressey, D., 2016. Paper that says human hand was 'designed by Creator' sparks concern. Nature 531, 143.

Csiszar, A., 2016. Peer review: troubled from the start. Nature 532, 306–308.

Curtis, M.J., Abernethy, D.R., 2015. Replication—why we need to publish our findings. Pharmacol. Res. Perspect. 3, e00164.

Curtis, M.J., Bond, R.A., Spina, D., Ahluwalia, A., Alexander, S.P.A., et al., 2015. Experimental design and analysis and their reporting: new guidance for publication in BJP. Brit. J. Pharmacol. 172, 3461–3471.

Cushing T. 2016. First Amendment Victorious: Protects Anonymous Critics On PubPeer. TechDirt. Available from: https://www.techdirt.com/blog/?tag=fazlul+sarkar.

Daniels, R.J., 2015. A generation at risk: young investigators and the future of the biomedical workforce. Proc. Natl. Acad. Sci. USA 112, 313–318.

Dansinger, M., 2017. Dear plagiarist: a letter to a peer reviewer who stole and published our manuscript as his own. Ann. Intern. Med. 166, 143.

Danthi, N., Wu, C.O., Shi, P., Lauer, M., 2014. Percentile ranking and citation impact of a large cohort of National Heart, Lung, and Blood Institute-funded cardiovascular R01 grants. Circ. Res. 114, 600–602.

Davis, P., 2011. Quoted in Mandavilli A. Peer review: trial by twitter. Nature 469, 286–287.

Davis, P., 2016. CiteScore–Flawed But Still A Game Changer. Scholary Kitchen. Available from: https://scholarlykitchen.sspnet.org/2016/12/12/citescore-flawed-but-still-a-game-changer/.

De Angelis, C., Drazen, J.M., Frizelle, F.A., Haug, C., Hoey, J., Horton, R., et al., 2004. Clinical trial registration: a statement from the International Committee of Medical Journal Editors. N. Engl. J. Med. 351, 1250–1251.

Degani, A., Weiner, E., 1993. Cockpit checklists: concepts, design, and use. Hum. Factors. 35, 345–359.

Dirnagl, U., Lauritzeb, M., 2010. Fighting publication bias: introducing the negative results section. J. Cereb. Blood Flow Metab. 30, 1263–1264.

Dolgos, H., Trusheim, M., Gross, D., Halle, J.-P., Ogden, J., Osterwalder, B., et al., 2016. *Translational Medicine Guide* transforms drug development processes: the recent Merck experience. Drug Discov. Today 21, 517–526.

DORA (Declaration on Research Assessment), 2012. San Francisco Declaration on Research Assessment. Putting science into the assessment of research. Available from: http://www.ascb.org/files/SFDeclaration-FINAL.pdf.

DORA (Declaration on Research Assessment), 2013. Letter to Thomson Reuters, Available from: http://www.ascb.org/a-letter-to-thompson-reuters/.

Drazen, J.M., 2016. Data sharing and the *Journal*. N. Engl. J. Med. 374, 19.

Drucker, D.J., 2016. Never Waste a Good Crisis: Confronting Reproducibility in Translational Research. Cell Metab. 24, 348360.

Dunn, A., Coiera, E., Mandl, K.D., Bourgeois, F.T., 2016. Conflict of interest disclosure in biomedical research: a review of current practices, biases, and the role of public registries in improving transparency. Res. Integr. Peer Rev. 1, 1.

Dzeng, E., 2014. How academia and publishing are destroying scientific innovation: a conversation with Sydney Brenner. King's Rev. Available from: http://kingsreview.co.uk/magazine/blog/2014/02/24/how-academia-and-publishing-are-destroying-scientific-innovation-a-conversation-with-sydney-brenner/.

Economist, 2013a. Peer to peer. Portable reviews look set to speed up the publication of papers. Economist. Available from: http://www.economist.com/news/science-and-technology/21578987-portable-reviews-look-set-speed-up-publication-papers-peer-peer.

Economist, 2013b. Looks good on paper. A flawed system for judging research is leading to academic fraud. Economist. Available from: http://www.economist.com/news/china/21586845-flawed-system-judging-research-leading-academic-fraud-looks-good-paper.

Economist, 2013c. Unreliable research. Trouble at the lab. Economist. Available from: http://www.economist.com/news/briefing/21588057-scientists-think-science-self-correcting-alarming-degree-it-not-trouble.

Economist, 2014. Metaphysicians sloppy researchers beware. A new institute has you in its sights. Economist. Available from: http://www.economist.com/news/science-and-technology/21598944-sloppy-researchers-beware-new-institute-has-you-its-sights-metaphysicians.

Economist, 2015. Spilling the beans. Failure to publish the results of all clinical trials is skewing medical science. Economist. Available from: http://www.economist.com/news/science-and-technology/21659703-failure-publish-results-all-clinical-trials-skewing-medical.

Economist, 2016. Schrödinger's panda. Fraud, bureaucracy and an obsession with quantity over quality still hold Chinese science back. Economist. Available from: http://www.economist.com/news/science-and-technology/21699898-fraud-bureaucracy-and-obsession-quantity-over-quality-still-hold-chinese.

Economist, 2017. Peer review is a thankless job. One firm wants to change that. Economist. Available from: http://www.economist.com/news/science-and-technology/21722822-publons-wants-scientists-be-rewarded-assessing-others-work-peer-review.

Eisen, J., 2014. Some notes on "Citations for Sale" about King Abdulaziz University offering me $$ to become an adjunct faculty. The Tree of Life Blog. Available from: https://phylogenomics.blogspot.com/2014/12/some-notes-on-citations-for-sale-about.html.

Eisen, M. 2017. Replace Francis Collins as NIH Director. It is NOT junk blog. Available from: http://www.michael-eisen.org/blog/?p=1967.

Eisen, M., 2013. PubMed Commons: post publication peer review goes mainstream. It is NOT junk blog. Available from: http://www.michaeleisen.org/blog/?p=1472.

Else, H., 2015. 'Sexist' peer review causes storm online. Times High Edu. Available from: https://www.timeshigher-education.com/news/sexist-peer-review-causes-storm-online/2020001.article.

Emerson, G.B., Warme, W.J., Wolf, F.M., Heckman, J.D., Brand, R.A., Leopold, S.S., 2010. Testing for the presence of positive-outcome bias in peer review: a randomized controlled trial. Arch. Intern. Med. 170, 1934–1939.

Engber, D., 2016. Cancer research is broken. Slate. Available from: http://www.slate.com/articles/health_and_science/future_tense/2016/04/biomedicine_facing_a_worse_replication_crisis_than_the_one_plaguing_psychology.html.

Erb, T.J., Kiefer, P., Hattendorf, B., Günther, D., Vorholt, J.A., 2012. GFAJ-1 is an arsenate-resistant, phosphate-dependent organism. Science 337, 467–470.

Errington, T.M., Iorns, E., Gunn, W., Tan, F.E., Lomax, J., Nosek, B.A., 2014. Science forum: an open investigation of the reproducibility of cancer biology research. eLife 3, e04333.

Etz, A., Vandekerckhove, J., 2016. A Bayesian perspective on the reproducibility project: psychology. PLoS One 11, e0149794.

Eyding, D., Lelgemann, M., Grouven, U., Härter, M., Kromp, M., Kaiser, T., et al., 2010. Reboxetine for acute treatment of major depression: systematic review and meta-analysis of published and unpublished placebo and selective serotonin reuptake inhibitor controlled trials. BMJ 341, c4727.

Eyre-Walker, A., Stoletzki, N., 2013. The assessment of science: the relative merits of post-publication review, the impact factor, and the number of citations. PLoS Biol. 11, e1001675.

Fan, J., de Lannoy, I.A.M., 2014. Pharmacokinetics. Biochem. Pharmacol. 87, 93–120.

Fanelli, D., 2012. Negative results are disappearing from most disciplines and countries. Scientometrics 90, 891–904.

Fang, F.C., Casadevall, A., 2009. NIH peer review reform—change we need, or lipstick on a pig? Infect. Immun. 77, 929–932.

Fang, F.C., Bowen, A., Casadevall, A., 2016. Research: NIH peer review percentile scores are poorly predictive of grant productivity. eLife 5, e13323.

Faulkes, Z., 2014. The vacuum shouts back: post-publication peer-review on social media. Neuron 82, 258–260.

Faulkes, Z., 2016. Mission creep in scientific publishing. NeuroDojo. Available from: http://neurodojo.blogspot.com/2016/02/mission-creep-in-scientific-publishing.html?m=1.

Federico, C.A., Carlisle, B., Kimmelman, J., Fergusson, D.A., 2014. Late, never or non-existent: the inaccessibility of preclinical evidence for new drugs. Br. J. Pharmacol. 171, 4247–4254.

Ferguson, C., Marcus, A., Oransky, I., 2014. Publishing: the peer-review scam. Nature 515, 480–482.

Ferreira, C., Bastille-Rousseau, G., Bennett, A.M., Ellington, E.H., Terwissen, C., Austin, C., et al., 2016. The evolution of peer review as a basis for scientific publication: directional selection towards a robust discipline? Biol. Rev. Camb. Phil. Soc. 91, 597–610.

Fersht, A., 2009. The most influential journals: impact factor and eigenfactor. Proc. Natl. Acad. Sci. USA 106, 6883–6884.

Flier, J.S., 2017. Irreproducibility of published bioscience research: Diagnosis, pathogenesis and therapy. Mol. Metab. 6, 2–9.

Foster, J.G., Rzhersky, A., Evans, J.A., 2015. Tradition and innovation in scientists' research strategies. Amer. Sociol. Rev. 80, 875–908.

Fox, J., Petchey, O.L., 2010. Pubcreds: fixing the peer review process by "privatizing" the reviewer commons. Bull. Ecol. Soc. Am. 91, 325–333.

Freedman, L.P., Cockburn, I.M., Simcoe, T.S., 2015. The economics of reproducibility in preclinical research. PLoS Biol. 13, e1002165.

Fulmer, T., 2012. The cost of reproducibility. SciBX 5(34). Available from: http://www.nature.com/scibx/journal/v5/n34/full/scibx.2012.888.html.

Garfield, E., 1955. Citation indexes for science: a new dimension in documentation through association of ideas. Science 122, 108–111.

Garfield, E., 2006. The history and meaning of the journal impact factor. JAMA 295, 90–93.

Garner, H.R., McIver, L.J., Waitzkin, M.B., 2013. Research funding: same work, twice the money? Nature 493, 599–601.

Gawande, A., 2009. The Checklist Manifesto How to Get Things Right. Metropolitan, New York.

GBSI (Global Biological Standards Institute), 2015. The case for standards in life science research. Available from: https://www.gbsi.org/gbsi-content/uploads/2015/10/The-Case-for-Standards.pdf.

Germain, R.N., 2015. Healing the NIH-funded biomedical research enterprise. Cell 161, 1485–1491.

Gilbert, D.T., King, G., Pettigrew, S., Wilson, T.D., 2016. Comment on "Estimating the reproducibility of psychological science". Science 351 (6277), 1037.

Glasziou, P., Altman, D.G., Bossuyt, P., Boutron, I., Clarke, M., Julious, S., et al., 2014. Research: increasing value, reducing waste 5. Reducing waste from incomplete or unusable reports of biomedical research. Lancet 383, 267–276.

Goldacre, B., 2012. Bad Medicine. Fourth Estate, London.

Goldacre, B., 2015. How to get all trials reported: audit, better data, and individual accountability. PLoS Med. 12, e100182.

Goldacre, B., 2016. Make journals report clinical trials properly. Nature 530, 7.

González-Pereira, B., Guerrero-Bote, V.P., Moya-Anegon, F., 2010. A new approach to the metric of journals' scientific prestige: the SJR indicator. J. Informat. 4, 379–391.

Goodlee, F., 2011. Evidence given to the UK House of Commons Science and Technology. Inquiry into Peer Review. HC856, Q97/Ev71. Available from: http://www.publications.parliament.uk/pa/cm201012/cmselect/cmsctech/856/856.pdf.

Goodman, F., 2010. Fortune's Fool: Edgar Bronfman, Jr., Warner Music, and an Industry in Crisis. Simon & Schuster, New York.

Goodman, S.N., Fanelli, D., Ioannidis, J.P.A., 2016. What does research reproducibility mean? Sci. Transl. Med. 8, 342ps12.

Goodstein, D., 1995. Conduct and misconduct in science. Ann. NY Acad. Sci. 75, 31–38.

Goodstein, D., 2000. How science works. US Federal Judiciary Reference Manual on Evidence, pp. 66–72,.

Goodyear, D., 2016. The stress test. Rivalries, intrigue, and fraud in the world of stem-cell research. New Yorker. Available from: http://www.newyorker.com/magazine/2016/02/29/the-stem-cell-scandal.

Green, M., 2013. The lost world of the London coffeehouse. Public Domain Rev. Available from: https://publicdomainreview.org/2013/08/07/the-lost-world-of-the-london-coffeehouse/.

Green, S.B., 2015. Can animal data translate to innovations necessary for a new era of patient-centred and individualised healthcare? Bias in preclinical animal research. BMC Med. Ethics. 16, 53.

Greshake, B., 2017. Looking into Pandora's box: The Content of Sci-Hub and its Usage. F1000 Research. doi: http://dx.doi.org./10.1101/124495.

Grivell, L., 2006. Through a glass darkly: the present and the future of editorial peer review. EMBO Rep. 7, 567–570.

Grootveld, M., van Egmond, J., 2012. Peer-reviewed open research data: results of a pilot. Inter. J. Dig. Curation 7, 81–91.

Gross, J., Ryan, J.C., 2015. Landscapes of research: perceptions of open access (OA) publishing in the arts and humanities. Publications 3, 65–88.

Guerrero-Bote, V.P., Moya-Anegon, F., 2012. A further step forward in measuring journals' scientific prestige: the SJR2 indicator. J. Infomet. 6, 674–688.

Hahnel, M., 2015. The year of open data mandates. figshare.com. Available from: https://figshare.com/blog/2015_The_year_of_open_data_mandates/143.

Hames, I., 2014. Peer review at the beginning of the 21st century. Sci. Ed. 1, 4–8.
Hardin, G., 1968. The tragedy of the commons. Science 162, 1243–1248.
Harold, S., 2013. Portable peer-review to prevent a pillar-to-post process. BMC Series Blog. Available from: http://blogs.biomedcentral.com/bmcseriesblog/2013/06/11/portable-peer-review-to-prevent-a-pillar-to-post-process/.
Harriman, S.L., Kowalczuk, M.K., Simera, I., Wager, E., 2016. A new forum for research on research integrity and peer review. Res. Integr.Peer Rev. 1, 5.
Hartung, T., 2013. Look back in anger—what clinical studies tell us about preclinical work. ALTEX 30, 275–291.
Harris, R., 2017a. Rigor Mortis: How Sloppy Science Creates Worthless Cures, Crushes Hope, and Wastes Billions. Basic Books, New York.
Harris, R., 2017b. What Does It Mean When Cancer Findings Can't Be Reproduced? NPR Morning Edition. Available from: http://www.npr.org/sections/health-shots/2017/01/18/510304871/what-does-it-mean-when-cancer-findings-cant-be-reproduced.
Hayden, E.C., 2012. Study challenges existence of arsenic-based life. Open-science advocates fail to reproduce controversial findings. Nature. Available from: http://www.nature.com/news/study-challenges-existence-of-arsenic-based-life-1.9861.
Helland, T., 2011. How to access science and medical research without paying an arm and a leg for it. Tanner.Helland.com. Available from: http://www.tannerhelland.com/3705/legally-access-medical-research-free/.
Henderson, V.C., Kimmelman, J., Fergusson, D., Grimshaw, J.M., Hackam, D.G., 2013. Threats to validity in the design and conduct of preclinical efficacy studies: a systematic review of guidelines for in vivo animal experiments. PLoS Med. 10, e1001489.
Hicks, D., Wouters, P., Waltman, L., de Rijcke, S., Rafols, I., 2015. Bibliometrics: the Leiden Manifesto for research metrics. Nature 520, 429–431.
Higgins, J.P.T., Altman, D.G., Gøtzsche, P.C., Jüni, P., Moher, D., Oxman, A.D., et al., 2011. The Cochrane Collaboration's tool for assessing risk of bias in randomised trials. BMJ 343, d5928.
Higgins, J.R., Lin, F.-C., Evans, J.P., 2016. Plagiarism in submitted manuscripts: incidence, characteristics and optimization of screening—case study in a major specialty medical journal. Res. Integr. Peer Rev. 1, 13.
Hirsch, J.E., 2005. An index to quantify an individual's scientific research output. Proc. Natl. Acad. Sci. USA 102, 16569–16572.
Hirsch, J.E., 2007. Does the h index have predictive power? Proc. Natl. Acad. Sci. USA 104, 19193–21918.
Hiyama, A., Nohara, C., Kinjo, S., Taira, W., Gima, S., Tanahara, A., Otaki, J.M., 2012. The biological impacts of the Fukushima nuclear accident on the pale grass blue butterfly. Sci. Rep. 2, 570.
Hooijmans, C.R., Leenaars, M., Ritskes-Hoitinga, M., 2010. A gold standard publication checklist to improve the quality of animal studies, to fully integrate the three Rs, and to make systematic reviews more feasible. Altern. Lab. Anim. 38, 167–182.
Hooijmans, C.R., de Vries, R., Leenaars, M., Curfs, J., Ritskes-Hoitinga, M., 2011. Improving planning, design, reporting and scientific quality of animal experiments by using the Gold Standard Publication Checklist, in addition to the ARRIVE guidelines. Br. J. Pharmacol. 162, 1259–1260.
Hooijmans, C.R., Rovers, M.M., de Vries, R.B.M., Leenaars, M., Ritskes-Hoitinga, M., Langendam, M.W., 2014. SYRCLE's risk of bias tool for animal studies. BMC Med. Res. Methodol. 14, 43.
Horrobin, D.F., 1990. The philosophical basis of peer review and the suppression of innovation. JAMA 263, 1438–1441.
Horrobin, D.F., 2001. Something rotten at the core of science? Trends Pharmacol. Sci. 22, 51–52.
Horton, R., 2000. Genetically modified food: consternation, confusion, and crack-up. Med. J. Aust. 172, 148–149.
House of Commons Science and Technology Committee, 2011. Peer review in scientific publication. Eighth Report of Session 2010–12. Available from: http://www.publications.parliament.uk/pa/cm201012/cmselect/cmsctech/856/856.pdf.
Hoyt, J., Binfield, P., 2013. Who killed the preprint, and could it make a return? Sci. Amer. Blog. Available from: http://blogs.scientificamerican.com/guest-blog/who-killed-the-preprint-and-could-it-make-a-return/.
Humphries, M., 2016. How a happy moment for neuroscience is a sad moment for science. Spike. Available from: https://medium.com/the-spike/how-a-happy-moment-for-neuroscience-is-a-sad-moment-for-science-c4ba00336e9c#.ufplnwqqc.
Hünig, T., 2012. The storm has cleared: lessons from the CD28 superagonist TGN1412 trial. Nat. Rev. Immunol. 12, 317–318.

Hunter, J., 2012. Post-publication peer review: opening up scientific conversation. Front. Comput. Neurosci. 6, 63.

Husten, L., 2015. Steven Nissen, conflicts of interest, and the new cholesterol drugs. Forbes. Available from: http://www.forbes.com/sites/larryhusten/2015/06/12/steven-nissen-conflicts-of-interest-and-the-new-cholesterol-drugs/#5f4ea9dc3a65.

Hutchins, B.I., Yuan, X., Anderson, J.M., Santangelo, G.M., 2015. Relative citation ratio (RCR): a new metric that uses citation rates to measure influence at the article level. bioRxiv preprint.

Hutchins, B.I., Yuan, X., Anderson, J.M., Santangelo, G.M., 2016. Relative citation ratio (RCR): a new metric that uses citation rates to measure influence at the article level. PLoS Biol 14, e1002541.

Hvistendahl, M., 2013. China's publication bazaar. Science 342, 1035–1039.

Ioannidis, JPA, 2005. Why most published research findings are false. PLoS Med 2, e124.

Ioannidis, J.P.A., 2014. How to make more published research true. PLoS Med. 11, e1001747.

Ioannidis, J.P.A., 2016a. The mass production of redundant, misleading, and conflicted systematic reviews and meta-analyses. Millbank Q. 94, 485–514.

Ioannidis, J.P.A., 2016b. Evidence-based medicine has been hijacked: a report to David Sackett. J. Clin. Epidemiol. 73, 82–86.

Ioannidis, J.P.A., Khoury, M.J., 2014. The PQRST of appraisal and reward. JAMA 312, 483–484.

Ioannidis, J.P.A., Nicholson, J.M., 2012. Research grants: conform and be funded. Nature 492, 34–36.

Ioannidis, J.P., Trikalinos, T.A., 2005. Early extreme contradictory estimates may appear in published research: the Proteus phenomenon in molecular genetics research and randomized trials. J. Clin. Epidemiol. 58, 543–549.

Ioannidis, J.P., Greenland, S., Hlatky, M.A., Khoury, M.J., Macleod, M.R., Moher, D., et al., 2014. Research: increasing value, reducing waste 2. Increasing value and reducing waste in research design, conduct, and analysis. Lancet 383, 166–175.

Iorns, E., Gunn, W., Erath, J., Rodriguez, A., Zhou, J., et al., 2014. Replication Attempt: "effect of BMAP-28 antimicrobial peptides on Leishmania Major Promastigote and Amastigote growth: role of Leishmanolysin in parasite survival". PLoS ONE 9, e114614.

Jackson, A., 2015. Fast-track peer review experiment: first findings. Nature Blog of schemes and memes. Available from: http://blogs.nature.com/ofschemesandmemes/2015/04/21/fast-track-peer-review-experiment-first-findings.

Jacobs, H., 2013. Howyland. EMBO Rep. 14, 48.

Jansen of Lorkeers, S.J., Doevendans, P.A., Chamuleau, S.A.J., 2014. All preclinical trials should be registered in advance in an online registry. Eur. J. Clin. Invest. 44, 892–1892.

Jarvis, M.F., Williams, M., 2016. Irreproducibility in preclinical biomedical research: perceptions, uncertainties, and knowledge gaps. Trends Pharmacol Sci 37, 290–302.

Jennings, C.G., 2006. Quality and value: the true purpose of peer review? Nature peer review blog. Available from: http://blogs.nature.com/peer-to peer/2006/06/quality_and_value_the_true_pur.html.

Johnson, V.E., 2008. Statistical analysis of the National Institutes of Health peer review system. Proc. Natl. Acad. Sci. USA 105 (32), 11076–11080.

Jubb, M., 2016. Peer review: the current landscape and future trends. Learn. Publ. 29, 13–21.

Kaiser, J., 2013. In 'insurrection,' scientists, editors call for abandoning journal impact factors. Science, Available from: http://www.sciencemag.org/news/2013/05/insurrection-scientists-editors-call-abandoning-journal-impact-factors.

Kaiser, J., 2015a. The cancer test. Science 348, 1411–1413.

Kaiser, J., 2015. NIH proposal to create grant for aging scientists hits a nerve. ScienceInsider. Available from: http://www.sciencemag.org/news/2015/02/nih-proposal-create-grant-aging-scientists-hits-nerve.

Kaplan, S., 2017. Scott Gottlieb preps for FDA's top post with a resume that cuts both ways. STATnews. Available from: https://www.statnews.com/2017/04/05/gottlieb-fda-profile/.

Karp, N.A., Meehan, T.F., Morgan, H., Mason, J.C., Blake, A., Kurbatova, N., et al., 2015. Applying the ARRIVE guidelines to an in vivo database. PLoS Biol. 13, e1002151.

Keefe, P.R., 2014. The empire of edge. New Yorker. Available from: www.newyorker.com/magazine/2014/10/13/empire-edge.

Kennison, R., 2016. Back to the future: (re)turning from peer review to peer engagement. Learn. Publ. 29, 69–71.

Khabsa, M., Giles, C.L., 2014. The number of scholarly documents on the public web. PLoS ONE 9, e93949.

Kilkenny, C., Browne, W.J., Cuthill, I.C., Emerson, M., Altman, D.G., 2010. Animal research: reporting *in vivo* experiments: the ARRIVE guidelines. Br. J. Pharmacol. 160, 1577–1579.

Kilkenny, C., Browne, W., Cuthill, I.C., Emerson, M., Altman, D.G., 2014. Improving bioscience research reporting: the ARRIVE guidelines for reporting animal research. Animals 4, 35–44.

Kimmelman, J., Anderson, J.A., 2012. Should preclinical studies be registered? Nat. Biotechnol. 30, 488–489.

Kimmelman, J., Mogil, J.S., Dirnagl, U., 2014. Distinguishing between exploratory and confirmatory preclinical research will improve translation. PLoS Biol. 12, e100186.

Kitchen, I., 1987. Statistics and pharmacology: the bloody obvious test. Trends Pharmacol. Sci. 8, 252–253.

Knoepfler, P., 2015. Reviewing post-publication peer review. Trends Genet 31, 221–223.

Kozak, M., Hartley, J., 2013. Publication fees for open access journals: different disciplines—different methods. J. Am. Soc. Inform. Sci. Technol. 64, 2591–2594.

Kraus, W.L., 2014. Editorial: do you see what i see? Quality, reliability, and reproducibility in biomedical research. Mol. Endocrinol. 38, 277–280.

Krebs, H.A., Johnson, W.A., 1937. The role of citric acid in intermediate metabolism in animal tissues. Enzymologia 4, 148–156.

Kriegeskorte, N., 2012. Open evaluation: a vision for entirely transparent post-publication peer review and rating for science. Front. Comput. Neurosci. 6, 79.

Kronick, D., 1984. Literature of the life sciences: the historical background. Ann. NY Acad. Sci. 60, 857–875.

Kronick, D.A., 1990. Peer review in 18th-century scientific journalism. JAMA 263, 1321–1322.

Kyriazis, M., 2013. Limitations of the peer review system and possible alternatives. J. Mol. Biochem. 2, 1-2.

Lai, Y.Y., Li, Y., Lang, J., Tong, X., Zhang, L., Fang, J., et al., 2015. Metagenomic human repiratory air in a hospital environment. PLoS ONE 10, e013904.

Laine, C., 2017. Scientific misconduct hurts. Ann. Intern. Med, doi:10.7326/M16-2550.

Lancho-Barrantes, B.S., Guerrero-Bote, V.P., Moya-Anegón, F., 2010. What lies behind the averages and significance of citation indicators in different disciplines? J. Info. Sci. 36, 371–382.

Landis, S.C., Amara, S.G., Asadullah, K., Austin, C.P., Blumenstein, R., Bradley, E.W., et al., 2012. A call for transparent reporting to optimize the predictive value of preclinical research. Nature 490, 187–191.

Lapchak, P.A., Zhang, J.H., Noble-Haeusslein, L.J., 2013. RIGOR guidelines: escalating STAIR and STEPS for effective translational research. Transl. Stroke Res. 4, 279–285.

Larivière, V., Ni, C., Gingras, Y., Cronin, B., Sugimoto, C.R., 2013. Bibliometrics: global gender disparities in science. Nature 504, 211–213.

Larivière, V., Haustein, S., Mongeon, P., 2015. The oligopoly of Academic Publishers in the digital era. PLoS ONE 10, e0127502.

Larivière, V., Kiermer, V., MacCallum, C.J., McNutt, M., Patterson, M., Pulverer, B., et al., 2016. A simple proposal for the publication of journal citation distributions. Preprint. Available from: bioRxiv doi: http://dx.doi.org/10.1101/062109.

Lauer, M.S., Danthi, N.S., Kaltman, J.R., Wu, C.O., 2015. Predicting productivity returns on investment: thirty years of peer review, grant funding, and publication of highly cited papers at the National Heart, Lung, and Blood Institute. Circ. Res. 117, 239–243.

Lăzăroiu, G., 2013. On citation ethics: editorial shenanigans to boost impact factor. Contemp. Readings Law Soc. Just. 5, 82–87.

Lee, K., Bacchetti, P., Sim, I., 2008. Publication of clinical trials supporting successful new drug applications: a literature analysis. PLoS Med. 5, e191.

Lee, C.J., Sugimoto, C.R., Zhang, G., Cronin, B., 2013. Bias in peer review. J. Am. Soc. Info. Sci. Technol. 64, 2–17.

Leek, J., 2016. Preprints are great, but post publication peer review isn't ready for prime time. Simply Statistics. Available from: http://simplystatistics.org/2016/02/26/preprints-and-pppr/.

Leslie, I., 2016. The sugar conspiracy. Guardian. Available from: http://www.theguardian.com/society/2016/apr/07/the-sugar-conspiracy-robert-lustig-john-yudkin.

Li, D., Agha, L., 2015. Research funding: big names or big ideas: do peer-review panels select the best science proposals? Science 348, 434–438.

Lin, S., 2013. Why serious academic fraud occurs in China. Learn. Publ. 26, 24–27.

Lin, S., Zhan, L., 2014. Trash journals in China. Learn. Publ. 27, 145–154.

Lindner, M.D., Nakamura, R.K., 2015. Examining the predictive validity of NIH peer review scores. PLoS One 10, e0126938.

Liu, M.-J., Xiong, C.-H., Xiong, L., Huang, X.-L., 2016. Biomechanical characteristics of hand coordination in grasping activities of daily living. PLoS ONE 11, e0146193.

Longo, D.L., Drazen, J.M., 2016. Data sharing. N. Engl. J. Med. 374, 276–277.

Lowe, D., 2014. Fazlul Sarkar Subpoenas PubPeer. In the pipeline. Sci. Transl. Med. Available from: http://blogs.sciencemag.org/pipeline/archives/2014/10/27/fazlul_sarkar_subpoenas_pubpeer.

Lowe, D., 2016. Crap, courtesy of a major scientific publisher. In the pipeline. Sci. Transl. Med. Available from: http://blogs.sciencemag.org/piline/archives/2016/06/10/crap-courtesy-of-a-major-scientific-publisher.

Ludbrook, J., 2007. Writing intelligible English prose for biomedical journals. Clin. Exp. Pharmacol. Physiol. 34, 508–514.

Lynn, M.A., Kindrachuk, J., Marr, A.K., Jenssen, H., Panté, N., Elliott, M.R., et al., 2011. Effect of BMAP-28 antimicrobial peptides on *Leishmania major* promastigote and amastigote growth: role of Leishmanolysin in parasite survival. PLoS Negl. Trop. Dis. 5, e1141.

Macdonald, S., Kam, J., 2007. Aardvark et al.: quality journals and gamesmanship in management studies. J. Info. Sci. 33, 702–717.

Macleod, M.R., van der Worp, H.B., Sena, E.S., Howells, D.W., Dirnagl, U., et al., 2008. Evidence for the efficacy of NXY-059 in experimental focal cerebral ischaemia is confounded by study quality. Stroke 39, 2824–2829.

Macleod, M.R., Fisher, M., O'Collins, V., Sena, E.S., Dirnagl, U., Bath, P.M.W., et al., 2009. Reprint: Good laboratory practice: preventing introduction of bias at the bench. J. Cerebral Blood Flow Metab. 29, 221–223.

Macleod, M.R., Michie, S., Roberts, I., Dirnagl, U., Chalmers, I., Ioannidis, J.P.A., et al., 2014. Biomedical research: increasing value, reducing waste. Lancet 383, 101–104.

Maher, B., 2015. Cancer reproducibility project scales back ambitions. Nature. Available from: http://www.nature.com/news/cancer-reproducibility-project-scales-back-ambitions-1.18938.

Mahoney, M.J., 1977. Publication prejudices: an experimental study of confirmatory bias in the peer review system. Cognit. Ther. Res. 1, 161–175.

Marcus, A., Oransky, I., 2016. Why fake data when you can fake a scientist? Nautilus. Available from: http://nautil.us/issue/42/fakes/why-fake-data-when-you-can-fake-a-scientist.

Marino, M., 2014. The use and misuse of statistical methodologies in pharmacology research. Biochem. Pharmacol. 87, 78–92.

Marsh, H.W., Jayasinghe, U.W., Bond, N.W., 2011. Gender differences in peer reviews of grant applications: a substantive-methodological synergy in support of the null hypothesis model. J. Informetr. 5, 167–180.

Marshall, E., 1998. Franz Ingelfinger's legacy shaped biology publishing. Science 282, 861.

Martin, B.R., 2016. Editors' JIF-boosting stratagems—which are appropriate and which not? Res. Policy 45, 1–7.

Masca, N.G.D., Hensor, E.M.A., Cornelius, V.R., Buffa, F.M., Marriott, H.M., et al., 2015. RIPOSTE: a framework for improving the design and analysis of laboratory-based research. eLife 4, e05519.

Matthews, D., 2016. Scientists offered $1 million in prizes to pre-register experiments. Times Higher Educ. Available from: https://www.timeshighereducation.com/news/scientists-offered-1-million-dollars-in-prizes-to-pre-register-experiments.

Mayernik, M.S., Callaghan, S., Leigh, R., Tedds, J., Worley, S., 2015. Peer review of datasets: when, why, and how. Bull. Am. Meterol. Soc. 96, 191–201.

McGrath, J.C., Lilley, E., 2015. Implementing guidelines on reporting research using animals (ARRIVE etc.): new requirements for publication in BJP. Br. J. Pharmacol. 172, 3189–3193.

McGrath, J.C., Drummond, G.B., McLachlan, E.M., Kilkenny, C., Wainwright, C.L., 2010. Guidelines for reporting experiments involving animals: the ARRIVE guidelines. Br. J. Pharmacol. 160, 1573–1576, 2010.

McGuire, J., 2014. Teaching of basic literacy skills is being eroded in our schools. South China Morning Post. Available from: http://www.scmp.com/lifestyle/family-education/article/1399083/teaching-basic-literacy-skills-being-eroded-our-schools.

McKenna, L., 2015. The convoluted profits of academic publishing. Atlantic Magazine. Available from: http://www.theatlantic.com/education/archive/2015/12/the-convoluted-profits-of-academic-publishing/421047/.

McNutt, M., 2014. Journals unite for reproducibility. Science 236, 679.

Meadows, A., 2015. ORCID peer review. ORCID blog. Available from: https://orcid.org/blog/2015/07/31/orcids-early-adopter-peer-review-program-progress-report-0.

Meadows, A., 2015. Peer review week—a celebration! The scholarly kitchen. Available from: http://scholarlykitchen.sspnet.org/2015/09/10/peer-review-week-a-celebration/.

Meadows, A., 2015. Peer review—recognition wanted! The scholarly kitchen. Available from: http://scholarlykitchen.sspnet.org/2015/01/08/peer-review-recognition-wanted/.

Meho, L.I., 2007. The rise and rise of citation analysis. Phys. World 202, 32–36.

Mervis, J., 2014. Peering into peer review. Science 343, 596–598.

Messerly, M., 2014. Citations for sale. Daily Californian. Available from: http://www.dailycal.org/2014/12/05/citations-sale/.

Michael, A., 2015. Ask the chefs: how can we improve the article review and submission process? The scholarly kitchen, Available from: http://scholarlykitchen.sspnet.org/2015/03/26/ask-the-chefs-how-can-we-improve-the-article-review-and-submission-process/.

Misteli, T., 2013. Eliminating the impact of the impact factor. J. Cell Biol. 201, 651–652.

Moher, D., Schulz, K.F., Altman, D., CONSORT Group (Consolidated Standards of Reporting Trials), 2001. The CONSORT statement: revised recommendations for improving the quality of reports of parallel-group randomized trials. JAMA 285, 1987–1991.

Moher, D., Avey, M., Antes, G., Altman, D.G., 2015. The National Institutes of Health and guidance for reporting preclinical research. BMC Med. 13, 34.

Moher, D., Glasziou, P., Chalmers, I., Nasser, M., Bossuyt, P.M.M., Korevaar, D.A., et al., 2016. Increasing value and reducing waste in biomedical research: who's listening? Lancet 397, 1573–1586.

Moore, J., 2006. Perspective: does peer review mean the same to the public as it does to scientists? Nature. Available from: http://www.nature.com/nature/peerreview/debate/nature05009.html.

Morey, R.D., Chambers, C.D., Etchells, P.J., Harris, C.R., Hoekstra, R., Lakens, D., et al., 2016. The peer reviewers' openness initiative: incentivizing open research practices through peer review. R. Soc. Open Sci. 3, 150547.

Motulsky, H., 2014. Editorial: common misconceptions about data analysis and statistics. J. Pharmacol. Exp. Ther. 351, 200–205.

Muhlhausler, B.S., Bloomfield, F.H., Gillman, M.W., 2013. Whole animal experiments should be more like human randomized controlled trials. PLoS Biol. 11, e1001481.

Mullane, K., Williams, M., 2015. Unknown unknowns in biomedical research: does an inability to deal with ambiguity contribute to issues of irreproducibility? Biochem. Pharmacol. 97, 133–136.

Mullane, K., Williams, M., 2017. Enhancing reproducibility: failures from reproducibility initiatives underline core challenges. Biochem. Pharmacol. 138, 7–18.

Mullane, K., Enna, S.J., Piette, J., Williams, M., 2015. Guidelines for manuscript submission in the peer-reviewed pharmacological literature. Biochem. Pharmacol. 97, 224–239.

Mulligan, A., Hall, L., Raphael, E., 2013. Peer review in a changing world: an international study measuring the attitudes of researchers. J. Am. Soc. Inform. Sci. Tech. 64, 132–161.

Mullis, K.B., Faloona, F.A., 1987. Specific synthesis of DNA in vitro via a polymerase-catalyzed chain reaction. Methods Enzymol. 155, 335–350.

Munafò, M., Noble, S., Browne, W.J., Brunner, D., Button, K., Ferreira, J., et al., 2014. Scientific rigor and the art of motorcycle maintenance. Nat. Biotechnol. 32, 871–873.

Munafò, M.R., Nosek, B.A., Bishop, D.V.M., Button, K.S., Chambers, C.D., et al., 2017. A manifesto for reproducible science. Nature Human Behav. 1, 0021.

Munch, T., Dufka, F.L., Greene, K., Smith, S.M., Dworkin, R.H., Rowbotham, M.C., 2014. RReACT goes global: perils and pitfalls of constructing a global open-access database of registered analgesic clinical trials and trial results. Pain. 55(7),1313–1317.

Murphy, F., 2016. An update on peer review and research data. Learn. Pub. 29, 51–53.

Murray, D.L., Morris, D., Lavoie, C., Leavitt, P.R., MacIsaac, H., Masson, M.E.J., et al., 2016. Bias in research grant evaluation has dire consequences for small universities. PLoS ONE 11, e0155876.

NASEM, The National Academies of Sciences, Engineering, and Medicine, 2017. Fostering Integrity in Research. The National Academies Press, Washington, DC, doi:https://doi.org/10.17226/21896.

Nature, 2005. Not-so-deep impact. Research assessment rests too heavily on the inflated status of the impact factor. Nature 435, 1003–1004.

Nature, 2006a. Nature's peer review trial. Nature. Avaialble from: http://www.nature.com/nature/peerreview/debate/nature05535.html.

Nature, 2006b. Peer review and fraud. Two assessments of the refereeing process highlight challenges for journals. Nature 444, 971–972.

Nature, 2008. Working double-blind. Nature 435, 605–606.

Nature, 2010. Response required. Nature 468, 867.

Nature, 2012a. John Maddox prize. Nature 491, 160.
Nature, 2012b. Must try harder. Nature 483, 509.
Nature, 2013. Time to talk. Online discussion is an essential aspect of the post-publication review of findings. Nature 502, 593–594.
Nature, 2014a. Journals unite for reproducibility. Nature 515, 7.
Nature, 2014b. The greater good. Nature 505, 5.
Nature, 2016a. Repetitive flaws. Nature 529, 256.
Nature, 2016b. Way of the dragon. Nature 534, 435.
Nature Neuroscience, 2008. Reducing the costs of peer review. Nat. Neurosci. 11, 375.
Nelson, L., 2015. Stat check: Is 98% of research in humanities and 75% in social science never cited again? Vox. Available from: http://www.vox.com/2015/11/30/9820192/universities-uncited-research.
Newby, K., 2014. Stanford launches center to strengthen quality of scientific research worldwide. Stanford Med. News Center. Available from: https://med.stanford.edu/news/all-news/2014/04/stanford-launches-center-to-strengthen-quality-of-scientific-research-worldwide.html.
Newman, L., 2015. The Biomedical Research Enterprise at a Crossroads. Taking on a broken system of too little funding and too many graduates. Clin. Lab. News. Available from: https://www.aacc.org/publications/cln/articles/2015/december/the-biomedical-research-enterprise-at-a-crossroads.
Neylon, C., Wu, S., 2009. Article-level metrics and the evolution of scientific impact. PLoS Biol. 7, e1000242.
Nicholson, J.M., Ioannidis, J.P.A., 2012. Research grants: conform and be funded. Nature 492, 34–36.
Nickerson, R.S., 1998. Confirmation bias: a ubiquitous phenomenon in many guises. Rev. Gen. Psychol. 2, 175–220.
Normile, D., Vogel, G., 2014. STAP cells succumb to pressure. Science 344, 1215–1216.
Nosek, B.A., Errington, T.M., 2017. Making sense of replications. eLife 6, e23383.
Nosek, B.A., Alter, G., Banks, G.C., Borsboom, D., Bowman, S.D., Breckler, S.J., et al., 2015. Promoting an open research culture. Science 348, 1422–1425.
NPG (Nature Publishing Group), 2015. Perceptions of open access publishing are changing for the better, a survey by Nature Publishing Group and Palgrave Macmillan finds. NPG Press Release. Available from: http://www.nature.com/press_releases/perceptions-open-access.html.
Obokata, H., Wakayama, T., Sasai, Y., Kojima, K., Vacanti, M.P., Niwa, H., et al., 2014a. Retracted. Stimulus-triggered fate conversion of somatic cells into pluripotency. Nature 505, 641–647.
Obokata, H., Sasai, Y., Niwa, H., Kadota, M., Andrabi, M., Takata, N., et al., 2014b. Retracted. Bidirectional developmental potential in reprogrammed cells with acquired pluripotency. Nature 505, 676–680.
Olefksy, J.M., 2007. The US's changing competitiveness in the biomedical sciences. J. Clin. Invest. 117, 270–276.
OSC (Open Science Collaboration), 2012. An open, large-scale, collaborative effort to estimate the reproducibility of psychological science. Perspect. Psychol. Sci. 7, 657–660.
OSC (Open Science Collaboration), 2015. Estimating the reproducibility of psychological science. Science 349, aac4716.
Oswald, A.J., 2008. Can we test for bias in scientific peer-review. IZA Discussion Paper 3665. Available from: http://ftp.iza.org/dp3665.pdf.
Pachter, L., 2014. To some a citation is worth $3 per year. Bits of DNA Blog. Available from: http://liorpachter.wordpress.com/2014/10/31/to-some-a-citation-is-worth-3-per-year/.
Pan, R.K., Fortunato, S., 2014. Author impact factor: tracking the dynamics of individual scientific impact. Sci. Rep. 4, 4880.
Parolo, P.D.B., Pan, R.K., Ghosh, R., Huberman, B.A., Kaski, K., Fortunato, S., 2015. Attention decay in science. J. Infometr. 9, 734–745.
Paulus, F.M., Rademacher, L., Schäfer, T.A.J., Müller-Pinzler, L., Krach, S., 2015. Journal impact factor shapes scientists' reward signal in the prospect of publication. PLoS ONE 10, e0142537.
Peres-Neto, P.R., 2016. Will technology trample peer review in ecology? Ongoing issues and potential solutions. Okios 125, 3–9.
Perrin, S., 2014. Preclinical research: make mouse studies work. Nature 507, 423–425.
Perry, G., Bertoluci, J., Bury, R.B., Hansen, R.W., Jehle, R., Measey, J., et al., 2012. The "peer" in "peer review". Herpetologica 68, 1–2.
Peters, D.P., Ceci, S.J., 1982. Peer-review practices of psychological journals: the fate of published articles, submitted again. Behav. Brain Sci. 5, 187–195.
Piwowar, H.A., 2013. Altmetrics: value all research products. Nature 493, 159.
Planck, M., 1950. Scientific Autobiography and Other Papers. Philosophical Library, New York, p. 97.

Plavén-Sigray, P., Matheson, G.R., Schiffler, B.J., Thompson, W.H., 2017. The readability of scientific texts is decreasing over time. bioRxiv. Available from: http://biorxiv.org/content/early/2017/03/28/119370.

Plenge, R.M., 2016. Disciplined approach to drug discovery and early development. Sci. Transl. Med. 8, 349ps15.

PLoS (Public Library of Science), 2015. Positively negative: a new PLOS ONE collection focusing on negative, null and inconclusive results. Available from: http://blogs.plos.org/everyone/2015/02/25/positively-negative-new-plos-one-collection-focusing-negative-null-inconclusive-results/.

PLoS ONE Editors, 2016. Retraction: metagenomic human repiratory air in a hospital environment. PLoS ONE 11, e0147243.

Pöschl, U., 2012. Multi-stage open peer review: scientific evaluation integrating the strengths of traditional peer review with the virtues of transparency and self-regulation. Front. Comput. Neurosci. 6, Art 33.

Powell, K., 2016. The waiting game. Nature 530, 149–151.

Priem, J., Taraborelli, D., Groth, P., Neylon, C., 2010. Altmetrics: a manifesto. Available from: http://altmetrics.org/manifesto.

Prinz, F., Schlange, T., Asadullah, K., 2011. Believe it or not: how much can we rely on published data on potential drug targets? Nat. Rev. Drug Discov. 10, 712–713.

Pulverer, B., 2015. Dora the brave. EMBO J. 34, 1601–1602.

Pusztai, L., Hatzis, C., Andre, F., 2013. Reproducibility of research and preclinical validation: problems and solutions. Nat. Rev. Clin. Oncol. 10, 720–724.

quantixed, 2015. Wrong number: a closer look at impact factors. Available from: https://quantixed.wordpress.com/2015/05/05/wrong-number-a-closer-look-at-impact-factors/.

quantixed, 2016. The great curve II: citation distributions and reverse engineering the JIF. Available from: https://quantixed.wordpress.com/2016/01/05/the-great-curve-ii-citation-distributions-and-reverse-engineering-the-jif/.

Rasko, J., Power, C., 2015. What pushes scientists to lie? The disturbing but familiar story of Haruko Obokata. Guardian. Available from: http://www.theguardian.com/science/2015/feb/18/haruko-obokata-stap-cells-controversy-scientists-lie.

Reaves, M.L., Sinha, S., Rabinowitz, J.D., Kruglyak, L., Redfield, R.J., 2012. Absence of detectable Arsenate in DNA from Arsenate-grown GFAJ-1 cells. Science 337, 470–473.

Remler, D., 2014. Are 90% of academic papers really never cited? Searching citations about academic citations reveals the good, the bad and the ugly. Available from: https://dahliaremler.com/2014/04/09/are-90-of-academic-papers-really-never-cited-searching-citations-about-academic-citations-reveals-the-good-the-bad-and-the-ugly/.

Rennie, D., 1986. Guarding the guardians: a conference on editorial peer review. JAMA 256, 2391–2392.

Rennie, D., 1999. Editorial peer review: its development and rationale. In: Goodlee, F., Jefferson, T. (Eds.), Peer Review in Health Sciences. BMJ Books, London.

Rennie, D., Flanagin, A., Smith, R., Smith, J., 2003. Fifth International Congress on peer review and biomedical publication. Call for research. JAMA 289, 1438.

Reuell, P., 2016. Study that undercut psych research got it wrong. Harvard Gazette. Available from: http://news.harvard.edu/gazette/story/2016/03/study-that-undercut-psych-research-got-it-wrong/.

Richmond, C., 2003. Obituary: David Horrobin. Br. Med. J. 326, 885.

RIN (Research Information Network), 2010. If you build it, will they come? How researchers perceive and use web 2.0. Research Information Network, London. Available from: http://www.rin.ac.uk/system/files/attachments/web_2.0_screen.pdf.

Ritskes-Hotinga, M., Leenaars, M., Avey, M., Rovers, M., Scholten, R., 2014. Systematic reviews of preclinical animal studies can make significant contributions to health care and more transparent translational medicine. Cochrane Database Syst. Rev. 3, ED000078 93.

Robertson, R.P., 2015. Research funding and ageism. J. Clin. Endocrinol. Metab. 100, 398–399.

Rockey, S. 2015. More data on age and the workforce. NIH Extramural Nexus. Available from: https://nexus.od.nih.gov/all/2015/03/25/age-of-investigator/.

Rosenblatt, M., 2016. An incentive-based approach for improving data reproducibility. Sci. Transl. Med. 8, 336ed5.

Rosenwald, M., 2016. This student put 50 million stolen research articles online. And they're free. Washington Post. Available from: https://www.washingtonpost.com/local/this-student-put-50-million-stolen-research-articles-online-and-theyre-free/2016/03/30/7714ffb4-eaf7-11e5-b0fd-073d5930a7b7_story.html.

Ross, J.S., Mulvey, G.K., Hines, E.M., Nissen, S.E., Krumholz, H.M., 2009. Trial publication after registration in clinicaltrials. Gov: a cross-sectional analysis. PLoS Med. 6, e1000144.

Rossner, M., Van Epps, H., Hill, E., 2007. Show me the data. J. Cell. Biol. 179, 1091–1092.

Roth, K.A., Cox, A.E., 2014. Science isn't science if it isn't reproducible. Am. J. Pathol. 185, 2–3.

Rothwell, P.M., Martyn, C., 2000. Reproducibility of peer review in clinical neuroscience—is agreement between reviewers any greater than would be expected by chance alone? Brain 123, 1964–1969.

Russell, W.M.S., Burch, R.L., 1959. The Principles of Humane Experimental Technique. 1959. London, Methuen. Republished Universities Federation for Animal Welfare, Wheathampstead, UK.

Sackett, D., 2004. Campaign to revitalize academic medicine: don't believe it. BMJ 329, 294.

Salman, R.A.-S., Beller, E., Kagan, J., Hemminki, E., Phillips, R.S., Savulescu, J., et al., 2014. Research: increasing value, reducing waste 3. Increasing value and reducing waste in biomedical research regulation and management. Lancet 383, 176–185.

Samie, N., Muniandy, S., Kanthimathi, M.S., Sadat Haerian, B.S., Azudin, R.E.R. Novel piperazine core compound induces death in human liver cancer cells: possible pharmacological properties. Sci. Rep. 6, 2417.

Santori, G., 2016. Research papers: journals should drive data reproducibility. Nature 535, 355.

Schekman, R., 2013a. How journals like nature, cell and science are damaging science. Guardian. Available from: http://www.theguardian.com/commentisfree/2013/dec/09/how-journals-nature-science-cell-damage-science.

Schekman, R., 2013. Quoted in "Nobel Laureate Schekman Offers NIH His First Post-Prize Talk". NIH Record LXV (24).

Schiermeier, Q., 2012. Arsenic-loving bacterium needs phosphorus after all. Nature. Available from: http://www.nature.com/news/arsenic-loving-bacterium-needs-phosphorus-after-all-1.10971.

Schiermeier, Q., 2015. Pirate research-paper sites play hide-and-seek with publishers. Nature. Available from: http://www.nature.com/news/pirate-research-paper-sites-play-hide-and-seek-with-publishers-1.18876.

Schmidt, C.W., 2014. Research wranglers: initiatives to improve reproducibility of study findings. Environ. Health Perspect. 122, A188–A191.

Schmitt, J., 2015. Can't disrupt this: Elsevier and the 25.2 billion dollar a year academic publishing business. Medium.com. Available from: https://medium.com/@jasonschmitt/can-t-disrupt-this-elsevier-and-the-25-2-billion-dollar-a-year-academic-publishing-business-aa3b9618d40a#.op0bgmnt7.

Schroter, S., Groves, T., Højgaard, L., 2010. Surveys of current status in biomedical science grant review: funding organisations' and grant reviewers' perspectives. BMC Med. 8, 6.

Schulz, K.F., Altman, D.G., Moher, D., CONSORT Group, 2010. CONSORT 2010 statement: updated guidelines for reporting parallel group randomised trials. J. Clin. Epidemiol. 63, 834–840.

Schweinsberg, M., Madan, N., Vianello, M., Sommer, S.A., Jordan, J., Tierney, W., et al., 2016. The pipeline project: pre-publication independent replications of a single laboratory's research pipeline. J. Exp. Soc. Psychol. 66, 55–67.

Scott, S., 2013. Pre-registration would put science in chains. Times Higher Educ. Available from: https://www.timeshighereducation.com/comment/opinion/pre-registration-would-put-science-in-chains/2005954.article.

Seglen, P.O., 1997. Why the impact factor of journals should not be used for evaluating research. BMJ 314, 498–502.

Sena, E., van der Worp, H.B., Howells, D., Macleod, M., 2007. How can we improve the preclinical development of drugs for stroke? Trends Neurosci. 30, 433–439.

Sena, E.S., van der Worp, H.B., Bath, P.M.W., Howells, D.W., Macleod, M.R., 2010. Publication bias in reports of animal stroke studies leads to major overstatement of efficacy. PLoS Biol. 8, e1000344.

Sena, E.M., Currie, G.L., McCann, S.K., Macleod, M.R., Howells, D.W., 2014. Systematic reviews and meta-analysis of preclinical studies: why perform them and how to appraise them critically. J. Cereb. Blood Flow Metab. 34, 737–742.

Sertkaya, A., Wong, H.-H., Jessup, A., Beleche, T., 2016. Key cost drivers of pharmaceutical clinical trials in the United States. Clin. Trials 13, 117–126.

Shaklee, P., Cousijn, H., 2015. Can data be peer-reviewed? Elsevier Connect. Available from: https://www.elsevier.com/connect/can-data-be-peer-reviewed.

Sharma, V.V., Young, L., Cavadas, M., Owen, K., 2016. Reproducibility Project: Cancer Biology. Registered report: COT drives resistance to RAF inhibition through MAP kinase pathway reactivation. eLife 5, e11414.

Shea, C., 2014. The new academic celebrity. Chron Higher Educ. Available from: http://chronicle.com/article/The-New-Academic-Celebrity/145845.

Shema, H., Bar-Ilan, J., Thelwall, M., 2014. Do blog citations correlate with a higher number of future citations? Research blogs as a potential source for alternative metrics. J. Assoc. Inform. Sci. Technol. 65, 1018–1027.

Shen, H., 2016. Brain-data gold mine could reveal how neurons compute. Nature 535, 209–210.

Shen, C., Björk, B.-C., 2015. 'Predatory' open access: a longitudinal study of article volumes and market characteristics. BMC Med. 13, 230.

Siegelman, S.S., 1998. The genesis of modern science: contributions of scientific societies and scientific journals. Radiol 208, 9–16.

Siler, K., Lee, K., Bero, L., 2015. Measuring the effectiveness of scientific gatekeeping. Proc. Natl. Acad. Sci. USA 112, 360–365.

Skeptico Blog, 2008. Extraordinary claims require extraordinary evidence. Available from: http://skeptico.blogs.com/skeptico/2008/01/extraordinary-c.html.

Smith, R., 1997. Peer review: reform or revolution? BMJ 315, 759–760.

Smith, R., 2006. Peer review: a flawed process at the heart of science and journals. J. R. Soc. Med. 99, 178–182.

Smith, R., 2010. Classical peer review: an empty gun. Breast Cancer Res. 12 (Suppl. 4), S13.

Smith, M.M., Clarke, E., Little, C.B., 2016. Considerations for the design and execution of protocols for animal research and treatment to improve reproducibility and standardization: "DEPART well-prepared and ARRIVE safely". Osteoarthritis Cartilage 25, 354–363.

Snodgrass, R.T., 2007. Editorial: single versus double-blind reviewing. ACM Transact. Database Syst. 32, 1–29.

Snyder, S.H., 2013. Science interminable: blame Ben? Proc. Natl. Acad. Sci. USA 110, 2428–2429.

Solomon, D., Björk, B.C., 2012. A study of open access journals using article processing charges. J. Am. Soc. Int. Sci. Technol. 63, 1485–1495.

Song, F., Hooper, L., Loke, Y.K., 2013. Publication bias: what is it? How do we measure it? How do we avoid it. Open Access J. Clin. Trials 2013, 71–81.

Sorokowski, P., Kulczycki, E., Sorokowska, A., Pisanski, K., 2017. Predatory journals recruit fake. Nature 543, 481–483.

Souder, L., 2011. The ethics of scholarly peer review: a review of the literature. Learn. Pub. 24, 55–74.

Spier, R., 2002. The history of the peer-review process. Trends Biotechnol. 20, 357–358.

Sterne, J.A.C., Egger, M., Mohe, D., 2011. Addressing reporting biases. In: Higgins, J.P.T., Green S. (Eds.), Cochrane Handbook for Systematic Reviews of Interventions. Version 5.1.0, Chapter 10. Available from: http://handbook.cochrane.org.

Stossel, T.P., 2015. Pharmaphobia: How the Conflict of Interest Myth Undermines American Medical Innovation. Rowman & Littlefield, Lanham, MD.

Stoye, E., 2015. Post publication peer review comes of age. Chemistryworld. Available from: http://www.rsc.org/chemistryworld/2015/01/post-publication-peer-review-stap-comes-age.

Suber, P., 2006. No-fee open-access journals SPARC Open Access Newsletter. Available from: http://legacy.earlham.edu/~peters/fos/newsletter/11-02-06.htm#nofee.

Suber, P., 2012. Open Access. MIT Press, Cambridge MA, pp. 7–8.

Swoger, B., 2014. Post publication peer-review: everything changes, and everything stays the same. Sci. Am. Available from: http://blogs.scientificamerican.com/information-culture/post-publication-peer-review-everything-changes-and-everything-stays-the-same/.

Taylor and Francis, 2015. Peer review in 2015. A global view. Available from: http://authorservices.taylorandfrancis.com/wp-content/uploads/2015/10/Peer-Review-2015-white-paper.pdf.

Taylor, C.F., Field, D., Sansone, S.-A., Aerts, J., Apweiler, R., Ashburner, M., et al., 2008. Promoting coherent minimum reporting guidelines for biological and biomedical investigations: the MIBBI project. Nat. Biotechnol. 26, 889–896.

Teixeira da Silva, J.A., Dobránszki, J., 2015. Problems with traditional science publishing and finding a wider niche for post-publication peer review. Accountability Res. 22, 22–40.

Tennant, J.P., Waldner, F., Jacques, D.C., Masuzzo, P., Collister, L.B., Hartgerink, C.H., 2016. The academic, economic and societal impacts of Open Access: an evidence-based review. [version 2; referees: 3 approved, 2 approved with reservations] F1000 Res. 5632.

ter Riet, G., Korevaar, D.A., Leenaars, M., Sterk, P.J., Van Noorden, C.J.F., et al., 2012. Publication bias in laboratory animal research: a survey on magnitude, drivers, consequences and potential solutions. PLoS ONE 7, e43404.

Thayer, K.A., Wolfe, M.S., Rooney, A.A., Boyles, A.L., Bucher, J.R., Birnbaum, L.S., 2014. Intersection of systematic review methodology with the NIH reproducibility initiative. Environ. Health Perspect. 122, A176–A177.

Theissen, S., 2015. Transferable peer review—breaking the cycle. BioMed Central blog. Available from: http://blogs.biomedcentral.com/bmcblog/2015/09/29/transferable-peer-review-breaking-cycle/.

Thrower, P., 2012. Eight reasons I rejected your article. Elsevier Connect. Available from: https://www.elsevier.com/connect/8-reasons-i-rejected-your-article\.

Toplansky, E.F., 2016. How can i possibly grade these students? Am. Thinker. Available from: http://www.americanthinker.com/articles/2016/03/how_can_i_possibly_grade_these_students.html.

Tracz, V., 2015. The five deadly sins of science publishing [version 1; referees: not peer reviewed]. F1000Res. 4, 112.

Travis, G.D.L., Collins, H.M., 1991. New light on old boys: cognitive and institutional particularism in the peer review system. Sci. Technol. Hum. Values 16, 322–341.

Triggle, D.J., Miller, K.W., 2002. Doctoral education: another tragedy of the commons? Am. J. Pharm. Educ. 66, 287–294.

Triggle, C.R., Triggle, D.J., 2007. What is the future of peer review? Why is there fraud in science? Is plagiarism out of control? Why do scientists do bad things? Is it all a case of: "All that is necessary for the triumph of evil is that good men do nothing?". Vasc. Health Risk Manag. 3, 39–53.

Triggle, C.R., Triggle, D.J., 2017. From Gutenberg to open science: an unfulfilled odyssey. Drug Dev. Res. 78, 3–23.

Van Bavel, J.J., Mende-Siedlecki, P., Brady, W.J., Reinero, D.A., 2016. Contextual sensitivity in scientific reproducibility. Proc. Natl. Acad. Sci. USA 113, 6454–6459.

van Dalen, H.P., Henkens, K., 2012. Intended and unintended consequences of a publish-or-perish culture: a worldwide survey. J. Am. Soc. Inform. Sci. Technol. 63, 1282–1293.

van Hilten, L.G., 2015. Why it's time to publish research "failures". Elsevier Connect. Available from: https://www.elsevier.com/connect/scientists-we-want-your-negative-results-too.

van Luijk, J., Bakker, B., Rovers, M.M., Ritskes-Hoitinga, M., de Vries, R.B.M., et al., 2014. Systematic reviews of animal studies; missing link in translational research? PLoS ONE 9, e89981.

Van Noorden, R., 2012. Global mobility: science on the move. Nature 490, 326–329.

Van Noorden, R, 2013a. Scientists join journal editors to fight impact-factor abuse. Nature. Available from: http://blogs.nature.com/news/2013/05/scientists-join-journal-editors-to-fight-impact-factor-abuse.html.

Van Noorden, R., 2013b. Open access: the true cost of science publishing. Nature 495, 426–429.

Van Noorden, R., 2014a. Publishers withdraw more than 120 gibberish papers. Nature.

Van Noorden, R., 2014b. The new dilemma of online peer review: too many places to post? Nature. Available from: http://blogs.nature.com/news/2014/03/the-new-dilemma-of-online-peer-review-too-many-places-to-post.html.

Van Noorden, R., 2014c. The scientists who get credit for peer review. Nature. Available from: http://www.nature.com/news/the-scientists-who-get-credit-for-peer-review-1.16102.

Van Noorden, R., 2014d. Parasite test shows where validation studies can go wrong. Nature. Available from: http://www.nature.com/news/parasite-test-shows-where-validation-studies-can-go-wrong-1.16527.

Van Noorden, R., 2016a. China by the numbers. Nature 534, 452–453.

Van Noorden, R., 2016b. Impact factor gets heavyweight rival. Nature 540, 325–326.

van Rooyen, S., Godlee, F., Evans, S., Black, N., Smith, R., 1999. Effect of open peer review on quality of reviews and on reviewers' recommendations: a randomised trial. BMJ 318, 23–27.

Vanclay, J.K., 2012. Impact factor: outdated artefact or stepping-stone to journal certification? Scientometrics. Available from: http://arxiv.org/pdf/1201.3076v1.pdf.

Voshall, L.B., 2012. The glacial pace of scientific publishing: why it hurts everyone and what we can do to fix it. FASEB J. 26, 3589–3593.

Wadman, M., 2013. NIH mulls rules for validating key results. Nature 500, 14–16.

Wagenmakers, E.-J., Wetzels, R., Borsboom, D., van der Maas, H.L.J., Kievit, R.A., 2012. An agenda for purely confirmatory research. Perspect. Psychol. Sci. 7, 632–638.

Walbot, V., 2009. Are we training pit bulls to review our manuscripts? J. Biol. 8, 24.

Walker, R., Rocha da Silva, P., 2015. Emerging trends in peer review—a survey. Front. Neurosci. 9, 160.

Walker, R., Barros, B., Conejo, R., Neumann, K., Telefont, M., 2015. Personal attributes of authors and reviewers, social bias and the outcomes of peer review: a case study [version 2; referees: 2 approved]. F1000Res. 4, 21.

Waltman, L., 2016. Q&A on Elsevier's CiteScore metric. CWRS Meaningful Metrics. Avilable from: https://www.cwts.nl/blog?article=n-q2y254.

Wang, D., Song, C., Barabási, A.-L., 2013. Quantifying long-term scientific impact. Science 342, 127–132.

Ware, M., 2008. Peer review: benefits, perceptions and alternatives. PRC Summary Paper 4 2008. Available from: http://www.publishingresearch.org.uk/documents/PRCsummary4Warefinal.pdf.

Ware, M., 2011. Peer review: recent experience and future directions. New Rev. Inform. Networking 16, 23–53.

Ware, M., Mabe, M., 2015. The STM Report, fourth ed. March 2015. Available from: http://www.stm-assoc.org/2015_02_20_STM_Report_2015.pdf.

Warne, V., 2014. Wiley pilots transferable peer review. Wiley Exchanges. Available from: http://exchanges.wiley.com/blog/2014/01/06/wiley-pilots-transferable-peer-review/.

Warren, J.R., Marshall, B., 1982. Unidentified curved bacilli on gastric epithelium in active chronic gastritis. Lancet 1, 1273–1275.

Whalen, J., 2012. An outcast among peers grains traction on Alzheimer's cure. Wall St. J. Available from: https://www.wsj.com/articles/SB10000872396390443624204578060941988428604.

Whittaker, V.P., 1979. The vesicular hypothesis. Trends Neurosci. 2, 55–56.

Wicherts, J.M., 2016. Peer review quality and transparency of the peer-review process in open access and subscription journals. PLoS ONE 11 (1), e0147913.

Willis, M., 2016. Why do peer reviewers decline to review manuscripts? A study of reviewer invitation responses. Learn. Pub. 29, 5–7.

Wilsdon, J., Allen, L., Belfiore, E., Campbell, P., Curry, S., Hill, S., et al., 2015. The metric tide: report of the independent review of the role of metrics in research assessment and management. HEFCE (Higher Education Funding Council for England). Available from: http://www.hefce.ac.uk/media/HEFCE,2014/Content/Pubs/Independentresearch/2015/The,Metric,Tide/2015_metric_tide.pdf.

Wilson, J., 2012. Standing up for Science 3. Peer review. The nuts and bolts. Sense About Science, London, 2012. Available from: http://www.senseaboutscience.org/data/files/resources/99/Peer-review_The-nuts-and-bolts.pdf.

Wilson, W.A., 2016. Scientific Regress. First Things, May 2016. Available from: http://www.firstthings.com/article/2016/05/scientific-regress.

Wines, M., 2012. Global Research Awards Showcase China's gains and efforts to retain scientists. New York Times January 24, 2012. Available from: http://www.nytimes.com/2012/01/25/world/asia/research-awards-showcase-chinese-science-and-technology-gains.html?_r=0.

Witt, S., 2015. How Music Got Free: The End of an Industry, the Turn of the Century, and the Patient Zero of Piracy. Viking, New York.

Wolfe-Simon, F., Switzer Blum, J., Kulp, T.R., Gordon, G.W., Hoeft, S.E., Pett-Ridge, J., et al., 2011. A bacterium that can grow by using arsenic instead of phosphorus. Science 332, 1163–1166.

Woodgett, J., 2015. The worst piece of peer review I've ever received. Times High Educ. Available from: https://www.timeshighereducation.com/features/the-worst-piece-of-peer-review-ive-ever-received.

Wynia, M., Borden, D., 2009. Better regulation of industry-sponsored clinical trials is long overdue. J. Law Med. Ethics 37, 410–419.

Xia, J., 2015. Predatory journals and their article publishing charges. Learn. Pub. 28, 69–74.

Xia, J., Harmon, J.L., Connolly, K.G., Donnelly, R.M., Anderson, M.R., Howard, H.A., 2015. Who publishes in "predatory" journals? J. Assoc. Inform. Sci. Technol. 66, 1406–1417.

Xie, Y., Zhang, C., Lai, Q., 2014. China's rise as a major contributor to science and technology. Proc Natl Acad Sci USA 111, 9437–9442.

Yahia, M., 2012. Are Saudi universities buying their way into top charts? House of Wisdom blog. Nature Middle East. Available from: http://blogs.nature.com/houseofwisdom/2012/01/are-saudi-universities-buying-their-way-into-top-charts.html.

Yandell, K., 2015. Riding out rejection. Scientist. Available from: http://www.the-scientist.com/?articles.view/articleNo/42261/title/Riding-Out-Rejection/.

Zijlstra, H., McCullough, R., 2016. CiteScore: a new metric to help you track journal performance and make decisions. Elsevier Editors' Update. Available from: https://www.elsevier.com/editors-update/story/journal-metrics/citescore-a-new-metric-to-help-you-choose-the-right-journal#contributors.

CHAPTER

6

Biomedical Research in the 21st Century: Multiple Challenges in Resolving Reproducibility Issues

Kevin Mullane, Michael J. Curtis and Michael Williams

OUTLINE

Research in the Biomedical Sciences. http://dx.doi.org/10.1016/B978-0-12-804725-5.00006-9

6.1 INTRODUCTION

Technologies that had their origins in the digital revolution of the last two decades of the 20th century have driven multiple paradigm shifts in the understanding of the causality, diagnosis/prognosis, and treatment of human diseases (Oellrich et al., 2016; Topol, 2012; Wachter, 2015). Conceptually, these advances have positioned the global biomedical research enterprise—academic, federal, and corporate—to develop a cornucopia of novel diagnostics and therapeutic interventions, in some instances personalized to the individual patient (PCAST, 2012), with the aspiration that "all diseases can be cured, prevented, or managed by the end of the 21st century" (Chayowski, 2016), a time at which society will also have to deal with the consequences of population growth (Haberman, 2015; Kweifo-Okai and Holder, 2016).

This optimistic vision for the future of the health of society is tempered, however, by multiple disconnects in the provision of healthcare many—if not the majority—of which are related to drug costs and availability. These include: a lack of access to essential therapeutics for patients in developing countries (Callahan, 2007; Chorev, 2016; Hunt and Khosla, 2010); de facto drug rationing in many societies due to high drug prices that limit patient access (Daniel, 2016; Goodman, 2016; Gornall et al., 2016; Kapp, 2010; Smith, 2016) and have the potential to jeopardize the financial well being and stability of healthcare systems (Economist, 2016; Kantarjian et al., 2014; Lawitz et al., 2013); the "postantibiotic apocalypse," where infectious diseases have become resistant to existing antibiotics due to over/inappropriate use (Review on Antimicrobial Resistance, 2016) that require new incentives to revitalize research antibiotic research (Outterson et al., 2016; Yong, 2016) and; slow progress in developing medications to treat chronic diseases—including Alzheimer's disease, Type I diabetes (T1D), pain, etc.

A tendency to overpromise on the effectiveness of therapeutic interventions and the hyping of new technologies to develop personalized or precision medicine and/or cures (Drucker, 2016; Duffy, 2016; Hughes, 2013; Freedman and Mullane, 2017; Le Fanu, 2011; Lowe, 2016a,b; Whitehouse, 2016) coupled with the issues in reproducibility extensively documented in this monograph have the potential to undermine the fabric and future of biomedical research and by extrapolation human health and well-being.

In the present chapter, the impact of a lack of reproducibility on human health is discussed in the context of its direct impact on the drug discovery process—precisely where the current concerns with the topic originated (Begley and Ellis, 2012; Jarvis and Williams, 2016; Prinz et al., 2011). It also addresses disruptive influences, including the emergence of the biotech industry in the last two decades of the 20th century and the evolving interest in healthcare from Silicon Valley technology-based companies like Alphabet (Google/Verily), Apple, and Microsoft (Tahir, 2016).

6.2 BACKGROUND

Advances in healthcare over the past 200 years—not just medicines but also improved sanitation and nutrition—have led to major improvements in individual and societal health. These advances include the eradication and/or prevention of many transmittable diseases that had routinely decimated society. Vaccines for diseases, such as smallpox, polio, measles, and tuberculosis, antibiotics for infections, and insulin for diabetes changed what were typically life-threatening and often fatal events into diseases with a generally good prognosis and a reduced impact on an individual's quality of life. Similarly, improvements in surgical procedures were facilitated by the use of anesthetics, antibiotics, aseptic techniques, and improved analgesia, making it less likely that any invasive procedure was worse than the condition requiring surgery.

With such precedents, society worldwide has come to expect that the biomedical research "ecosystem" (Alberts et al., 2014), that includes researchers in academia, government and the diagnostic, medical device, bioengineering and biopharmaceutical industries, will continue to provide innovative, effective and timely diagnostics, devices and therapeutics, the latter encompassing both small molecules and biologics, that can diagnose and safely treat acute and chronic disease states (De George, 2009; Juliano, 2013).

An inevitable outcome of improving human health and consequent longevity is that as one disease is effectively treated, another condition emerges and becomes the focus of treatment. For each disease that becomes treatable, transitioning from acutely fatal to chronic—a shift "from premature death to years lived with disability" (Global Burden of Disease Study 2013 Collaborators, 2015)—the population lives longer. While a laudable outcome from a societal perspective, this adds to the financial burden for society, representing a vicious cycle when resources are limited, unavailable, or restricted on economic grounds (Gornall et al., 2016; Goodman, 2016). Decreases in the incidence and severity of heart disease, cancer, stroke, and T2D (Type II diabetes), the result of changes in lifestyle (Ma et al., 2015), have thus created a "silver tsunami"—an increase in the numbers of elderly in the population—with a change in life expectancy from an average of 50 years in 1910 to 80 or more in 2013 (OECD, 2016). In Western society 72 years old is now viewed as being the "new 30" (Burger et al., 2013).

Not all members of society, however, enjoy increased longevity owing to inequities in income, conflict, social and political unrest, urban violence, and poor lifestyle choices. The latter, which is the least controversial to address, has been the global spread of a hypercaloric, "Western"-style diet in societies that had previously consumed their own traditional diet, and a decrease in physical activity that has reversed gains in life expectancy. In 2015, China and India had the highest total number of diabetics, 110 million and 69 million, respectively with the highest population prevalence of diabetes, 30% and 24% of adults, being found in the Pacific Islands of Tokelu and Nauru (Tucker, 2015). In the United States, the acronym, HONDA (Hypertensive, Obese, Noncompliant, Diabetic Adult/Asthmatic) describes a sedentary, obese population that is expected to double in number over the next decade (Westphal, 2013).

Clearly, the cost effective, universal provision of healthcare is a major challenge in the 21st century society. One outcome of this success is the shift from treating an illness to the ``promotion of wellness," which involves engaging proactively in various activities and life style changes that avoid the HONDA syndrome that are discussed further. Among the numerous solutions to enhance wellness are measures to eradicate poverty, hunger and provide affordable housing, promoting healthy diets and exercise. Society however, inevitably expects the answer to be in the form of a homeopathic medicine, an approved drug or a dietary supplement all of which produces beneficial effects with the mere act of ingestion. Throughout the history of medicine, natural products and various herbal concoctions have been imbued with the power to improve brain function, improve sexual performance, arrest aging etc. In the 20th century, these ingestibles included Coca Cola as a brain tonic (a property, contrary to popular belief, that was due to caffeine rather than cocaine) and the stimulant amphetamines that were widely used as slimming pills. Some confused wellness with a sense of well-being or euphoria and ingested any manner of psychoactive and addictive substances. Most recently, an "exercise pill" or exercise mimetic, a glucose sparing PPARδ activator has been reported that may be useful in "The treatment of metabolic disease [and] dystrophies" and "unavoidably, the enhancement of athletic performance" (Fan et al., 2017). The provision of healthcare is a financially and emotionally complex issue that is inevitably colored by commercial and political interests. Such issues lie outside the direct daily activities of the biomedical research scientist who is responsible for the quality, relevance, and reproducibility of the basic science that is key to success in the drug discovery process (Mullane and Williams, 2015; Williams, 2011), the topic of this monograph.

6.3 BIOMEDICAL RESEARCH, DRUG DISCOVERY, AND DEVELOPMENT

The search for therapeutics from natural drug sources to treat human disease states can be traced back some 5,000 years (Ravina, 2011; Sneader, 1985, 2005) and is well documented in the Chinese *Materia Medica* (*Traditional Chinese Medicine;* Hanson, 2015; Qiu, 2015) Ayurvedic Medicine (Pole, 2013) and the writings of Dioscorides, Galen, Ishaq, Ibn Sina (Avicenna), al-Zahrawi, and Paracelsus from the Greco-Roman, Greco-Arabic, and Renaissance eras (Sneader, 2005).

It was not however, until the mid-late 19th century that the framework of modern day biomedical research emerged from the seminal work variously and collectively attributed to Bernard, Ehrlich, Koch, Pasteur, Schmiedeberg, and von Behring (Koch-Weser and Schecter, 1978; Sanchez-Serrano, 2011). This led to a more rational, and informed approach to the characterization and use of natural products, the latter involving systematic efforts to identify, isolate, and/or synthesize their active ingredients—that included aspirin and morphine—drugs that still remain a key part of the pharmacopeia to treat pain and inflammation. This process was complemented by the synthesis of small molecules by the established German and Swiss dyestuffs industries (Chandler et al., 1998) as potential therapeutics for the treatment of human diseases. Together with contributions from the emerging industries in England, France, Switzerland, and the United States, this led to the establishment of the ethical pharmaceutical industry (Chandler, 2005; Chast, 2008; Sneader, 2005), which has now evolved into the biopharmaceutical industry (Hopkins et al., 2007; Pisano, 2006), an enterprise the present operational paradigm of which is in flux (Munos and Orloff, 2016) and has been for the last 3 decades.

6.3.1 Public Expectations From Biomedical Research

Before the advent of modern medicine, individuals sought to improve their health and quality of life by the use of natural products (not the least of which was alcohol) together with medications containing mercury, cocaine, or heroin and interventions from shamans, apothecaries, and the like. The use of primitive surgical interventions (e.g., bloodletting, trepanation) were in time supplemented by elixirs, diets, dietary supplements, lifestyles, etc. that anecdotal evidence suggested would improve personal well-being and lifespan.

This trend remains unabated in the 21st century, with the mainstream media—at the behest of their readership—devoting considerable attention to reports indicating that certain foods, diets, nutritional supplements (vitamins, amino acids, fiber, chondroitin, etc.), and daily activities (sleeping, walking, cycling, sex, yoga) will, based on limited anecdotal and often conflicting evidence, increase (red wine, red meat, chocolate, fish and fish oil, olive oil, blueberries, garlic, too much or too little sleep and/or exercise/sex, etc.) or decrease (white wine, red meat, chocolate, olive oil, garlic, flying, too much or too little sleep and/or exercise/sex, cellphone usage etc.) health and life expectancy—towards the Holy Grail of immortality. The media also routinely hypes early stage research findings—often fueled by press releases from academic and commercial research institutions (Drucker, 2016; Freedman and Mullane, 2017; Sumner et al., 2014), and by individual researchers hoping for fame and glory—or at least additional leverage to increase their chances of research funding. Thus in the areas of cancer (Abola and Prasad, 2016) and neurodegenerative diseases (Wray and Fox, 2016), new options, therapeutic or lifestyle, are reported using qualifiers like "breakthrough," "game changer," "miracle," "cure," "home run," "revolutionary," "transformative," "life saver," and "ground breaking" however, sketchy, contradictory, distorted, implausible, or transitory such options may be. Many of these claims are contradicted within days or weeks or merely disappear from view to be replaced by new breakthroughs (Freedman and Mullane, 2017). A website, *Kill or Cure?* (http://kill-or-cure.herokuapp.com) has as its goal to "Help to make sense of the Daily Mail's ongoing effort to classify every inanimate object into those that cause cancer and those that prevent it." For readers unfamiliar with the *Daily Mail*, it is a UK tabloid newspaper, that along with the U.K. *Telegraph*, has "a tendency to jump from the slightest indication of efficacy in small groups [or even a press release of a test tube finding]...to wondering when the new cure will be approved... a cruel hoax on... patients and their families" (Carroll, 2016a).

Despite this, neither the media nor the public shows any indication of abandoning the firm conviction that technological advances in biomedical research bolstered by information technologies, collected under the rubric of "innovation," will provide relief for each and every disease state and human infirmity regardless of individual personal lifestyles (alcohol and drug abuse, obesity, etc.). The preoccupation with improving personal health via external factors led the philosopher, Alain de Botton to note that "the news keeps drawing our attention, with deranged zeal, to the.....latest findings about red wine, gene therapy and the benefits of eating walnuts with a superstitious reverence not dissimilar to that which might once have inspired a devout Catholic pilgrim to touch the shin bone of Mary Magdalene—in the hope of securing ongoing divine protection" (de Botton, 2014). In this same context, politicians—a societal group long on optimistic hyperbole and parroting "facts" (e.g., an Alzheimer's "cure" by 2018) that they cannot conceivably have taken time to appreciate and understand as sound bites, and short on ethics, logic, common sense, and an appreciation of marketplace realities—have concluded that the "past quarter century has seen stunning progress in basic biomedical research, propelled by powerful research technologies... [that reveal]... fundamental information about the biologic basis of disease. The opportunities for biomedical advances have never been brighter" (PCAST, 2012). This assessment may be viewed as somewhat premature given regular and highly controversial increases in drug development costs and a short term debatable decline in the number of new drug approvals (Carroll, 2016b; DiMasi et al., 2016; Scannell and Bosley, 2016).

The hype that is inherent when details of early stage preclinical research are exaggerated has generated a public expectation level that experience shows is far beyond what is feasible, setting biomedical research up for failure, raising false hopes for both patients and caregivers, and undermining confidence in biomedical science (Bowen and Casadevall, 2015; Carroll, 2016a; Gittelman, 2016). There is already a growing awareness that in the future "[i] nvestments in biomedical research must result in improvements in quality of life in the future" (Thayer et al., 2014)

6.3.2 Healthcare Costs

6.3.2.1 Healthcare as a Birth Right

"The fundamental right to health of every human being" is enshrined in the preamble to the 1948 Constitution of the World Health Organization that also notes that "the responsibility of governments for the health of their people can be met only by the provision of adequate health and social measures" (Grad, 2002). A nationwide infrastructure for the provision of healthcare is currently provided by a variety of iterations on essentially two systems depending on the country (Reed, 2010): (1) *single payer* national health insurance that is funded by taxation or social insurance models, with healthcare delivery provided either by government owned clinics and hospitals (the UK's National Health Service and healthcare systems throughout Europe) or via social insurance models that involve a mixture of government and private providers (Canada, South Korea) and; (2) the *"out-of-pocket"* model—epitomized in the highly dysfunctional and ineffective US healthcare model (Du and Lu, 2016) where patients are required to contribute toward their healthcare at point of delivery via copays. The US system remains the most expensive per capita of any advanced country—$9,451 per annum in 2015, 3 times that of the average of $3,814 in 19 other industrialized nations (OECD, 2016). In the United States in 2007, healthcare costs were associated with 62% of all personal bankruptcies (Himmelstein et al., 2009).

Shortcomings in healthcare provision, irrespective of the system type, are legion (Shah, 2011; Deloitte, 2016) and provide ample grist for the mainstream media debate and critics whose agenda is frequently antagonistic to the socialist/collectivist elements of healthcare provision. This is fueled by considerations that include: *"what is* healthcare?" (Roy, 2013); whether healthcare is a privilege or a right (Davis, 2013; McArdle, 2016) and how the absence of "an individual's responsibility for their own health by pursuing a healthy lifestyle" (Berkeley, 1999) can lead to self-inflicted diseases, for example, chronic obstructive pulmonary disease in a chronic smoker, that unnecessarily contribute to the burden of healthcare (Roy, 2013). These topics, while of paramount importance to the future of society, require far more space and expert consideration than the authors can provide and will only be tangentially addressed here in the context of the provision of new therapeutics, biomedical research, and the reproducibility issue.

6.3.2.2 The Concept of the Illness Profit Industry

In the United States, the totality of the healthcare—hospitals, physicians' offices, device and biopharmaceutical companies, and insurance companies—has been captured by Schwartz (2016) under the rubric, the "Illness Profit Industry". Other descriptors are "an impenetrable morass of regulations and acronyms" (Baicker et al., 2012), "overpriced" (Brill, 2013), "big business" (Rosenthal, 2017), and a political system with multiple "choke points" that permit special interest groups (Carpenter, 2014; Freudenberg, 2014) to block any concrete reforms to improve healthcare (Fuchs, 2013). Thus, while viewed as a free market system that ostensibly conforms to core Darwinian principles, for example, survival of the fittest, the US healthcare system has become a byzantine and inefficient bureaucracy similar—if not worse based on its per capita costs and unintended consequences—to that present in the healthcare systems of other nations, federal bureaucracies and many large, dysfunctional entities in the private sector. This is perhaps inherent in all complex human systems (including healthcare, and indeed the more ephemeral "scientific community") with their complexity almost guaranteeing system failure (Cook, 1998). Nonetheless, single care systems are not viewed that much differently, for example, in 2011, Dalrymple (2011) noted that the much revered and maligned (in equal measure; Timmins, 2013) NHS system in the United Kingdom "is now capable of absorbing infinite amounts of money with minimal benefit to the health of the population, though with great benefit to the pocketbooks of those who work in it," a sentiment that applies equally to other healthcare systems in the world. In the case of biomedical research and healthcare, it is hoped that tinkering with these systems will have better success than the 70 years of tinkering applied by its leaders to Soviet Union-style communism (Beissinger, 2009).

The reader may be concerned why the discussion on drugs is treated as inevitably "US-centric" given that healthcare is far from unique to the US. This can be justified for the following reasons. Firstly, the research efforts of the biopharmaceutical industry, whether US or foreign owned, have become geographically focused in the United States, primarily in Boston and San Francisco, and secondly, the costs of drug discovery and development are indirectly subsidized by the US taxpayer via drug prices which, in 2013, were on average twice ($858 vs. $400 Per capita per annum) that of 19 other industrialized nations (Kesselheim et al., 2016). This has been argued to be due to artificial government-protected monopolies (e.g., Medicare) for the products of biopharmaceutical companies (an argument cogently presented by Alexander (2016) who compared and contrasted a hypothetical *real* free market situation in high priced chairs with that of prescription drugs) that distort the free market for drugs (McArdle, 2016). When combined with an ineffective process for approving cheaper generic drugs that is heavily influenced by corporate lobbyists and cronyism, drug prices, as described by Alexander (2016), will remain at what the market will bear in much the same way that Apple, Rolex, and Hermes can charge, and receive, premium prices for their product line because of the perceived (or real) added value. These considerations should also not ignore the fact that the biopharma industry is a for-profit entity (Booth, 2016) that takes the risks to discover and develop drugs to provide a return on investment for its shareholders. Inadequate returns—with "adequate" being the contentious point under debate—will restrict the ability of the pharmaceutical industry to take the risks required for innovation. In addition, drug costs to US taxpayers also reflect the impact and cost of mandatory government price controls in many countries outside of the United States. The United Kingdom has no mandatory price controls as such, but the National Institute for Health and Clinical Excellence (NICE) will not designate a drug as one that can be bought by the NHS if NICE deems that benefit does not justify the cost (Claxton et al., 2015). However with some counterintuitive exceptions, for example, the UK Cancer Drugs Fund (https://www.nice.org.uk/about/what-we-do/our-programmes/nice-guidance/nice-technology-appraisal-guidance/cancer-drugs-fund#whatis) that provides patient access to drugs which are deemed to have "clinical uncertainty," and are not approved by NICE, the decision-making is binary. A pharmaceutical company can however, influence the decision by adjusting the price to a level lower than it charges in the United States.

6.3.2.3 *Drug Prices*

A focal point for many of the concerns regarding effective healthcare provision is prescription drug costs. While representing only 10% (http://www.cdc.gov/nchs/fastats/

health-expenditures.htm) to 17% (Kesselheim et al., 2016) of healthcare expenditures, drugs are often the most frequently encountered form of healthcare for the majority of patients and are often subject to real time copayments. This inevitably influences public opinion and in the United States, over 70% of the population think that drug costs are unreasonable and want government to "Mak[e] sure that high-cost drugs for chronic conditions, such as HIV, hepatitis, mental illness, and cancer, are affordable to those who need them" (Kaiser Foundation, 2016) squarely placing the blame on drug manufacturers for setting prices too high (DiJulio et al., 2015).

Calculating the cost to develop a drug and setting its eventual price in the marketplace are highly contentious issues, and the debate surrounding them is polarized with little in the way of any middle ground (Angell, 2009; DiMasi et al., 2016; Light and Warburton, 2011). To many, the pharmaceutical companies developing the drugs are perceived as being exclusively interested in profits rather than patients with "blockbuster drugs... [being] ... regarded by critics of big pharma as emphasizing sales volume over whether patients receiving a drug actually derive any benefit from it" (Kessel, 2015). In 2013, this led a group of 100 physicians and researchers working in the chronic myelogenous leukemia (CML) area to argue (Kantarjian et al., 2013) that drug prices were causing harm to cancer patients by restricting drug use and that prices were at levels viewed as "profiteering." Thus while most drugs in the pharmacopoeia are "off patent" and theoretically inexpensive when manufactured as generics, many new drugs including a majority of those in the areas of cancer and orphan diseases cost in excess of $100,000 per patient annually (Kantarjian et al., 2014; Senior, 2014). When expensive new drugs are paid for by the patient as occurs when the coverage is denied by the national healthcare or private insurance schemes—an event that happens all too frequently—the personal out-of-pocket costs have the potential to "wipe patients out" financially (Walker, 2015). One of these therapeutics, alipogene tiparvovec (Glybera), a gene therapy approach to lipoprotein lipase deficiency, an orphan disease affecting one in a million individuals, is actually priced above $1 million per course of therapy per patient making it the most expensive medicine the world. Perhaps unsurprisingly, it has been a commercial failure, as it apparently has only been used once since its regulatory approval in the EU in 2012 (Regalado, 2016) and will be effectively withdrawn from the market in October, 2017 (Carroll, 2017). The inevitable inference is that it is not worthwhile for a company to seek to invent a drug that is so expensive to develop and/or manufacture that it cannot be sold without the company losing money. Although Public Private Partnerships are an alternative model to develop drugs for orphan or third-world diseases, the problem is that the US taxpayers who help foot the bill and take the risk (e.g., via NIH funding), see no recouping of costs downstream or even reduced drug prices in the United States to recognize their contribution.

Added to the high costs of the new cancer and orphan disease drugs are the recent egregious activities of what have been termed "dark pool" or "rent-seeking" pharmaceutical companies that acquire generic drugs and through various legal manoeuvres (Portteus, 2016) raise prices (Kesselheim et al., 2016). In the case of the EpiPen, a device for precisely delivering the generic drug, epinephrine in cases of life-threatening allergic reactions the increase was five-fold. For pyrimethamine (Daraprim), a treatment for toxoplasmosis this was 55-fold, from $13.50 to $750 a pill (LaMattina, 2014; Lorenzetti, 2015; Lowe, 2015a; Smith, 2016). Excessive increases in drug prices are not the sole purview of the "rent seeking" sector as Pfizer was recently fined $108 million in the UK for an approximately threefold increase in the price of the generic anticonvulsant, phenytoin (Monaghan, 2016). This has continued to drive public outrage with calls for drug pricing reform (Sachs, 2016) that appears inevitable over the next 5 years.

A consequence of the perception that companies overcharge for their products is that regulators will not approve the use of new drugs that are expensive, but benefit the patient, sometimes making false calculations and overlooking actual financial and medical benefits. An example of this is sofosbuvir, a breakthrough drug that can *cure* over 90% of the 185 million individuals worldwide infected with hepatitis C virus (HCV; Lawitz et al., 2013) replacing an existing three drug combination (telaprevir, peginterferon, ribavirin) that costs approximately $190,000 per patient annually with a cure rate of only 44% while eliminating the annual cost for palliative healthcare ($24,000–$60,000) and the inevitable need for a liver transplant (a cost in excess of $500,000; Reau and Jensen, 2014; Smith, 2014). Rather extraordinarily, the $84,000 price for a 12-week course of treatment of sofosbuvir has been described as "a blank check.... [that] .....will blow up employer benefit costs... and wreak havoc on the federal debt" (LaMattina, 2014) while a federal investigation (Wyden and Grassley, 2015) concluded that "fostering broad, affordable access was not a key consideration in the process of setting the wholesale prices" by its manufacturer, adding further to the furor over drug pricing.

However, this needs to be viewed in a broader context, namely the willingness for society as a whole to realistically address the issue of the precise value of individual health (Dillon, 2015). The push back on sofosbuvir (LaMattina, 2014) and high priced cancer and orphan drugs (Kantarjian et al., 2014; Senior, 2014) is a variation on the rhetorical question "what's worse than not getting what you want?" where a drug that cures HCV is viewed as too expensive even when it costs far less than the existing chronic palliative treatment for the disease.

The sofosbuvir case may also represent the tip of the iceberg for public reaction to drug pricing in other chronic disease states, for example, Alzheimer's disease (AD) where the current global population with dementia is approximately 47 million and is estimated to be in excess of 130 million by 2050 (https://www.alz.co.uk/research/statistics) with costs in 2016 being $605–$808 billion projected to increase to $2–$3 trillion by 2050.

At the time an *effective* treatment for AD is approved, hopefully within the next decade, the anticipated cost may be deemed excessive because of the huge numbers of patients requiring treatment. Will this be considered in terms of another "blank check.... [that] .....will blow up employer benefit costs... and wreak havoc on the federal debt" (LaMattina, 2014)? If so, the company developing the compound will be subject to vilification and be disincentivized to further invest in high-risk drug research for chronic diseases. When the biopharmaceutical industry develops viable stem cell transplants to cure T1D (Type 1 diabetes) and new generations of sophisticated pain killers to replace morphine and other opioid-based analgesics and curtail the opioid epidemic (Alexander et al., 2015), will society's response be "we can't afford it, the drug companies charge too much?" If the industry cannot provide a return on investment to its shareholders, it cannot sustain its business with no viable alternative source of new drug products on the horizon despite various claims to the contrary. The economic arguments that are accepted for Apple, Boeing, or BMW become ethically inoperable when the merchandise is drugs, clearly a moral dilemma that society at large, not just the biopharmaceutical industry, must address in a constructive manner.

6.3.2.3.1 COSTS FOR DISCOVERING AND DEVELOPING A DRUG

The current debate on drug prices and "profiteering" with the negative impact on the reputation of the pharmaceutical industry are a far cry from the often quoted statement made by George W. Merck, the President of Merck in 1950, "We try never to forget that medicine is for the people. It is not for the profits. The profits follow, and if we have remembered that, they have never failed to appear. The better we have remembered it, the larger they have been" (Collins and Porras, 2007).

In the 1950s, the drug industry was held in high regard, but of late the reputation and public trust in the industry has declined (Kessel, 2015; Ramsey, 2016) with the cost of discovering and developing an approved drug being estimated as $2.6 billion in 2016 (DiMasi et al., 2016). This has been disputed, along with previous estimates, as being a "real number" (Light and Warburton, 2011; Maxmen, 2016), the implication being that the industry is exaggerating its costs in order to justify the market entry price of new drugs. This conclusion is supported by comments from the one-time CEO of GlaxoSmithKline, Andrew Witty, who characterized a figure "as low as $1 billion" to develop a drug as a "myth" noting that it was "entirely achievable" to make research more efficient with the rather obvious and naive comment "If you stop failing so often you massively reduce the cost of drug development" (Hirschler, 2013). This statement may be considered as generic CEO hubris since no one actually involved in drug R&D knows precisely how to avoid failure despite their dedication, experience, and skill in doing the appropriately relevant and good science (Booth, 2012).

This high attrition rate in drug discovery has led to questions as to whether there are alternative, more effective ways to develop drugs (Maxmen, 2016; Munos and Orloff, 2016), including new technologies which somehow always end up being more expensive than what they are intended to replace. The Médecins Sans Frontières' Drugs for Neglected Diseases Initiative (DNDi; Maxmen, 2016) claimed that by being more efficient than biopharma, they were able to produce 5 drugs/drug combinations over a decade at a cost of $262 million, an average of $52 million per therapeutic, 50-times less than the $2.6 billion average in the industry (DiMasi et al., 2016). While chastising pharma's opacity on drug costs (Malpani, 2014), DNDi has its own opacity issues with respect to how many of the drugs they produced were repurposed from other disease indications with the reduced costs reflecting their partial development elsewhere, the cost of which was not included in the $52 million (Lowe, 2016c).

A more positive view of the biopharmaceutical industry is that it is a key economic asset with investment in innovative drug discovery being a national economic priority (PCAST, 2012). In the United States, of the 30 companies comprising the Dow Jones Industrial Average, 3 are pharmaceutical companies—Johnson and Johnson, Pfizer and Merck. In 2013, the biopharmaceutical industry was responsible for the creation of 3.4 million jobs (Battelle Technology Partnership Practice, 2013).

Thus the societal view of the biopharmaceutical industry is schizoid. On the one hand, it is vilified as a major source of increased healthcare costs (Sachs, 2016) while on the other, it is a key player in improving patient health and well-being by developing drugs (De George, 2009) while also maintaining innovation and global competitiveness for developed nations (PCAST, 2012). The conflict in these perspectives needs to be reconciled to prevent the former viewpoint from slowly asphyxiating a key golden goose in the provision of effective healthcare.

6.3.2.3.2 DRUG COSTS

Drug costs raise an unresolved—and perhaps unresolvable—debate as to what precisely is the appropriate price for true innovation in drug discovery (Berndt et al., 2015). This also requires a balanced consideration of whether price controls will stifle drug innovation and therefore negatively impact the health of future generations of patients (Howard, 2015; Lakdawalla, 2015). Arguments against price controls are: the limited time to recoup the investment in developing a new drug (a concept dismissed by Scannell, 2015) due to short patent lives; high prices apply only to new drugs that are patented; that the price is set by the market (albeit inefficently) and; that introducing price control for this business sector is inherently unfair, and counterproductive. If a company cannot earn a profit inventing drugs then it will reinvent itself as a purveyor of OTC medicines, nutraceuticals, confectionary, or whatever else might conveniently turn a better, and more assured, profit. A logical initiative to address this is value-based drug pricing that is informed by clinical evidence of how much a drug improves patients' lives (ICER, 2016) using comparative effectiveness research (CER; Sox and Greenfield, 2009). Additionally, a therapeutic area has often moved on beyond the drugs being evaluated by the time that the CER data has been analyzed and reported. As a result, these types of initiative will also require a more inclusive discussion given the negative reaction to sofosbuvir.

6.4 DRUG DISCOVERY: TECHNOLOGICAL ADVANCES AND ALTERNATIVE APPROACHES

Historically, biomedical research has been a major beneficiary of advances in technology, if not a perennial early adopter. Throughout the four historical eras of drug discovery (Table 6.1; Williams, 2007): empirical/serendipitous/physiological (1820–1948); biochemical (1948–1987), molecular biology/biotechnological (1987–2001); and genomic/postgenomic (2001–present), basic biological research has embraced key enabling technologies in the physical, chemical, and biological sciences that once proven useful aided in drug discovery.

This technology-enabled approach has not only led to the development of new concepts and paradigms in understanding disease states, drug action, and drug discovery, for example, systems biology (Palsson, 2015) and organoid-on-a-chip and body-on-a-chip systems (Skardal et al., 2016). This has expedited data generation (Biesecker, 2013) that together with facile access to information technology (IT) infrastructure and large scale computational power (databases in the exabyte range) has led to a big data revolution in biomedical research (Auffray et al., 2016; Costa, 2014; Mullane and Williams, 2017) and the NIH Big Data to Knowledge (BD2K) initiative (Margolis et al., 2014), "data so large, varied, and generated at such an alarming rate that is too challenging for the conventional methods, tools, and technologies to handle it" (Ramannavar and Sindai, 2015) describing much of the output of omics-based experimentation in basic research. Like other aspects of biomedical research funding (Alberts et al., 2014), the costs of big data and its maintenance are challenging available budgets (Kaiser, 2016a).

TABLE 6.1 Biomedical Research Phases

Era	Concepts/technologies	Representative drug classes and drugs
Empirical/serendipitous/physiological—1820–1948	Receptor concepts • "Lock and key" theory of drug action • Magic bullets • Dose/concentration response curve Isolation and synthesis of active ingredients of natural products Medicinal chemistry Bioassays in cells, tissues, and animals Holistic, phenotypically based drug discovery	*Analgesics/anesthetics*—opium, morphine, heroin, cocaine *Volatile:* ether, nitrous oxide, chloroform, cyclopropane, halothane *Local:* Benzocaine, amylocaine, procaine *Antibiotics:* sulfonamides, penicillin, methicillin, cephalosporins, streptomycin, erythromycin *Anticancer*—nitrogen mustards, cyclophosphamide *Antimalarial/antipyretics:* quinine. paracetamol, aspirin *Cardiovascular/metabolic* *Antianginal*—Lidoflazine, papaverine, verapamil *Antihypertensives*—Methyldopa, thiazides, hydralazine, mecamylamine, reserpine, propranolol, sotalol, practolol *Antidiabetic*—Insulin *Antiarrhythmic*—Quinine *Congestive heart failure/cardiac stimulants*—digitalis *CNS*— *Anxiolytics*—Alcohol, cannabinoids *Hypnotics/anticonvulsants*—Barbiturates—barbitone, chloral hydrate, chloralose, ethosuximide, hexobarbitone, phenobarbitone, pentothal, phenytoin, thiopentone, valproic acid *Antidepressants*—Lithium *Stimulants:* Caffeine

Biochemical—1948–1987	Pharmacological classification of receptor subtypes Randomized clinical trials Drug targets/drug mechanism of actions Growth of big pharma	*Analgesics/anesthetics*—Alfentinil, bupivacaine, meptazinol, desflurane, ropivacaine *Antiallergy*—Cyclizine, diphenhydramine, pheniramine, cetirizine, repirinast *Antiarthritic*—Auranofin, diclofenac, diflunisal, ibuprofen, indomethacin, hydroxychloroquine, ketoprofen, naproxen, proxicam, sulindac *Antiasthmatic*—Sodium cromoglycate, amlexanox, doxofylline, levalbuterol, oxitropium *Antibiotics*: Macrolides, tetracyclines, rifamycins, quinolones *Anticancer*—Dactinomycin, doxorubicin, vinblastine, etoposide, cisplatin, taxol, etoposide, amsacrine, camostat, enocitabine, flutamide, mitoxantrone, nilutamide, ranimustine *Antiulcer*—Cimetidine *Antivirals*—Rimantadine, zidovudine *Cardiovascular/metabolic* *Antianginal*—Gallopamil, nicorandil, encainide *Antihypertensives*—Hydralazine, b-adrenoceptor blockers, ACE-inhibitors, calcium entry blockers *Antidiabetic*—Tolbutamide, glibenclamide, metformin *Antiarrhythmic*—Amiodarone, procainamide *Congestive heart failure*—Amiodarone *CNS*— *Antipsychotics*—Halperidol, chlorpromazine, amisulpride *Anxiolytics*—Diazepam, chlordazepoxide, buspirone, tandospirone *Anticonvulsants*—Carbamazepine, progabide, valproate *Antidepressants*—Amitriptyline, desimpramine, fluoxetine, fluvoxamine, lithium, iproniazide, imipramine, mianserin *Neurodegeneration* *Multiple Sclerosis*—Glatiramer *Parkinson's*—L-dopa *Oral contraceptives*—Norprogesterone, normegesterol, promegestrone *Immunosuppressant*: Cyclosporine

(*Continued*)

TABLE 6.1 Biomedical Research Phases (*cont.*)

Era	Concepts/technologies	Representative drug classes and drugs
Molecular biology/ biotechnology phase (1987–2001)	Recombinant cloning Monoclonal antibody production monoclonal antibodies as drugs Synthetic biology High throughput screening (HTS) and organic synthesis (HTOS) Target based drug discovery/reductionism Computational-based/bioinformatics Life style drugs Venture capital funding and the biotechnology industry	*Analgesics/anesthetics*—Ketorolac, mofezolac, propofol, remifentanil, ropivacaine, sevoflurane *Antiallergy*—Astemizole, levocabastine, loratridine *Antiarthritic*—Bromfenac, celexoxib, meloxicam, methotrexate, rofecoxib, lefunomide *Antiasthmatic*—Ibudilast, montelukast, pranlukast, salmeterol, zafirlukast *Antibiotics*—Azithromycin, carboplatin, ciprofloxacin, clarithromycin, spafloxacin, temafloxacin, telithromycin, trovafloxicin *Anticancer*—Bexarotene, docetaxol, flutamide, gemcitabine, interleukin-2, irinotecan, paclitaxel, pentostatin, temozolomide, topotecan, *Antiobesity*—Orlistat, silbutramine *Antiinflammatory*—Canakinumab, etodolac, fluticasone, piroxicam *Antiulcer*—Omeprazole, ranitidine *Antivirals*—Cidofovir, famciclovir, foscarnet, indinavir, ganciclovir, lamivudine, lopinavir, nevirapine, ritonovir *Cardiovascular/metabolic* *Antianginal*—Ranolazine, dofetilide *Antihypercholesterolemic*—Atorvastatin, cerivastatin, lovastatin, simvastatin *Antidiabetic*—Acarbose, epalrestat, glimepiride, glucagon, pioglitazone, rosiglitazone, troglitazone, vildagliptin *Antihypertensives*—Amlodipine, aranidipine, carvediol, fenoldopam, irbesartan, lisinopril, losartan, mibefradil, telmisartan, valsartan *Antiarrhythmic*—Nicorandil, nifekalant, encainide *Antiobesity*—Orlistat, silbutramine *Congestive heart failure (CHF)*—Levosimendan, milnirone *CNS*— *Antipsychotics*—Olanzapine, quietapine, remoxipride, risperidone, sertinodole, ziprasidone. *Anxiolytics*—Alpidem, tandospirone *Anticonvulsants*—Felbamate, fosphenytoin, lamtrogine, levetiracetam, oxcarbazepine, tiagabine, topirmate, vigabatrin *Antidepressants*—Buproprion, citalopram, milnacipran, mirtazapine, nefazodone, paroxetine, reboxetine, sertraline, tianeptine, venlafaxine *Antimigraine*—Sumatriptan, almotriptan, naratriptan, riztriptan, zomitriptan, *Neurodegeneration* *Alzheimer's*—Tacrine, donepezil, rivastigmine, *Multiple Sclerosis*—Glatiramer *Parkinson's*—Entapacone, pergolide, pramipexole, ropinirole, talipexole, tolcapone *Sleep disorders*—Modafinil, zolpidem, zopiclone *Male sexual dysfunction*—Sildenafil *Immunosuppressant*—Tacrolimus (FK-506), sirolimus (rapamycin)

Genomic—2001–present	Mapping of the human genome completed in 2003 Concept of disease-associated genomic targets Drugs to cure human disease—HCV HONDA	*Analgesics/anesthetics*—Parecoxib, tapentadol, ziconitide, zucapasicin *Antiallergy*—Rupatadine *Antiarthritic*—Adalimumab, anakinra, tofacitinib, vadecoxib *Antiasthmatic*—Arformoterol, ciclesonide, mepolizumab, omalizumab *Antibiotics*—Besifloxacin, biapenem, ceftaroline, ceftazidime-avibactam, dalbavancin , fidaxomicin, gemifloxacin, oritavancin, pazufloxacin, retapamulin, sitafloxacin, telvancin *Anticancer*—Abarelix, afatinib, alectinib, alemtuzmab, axitinib, belinostat, bevacizumab, blinatumomab, bortezomib, bosutinib, carfilzomib, ceritinib, cetuximab, cobimetinib, daratumumab, dasatinib, dinutuximab, elotuzumab, erlotinib, gefitinib, ibrutinib, idelalisib, imatinib, ixazomib, lenvatinib, mogamulizumab, necitumumab, nivolumab, obinutuzumab, olaparib, osimertinib, palbociclib, ponatinib, panobinostat, pembrolizumab, ramucirumab, siltuximab, sonidegib, sorafinib, sunitnib, trabectedin, vemurafenib, vorinostat *Antiinflammatory*—Canakinumab, etodolac, fluticasone, piroxicam *Antiulcer*—Dexlansoprazole *Antivirals*—Adefovir, boceprevir, clevudine, daclatasvir, darunavir, dolutegravir, elvitegravir, fosamprenavir, peramivir, permivir, rilpvirine, simeprevir, sofosbuvir, *Cardiovascular/metabolic* *Antianginal—Ranolazine, dofetilide* *Antihypercholesterolemic*—Alirocumab, choline fenofibrate, evolocumab, ezetimibe lomitapide, mipomersen, pitavastatin, rosuvastatin *Antidiabetic*—Albiglutide, alogliptin, anagliptin, canagliflozin, dapaglifozin, dulaglutide, empagliflozin, exanatide, liraglutide, linagliptin, lixisenatide, pramlintide, saxagliptin, sitagliptin, tenelgliptin, troglitazone, vildagliptin *Antihypertensives*—Aliskiren, azelnidipine, azilsartan, bosentan, clevidipine, *Antiarrhythmic*—Dofetilide, dronedarone, landiolol, vernakalant *Antiobesity*—Locaserin *Congestive heart failure (CHF)*—Ivabradine, sacubitril/valsartan, nesiritide

(*Continued*)

TABLE 6.1 Biomedical Research Phases (*cont.*)

Era	Concepts/technologies	Representative drug classes and drugs
		CNS— *Antipsychotics*—Ariperazole, asenapine, blonanserin, brexpiprazole, cariprazine, lurasidone, paliperidone, *Anxiolytics*—Alpidem, tandospirone *Anticonvulsants*—Eslicarbazepine, lacosamide, levetiracetam, perampanel, retigabine, rufinamide, *Antidepressants*—Desvenlafaxine, duloxetine, escitalopram, vilazodone, vortioxetine *Antimigraine*—Eletriptan, frovatriptan *Neurodegeneration* *Alzheimer's*—Memantine *Multiple Sclerosis*—Dalfampridine, dimethyl fumarate, fingolimod, natalizumab, peginterferon beta-1a, teriflunomide *Sleep disorders*—eszopiclone, ramelteon, suvorexant, tasimelteon *Sexual dysfunction*—avanafil, flibanserin, tadalafil, sudenafil, vardenafil *Immunosuppressant*—pimecrolimus

The list of approved drugs of note listed in this table is both selective and incomplete. A comprehensive timeline of approved drugs is extremely challenging to collate since the available public databases—with the exception of the yearly FDA NME/BLA lists that only dates back to 1999 http://www.fda.gov/Drugs/DevelopmentApprovalProcess/HowDrugsareDevelopedandApproved/DrugandBiologicApprovalReports/NDAandBLAApprovalReports/ucm373420.htm—are variously focused on the chemical classes, therapeutic categories/disease indications, commercial sources, sales, etc. of approved drugs with their year of introduction being of secondary concern. Of additional note is the fact that web searches on the topic "prescription drug introductions" inevitable pulls up links to drug abuse. The most comprehensive lists are those collated by Kennewell (2007) that covers new drug introductions, a total of 339 over the years 1993–2002 and a cumulative list of new drug introductions worldwide by indication from 1983 to 2013 in Annual Reports in Medicinal Chemistry (2014). Both are however based on therapeutic categories/disease indications and are consequently a challenge to search on a yearly basis.

Source: Sneader, 1985. Sneader, 2005. Ravina, 2011. Kennewell, 2007. Annual Reports in Medicinal Chemistry, 2014. http://www.fda.gov/Drugs/DevelopmentApprovalProcess/DrugInnovation/ucm381263.htm.

6.4.1 Are There Viable Alternatives to the Present Drug Discovery and Development Models?

Given issues with productivity, drug costs and unmet medical need there has been considerable discussion of late in assessing whether the current model of drug discovery and development is fit for its intended purpose, namely effectively contributing to human health (Bowen and Casadevall, 2015; Munos and Orloff, 2016). This has led to continuing initiatives from the Federal Government and from Silicon Valley entrepreneurs/philanthropists that have the potential, if successful, to disrupt the current model of drug discovery and development in a manner similar to the irrevocable impact of digital photography on Eastman Kodak and Über on the taxi and automotive industries (Kapeller, 2016). Before addressing these recent initiatives, it is useful to assess the impact of the biotechnology industry on drug discovery over the past 30 years.

6.4.1.1 *The Impact of Biotechnology*

From the perspective of the pharmaceutical industry, the most disruptive event in recent time was the biotech revolution. While the formation of Genentech in 1976 is widely viewed as the beginning of the biotech industry with its first success being the cloning and expression of human insulin in collaboration with Lilly (Hughes, 2011; Rasmussen, 2014), it is generally considered as an event waiting to happen. The existence of Genentech had been preceded by the formation of Cetus (Emeryville, CA) and Collaborative Research (Lexington MA) and was accompanied by a multitude of molecular biology/genetics-based companies (Amgen, Biogen, Celgene, Centocor, DNAX, Genetics Institute, Genetics Systems, Genzyme, Hybritech, and Xoma) within 5 years (Pfeffer, 2012).

The stories surrounding the science of biotechnology and the growth of the industry make for fascinating—if not always inspirational—reading with multiple books on the topic (Madrigal, 2013). Other interesting reading is that on the activities of some of biotechnology's highflyers, some of whom ran afoul of the US regulatory authorities and ended up being incarcerated (Hodgson, 2016; Prud'Homme, 2004; Pollack, 2013).

Of greater relevance however, is the financial impact of the industry—beyond wealth creation for company founders (Walker and McGinty, 2016), early employees and investors—which is unclear. In 2004, Hamilton (2004) noted that biotech had had \$40 billion in cumulative losses with "negative financial returns for decades," leading him to describe the sector as being "the ultimate roulette game." In his seminal study of the biotech industry, "*Science Business. The Promise, the Reality, and the Future of Biotech*," Pisano (2006) while acknowledging that Amgen, Genentech, and Genzyme were successful companies, noted that "the economic performance of the sector overall had been disappointing by any objective standard." And from either the Schumpeterian concept of "creative destruction" or Christensen's disruptive innovation concept (Bower and Christensen, 1995), the perceived high impact of the biotech revolution on the established pharmaceutical industry has been viewed as a myth (Hopkins et al., 2007). Rasmussen (2014) further noted that "Great science aside, biotechnology pharmaceuticals development has translated into pharmaceuticals business as usual, quantitatively speaking", becoming a "me-too" part of the drug industry. Thus many of the still existent "biotech" companies are indistinguishable in their operational model, products and, in some instances, leadership, from legacy pharma often having little in common with their original "visionary" business plan. An analysis in 2005 of 61 "biotech medicines" approved by the European Medicine Evaluation Agency (EMEA) from 1995 to 2003, revealed that only 15 represented therapeutic innovation, defined as a drug for a disease without effective treatment, that is more effective than existing treatment, or is active in patients resistant to current treatment. A further 22 had limited nontherapeutic advantages over existing drugs (10 in terms of safety and 12 in terms of convenience) while 24 were "me too" products (Joppi et al., 2005).

Based on an analysis of therapeutic products approved by the FDA on a priority basis from 1998–2012, Drakeman (2014) concluded that biotech had been more productive (89 vs. 73 products) than pharma and at lower cost, $6.67 billion per compound for pharma and $1.84 billion for biotech (neither number of which tallies with the recent $2.6 billion figure from DiMasi et al. (2016), a reconciliation challenge for another day). Drakeman has attributed this to biotech companies being perpetually short of cash—the survival index or cash runway typically being a maximum of 3 years—and therefore conducting R&D at a "bare bones level" eschewing big pharma's "nice to have" studies. This approach had been viewed by Horrabin (2001) as "methodologically problematic" as smaller studies may lack the power to detect side effects that may be overlooked with the mistaken assumption that biologics, being "natural," are predictably less toxic (Pollack, 2002).

None of these retrospective studies on the impact of biotech adequately account for the private biotech companies. These may have been started as quasi-Ponzi schemes or as tax shelters for investors who were as interested in the societal impact of drug discovery as they were in the profitability of coin-operated laundries. Annual reports from Ernst and Young that were cited in Hopkins et al. (2007) and Pfeffer (2012) estimated that in 2004 there were 4,000–5,000 biotech companies with only 1750 or so remaining in 2008 (Pfeffer, 2012), a huge attrition rate. While successful biotechs remain, for example, Amgen, Biogen IDEC, Celgene, Genentech, Gilead, Regeneron, and Vertex (to name just a few), thousands of biotech startups have failed over the past three decades. Detailed information on this phenomenon to assess the probable reasons is surprising difficult to access on public databases (but probably exists in databases in financial accounting and consulting companies that is accessible for a fee and in any event cannot be publically shared) if indeed such information has been collected (Carlson, 2016). What information that is publically available is often in the form of personal interest stories in the mainstream media related to the impact of failed biotech companies on the economic well-being of their former employees (Nordrum, 2015) or is conflated with compound failures.

The view that biotech era may be less of a revolution and more of an incremental change in technology (Hopkins et al., 2007) is no doubt controversial, but it is a reasonable rebuttal of the public relations/media narrative of the biotech sector that tends to studiously ignore failures and extol successful companies in the form of the visionary logic, genius, and ruthless determination of their legendary leader. This was succinctly captured by Martinez (2016) in the context of the Facebook success story—"The amnesiac tech press weaves the narrative fallacy around the proceedings, fabricating a make-believe dramatic arc from steely-eyed product ideation to flawless and unhesitating technical execution. What was an improbable bonanza at the hands of the flailing half-blind becomes the inevitable coup of the assured visionary. The world crowns you a genius and you start acting like one."

6.4.1.1.1 BIOTECH—CHANGING THE FRAMEWORK OF APPLIED BIOMEDICAL SCIENCE

Biotechnology industry models—whether building a product in-house or licensing and/or selling/divesting to ensure/enhance survival (Tsai and Erickson, 2006; Shin et al., 2017)—have had a major impact on the structure of mainstream pharmaceutical industry. The core technologies of biotechnology were developed and reduced to practice far more quickly as a result of (1) venture capital funding and (2) the absence of a NIH (not invented here) risk aversive bureaucracy than would have been the case in the pre-1975 legacy industry. The existence of separate companies owning the output of basic academic research—often before it had been published in the peer review literature—restricted the access of traditional pharma to information that included novel drug targets, new technologies including chemistry and biological IP and potential drug candidates—necessitatitating strategic alliances and the outsourcing of preclinical research. This irrevocably changed the infrastructure of drug discovery (Hopkins et al., 2007; Gittelman, 2016) such that, in addition to their day jobs conducting research or running clinical trials, preclinical researchers and company clinicians were tasked with assessing strategic alliances and company/compound acquisitions, diminishing the time, and personal investment spent on internal projects while enhancing personal networks and increasing the accumulation of frequent flyer miles. With defined portions of pharma R & D budgets being mandated for use in external activities and with many large companies repetitively (perhaps addictively) acquiring products, some have questioned the logic of maintaining internal research and development capabilities other than for the oversight of external opportunities and collaborations, in effect making their R&D function virtual. This led Dixon et al., (2009) to consider vertical disintegration of legacy pharma as a means to enhance productivity. In 2014, an article in the Economist (2014) noted that mainstream pharma "got more than 70% of their revenues from products that were not developed in-house." The success in in-licensing technology from, or acquiring, biotechs was often in contrast with outright mergers that " rarely produce significant advances in innovation or research productivity," no doubt the result of creating organizational entropy and a decrease in risk avoidance (Booth, 2012). Not that acquisitions have always been beneficial for the acquired biotech beyond the initial financial transaction. Acquisitions that were heralded with press releases that highlighted the multiple opportunities in terms of compounds, technology, and world class researchers were reinforced with promises that the acquired company would have a major influence on the culture and operations of the larger/acquiring entity. These were often followed within 12 months—or sooner—with the Orwellian disappearance of the acquisition and its products due to either the compounds/technologies not living up to expectations or cultural/philosophical disconnects.

Another more subtle aspect of biotech was the creation, for want of a better term, of research sandboxes. Thus, the biotech industry created far more avenues for conducting innovative research via the influx of both money and ideas than had ever existed for legacy pharma. While some of the latter justifiably fell by the wayside, others led to new opportunities for drug discovery, the application of which was driven by an admix of a culture unencumbered by the aforementioned risk aversive "NIH"-culture and the considerable pressure exerted by VCs to ensure a timely return on their investment.

6.4.1.2 Moonshots—Federal and Otherwise

The moonshot metaphor for cancer research is a reference to the successful 1969 NASA Apollo 11 mission to land on the moon, the conceptual catalyst for the *War on Cancer*—the National Cancer Act of 1971—(Mukherjee, 2010). The success of the War—aka cancer moonshot v.1.—has been viewed ambivalently (DeVita and DeVita-Raeburn, 2015; Spector, 2010; Watson, 2009; Weinberg, 2014) with many of the recent advances in cancer therapeutics viewed as being independent of the War—being more incremental than revolutionary.

The most recent "cancer moon shot"—v.4—was an initiative announced at the 2016 Presidential State of The Union Address "to make more therapies available to more patients, while also improving our ability to prevent cancer and detect it at an early stage" (https://www.whitehouse.gov/the-press-office/2016/02/01/fact-sheet-investing-national-cancer-moonshot). The resultant *National Cancer Moonshot* (NCM) goals were subsequently refined by the inevitable Cancer Moonshot Task Force (Singer et al., 2016) "to dramatically accelerate efforts to prevent, diagnose, and treat cancer—to achieve a decade's worth of progress in 5 years" (https://medium.com/cancer-moonshot/report-of-the-cancer-moonshot-task-force-executive-summary-e711f1845ec#.bd8mt3kiz). In the context of: (1) the project management "iron triangle"—time, cost, and scope," more familiar as "fast, good, cheap—pick two" and (2) valid concerns as to where the $1 billion to jump start the NCM initiative would come from (Kaiser, 2016b)—it is unlikely that "good" will be part of the outcome. While halving the time to achieve unknown goals (zero divided by two still equals zero) is a very fuzzy target, the NCM has managed to catalyze considerable interest with several entrepreneurs funding new academic research initiatives in the cancer area (Lowy et al., 2016; Nature, 2016) that include The Ellison Institute for Transformative Medicine at USC, the Parker Institute for Cancer Immunotherapy sites and the Bloomberg–Kimmel Institute for Cancer Immunotherapy at Johns Hopkins.

Concomitant with the announcement of the NCM, a similar initiative, the *Cancer MoonShot 2020 Program* (cancer moon shot v. 3) was also announced that was intended to bring together stakeholders from industry, academia, community oncologists, and government to produce vaccine-based immunotherapies against cancer (Lowy et al., 2016). This had been preceded in 2012 by the MD Anderson Cancer Center in its *Moon Shots Program* (cancer moon shot—v. 2), a large scale, focused, well-coordinated, and well-funded research effort where multiple teams were focused on the goal of dramatically improving cancer survival (Kaiser, 2012). The existence of three temporarily congruent Moon Shots programs has raised concerns as to how these groups will work together to avoid duplication of effort (Hayden, 2016). Indeed, the MD Anderson Cancer Center has sued the Moonshot 2020 Program for using the "Moonshots" tagline (Sagonowsky, 2016) while both parties have been accused of conflicts of interest (Ross, 2017; Lowe, 2017).

The NCM concept has thus been described as a *"vaguely thought-out… terrible idea"* (Gitlin, 2016) with considerable potential to duplicate efforts in the private sector with "…a lack of overt leadership, and in the absence of a logical strategy ….a tendency to throw plates of spaghetti against the wall and hope it sticks," (Hayden, 2016).

The repeated analogy of drug discovery with the Apollo 11 mission has led to widespread criticism from experienced drug discovery scientists. While universally acknowledging the NASA achievement, these individuals are unable—for good reason—to reconcile the comparison of what was a successful engineering challenge with the complex biological hurdles of the drug discovery process. This viewpoint has been encapsulated by Lowe (2016a) in his *In the Pipeline* blog (http://blogs.sciencemag.org/pipeline/). Lowe noted that "going to the moon was easier [than discovering a drug as]....the basic principles were well known. Trying to cure cancer .... would be like trying to go to the moon without really knowing how rocket engines actually work, without being quite sure if Newton's laws of motion would hold up, and with some real uncertainty in the position of the moon. It can be hard to explain this to people who haven't done such research, and it's probably impossible to explain it without sounding like you're ..... making excuses. ...There are just far too many factors at work, many of which depend on each other, and many of which were not even clear about yet. Forget going to the moon—you wouldn't want to jump off a chair if our knowledge of physics was as inexact as our knowledge of human biology." Also, as several experts have pointed out, the Apollo 11 mission was a single defined event while cancer is not a single disease (Kaiser, 2012) not even within a single tumor (Mosoyan et al., 2013) but instead several hundred diseases each different in their origin, progress, drug responsiveness, and outcome.

To many observers, the 2016 NCM is reminiscent of a series of other recent White House initiatives related to healthcare: the 2011 National Alzheimer's Project Act (to accelerate the development of treatments that would prevent, halt or reverse the course of Alzheimer's Disease); the 2013 BRAIN Initiative (revolutionizing our understanding of the human brain); and the 2015 Precision Medicine Initiative (to enable a new era of medicine through research, technology, and policies that empower patients, researchers, and providers to work together toward development of individualized care). To an NIH insider, the latter two followed a similar pattern (Gitlin, 2016)—"flashy announcement first, followed by a year or more of meetings, workshops, and conference calls where researchers and policy makers had to sit down and work out what the actual scientific questions were supposed to be and what could they actually accomplish with the amounts of money on offer." This appears to be a recurrent process despite repeated concerns regarding access to funding that inevitably impact existing research plans that, unlike the White House initiatives, have, already gone through an objective peer review process that is informed by expertise in the sector. This contrasts with the knee jerk tenor—"more ambition than attention span" (Marcus, 2015)—of these governmental initiatives that is more concerned with self-serving, self-aggrandizing press conferences than advancing science. In 2016, the outgoing White House unveiled the National Microbiome Initiative, yet another whimsical and ill-defined exercise, in this instance "to advance understanding of microbiome behavior and enable protection and restoration of healthy microbiome function."

6.4.1.3 *Tech Titans and Healthcare*

The announcement for a workshop entitled, "The Search for Cures Leads to Silicon Valley" (http://www.milkeninstitute.org/events/conferences/global-conference/2016/panel-detail/6163) at the Milken Institute Global Conference in Santa Monica in May, 2016 read as follows; "The *tech titans of Silicon Valley are joining the search for cures*. Leading companies such as Google, Apple and Intel are helping refine methods for collecting and analyzing vast amounts of medical data to assist medical researchers. ......Could this be a disruptive force in medical research, helping shrink the time and cost of bringing treatments to patients? Can we make Moore's Law work in the search for cures?"

For those who are not especially computer savvy, Moore's Law—named after Gordon Moore, a cofounder of the computer chip maker, Intel in 1965—states that the number of transistors in a dense integrated circuit doubles approximately every 2 years along with the performance of the chip—"a bucolic vision of technological progress" (Jogalekar, 2012). Outside the computer industry, Moore's Law is frequently used as a metaphor for innovation, change, improved efficiency, and progress even though its actual days appear numbered due to unsolvable issues with heat generation in new chip design (Waldrup, 2016).

To the titans of technology in Silicon Valley, the pace of drug discovery and development is viewed as "glacial" and "the opposite of the way information technology works" and requires "re-thinking the current norms of biomedical research and drug discovery" (Suennen, 2016), which Lowe (2015b) and Piller (2016) have equated with an attitude of "Clearly I'm smart and successful, so clearly I have something to offer in this other field over here" and "Silicon Valley arrogance", respectively.

6.4.2 Reductionism—Of Transistor Radios and Donkey Kong

The tech titan approach to biomedical research is, to major extent, a reflection of a reductionist approach that, while useful when placed in the context of the inherent hierarchical complexity of biology, logic and experience is, on its own, not especially useful as Lazebnik (2002) and Jonas and Kording (2017) have argued.

6.4.2.1 Can a Biologist Fix a Radio?

In a widely read paper entitled, "*Can a biologist fix a radio?—Or, what I learned while studying apoptosis*" that has become widely appreciated by the skeptics in drug discovery who tend to regard the present nonhierarchical, reductionistic approach to research as both naïve and intellectually sloppy, Lazebnik (2002), provided an interesting perspective on absolutes in biomedical research. In his example, an unexpected research finding in the field of apoptosis resulted in a "Klondike gold rush with all the characteristic dynamics, mentality, and morals. A major driving force becomes the desire to find the nugget that will secure a place in textbooks, guarantee an unrelenting envy of peers, and, at last, solve all financial problems… [resulting]…. in crystal clear models that often explain everything and point at targets for future miracle drugs." This is inevitably followed by: obvious predictions falling apart; a "sense of frustration at the complexity of the process"; and the conclusion that the more that is learnt the less that is understood regarding the problem being studied.

Lazebnik then suggested that there was a fundamental flaw in how biologists approached complex problems and posited that this could be addressed by testing the value of an approach by applying it to a problem with a known solution which, in this instance, was a hypothetical problem of a biologist (a nonengineer) repairing a broken transistor radio. By determining the function of its component parts, approximately 100 capacitors, resistors, and transistors, those critical to functionality could conceptually be identified. This could be done by removing these components one at a time to see what effect they had on radio function. Some components—often more than one—were deemed critical for a "normal radio phenotype," for example, playing music. For the biologist to be able to repair the radio was deemed to be largely dependent on serendipity as the language used by biologists was viewed as "useless for a quantitative analysis, which limits its predictive or investigative value to a very narrow range." In contrast, an engineer would be expected to easily repair a dysfunctional radio as the result of an unambiguous, standardized and quantitative means of communication, invariable rules and a series of systematic, formal analytical tools—the concept of engineering modularity discussed by Pisano (2006).

In an early comparison of the semiconductor and biotechnology industries, Pisano had noted that the former is an industry that integrates know-how, is modular in nature, uses broadly accepted platforms, e.g., operating systems, and is also codified, which allows developers to easily collaborate at low cost. In contrast, biology (and by extrapolation drug discovery) lacks modularity or a standard operating platform, with processes that are "iterative and messy" such that each project is unique. In the context of the latter, system complexity can be overwhelming, especially a cell or an organ, with biological thinking tending toward oversimplification and a return to reductionism in a nonhierarchical, context-free setting (DeLisi, 2004). A further extension of the Lazebnik parable of the transistor radio was the "function from component parts" argument where the map of the human genome has been compared to the basic parts list for a Boeing 747 with none of the component parts on their own providing any precise understanding of which are critical for function, life, and safe flight, respectively (Winquist et al., 2014).

6.4.2.2 Donkey Kong

A second paper in the vein of the biologist transistor radio was "Could a neuroscientist understand a microprocessor?" (Jonas and Kording, 2017). Directly acknowledging inspiration by the Lazebnik paper, this paper focused on "big data" approaches in neuroscience (e.g., the US BRAIN and the EU Human Brain Mapping initiatives; Sejnowski et al., 2014) that are intended to provide fundamental insights as to how the brain processes information. Like Labeznik, these authors assessed whether the current approaches used for understanding brain function would allow a biologist to understand the processing systems of the MOS 6502 microchip processor that was widely used in a variety of early computers—for example, the Apple I and Commodore 64. While the 86 billion neurons in the brain dwarf the 3,510 transistors in the MOS 6502 chip, this microprocessor, like the brain, is a complex information processing system. Unlike the brain however, its function has been well established and has been reverse-engineered by "microchip archeologists" who recreated its function based on connectivity modeling (Jonas and Kording, 2017). Using this microchip as a model system, the authors applied data analysis methods used to study brain function to perform a reality check as to whether these were able to elucidate how information was processed by the chip. They reported that their results failed to produce any meaningful understanding of the microprocessor. Furthermore, using the videogame *Donkey Kong* in combination with the "lesioning" (e.g., removal) of individual transistors as behavioral outputs for the 6502 chip, the authors searched for transistors that could be uniquely responsible for an individual videogame—in essence "the *Donkey Kong* transistor"—which in the context of brain research would be *the* gene for depression, schizophrenia, autism, etc—the Holy Grail of neuroscience research. The outcomes were deemed "incredibly superficial [with] the transistors… not specific to any one behavior or game but rather implement simple functions" echoing the findings from Lazebnik's transistor radio example. In testing the approaches used to understand informational processing in the brain to a problem with a known solution, Jonas and Kording further focused additional attention on the limitations of the reductionist approach to studying brain function and the current tools available, reinforcing the naivety and lack of hierarchal integration of function that is the current *modus operandi* associated with "big data" approaches in neuroscience (O'Neil, 2015).

6.4.2.3 The Andy Grove Fallacy

The perceived shortcomings of biomedical research in the eyes of the titans of technology in Silicon Valley were also addressed by another Intel cofounder, the late Andy Grove, who argued that "pharma could learn something from how tech industries learn from their successes and failures" (Begley, 2007) implying that "all that drug industry really needs is more gumption about getting products to market quickly" (Hamilton, 2007), a perspective that is focused on "getting incomplete products into the hands of key user groups for beta testing and refinement" (Mullane and Williams, 2017). Grove also made an argument for abolishing the current clinical trial model in an editorial in *Science* entitled "Rethinking Clinical Trials" (Grove, 2011). In it, he advocated taking success stories from the Silicon Valley computer industry as examples to improve the speed, cost and efficiency of clinical trials. In his view, an "e-trial" system would determine efficacy on a patient-by-patient basis once safety had been established in a traditional Phase I trial. This information would then be used to develop databases similar to those used by Amazon.com that would allow meta-analyses to be conducted to "liberate drugs from the tyranny of the averages" Grove's insights to improving clinical trials were taken one step further by another Silicon Valley entrepreneur, Balaji Srinivasan, who in 2014 had tweeted "we can do vastly better than FDA w/a Yelp for drugs, including MD star ratings," another free market view of the drug approval process that was extended to eBay, Uber, and Airbnb as models (Orac, 2017).

Grove's suggestions led to some interesting comments, one of which correctly noted that drug candidate safety was a function of the therapeutic index, which required a measure of human efficacy that is typically unknown until Phase IIa trials are completed. In an e-trial this could not occur because the suggested physician-based efficacy trials were likely to be one-offs (e.g., single patients for an $n = 1$) and there would probably be no defined endpoints or controls. This led to the considered conclusion that "Groves clearly *doesn't understand clinical trials*" (Gorski, 2011). As with the simplistic oversimplification of drug discovery in the context of the 4th version of the cancer moonshot, the ever insightful Derek Lowe (2007) immortalized Grove's insights into drug discovery as the "Andy Grove Fallacy." Noting that Grove had a "warped... worldview", Lowe's blog provided a commentary that remains highly relevant to the Silicon Valley misperceptions of the drug discovery process and is therefore quoted in detail: "*.....medical research is different than semiconductor research.* It's harder. Ever seen one of those huge blow-ups of a chip's architecture? It's awe-inspiring, the amount of detail that's crammed into such a small space. And guess what—it's nothing, it's the instructions on the back of a shampoo bottle compared to the complexity of a living system. That's partly because we didn't build them. Making the things from the ground up is a real advantage when it comes to understanding them, but we started studying life after it had a few billion years head start. What's more, Intel chips are (presumably) actively designed to be comprehensible and efficient, whereas living systems—sorry, Intelligent Design people—have been glued together by relentless random tinkering. Mr. Grove, you can print out the technical specs for your chips. We don't have them for cells. And believe me, there are a lot more different types of cells than there are chips. Think of the untold number of different bacteria, all mutating and evolving while you look at them. Move on to all the so-called simple organisms, your roundworms and fruit flies, which have occupied generations of scientists and still not given up their biggest and most important mysteries. Keep on until you hit the lower mammals, the rats and mice that we run our efficacy and tox models in. Notice how many different kinds there are, and reflect on how much we really know about how they differ from each other and from us. Now you're ready for human patients, in all their huge, insane variety. Genetically we're a mighty hodgepodge, and when you add environment to that it's a wonder that any drug works at all."

Among the comments associated with this blog, one cited Pisano's book "Science Business" (Pisano, 2006) with the comment, "Changing the socket on a motherboard is a lot easier than altering a gene to modify a kinase's active site to fit your drug." As already discussed, Pisano had made the point that unlike engineering, biotechnology/drug discovery lacks modularity or a standard operating platform so that the question, "What is known about a target or a molecule or the behavior of a drug inside the body cannot be completely codified or reduced to precise rules (if X, then Y)." In short, unlike semiconductor engineering "there is still an 'art' to drug discovery that relies on judgment, instinct, and experience" (Pisano, 2006). Gittelman (2016) has further emphasized the unpredictability embedded in the complexity of human physiology that is an impediment to identifying the "causal triggers" that are viewed as being the rational means by which to identify drugs and cure disease. Little has changed in the decade since Andy Grove suggested how biopharma could improve its success, in fact the pharmaceutical industry has managed to rewrite Moore's Law, as Eroom's Law—Moore in reverse.

6.4.2.4 *Eroom's Law*

In 2012 Moore's Law was restated in the context of productivity in the pharmaceutical industry as Eroom's Law (Scannell et al., 2012) and was loosely defined in terms of the cost of developing a new drug doubling every 9 years such that drug discovery is becoming both slower and more expensive. Among the identified reasons for this, two are interlinked—"throwing money" at a drug discovery problem and relying on a "basic science-brute force" paradigm. In both instances, these approaches fail to address the "complex, multifaceted puzzle like the discovery of a new drug" (Jogalekar, 2012). The problem is neither amenable to adding more financial and human resources nor to being addressed via the molecular reductionism involved in industrial-level high throughput screening with hundreds of thousands (if not millions) of compounds produced using combinatorial/parallel chemical synthesis to optimize in vitro target interactions.

6.4.2.5 *Not Knowing What You Don't Know*

The criticism both from and to the tech titans can easily be categorized as patronizing, complacent, hubristic, etc., from either side. Accordingly, the negative receptivity from the drug discovery community needs to be put in the context of how many well-meaning "disruptive forces," technologies, cultures and management concepts, and organizations, have been directed at the drug discovery process over the past 40 years—much of it computationally enabled—with questionable success. Technologies including molecular biology, genomics, high throughput screening, combinatorial/parallel chemical synthesis, CAMD (computer assisted molecular design) and genomics, transcriptomics, metabolomics etc. have accordingly been integrated into diverse operational structures, for example, therapeutic, scientific discipline and strategic business unit (Williams, 2007), and concepts like systems biology and translational medicine, all buffeted by biotech collaborations and compound acquisitions without the emergence of any real paradigm for success. The most obvious disruptive force in traditional drug discovery that has already been discussed is the biotech industry, which has markedly changed the way in which legacy pharma operates while offering an alternative source for drugs albeit using the same technical approaches as pharma, blunting the perception of its disruptive potential if any such potential did indeed exist.

Despite a high level of pessimism, there have been positive outcomes related to the use of new, disruptive technologies enabled by the Internet and crowdsourcing. One involved 57,000 "citizen scientists" using the online video game "Foldit" to solve the structure of the M-PMV retroviral protease, an AIDS-causing monkey virus in 3 weeks, a problem that had not been solved using conventional approaches in 15 years (Khatib et al., 2011). The second was the "massive open laboratory" model using the online videogame, EteRNA (Treuille and Das, 2014) where 37,000 users developed design rules to improve the experimental accuracy of RNA structures (Lee et al., 2014).

6.4.2.6 App-Enabled Disruptions in Biomedical Research

Several Internet/technology companies have established a visible and highly disruptive presence in the healthcare management space. Thus Amazon Web Services (AWS), Apple, Facebook, Google, IBM Watson Health, Microsoft, and Qualcomm Life have, to varying degrees, become involved in initiatives in partnership with established healthcare providers, device manufacturers, hospitals, insurance companies, and directly with patients via apps, health and fitness devices, and online communities. These multiple initiatives are intended to facilitate health and wellness programs, oversee chronic care management, prescription drug compliance, genetic testing, etc. in addition to collecting healthcare metrics and providing clinical decision support, activities that may in time extend to the challenge of developing interactive and shareable electronic health records (EHRs) to replace paper files (Freudenheim, 2012).

Apple, Google, and Microsoft also have initiatives that have the potential to directly address healthcare in terms of devices and therapeutics. In 2013, Google (now Alphabet) established Calico (California Life Company), a biotech focused on treatments for aging and age-related diseases, including cancer and neurodegeneration. Calico, headed by the former CEO of Genentech, has established collaborations with AbbVie and a number of academic research centers.

A second Alphabet initiative is Verily Life Sciences (formerly Google X), an "ambitious and extravagantly funded biotech venture," (Piller, 2016) focused on developing the "Google Baseline" database to define "healthy" that is intended to provide a resource in identifying early signs of disease and a nanodiagnostic computer and a contact lens to read glucose levels to monitor diabetes. The Verily initiative has been described as "quixotic" applying "the confident impatience of computer engineering, along with extravagant hype, to biotech ideas that demand rigorous peer review and years or decades of painstaking work" with its devices, like Theranos' diagnostic tests, having no peer-review validation, instead being viewed as vaporware—"technology products that don't actually exist" (Piller, 2016).

In addition to developing predictive algorithms for more accurate online disease diagnosis, Microsoft has launched itself into the drug discovery space via a collaboration with Cambridge University in the United Kingdom to "solve the problem of cancer" within a decade (Knapton, 2016). By treating cancer as either an information processing system or a computer virus (depending on the media headline), Microsoft intends to map the code of cancer cells to "debug" them, so they can be reprogrammed back to a healthy state. The merit of this highly novel, blue sky approach to treating cancer has inevitably been questioned (Lowe, 2016b) and it, like Verily's glucose-sensing contact lens, is difficult to assess in the absence of any product or data. Many experienced drug researchers view the Microsoft approach as naïve and worthy of derision questioning how Microsoft intends to "debug" cancer cells when it cannot debug its frequently dysfunctional Office software suite and has failed to address issues with its Excel program that introduce errors into gene-related research (Ziemann et al., 2016).

In addition to its efforts in healthcare IT and CareKit apps Apple, like Verily is working on a noninvasive glucose monitor, in this instance using an *iWatch* based approach (Taylor, 2017), which, if the technical challenges can be overcome by the 200 Ph.D.s Apple has allocated to the project, would revolutionize diabetes healthcare.

Entering the pharmaceuticals business can be challenging—even when a company buys existing expertize and products in the form of another drug company as both Kodak and Proctor and Gamble did in the last decades of the 20th century - without success (Chandler, 2005). The future of the Alphabet/Google and Microsoft initiatives will be interesting to say the least especially with the "Silicon Valley arrogance" (Piller, 2016) and an apparent disdain for regulatory guidelines that no doubt resulted in the lack of any peer reviewed validation of the potentially disruptive diagnostic and device technologies emerging from Theranos (Bilton, 2016) and Verily (Piller, 2016), respectively.

6.4.2.7 *Silicon Valley Philanthropy and Biomedical Research*

An additional aspect of the tech titan interest in healthcare is the $3 billion Chan Zuckerberg Initiative recently announced by Facebook's founder, Mark Zuckerberg and his wife, Priscilla Chan, that has the ambitious goal "to eliminate, manage, or prevent all major disease by 2100" (Chayowski, 2016), a tall order that will need to take into account how society will be able to afford a world where disease has been eliminated or minimized and humans have been replaced by robots. This initiative follows in the tradition of the Rockefeller Foundation, Howard Hughes Medical Institute, and the Stowers, Broad, and Gates Foundations, The Michael J. Fox Foundation for Parkinson's Research all of which have historically funded biomedical research and the Theil Breakout Labs program that supports the formation of new biotech startup companies (Parthasarathy and Fishburne, 2015). This has led to the view that "the practice of science in the 21st century is becoming shaped less by national priorities or by peer-review groups and more by the particular preferences of individuals with huge amounts of money," (Broad, 2014) which is seen as a mixed blessing.

6.4.2.8 *Philanthropy Supports Research or Research Infrastructure?*

Investments in biomedical research initiatives whether arising from Federal, commercial, or philanthropic interest have, of late, tended to involve numerous meetings, workshops, etc. (Gitlin, 2016) with no clear direction (Hayden, 2016) at the end of which much of the investment has disappeared with debatable outcomes beyond the meetings and workshops (Marcus, 2015).

6.5 RESEARCHERS—THE IMAGES, THE REALITY, THE INCENTIVES

The majority of the proposed solutions intended to address and resolve issues with reproducibility in biomedical research (Begley and Ellis, 2012; Begley and Ioannidis, 2015; Jarvis and Williams, 2016) are directed towards the tools used to execute, analyze, and report experimental findings. These include: planning and execution (Chapter 2), statistics (Chapters 2 and 3) and reporting (Chapter 4) that have been the focus of high level attention with only oblique references (Alberts et al., 2014; Ioannidis, 2014a) to the actual vehicle for ensuring tangible outcomes—the biomedical researcher.

Even though society is becoming increasingly automated, the biomedical research enterprise remains highly dependent on the continued recruitment of highly trained, experienced, and motivated researchers who during their early career training and mentoring, acquire and subsequently provide, the expertize and decision-making experience to ensure that the output from basic research has value and is relevant to the needs of society. Such individuals are in marked contrast to Thiel's uncommitted "high-salaried, unaligned lab drones" who successfully populate biotech (Thiel and Masters, 2014).

6.5.1 The Very Model of a Modern Biomedical Researcher

6.5.1.1 Creating an Image

The general image of a biomedical researcher, and more often than not the reality, is of a gifted, well-trained, dedicated, and altruistic scientist whose major interest in working in research is solving a problem that will benefit societal health. To many, the archetypes of this type of individual are the eponymous hero of Sinclair Lewis' novel, "Arrowsmith" (Lewis, 1925) and, in real life, Rosalind Franklin, the unsung heroine of the Double Helix (Maddox, 2002) and Robert Langer, the scientist-entrepreneur, whose lab at MIT has spun out more than 40 companies, the products of which are estimated to have improved the lives of more than 2 billion people (Fleming, 2016; Prokesch, 2017)

Using Lewis' fictionalized version of the researcher as a reference point, the latter often works on their own, exhibits varying degrees of assertiveness and communication skills, is self-motivated, self-effacing and self-critical with high standards in his/her research work, and finds personal rewards in the intellectual challenge of doing research and sharing the outcomes with their peers. In addition, they may also suffer from an inflated ego, reflecting aspects of Kruger-Dunning Syndrome (Kruger and Dunning, 1999), be brusque and abrasive, ill tempered, with a marked inability to suffer fools gladly, an inability or refusal to work well with others and an inclination to take criticism either badly or as a measure of the intrinsic importance of their work. In fact, the more persistent the criticism, the greater the conviction that the ghost of Alfred Nobel is avidly watching over them. However, as long as these individuals do good science, character flaws are a minor distraction to deal with.

In contrast, defining the bad scientist involves the description of core character deficits that include questionable ethics, incompetence, poor work ethic, hubris, and/or taking full advantage of the knowledge that colleagues will tolerate many divergent viewpoints and approaches that are not and cannot necessarily viewed as right or wrong. In 1975, Feyerabend (1975) noted in the context of science becoming a career rather than a vocation that "Most scientists today are devoid of ideas, full of fear, intent on producing some paltry result so that they can add to the flood of inane papers that now constitutes 'scientific progress' in many areas." By operating in gray areas and keeping a low profile, the activities of a less than stellar scientist are often overlooked, ignored, or avoided.

There is also what might be viewed as a code of conduct in current biomedical research that places a higher value on the social interactions that contribute to a research environment, the academic circus (Koestler, 1973), than on the science itself, that facilitates bad science and fraud. Thus while reserving criticism of their peers to anonymous peer review, in their immediate sphere of research, researchers dismiss, or ignore the possibility of fraud, malicious intent, incompetence and hubris in their colleagues, thus ensuring that that the latter have no need—or opportunity—to question their own behavior.

Building on the psychological traits of desperation, perfectionism, flawed ethics, grandiosity, and sociopathistic behavior identified by Kornfeld (2012) in his metaanalysis of 146 reports of individual scientific misconduct identified by the US Federal Office of Research Integrity (ORI), bad scientists can be further categorized (Jarvis and Williams, 2016) as:

- *Inept*—Researchers with inadequate/poor training and mentoring that results in poorly designed, hypothesis-free, underpowered experiments that are often poorly executed, analyzed and reported, with the results being irreproducible.
- *Incompetent*—Researchers who despite appropriate training and mentoring lack the necessary skill sets, intellectual and physical, to conduct research successfully.
- *Indifferent*—Researchers who are competent, have been appropriately trained and mentored but whose contributions to irreproducibility are the result of laziness and an indifferent work ethic—the result of a high level of comfort with the *status quo* and an absence of any tangible consequences to affect their lifestyle; and
- *Ingenious*—Individuals who possess "enlightened" self-interest and hubris and whose behavior is driven by the "perverse incentives" of career advancement and/or financial gain that "...pressure ... authors to rush into print, cut corners, exaggerate ... and overstate the significance of their work" (Alberts et al., 2014). That such self-deluded dishonesty can thrive reflects a "system ... willing to overlook and ignore lack of scientific rigor and instead reward(s) flashy results that generate scientific buzz and excitement" (Begley and Ioannidis, 2015).

This final group are perhaps the most dangerous since it inevitably includes charlatans, narcissists (Lemaitre, 2016), and sociopaths who crave fame, glory and fortune, and abandon all scientific norms to accomplish these ends by generating findings that are eye-catching and thus have a higher probability of appearing in high Impact Factor journals.

6.5.1.2 Individualism in the era of "Big Science"

The image of the individual biomedical researcher altruistically searching for cures for disease, working 24/7 in a self-motivated quest for uncertain rewards beyond the prestige of peer recognition and the Holy Grail of a Nobel prize has become increasingly untenable. On the one hand there are far fewer opportunities for an individual to embark on their own career. On the other the advent of "big science" like the NIH's Big Data to Knowledge (BD2K) Initiative subsumes the role of the individual to that of a consortium, the members of which rely on one another for access to expertize in disparate disciplines. Researchers' insights and responsibility for consortia outcomes in such situations become diffuse to the point of opacity with the biostatistician collating the data and preparing the paper being the only individual with oversight and insight for the whole. While individual scientists to a maximum of three have historically been awarded the annual Nobel Prize in Physiology or Medicine, future Nobelists may be the representative(s) of the consortia (the biostatistician?) or to a diffuse and ultimately anonymous group as occurred in 2012 when the Nobel Peace prize was awarded to the European Union (http://www.nobelprize.org/nobel_prizes/peace/laureates/2012/).

6.5.1.3 Perpetuating the Image—Narratives in Reporting and Rewarding

Manuscripts submitted for peer review and subsequently published are by their nature and format an opportunity to present a measured, usually logical, progression from the formulation of an experimental hypothesis to the generation of the experimental data to the analysis and conclusions. Such manuscripts are however, infrequently a "true account" of how the research reported in the paper was actually done (Casadevall and Fang, 2015).

In a seminal 1964 paper entitled "Is The Scientific Paper a Fraud?" (Medawar, 1996), Peter Medawar, corecipient of the 1960 Nobel Prize in Physiology or Medicine "for discovery of acquired immunological tolerance" argued that a scientific paper typically "misrepresents the processes of thought that accompanied or gave rise to the work that is described in the paper." In providing additional detail of this contention, he characterized the results section of a paper as "a stream of factual information in which it is considered extremely bad form to discuss the significance of the results you are getting. You have to pretend that your mind is.... an empty vessel, for information, which floods into it from the external world for no reason, which you yourself have revealed." The appraisal of the factual information does not occur until the discussion section, where "you adopt the ludicrous pretense of asking yourself if the information you have collected actually means anything," a style of writing that Medawar characterizes as demonstrating the inductive process of scientific discovery.

Medawar's paper was recently revisited by Howitt and Wilson (2014), who noted that the IMRAD reporting format (Chapter 4) reinforces the impression that a research scientist is a highly logical and rational individual who in "doggedly adhering" to the scientific method, creates an image of infallibility that is often self-internalized (Lemaitre, 2016) and precludes the opportunity for serendipity, failure, and unexpected results that have been so critical in advancing research (Schwarz, 2017). Howitt and Wilson thus contrasted the "sanitized" textbook account of Watson and Crick's discovery of the Double Helix with Watson's account in his book of the same name (Watson, 1968) where his own mistakes and those of Linus Pauling were redeemed by input from a chemist, Jerry Donohue. The take home was that high school students and undergraduates, in being taught "sanitized" science with an overt factual and authoritarian focus, did not understand the intrinsic uncertainty of biomedical science and the fact that mistakes were inherent in experimentation and that these could nevertheless be resolved by intellectual engagement and experienced judgement and that "doing science and communicating science are quite different things." Casadevall and Fang (2013) have also noted that the Nobel Prize has distorted the history of science promulgating a "winner-take-all" culture to the exclusion of many scientists who made significant contributions (including Rosalind Franklin), additionally noting (Casadevall and Fang, 2015) that the scientific literature by "idealizing" the research process drained "the passion of discovery" that like serendipity and failure is another cornerstone in the discovery process.

6.6 REPRODUCIBILITY AND THE RESEARCH CULTURE

Researcher training, motivation and incentivization contribute in a variety of ways to the reproducibility issues that have been considered in this monograph.

In asking what motivation there would be for an individual to enter into a career in research beyond the actual science, the interrogator must squarely focus on the systematic flaws in the system that reflect the unintended but inevitable consequences of funding, that is inadequate and permanent positions for the number of individuals seeking them. This results in the normally competitive culture of research—first to publish, next in line for the Lasker or Nobel Prize, etc., - becoming hypercompetitive, with competition morphing into Darwinian survival, with citations as the criteria for tenure and grants (Ioannidis, 2014b). The resultant unintended consequences include "supress[ing] the creativity, cooperation, risk-taking, and original thinking required to make fundamental discoveries," "pressure on authors to rush into print, cut corners, exaggerate their findings, and overstate the significance of their work" (Alberts et al., 2014), with an associated decline in the quality of the science, its relevance and reproducibility.

6.6.1 Science and Money

Stories abound where the "grantsmanship" of an individual scientist can be valued far more than the credibility or novelty of their research. This can play a major role in career advancement and rewards that include tenure and space with those successful in attracting grants occupying whole floors of research buildings and enhancing the visibility and ranking of research institutes/universities and their ability to recruit new students and faculty. For those less gifted in their "grantsmanship," greater than 50% of their time can be spent in writing grants instead of doing research, leading inevitably to reduced data productivity and fewer publications that further diminishes the likelihood of being funded. In one widely reported instance, a senior researcher in the United Kingdom committed suicide (Parr, 2014) after his supervisor informed him that his grants were "not enough" and that he would "soon be sacked" (Colquhoun, 2014). While the ability to raise money endears such individuals to administrators, it can encourage: (1) shortcuts that lead to irreproducible research with "no-one [being] incentivized to be right. Instead, scientists are incentivized to be productive and innovative" (Horton, 2015) with "outputs" that are currently measured almost exclusively by citations, and: (2) administrative "indecisiveness" when a researcher in their institution is accused of research fraud, plagiarism, etc., and their initial response is a concern for losing the funding that the researcher brings to the institution.

6.6.2 The "Businessification" of Science

The leadership, structure, and infrastructure of biomedical research in the 21st century has been encapsulated by Lazebnik (2015) as the "businessification" of science. Like his seminal and thought provoking reflections on biologists fixing transistor radios, Lazebnik took issue with some of the conclusions in the Alberts et al. (2014) report on the challenges facing the US biomedical research ecosystem and contrasted it with the *Endless Frontier* report (Bush, 1945) that had underpinned the creation of the ecosystem.

Lazebnik noted that, in his view, young researchers, rather than being respected as "unique creative individuals," have increasingly been treated as an "economical workforce....maintained at a low cost, propagated in needed quantities, and trained for use in the laboratory" and found numerous instances where this change in status could not be ascribed to the current funding shortage. Instead, the ecosystem had "changed from one of advisors, trainees, and colleagues" to that of "a workforce and its users" with career administrators transitioning themselves from managing their institutions for the benefit of the scientists to becoming the leaders to whom the scientists were answerable. Lazebnik also noted that over the past 30 years while the tenured or tenure-track faculty at institutions of higher education increased 23%, the number of administrators grew 369%, 16 times greater with salaries that grew 2–3 times more than those of professors. Rather than doing science, the priority became raising money—especially grant overheads—that could support an ever-growing administration and facilities infrastructure with the director of a scientific institute purloining the title of Chief Operating Officer or Chief Executive Officer from the business environment. This is explained by the comment from Alberts et al. (2014), "Salaries paid by grants …[being]… subject to indirect cost reimbursement, creating a strong incentive for universities to enlarge their faculties by seeking as much faculty salary support as possible on government grants. This has led to an enormous growth in "soft money" positions, with stagnation in the ranks of faculty who have institutional support." These soft money positions in creating "fixed-term, precarious employment" contrasted with the "insiders in secure, stable employment" and led observers to compare the current situation of a new postdoc in academic biomedical research with that of being a member of a drug gang (Afonso, 2013) not perhaps the *Endless Frontier* that Bush had envisaged. Lazebnik further argued that academia in importing business more and superimposing these on the existing scientific research structure has created a dysfunctional hybrid business system with "merging love… [science]…and profit …[having]…..a danger of leading to prostitution" with the debasing of the ethic and passion that underlies the scientific method.

The converse of this argument was a period in time in the biopharma industry when "academic superstars" were hired into industry based on the successes of Pedro Cuatrecasas, Phil Needleman and Roy Vagelos at Burroughs Wellcome, Monsanto/Searle and Merck, respectively to motivate in-house researchers and enhance innovation. Many of these individuals did not last too long when their efforts resulted in nothing more than a string of high profile publications, extensive travel to exotic venues, a disgruntled staff, and no drug candidates.

The argument that something more than a lack of funds, insufficient positions, and too many scientists is responsible for the systemic flaws in US biomedical research (Alberts et al., 2014) was further developed by Smaldino and McElreath (2016), who argued based on a 60-year metaanalysis of statistical power in the behavioral sciences and dynamic modeling of scientific communities that the "most powerful incentives in contemporary science actively encourage, reward and propagate poor research methods and abuse of statistical procedures" leading to "the natural selection of poor science...[where]...exploitable quantitative metrics are used as proxies to evaluate and reward scientists," the obvious self-inflicted curse of citations and Impact Factors.

The concept of dysfunctional/perverse incentives was the topic of the management paper, "*On the Folly of Rewarding A, While Hoping for B*" (Kerr, 1975). In this seminal paper, Kerr noted in the context of science that "Society *hopes* that professors will not neglect their teaching responsibilities but *rewards* them almost entirely for research and publications," which in the context of this monograph can be restated as "Society *hopes* that biomedical researchers will conduct their research activities ensuring that the reagents they use in their experiments have been validated and/or authenticated and that the experiments are appropriately designed, powered, executed, and reported in accord with established best practices so that they can be facilely reproduced but *rewards* them almost entirely for publications in journals with a high Impact Factor that are highly cited and often cannot be reproduced."

6.6.3 Superstars and the Cult of Science

An additional facet of the "businessification" of science that tends to conflate the image of a research scientist is a mixture of what has been described as the "cult of science" (Wilson, 2016), and celebrity or scientific superstars.

The cult of science is associated with a phenomenon known as "scientism" (Hughes, 2012), a philosophical worldview that "scientific knowledge [is] a holy book .... that offers simple and decisive resolutions to deep questions. But it adds to this a pinch of glib frivolity and a dash of unembarrassed ignorance. Its rhetorical tics include a forced enthusiasm. Some of the Cult's leaders like to play dress-up as scientists .... but hardly any of them have contributed any research results of note" (Wilson, 2016) with "Cult leadership trend[ing] heavily in the direction of educators, popularizers, journalists." This also includes the superstars, a term that has long been used in science in the context of Nobel laureates, Lasker Awardees, etc., and also to describe up and coming researchers, often preceded by a copulatory adjective. The superstar concept of researchers has now part of the mainstream media including the world of entertainment and philanthropy. Prominent cancer researchers, including Nobel laureates, have been photographed in black leather jackets as GQ's "Rock Stars of Science" (Minogue, 2010) alongside and compared to, perhaps unfairly, "rock gods who don't follow orders" (https://www.mskcc.org/blog/geoffrey-beene-rock-stars-science-campaign-features-msk-researchers). While Robbins (2010) comments "if you have to resort to rockstars... [to] make science cool, you're really not very good at communicating science," the concept has become a facile, if somewhat embarrassing means, along with TV pundits like Dr. Oz and TED speakers, to help the public understand the science of healthcare and also raise funds for research.

6.7 CONSEQUENCES——NOT

Another perverse incentive within the culture of research is a lack of consequences for doing bad science. In 2008, Nardone (2008) proposed a zero tolerance policy regarding cell line authentication—"No cell line authentication, no grant; no cell line authentication, no access to journals as publication outlets"—which he considered "would … harmonize with the missions of granting agencies, journals, and professional societies—all of them run by intelligent, dedicated, hard-working people." The response he received from the NIH was that "mandating testing—and specifying particular tests—would conflict with the spirit of the grants program, which encourages individual responsibility for the conduct and direction of sponsored research." Similarly, part of the NIH response to the Begley and Ellis (2012) concerns was to introduce training modules in experimental design and record keeping (Wadman, 2013). However, the Deputy Director of the NIH noted (Fishburn, 2014) that their emphasis was "on recommendations rather than rules" with no expectation of "punitive actions for failure to comply. We prefer to use the bully pulpit to convince people to do the right thing and make it obligatory when there's no alternative," a rather puzzling position given estimates that 85% of the investment across the entirety of biomedical research (e.g., clinical, health services, and basic science) was being avoidably wasted (Chalmers and Glasziou, 2009). A later study (Freedman et al., 2015) estimated that more than 50% of preclinical research was irreproducible at a cost of $28 billion per year. With concerns about reduced funding and its decidedly negative impact on the future of biomedical research, one would think that avoiding egregiously wasting what money there was available under the rubric of "recommendations rather than rules" could be considered the height of irresponsibility, consistent with the continued *laissez faire* attitude to research misconduct (Kaiser, 2014).

6.8 CONCLUSIONS

Cataloging of the many errors and shortcomings in the conduct and reporting of scientific research—research on the research and metaanalysis of published studies (Ioannidis, 2014a; Ioannidis et al., 2015)—coupled with the concerns that this will inevitably impact human health outcomes has become a cottage industry (much of it published in the open access and mainstream media literature) that is progressively vocal and increasingly critical, a reflection of frustration with the *status quo* and the absence of any concrete suggestions or a demonstrated willingness to *proactively* rectify the situation. There is thus an urgent need to take steps to rectify this situation in order to improve the quality and validity of papers being published and the outcomes of peer review as the primary gatekeeper of quality.

This raises a number of issues, the resolutions to which remain elusive despite the motivation, passion, and acrimony that is brought to their debate, despite an extensive literature showing the consequences of:

- not validating antibodies,
- not authenticating cell lines,
- not powering experiments,
- not defining endpoints,
- not blinding and randomizing experiments,
- not using appropriate statistics,
- inappropriately normalizing data to control or baselines, and
- P-Hacking.

This leads to the following questions:

1. Why do researchers complacently continue with "business as usual" ignoring the issues bullet pointed above and wasting resources on research that not only lacks relevance and cannot be reproduced but is just plainly irresponsible from the get go?
2. Why has inadequate training and mentoring of young scientists in biomedical research become a facet of the present system and what is being done to address this given the time lag to rectify the situation (Rockey, 2014; Meadows, 2015; Redman and Caplan, 2015; Jarvis and Williams, 2016)?
3. Why do the guidelines for addressing shortcomings in reproducibility—most clearly and economically encapsulated in Begley's Rule of Six (Begley, 2013)—routinely appear to be treated as "suggestions" or "recommendations" rather than being mandatory with real consequences for their lack of use?
4. When will a statistical consensus view point on the generation and use of p-values and/or effect sizes that is coherent, consistent, and applicable for use by biomedical researchers replace the current perceived schadenfreude-like position, which is in effect little more than various statements that reflect diverse and often strongly held opinions that add confusion without tangible solutions?
5. Why does the scientific community confuse the translatability of preclinical findings to the clinic, a function of long standing challenges with the predictability of animal models, with reproducibility?
6. Why do concerns regarding the perceived shortcomings of the *CPR* process result in ever more convoluted and less transparent substitutes (e.g., *postpublication peer review, postpublication commentary*), the majority of which ignore the expertise, experience, time, and concerns of the long suffering, altruistic, unpaid, and disappearing, peer reviewer?
7. Why does the scientific community continue to dismiss the Journal Impact Factor (JIF) as "*useless or even destructive to the scientific community*" and a "tragedy of the commons" (Casadevall and Fang, 2014) then advocate it by using "it not only for deciding where to submit research papers, but for judging their peers, as well as influencing who wins jobs, tenure, and grants" (Bohannon, 2016)?
8. Why is the scientific community currently obsessed with reproducibility initiatives to replace the historical process of informal replication/reproducibility that have been singular and utter failures? And why, given that the original study plus the reproducibility initiative comprise n = 2, is money being wasted on such ill-judged initiatives? (Hint—the failure to confirm the studies on the STAP phenomenon numbered 133 in 7 labs within 3 months of its publication).

Until the reproducibility debate turns to prioritizing action over research on research voyeurism, the volume of which is increasing as its relevance and quality declines (Ioannidis, 2016), these questions will not receive logical, if any, answers.

6.8.1 Frameworks for Improving Reproducibility

Two series of frameworks have been recently published that provide a series of recommendations to address the high profile issue of reproducibility and its underlying causes, both perceived and real.

6.8.1.1 The American Academy of Microbiology (AAM) Framework

The American Academy of Microbiology (AAM) has created "A Framework for Improving the Quality of Research in the Biological Sciences" (Casadevall et al., 2016) that comprises six recommendations to provide an actionable framework to improve the quality of biological research. These are:

1. Design rigorous and comprehensive evaluation criteria to recognize and reward high-quality scientific research.
2. Require universal training in good scientific practices, appropriate statistical usage, and responsible research practices for scientists at all levels, with training content regularly updated and presented by qualified scientists.
3. Establish open data as the standard operating procedure throughout the scientific enterprise.
4. Encourage scientific journals to publish negative data that meet standards of quality.
5. Agree upon common criteria among scientific journals for retraction of published papers, to provide consistency and transparency.
6. Strengthen research integrity, oversight, and training.

6.8.1.2 The Committee on Responsible Science—the National Academies of Sciences, Engineering, and Medicine (NASEM) Recommendations—NASEM (2017)

As a follow-on to a 1992 panel report published by the Committee on Science, Engineering, and Public Policy of the National Academy of Sciences (NAS), the National Academy of Engineering (NAE), and the Institute of Medicine (IOM), "Responsible Science: Ensuring the Integrity of the Research Process" which recommended steps for reinforcing responsible research practices (NAS-NAE-IOM, 1992), NASEM (2017) published "Fostering Integrity in Research", that comprised the 11 recommendations listed below that have been edited/paraphrased for conciseness:

1. All stakeholders in the research enterprise—researchers, research institutions, research sponsors, journals, and societies—should improve/update their practices and policies to respond to the threats to research integrity identified in the "Fostering Integrity in Research" report.
2. Since research institutions play a central role in fostering research integrity and addressing current threats, they should maintain the highest standards for research conduct, going beyond simple compliance with federal regulations in undertaking research misconduct investigations and in other areas.
3. Research institutions and federal agencies should work to ensure that good-faith whistleblowers are protected and that their concerns are assessed and addressed in a fair, thorough, and timely manner.
4. A Research Integrity Advisory Board (RIAB) should be established as an independent nonprofit organization to provide a continuing organizational focus for fostering research integrity across research disciplines and sectors.
5. Clear disciplinary authorship standards should be developed by societies and journals (Chapter 4.5.1.1).
6. Publishers, via their policies and support infrastructure, should ensure that information sufficient for a researcher knowledgeable about the field and its techniques to reproduce reported results is available at the time of publication or as soon as possible thereafter.

7. Funds should be allocated by funding agencies and research sponsors to facilitate the long-term storage, archiving, and access of datasets and code necessary for reproducing published findings.
8. Researchers should routinely disclose all statistical results, including negative findings, to avoid unproductive duplication of research and to permit effective judgments on the statistical significance of findings. Key stakeholders (research sponsors, research institutions, and journals) should support this level of transparency.
9. Funding agencies and research sponsors in the US should fund research to quantify, and develop responses to, conditions in the research environment that may be linked to research misconduct and detrimental research practices using the data accumulated to monitor and modify existing policies and regulations.
10. Key stakeholders should continue to develop and assess more effective education and other programs that support research integrity. These improved programs should be globally adopted across disciplines.
11. Key stakeholders that participate in and support international collaborations should leverage these to foster research integrity through mutual learning and sharing.

A common theme in both the AAM Framework and the NASEM Recommendations is a clear focus on the responsibility of key shareholders especially research institutions and funding bodies to participate in developing and maintaining standards for research integrity. This contrasts with a trend to shift the responsibility exclusively to publishers and the much maligned peer review process. They also emphasize improved training in best practices in responsible research.

The AAM Framework, the NASEM recommendations with the Good Institutional Practices (Begley et al., 2015) and be supported by all researchers interested in a sustainable and relevant future for biomedical research.

In closing, Kraus (2014) has commented that "Science… is built upon a foundation of trust and verification—trust among scientists, who rely on each other's data and conclusions, and trust between scientists and the public, which funds much of the science." Unless actionable and transparent initiatives along the lines of the AAM Guidelines are taken seriously rather than discussing how to rearrange the deck chairs via moonshots-like initiatives and the like—borrowed from the inevitable *Titanic* analogy (Bourne and Lively 2012)—to restore this trust, the credibility and relevance of basic biomedical research will continue to erode with negative consequences for healthcare. An argument made against such overt pessimism is that the glass of biomedical research is being inappropriately viewed as half empty rather than half full, the latter being the result of the incredible wealth of knowledge and technologies that 21st century biomedical research has at its disposal. The issues that contribute to a pessimistic viewpoint are, like the negative views on peer review, less a comment on the immense potential for biomedical research per se but rather the way in which the science is practiced. Few, if any, of the issues raised in this monograph on research reproducibility can be considered novel. In contrast, the solutions to detrimental research practices are varied and reflect the difficulty in their implementation. These range from the relatively simple and inexpensive—authenticating cell lines, appropriately powering and randomizing experiments, etc.—to dealing with the widespread cultural issues that include the perverse incentives in 21st century research that presents researchers the choice between doing "what's best for medical advancement by adhering to the rigorous standards of science, or…what they perceive is necessary to maintain a career in the hypercompetitive environment of biomedical research" (Harris, 2017).

While new technologies and research discoveries will continue to provide opportunities to advance biomedical research and the drug discovery process, shortcomings in reproducibility will continue to undermine progress until there is a more focused, concerted, and responsible effort to restore best practices in research. These involve improved training and mentoring, the restructuring of incentives, personal and institutional, with stronger ethical norms that include real-time accountability and consequences for poor behaviors, together with the political will to impose them (Redman, 2015) and efforts to reverse the unnecessary, unproductive and unethical "businessification" of academic research (Lazebnik, 2015). To continue to condone the status quo in biomedical research, its funding and conduct, while equating its widespread criticism as misplaced pessimism is little more than a research-based form of the current societal addiction to political correctness, an activity that has been appropriately compared to necrotizing fasciitis (Deresiewicz, 2017) and is no solution to what ails 21st century biomedical research and its contributions to societal health and well being.

References

Abola, M.V., Prasad, V., 2016. The use of superlatives in cancer research. JAMA Oncol. 2, 139–141.

Afonso, A., 2013. How academia resembles a drug gang. Afonso Blog. Available from: https://alexandreafonso.me/2013/11/21/how-academia-resembles-a-drug-gang/.

Alberts, B., Kirschner, M.W., Tilghman, S., Varmus, H., 2014. Rescuing US biomedical research from its systemic flaws. Proc. Natl. Acad. Sci. USA 111, 5573–5777.

Alexander, S., 2016. Reverse voxsplaining: drugs vs chairs. Slate Star Codex. Available from: http://slatestarcodex.com/2016/08/29/reverse-voxsplaining-drugs-vs-chairs/.

Alexander, G.C., Frattaroli, S., Gielen, A.C., 2015. The Prescription Opioid Epidemic: An Evidence-Based Approach. Johns Hopkins Bloomberg School of Public Health, Baltimore, Maryland. http://www.jhsph.edu/research/centers-and-institutes/center-for-drug-safety-and-effectiveness/opioid-epidemic-town-hall-2015/2015-prescription-opioid-epidemic-report.pdf.

Angell, M., 2009. Drug Companies & Doctors: A Story of Corruption. Rev Books, New York. http://www.nybooks.com/articles/2009/01/15/drug-companies-doctorsa-story-of-corruption/.

Annual Reports in Medicinal Chemistry, 2014. Cumulative NCE Introduction Index, 1983–2013. Annu. Rep. Med. Chem. 49, 569–594.

Auffray, C., Balling, R., Barroso, I., Bencze, L., Benson, M., Bergeron, J., et al., 2016. Making sense of big data in health research: towards an EU action plan. Genome Med. 8, 71.

Baicker, K., Chandra, A., Skinner, J.S., 2012. Saving money or just saving lives? Improving the productivity of US health care spending. Ann. Rev. Econ. 4, 33–56.

Battelle Technology Partnership Practice, 2013. The Economic Impact of the U.S. Biopharmaceutical Industry, Batelle, Columbus, OH. http://phrma.org/sites/default/files/pdf/The-Economic-Impact-of-the-US-Biopharmaceutical-Industry.pdf.

Begley, S., 2007. Andy Grove: where are the cures?! Newsweek. Available from: http://www.newsweek.com/andy-grove-where-are-cures-221880.

Begley, C.G., 2013. Reproducibility: six red flags for suspect work. Nature 497, 433–434.

Begley, C.G., Ellis, L.M., 2012. Raise standards for preclinical cancer research. Nature 483, 531–533.

Begley, C.G., Ioannidis, J.P.A., 2015. Reproducibility in science: Improving the standard for basic and preclinical research. Cir. Res. 116, 116–126.

Begley, C.G., Buchan, A.M., Dirnagl, U., 2015. Robust research: institutions must do their part for reproducibility. Nature 525, 25–27.

Beissinger, M.R., 2009. Nationalism and the collapse of Soviet Communism. Contemp. Eur. History 18, 331–347.

Berkeley, J., 1999. Health care is not a human right. BMJ 319, 32.

Berndt, E.R., Nass, D., Kleinrock, M., Aitken, M., 2015. Decline in economic returns from new drugs raises questions about sustaining innovations. Health Aff. 34, 245–552.

Biesecker, L.G., 2013. Hypothesis-generating research and predictive medicine. Genome Res. 23, 1051–1053.

Bilton, N., 2016. Exclusive: how Elizabeth Holmes's house of cards came tumbling down. Vanity Fair. Available from: http://www.vanityfair.com/news/2016/09/elizabeth-holmes-theranos-exclusive.

Broad, W.J., 2014. Billionaires with big ideas are privatizing American science. New York Times. Available from: http://www.nytimes.com/2014/03/16/science/billionaires-with-big-ideas-are-privatizing-american-science.html.

Bohannon, J., 2016. Hate journal impact factors? New study gives you one more reason. *Science* Insider. Available from: http://www.sciencemag.org/news/2016/07/hate-journal-impact-factors-new-study-gives-you-one-more-reason.

Booth, B., 2012. Culture as a culprit of the Pharma R & D Crisis. Forbes. Available from: http://www.forbes.com/sites/brucebooth/2012/04/19/culture-as-a-culprit-of-the-pharma-rd-crisis/#6e41977176dd.

Booth, B., 2016. Innovators vs exploiters: drug pricing and the future of pharma, 2016;. Life Sci. VC. Available from: https://lifescivc.com/2016/08/innovators-vs-exploiters-drug-pricing-future-pharma/.

Bourne, H.R., Lively, M.O., 2012. Iceberg alert for NIH. Science 337, 390.

Bowen, A., Casadevall, A., 2015. Increasing disparities between resource inputs and outcomes, as measured by certain health deliverables, in biomedical research. Proc. Natl. Acad. Sci. USA 112, 11335–11340.

Bower, J.L., Christensen, C.M., 1995. Disruptive technologies: catching the wave. Harvard Bus Rev. Available from: https://hbr.org/1995/01/disruptive-technologies-catching-the-wave.

Brill, S., 2013. Bitter pill: why medical bills are killing us. Time Magazine. Available from: http://healthland.time.com/2013/02/20/bitter-pill-why-medical-bills-are-killing-us/print/.

Burger, O., Baudischa, A., Vaupela, J.W., 2013. Human mortality improvement in evolutionary context. Proc. Natl. Acad. Sci. USA 109, 18210–18214.

Bush, V., 1945. Science, the Endless Frontier: a Report to the President. U.S. Government Printing Office, Washington, D.C.

Callahan, D., 2007. How much is enough or too little: assessing healthcare demand in developed countries. Comp. Med. Chem. III 1, 627–635.

Carlson, R., 2016. Estimating the biotech sector's contribution to the US economy. Nat. Biotechnol. 34, 247–255.

Carpenter, D., 2014. Corrosive capture? The dueling forces of autonomy and industry influence in FDA pharmaceutical regulation. In: Carpenter, D., Moss, D.A. (Eds.), Preventing Regulatory Capture. Special Interest Influence and How to Limit It. Cambridge University Press, New York, pp. 152–172.

Carroll, J., 2016a. Hillary Clinton's war on pharma II. A call for common sense reporting on Alzheimer's R&D. Endpoints News. Available from: https://endpts.com/hillary-clintons-war-on-pharma-ii-a-call-for-common-sense-reporting-on-alzheimers-rd/?utm.

Carroll, J., 2016b. Hold the applause: Big Pharma has little to brag about regarding 2016 R&D productivity. Endpoints News. Available from: http://endpts.com/hold-the-applause-big-pharma-has-little-to-brag-about-regarding-2016-rd-productivity/.

Carroll, J., 2017. The list of the top 10 most expense drugs on the planet will soon have a new opening. Endpoints News. Available from: https://endpts.com/the-list-of-the-top-10-most-expensive-drugs-on-the-planet-will-soon-have-a-new-opening/?utm.

Casadevall, A., Fang, F.C., 2013. Is the Nobel Prize good for science? FASEB J. 27, 4682–4690.

Casadevall, A., Fang, F.C., 2014. Causes for the persistence of impact factor mania. mBio 5, ee00065-14.

Casadevall, A., Fang, F.C., 2015. (A) historical science. Infect. Immunol. 83, 4460–4464.

Casadevall, A., Ellis, L.M., Davies, E.W., McFall-Ngai, M., Fang, F.C., 2016. A framework for improving the quality of research in the biological sciences. mBio 7, e01256-16.

Chalmers, I., Glasziou, P., 2009. Avoidable waste in the production and reporting of research evidence. Lancet 374, 86–89.

Chandler, Jr., A.D., 2005. Shaping the Industrial Century. The Remarkable Story of the Evolution of the Modern Chemical and Pharmaceutical Industries. Harvard University Press, Cambridge, MA.

Chandler, Jr., A.D., Hikino, T., Mowery, D., 1998. The evolution of corporate capability and corporate strategy and structure within the world's largest chemical firms: the twentieth century in perspective. In: Arora, A., Landau, R., Rosenberg, N. (Eds.), Chemicals and Long-Term Economic Growth: Insights from the Chemical Industry. Wiley Interscience, New York, pp. 415–458.

Chast, F., 2008. A history of drug discovery. In: Wermuth, C.G. (Ed.), The Practice of Medicinal Chemistry. third ed. Academic Press, Burlington, MA, pp. 3–62.

Chayowski, K., 2016. Chan Zuckerberg initiative promises to spend $3 billion to research and cure all diseases. Forbes. Available from: http://www.forbes.com/sites/kathleenchaykowski/2016/09/21/chan-zuckerberg-initiative-invests-3-billion-to-cure-disease/#112ec682706c.

Chorev, B., 2016. Good drugs, good intentions, and the (bumpy) road to development. World Dev. Perspect. 1, 405.

Claxton, K., Martin, S., Soares, M., Rice, N., Spackman, E., Hinde, S., et al., 2015. Methods for the estimation of the National Institute for Health and Care Excellence cost-effectiveness threshold. Health Technol. Assess. 19, 1–503.

Collins, J.C., Porras, J.I., 2007. Built To Last. Successful Habits of Visionary Companies. Harper Business, New York, p. 48.

Colquhoun, D., 2014. Publish *and* perish at Imperial College London: the death of Stefan Grimm. DC's Improbable Science. Available from: http://www.dcscience.net/2014/12/01/publish-and-perish-at-imperial-college-london-the-death-of-stefan-grimm/.

Cook, R.I., 1998. How Complex Systems Fail. Cognitive Technologies Laboratory, University of Chicago, Chicago. http://web.mit.edu/2.75/resources/random/How%20Complex%20Systems%20Fail.pdf.

Costa, F.F., 2014. Big data in biomedicine. Drug Discov. Today 19, 433–440.

Dalrymple, T., 2011. New efficiencies in health care? Not likely. Wall St. J. Available from: http://online.wsj.com/article/SB10001424052748704116404576262943694897016.html.

Daniel, H., 2016. Stemming the escalating cost of prescription drugs: a position paper of the American College of PHYSICIANS. Ann. Intern. Med. 165, 50–52.

Davis, M.M., 2013. Right, Privilege—or Tragedy of the Commons? Culture of Health. Robert Wood Johnson Foundation. http://www.rwjf.org/en/culture-of-health/2013/08/right_privilege_or.html.

de Botton, A., 2014. The News: A User's Manual. Pantheon, New York, pp. 220–221.

De George, R., 2009. Two cheers for the pharmaceutical industry. In: Arnold, D. (Ed.), Ethics and Business of Biomedicine. Cambridge University Press, Cambridge, pp. 169–197.

DeLisi, C., 2004. Guest editorial: systems biology, the second time around. Environ. Health Perspect. 112, A926–A927.

Deloitte, 2016. 2016 Global health care outlook. Battling costs while improving care. Available from: https://www2.deloitte.com/content/dam/Deloitte/global/Documents/Life-Sciences-Health-Care/gx-lshc-2016-health-care-outlook.pdf.

Deresiewicz, W., 2017. On Political Correctness. Amer. Scholar, Available from: https://theamericanscholar.org/on-political-correctness/#.

DeVita, Jr., V.T., DeVita-Raeburn, E., 2015. The Death of Cancer. Farrar, Straus and Giroux, New York.

DiJulio, B., Firth, J., Brodie, M., 2015. Kaiser health tracking poll: June 2015. Available from: http://kff.org/health-costs/poll-finding/kaiser-health-tracking-poll-june-2015/.

Dillon, A., 2015. Carrying NICE over the threshold. National Institute for Health and Clinical Excellence (NICE) Blog February 19, 2015. Available from: https://www.nice.org.uk/news/blog/carrying-nice-over-the-threshold.

DiMasi, J.A., Grabowski, H.G., Hansen, R.W., 2016. Innovation in the pharmaceutical industry: new estimates of R&D costs. J. Health Econ. 47, 20–33.

Dixon, J., Lawton, G., Machin, P., 2009. Vertical disintegration: a strategy for pharmaceutical businesses in 2009? Nature Rev. Drug Discov. 8, 435.

Drakeman, D.L., 2014. Benchmarking biotech and pharmaceutical product development. Nat. Biotechnol. 32, 621–625.

Drucker, D.J., 2016. Never waste a good crisis: confronting reproducibility in translational research. Cell Metab. 24, 348360.

Du, L., Lu, W., 2016. U.S. Health-care system ranks as one of the least-efficient. Bloomberg. Available from: https://www.bloomberg.com/news/articles/2016-09-29/u-s-health-care-system-ranks-as-one-of-the-least-efficient.

Duffy, D.J., 2016. Problems, challenges and promises: perspectives on precision medicine. Brief Bioinform. 17, 494–504.

Economist, 2014. Invent it, swap it or buy it. Economist. Available from: http://www.economist.com/news/business/21632676-why-constant-dealmaking-among-drugmakers-inevitable-invent-it-swap-it-or-buy-it.

Economist, 2016. Why drug prices in America are so high. Economist. Available from: http://www.economist.com/blogs/economist-explains/2016/09/economist-explains-2.

Fan, W., Waizenegger, W., Lin, C.S., He, M.-X., et al., 2017. PPARd promotes running endurance by preserving glucose. Cell Metabolism 25, 1186–1193.

Feyerabend, P., 1975. How to defined society against science. Radical Philosp. 11, 3–8.

Fishburn, C.S., 2014. Repairing reproducibility. SciBx 7. Available from: http://www.nature.com/scibx/journal/v7/n10/full/scibx.2014.275.html.

Fleming, A., 2016. From super-pills to second skin: meet the Willly Wonka revolutionising medicine. Guardian. Available from: https://www.theguardian.com/lifeandstyle/2016/oct/17/robert-langer-nanotechnology-pioneer-willy-wonka-revolutionising-medicine.

Freedman, L.P., Cockburn, I.M., Simcoe, T.S., 2015. The economics of reproducibility in preclinical research. PLoS Biol. 13, e1002165.

Freedman, S., Mullane, K., 2017. The academic-industrial complex: navigating the translational and cultural divide. Drug Discov. Today, Available from: http://dx.doi.org/10.1016/j.drudis.2017.03.005.

Freudenberg, N., 2014. Lethal but legal: corporations, consumption, and protecting public health. Oxford University Press, Oxford UK.

Freudenheim, M., 2012. The ups and downs of electronic medical records. New York Times. Available from: http://www.nytimes.com/2012/10/09/health/the-ups-and-downs-of-electronic-medical-records-the-digital-doctor.html.

Fuchs, V.R., 2013. How and why us health care differs from that in other OECD countries. JAMA 309, 33–34.

Gitlin, J.M., 2016. Dear Mr. President: please stop with these science "moonshots". Ars Technica. Available from: http://arstechnica.com/science/2016/01/dear-mr-president-please-stop-with-these-science-moonshots/.

Gittelman, M., 2016. The revolution re-visited: clinical and genetics research paradigms and the productivity paradox in drug discovery. Res. Policy 45, 1570–1585.

Global Burden of Disease Study 2013 Collaborators, 2015. Global, regional, and national incidence, prevalence, and years lived with disability for 301 acute and chronic diseases and injuries in 188 countries, 1990-2013: a systematic analysis for the Global Burden of Disease Study 2013. Lancet 386, 743–800.

Goodman, J.C., 2016. The US is rationing life-saving drugs. Forbes. Available from: http://www.forbes.com/sites/johngoodman/2016/02/02/the-us-is-rationing-life-saving-drugs/#525fef3b8709.

Gornall, J., Hoey, A., Ozieranski, P., 2016. A pill too hard to swallow: how the NHS is limiting access to high priced drugs. BMJ 354, 4117.

Gorski, D., 2011. The wrong way to "open up" clinical trials. Science-Based Medicine. Available from: https://www.sciencebasedmedicine.org/the-wrong-way-to-open-up-clinical-trials/.

Grad, F.P., 2002. The preamble of the constitution of the World Health Organization. Bull. World Health Org. 80, 981–982.

Grove, A., 2011. Rethinking clinical trials. Science 333, 1679.

Haberman, C., 2015. The unrealized horrors of population explosion. New York Times. Available from: http://www.nytimes.com/2015/06/01/us/the-unrealized-horrors-of-population-explosion.html.

Hamilton, D.P., 2004. Biotech's dismal bottom line: more than $40 billion in losses. Wall St. J. Available from: http://www.wsj.com/articles/SB108499868760716023.

Hamilton, D.P., 2007. What Andy Grove still gets wrong about the life sciences—and how he could help fix them Venture Beat. Available from: http://venturebeat.com/2007/11/08/what-andy-grove-still-gets-wrong-about-the-life-sciences-and-how-he-could-help-fix-them/.

Hanson, M., 2015. Is the 2015 Nobel Prize a turning point for traditional Chinese medicine? The Conversation. Available from: http://theconversation.com/is-the-2015-nobel-prize-a-turning-point-for-traditional-chinese-medicine-48643.

Harris, R., 2017. Rigor Mortis: How Sloppy Science Creates Worthless Cures, Crushes Hope, and Wastes Billions. Basic Books, New York, p. 3.

Hayden, E.C., 2016. Scientists worry as cancer moonshots multiply. Nature 532, 424–425.

Himmelstein, D.U., Thorne, D., Warren, E., Woolhandler, S., 2009. Medical bankruptcy in the United States, 2007, results of a national study. Am. J. Med. 122, 741–746.

Hirschler, B., 2013. GlaxoSmithKline boss says new drugs can be cheaper. Reuters. Available from: http://www.reuters.com/article/us-glaxosmithkline-prices-idUSBRE92D0RM20130314.

Hodgson, J., 2016. When biotech goes bad. Nat. Biotechnol. 34, 284–291.

Hopkins, M.M., Martin, P.A., Nightingale, P., Kraft, A., Surya Mahdi, S., 2007. The myth of the biotech revolution: an assessment of technological, clinical and organisational change. Res. Policy 36, 566–589.

Horrabin, D.F., 2001. Realism in drug discovery—could Cassandra be right? Nat. Biotech. 19, 1099–1100.

Horton, R., 2015. Offline: what is medicine's 5 sigma? Lancet 285, 1380.

Howard, P., 2015. To lower drug prices, innovate, don't regulate. New York Times. Available from: http://www.nytimes.com/roomfordebate/2015/09/23/should-the-government-impose-drug-price-controls/to-lower-drug-prices-innovate-dont-regulate.

Howitt, S.M., Wilson, A.N., 2014. Revisiting "is the scientific paper a fraud?". EMBO Rep. 15, 481–484.

Hughes, S.S., 2011. Genentech. The Beginnings of Biotech. University of Chicago Press, Chicago.

Hughes, A.L., 2012. The folly of scientism. New Atlantis 37, 32–50.

Hughes, V., 2013. The big sell. Slate. Available from: http://www.slate.com/articles/health_and_science/human_genome/2013/10/brain_mapping_project_how_the_human_genome_project_mastered_big_science.html.

Hunt, P., Khosla, R., 2010. Are drug companies living up to their human rights responsibilities? The perspective of the former United Nations Special Rapporteur (2002–2008). PLoS Med. 7, e1000330.

ICER (Institute for Clinical and Economic Review), 2016. Addressing the myths about ICER and value assessment. Available from: https://icer-review.org/myths/.

Ioannidis, J.P.A., 2014a. Modeling and research on research. Clin. Chem. 60, 1238–1239.

Ioannidis, J.P.A., 2014b. How to make more published research true. PLoS Med 11, e1001747.

Ioannidis, J.P.A., 2016. The mass production of redundant, misleading, and conflicted systematic reviews and meta-analyses. Millbank Quart. 94, 485–514.

Ioannidis, J.P.A., Fanelli, D., Dunne, D.D., Goodman, S.N., 2015. Meta-research: evaluation and improvement of research methods and practices. PLoS Biol. 13, e1002264.

Jarvis, M.F., Williams, M., 2016. Irreproducibility in preclinical biomedical research: perceptions, uncertainties and knowledge gaps. Trends Pharmacol. Sci. 37, 290–302.

Jogalekar, A., 2012. The unstoppable Moore hits the immovable Eroom. Curious Wavefunction. Available from: http://wavefunction.fieldofscience.com/2012/03/unstoppable-moore-hits-immovable-eroom.html.

Jonas, E., Kording, K. P., 2017. Could a neuroscientist understand a microprocessor? PLoS Comput. Biol. 13, e1005268.

Joppi, R., Bertele, V., Garattini, S., 2005. Disappointing Biotech. BMJ 331, 895.

Juliano, R.L., 2013. Pharmaceutical innovation and public policy: the case for a new strategy for drug discovery and development. Sci. Pub Policy 42, 1–13.

Kaiser, J., 2012. MD Anderson launches $3 billion 'moon shot' to fight cancer. ScienceInsider. Available from: https://www.sciencepubs.com/news/2012/09/md-anderson-launches-3-billion-moon-shot-fight-cancer.

Kaiser, J., 2014. Top U.S. scientific misconduct official quits in frustration with bureaucracy. ScienceInsider Available from: http://news.sciencemag.org/people-events/2014/03/top-u.s.-scientific-misconduct-official-quits-frustration-bureaucracy.

Kaiser, J., 2016a. Funding for key data resources in jeopardy. Science 351, 14.

Kaiser, J., 2016b. White House wants $1 billion for Vice President Biden's cancer moonshot. Where will it come from? Science. Available from: http://www.sciencemag.org/news/2016/02/white-house-wants-1-billion-vice-president-biden-s-cancer-moonshot-where-will-it-come.

Kaiser Foundation, 2016. Kaiser health tracking poll. Available from: http://kff.org/health-reform/poll-finding/kaiser-health-tracking-poll-october-2016/?utm_campaign=KFF-2016-October-TrackingPoll&utm_source=hs_email&utm_medium=email&utm_content=36540781&_hsenc=p2ANqtz9y4XyWbnH2EByO5oZUOm602gYAR4P8H5SXRNCJHtMH3Hd3MKgfHUoxkcci1a9XhGmJREOPFVA8z14CwZiuoa-pi1Ya9Q&_hsmi=36540781.

Kantarjian, H.M., Fojo, T., Mathisen, M., Zwelling, L.A., 2013. Cancer drugs in the United States: *Justum Pretium*—the just price. J. Clin. Oncol. 31, 3600–3604.

Kantarjian, H., Steensma, D., Sanjuan, J.R., Elshaug, A., Light, D., 2014. High cancer drug prices in the United States: reasons and proposed solutions. J. Oncol. Pract. 10, e208–e211.

Kapeller, R., 2016. Is biotech ready for an Über disruption? Life Sci. VC. Available from: https://lifescivc.com/2016/03/biotech-ready-uber-disruption/.

Kapp, M.B., 2010. Health care technology, health care rationing, and older Americans: enough already? FSU College of Law, Public Law Research Paper No. 423. Available from: http://fsu.digital.flvc.org/islandora/object/fsu%3A209891.

Kennewell, P.D., 2007. Major drug introductions. Comp. Med. Chem. II 1, 105–249.

Kerr, S., 1975. On the folly of rewarding A while hoping for B. Acad. Mang. J. 18, 769–783.

Kessel, M., 2015. Restoring the pharmaceutical industry's reputation. Nat. Biotechnol. 32, 983–990.

Kesselheim, A.S., Avorn, J., Sarpatwari, A., 2016. The high cost of perscription drugs in the United States: origins and prospects for reform. JAMA 316, 858–871.

Khatib, F., Dimaio, F., Foldit Contenders Group, Foldit Void Crushers Group, Cooper, S., Kazmierczyk, M., et al., 2011. Crystal structure of a monomeric retroviral protease solved by protein folding game players. Nat. Struct. Mol. Biol. 18, 1175–1177.

Knapton, S., 2016. Microsoft will 'solve' cancer within 10 years by 'reprogramming' diseased cells. Telegraph. Available from: http://www.telegraph.co.uk/science/2016/09/20/microsoft-will-solve-cancer-within-10-years-by-reprogramming-dis/.

Koch-Weser, J., Schecter, P.J., 1978. Schmiedeberg in Strassburg 1872-1918: the making of modern pharmacology. Life Sci. 22, 1361–1371.

Koestler, A., 1973. The Call Girls. Random House, New York.

Kornfeld, D.S., 2012. Perspective: research misconduct: the search for a remedy. Acad. Med. 87, 877–882.

Kraus, W.L., 2014. Editorial: do you see what i see? quality, reliability, and reproducibility in biomedical research. Mol. Endocrinol. 38, 277–280.

Kruger, J., Dunning, D., 1999. Unskilled and unaware if it: how difficulties in recognizing one's own incompetence lead to inflated self-assessment. J. Personal. Soc. Psychol. 77, 1121–1134.

Kweifo-Okai, C., Holder, J., 2016. Over-populated or under-developed? The real story of population growth. Guardian. Available from: https://www.theguardian.com/global-development/datablog/2016/jun/28/over-populated-or-under-developed-real-story-population-growth.

Lakdawalla, D., 2015. Drug price controls end up costing patients their health. New York Times. Available from: http://www.nytimes.com/roomfordebate/2015/09/23/should-the-government-impose-drug-price-controls/drug-price-controls-end-up-costing-patients-their-health.

LaMattina, J., 2014. What price innovation? The Sovaldi Saga. Forbes.com. Available from; http://www.forbes.com/sites/johnlamattina/2014/05/29/what-price-innovation-the-sovaldi-saga/.

Lawitz, E., Mangia, A., Wyles, D., Rodriguez-Torres, M., Hassanein, T., Gordon, S.C., et al., 2013. Sofosbuvir for previously untreated chronic hepatitis C infection. NEJM 368, 1878–1887.

Lazebnik, Y., 2002. Can a biologist fix a radio?—Or, what I learned while studying apoptosis. Cancer Cell 2, 179–182.

Lazebnik, Y., 2015. Are scientists a workforce? Or, how Dr Frankenstein made biomedical research sick. EMBO Rep. 16, 1592–1600.

Le Fanu, J., 2011. The Rise and Fall of Modern Medicine. Abacus, London.

Lee, J., Kladwang, W., Lee, M., Cantu, D., Azizyan, M., Kim, H., et al., 2014. RNA design rules from a massive open laboratory. Proc. Natl. Acad. Sci. USA 111, 2122–2127.

Lemaitre, B., 2016. An Essay On Science And Narcissism: how do high-ego personalities drive research in life sciences? brunolemaitre.ch, Switzerland, Lausanne.

Lewis, S., 1925. Arrowsmith. 1925. New York, Harcourt, Brace. Republished, New York, Library of America, 133, 2002.

Light, D.W., Warburton, R., 2011. Demythologizing the high costs of pharmaceutical research. BioSocieties 6, 34–50.

Lorenzetti, L., 2015. Here's why Turing Pharmaceuticals says 5,000% price bump is necessary. Fortune. Available from; http://fortune.com/2015/09/21/turing-pharmaceuticals-martin-shkreli-response/.

Lowe, D., 2007. Andy Grove: rich, famous, smart and wrong. In the Pipeline. Available from; http://blogs.sciencemag.org/pipeline/archives/2007/11/06/andy_grove_rich_famous_smart_and_wrong.

Lowe, D., 2015a. What to do about Turing (and the others). In the Pipeline. Sci. Transl. Med. Available from; http://blogs.sciencemag.org/pipeline/archives/2015/09/21/what-to-do-about-turing-and-the-others.

Lowe, D., 2015b. Silicon valley sunglasses. In the Pipeline. Available from: http://blogs.sciencemag.org/pipeline/archives/2015/04/02/silicon_valley_sunglasses.

Lowe, D., 2016a. Moonshot, they say. In the Pipeline. Available from; http://blogs.sciencemag.org/pipeline/archives/2016/01/15/moonshot-they-say.

Lowe, D., 2016b. Better, faster, more comprehensive manure distribution. In the Pipeline. Available from; http://blogs.sciencemag.org/pipeline/archives/2016/09/21/better-faster-more-comprehensive-manure-distribution.

Lowe, D., 2016c. Gosh, fellows, there's a better way, in the pipeline. Available from; http://blogs.sciencemag.org/pipeline/archives/2016/08/26/gosh-fellows-theres-a-better-way.

Lowe, D., 2017. The Soon-Shiong hype machine. In the Pipeline. Available from: http://blogs.sciencemag.org/pipeline/archives/2017/02/14/the-soon-shiong-hype-machine.

Lowy, D., Singer, D., DePinho, R., Simon, G.C., Soon-Shiong, P., 2016. Cancer moonshot countdown. Nat. Biotechnol. 34, 596–599.

Ma, J., Ward, E.M., Siegel, R.L., Jemal, A., 2015. Temporal trends in mortality in the United States, 1969-2013. JAMA 314, 1731–1739.

Maddox, B., 2002. Rosalind Franklin: The Dark Lady of DNA. HarperCollins, London.

Madrigal, C., 2013. The best books about biotech. Atlantic. Available from; http://www.theatlantic.com/technology/archive/2013/03/the-best-books-about-biotechnology/273683/.

Malpani, R., 2014. R&D cost estimates—MSF response to Tufts CSDD study on cost to develop a new drug. Médecins Sans Frontières Access Campaign. Available from; http://www.msfaccess.org/content/rd-cost-estimates-msf-response-tufts-csdd-study-cost-develop-new-drug.

Marcus, R., 2015. We now know more about how Newark schools partially squandered Mark Zuckerberg's $100 million donation. Washington Post. Available from: http://www.businessinsider.com/we-now-know-more-about-how-newark-schools-partially-squandered-mark-zuckerburgs-100-million-donation-2015-10.

Margolis, R., Derr, L., Dunn, M., Huerta, M., Larkin, J., Sheehan, J., et al., 2014. The National Institutes of Health's Big Data to Knowledge (BD2K) initiative: capitalizing on biomedical big data. J. Am. Med. Inform. Assoc. 21, 957–958.

Martinez, A.G., 2016. Chaos Monkeys. Ebury Press, London, p. 490.

Maxmen, A., 2016. Big Pharma's—cost-cutting challenger. Nature 536, 388–390.

McArdle, M., 2016. Health care is a business, not a right. Bloomberg View. Available from; https://www.bloomberg.com/view/articles/2016-08-23/health-care-is-a-business-not-a-right.

Meadows, A., 2015. Meadows, A. quoted in Michael A. Ask the chefs: how can we improve the article review and submission process? The Scholarly Kitchen. Available from; http://scholarlykitchen.sspnet.org/2015/03/26/ask-the-chefs-how-can-we-improve-the-article-review-and-submission-process/.

Medawar, P., 1996. Is The scientific paper a fraud? In: Medawar, P., Gould, S.J. (Eds.), The Strange Case of the Spotted Mice: and Other Classic Essays on Science. Oxford University Press, UK, pp. 33–39.

Monaghan, A., 2016. Pfizer fined record £84.2m over NHS overcharging. Guardian. Available from: https://www.theguardian.com/business/2016/dec/07/pfizer-fined-nhs-anti-epilepsy-drug-cma.

Minogue, K., 2010. Can 'rock stars of science' cut through the noise? ScienceInsider Available from; http://www.sciencemag.org/news/2010/11/can-rock-stars-science-cut-through-noise.

Mosoyan, G., Nagi, C., Marukian, S., Teixeira, A., Simonian, A., Resnick-Silverman, L., et al., 2013. Multiple breast cancer cell-lines derived from a single tumor differ in their molecular characteristics and tumorigenic potential. PLoS ONE 8, e55145.

Mukherjee, S., 2010. The Emperor of All Maladies: A Biography of Cancer. Scribner, New York.

Mullane, K., Williams, M., 2015. Unknown unknowns in biomedical research: does an inability to deal with ambiguity contribute to issues of irreproducibility? Biochem. Pharmacol. 97, 133–136.

Mullane, K., Williams, M., 2017. Enhancing reproducibility: failures from reproducibility initiatives underline core challenges. Biochem. Pharmacol. 138, 7–18.

Munos, B., Orloff, J.J., 2016. Disruptive innovation and transformation of the drug discovery and development enterprise. Discussion Paper, National Academy of Medicine, Washington, DC. Available from; https://nam.edu/wp-content/uploads/2016/07/Disruptive- Innovation-and-Transformation-of-the-Drug-Discovery-and-Development-Enterprise.pdf.

Nardone, R.M., 2008. Curbing rampant cross-contamination and misidentification of cell lines. BioTechniques 45, 221–227.

NAS-NAE-IOM (National Academy of Sciences, National Academy of Engineering and Institute of Medicine), 1992. Responsible Science: Ensuring the Integrity of the Research Process. National Academies Press, Washingon, DC.

NASEM (National Academies of Science, Engineering and Medicine), 2017. Fostering Integrity in Research. The National Academies Press, Washington, DC, Available from: http://www.nap.edu/21896.

Nature, 2016. Tech billionaires fund new cancer centers. Nat. Biotechnol. 34, 583.

Nordrum, A., 2015. The fallout: what happens when biotech companies fail? Inter Business Times. Available from; http://www.ibtimes.com/fallout-what-happens-when-biotech-companies-fail-2148850.

O'Neil, C., 2015. Guest post: dirty rant about the human brain project. Available from; https://mathbabe.org/2015/10/20/guest-post-dirty-rant-about-the-human-brain-project/.

OECD, 2016. Health Statistics 2016. Available from; http://www.oecd.org/els/health-systems/health-data.htm.

Oellrich, A., Collier, N., Groza, T., Rebholz-Schuhmann, D., Shah, N., Bodenreider, O., et al., 2016. The digital revolution in phenotyping. Briefings Bioinform. 17, 819–830.

Orac, 2017. Next up on the Trump crazy train: A man who thinks that a "Yelp for drugs" will do a better job than the FDA. Respectful Insolence. ScienceBlogs. Available from: http://scienceblogs.com/insolence/2017/01/16/next-up-on-the-trump-fda-crazy-train-a-man-who-thinks-that-a-yelp-like-system-will-do-better-than-the-fda-at-maintaining-drug-safety/.

Outterson, K., Rex, J.H., Jinks, J., Jackson, P., Hallinan, J., Karp, S., et al., 2016. Accelerating global innovation to address antibacterial resistance: introducing CARB-X. Nat. Rev. Drug Discov. 15, 589–590.

Palsson, B.O., 2015. Systems Biology. Cambridge University Press, Cambridge.

Parr, C., 2014. Imperial College London to 'review procedures' after death of academic. Times Higher Edu. Available from; https://www.timeshighereducation.com/news/imperial-college-london-to-review-procedures-after-death-of-academic/2017188.article.

Parthasarathy, H., Fishburne, L., 2015. Philanthropy's role in translating scientific innovation. Nat. Biotechnol 33, 1022–1025.

PCAST (President's Council of Advisors on Science and Technology), 2012. Report to the President on propelling innovation in drug discovery, development, and evaluation. Available from; http://www.whitehouse.gov/sites/default/files/microsites/ostp/pcast-fda-final.pdf.

Pfeffer, G.G., 2012. The biotechnology sector: therapeutics. In: Burns, L.R. (Ed.), The Business of Healthcare Innovation. second ed. Cambridge University Press, Cambridge UK, pp. 194–345.

Piller, C., 2016. 'Silicon Valley arrogance'? Google misfires as it strives to turn Star Trek fiction into reality. STAT News. Available from; https://www.statnews.com/2016/06/06/google-star-trek-fiction/.

Pisano, G., 2006. Science Business. The Promise, the Reality, and the Future of Biotech. Harvard Business School Press, Boston, MA.

Pole, S., 2013. Ayurvedic Medicine: The Principles of Traditional Practice. Singing Dragon/Kingsley, London.

Pollack, A., 2002. Rebellious bodies dim the glow of 'natural' biotech drugs. New York Times. Available from; http://www.nytimes.com/2002/07/30/health/rebellious-bodies-dim-the-glow-of-natural-biotech-drugs.html?_r=0.

Pollack, A., 2013. A biotech king, dethroned. New York Times. Available from; http://www.nytimes.com/2013/09/06/business/david-blech-a-biotech-king-dethroned.html?_r=0.

Portteus, K., 2016. EpiPen debate sheds light on ruling class. Forbes Capital Flows. Available from; http://www.forbes.com/sites/realspin/2016/09/13/epipen-debate-sheds-light-on-ruling-class/#571365ba6334.

Prinz, F., Schlange, T., Asadullah, K., 2011. Believe it or not: how much can we rely on published data on potential drug targets? Nat. Rev. Drug Discov. 10, 712–713.

Prokesch, S., 2017. The Edison of Medicine. Harvard Business Review, pp. 134–143.

Prud'Homme, A., 2004. The Cell Game. Sam Waksal's Fast Money and False Promises—and the Fate of Imclone's Cancer Drug. Harper Business, New York.

Qiu, J., 2015. When the East meets the West: the future of traditional Chinese medicine in the 21st century. Natl. Sci. Rev. 2, 377–380.

Ramannavar, M., Sindai, N.S., 2015. Big Data and analytics—a journey through basic concepts to research issues. Adv. Intelligent Syst. Comput. 398, 291–306.

Ramsey, L., 2016. RANKED: These are the most and least reputable drug companies in the world, BusinessInsider. Available from; http://www.businessinsider.com/pharmaceutical-companies-reputation-rankings-2016-5/#14-pfizer-reptrak-points-659-1.

Rasmussen, N., 2014. Gene Jockeys. Life Science and the Rise of the Biotech Enterprise. Johns Hopkins University Press, Baltimore MD.

Ravina, E., 2011. The Evolution of Drug Discovery. Wiley-VCH, Weinheim, Germany.

Reau, N.S., Jensen, D.M., 2014. Sticker shock and the price of new therapies for hepatitis C: Is it worth it? Hepatology 59, 1246–1249.

Redman, B.K., 2015. Are the Biomedical Sciences Sliding Toward Institutional Corruption? And Why Didn't We Notice It? Edmond J. Safra Working Papers, No. 59. Harvard University, 2015. Available from: http://papers.ssrn.com/sol3/papers.cfm?abstract_id=2585141.

Redman, B.K., Caplan, A.L., 2015. Closing the barn door: coping with findings of research misconduct by trainees in the biomedical sciences. Res. Ethics 11, 124–132.

Reed, T.R., 2010. The Healing of America: A Global Quest for Better, Cheaper, and Fairer Health Care. Penguin, New York.

Regalado, A., 2016. The world's most expensive medicine is a bust. MIT Tech. Rev. Available from: https://www.technologyreview.com/s/601165/the-worlds-most-expensive-medicine-is-a-bust/.

Review on Antimicrobial Resistance, 2016. Tackling drug-resistant infections globally: final report and recommendations. Wellcome Trust HM Government. Available from; https://amr-review.org/sites/default/files/160525_Final%20paper_with%20cover.pdf.

Robbins, M., 2010. 'Rockstars of science' should be 'scientists of rock'. Guardian. Available from: https://www.theguardian.com/science/the-lay-scientist/2010/nov/20/1.

Rockey, S., 2014. Mentorship matters for the biomedical workforce. Nat. Med. 20, 575.

Roy, A., 2013. Yes, health care is a right—an individual right. Forbes. Apothecary. Available from: https://www.forbes.com/sites/theapothecary/2013/03/28/yes-health-care-is-a-right-an-individual-right/.

Rosenthal, E., 2017. An American Sickness: How Healthcare Became Big Business and How You Can Take It Back. Penguin, New York.

Ross, C., 2017. Ronald DePinho, embattled chief of MD Anderson Cancer Research, resigns. STATnews. Available from: https://www.statnews.com/2017/03/08/md-anderson-depinho-resigns/.

Sachs, R., 2016. Drug prices: where do we go after the election? Conversation. Available from: http://theconversation.com/drug-prices-where-do-we-go-after-the-election-67812?.

Sagonowsky, E., 2016. Who owns 'moonshot'? MD Anderson says it does in suit against biotech billionaire Soon-Shiong. FiercePharma. Available from: http://www.fiercepharma.com/pharma/md-anderson-files-lawsuit-over-alleged-moonshot-misuse.

Sanchez-Serrano, I., 2011. The World's Healthcare Crisis. Elsevier, London, pp. 47–70.

Scannell, J., 2015. Four reasons drugs are expensive, of which two are false. Forbes. Available from: http://www.forbes.com/sites/matthewherper/2015/10/13/four-reasons-drugs-are-expensive-of-which-two-are-false/#6e5e3ae848a5.

Scannell, J.W., Bosley, J., 2016. When quality beats quantity: decision theory, drug discovery, and the reproducibility crisis. PLoS ONE 11, e0147215.

Scannell, J.W., Blanckley, A., Boldon, H., Warrington, B., 2012. Diagnosing the decline in pharmaceutical R&D efficiency. Nat. Rev. Drug Discov. 11, 191–200.

Schwartz, S.A., 2016. American healthcare: a profile in shortages. Explore 12, 167–170.

Schwarz, H., 2017. On the usefulness of useless knowledge. Nature Rev. Chem. 1, 0001.

Sejnowski, T.J., Churchland, P.S., Movshon, J.A., 2014. Putting big data to good use in neuroscience. Nat. Neurosci. 17, 1440–1441.

Senior, M., 2014. Sovaldi makes blockbuster history, ignites drug pricing unrest. Nat. Biotech. 32, 501–502.

Shah, A., 2011. Health care around the world. Global Issues. Available from; http://www.globalissues.org/article/774/health-care-around-the-world.

Shin, K., Park, G., Choi, J.Y., Choy, M., 2017. Factors affecting the survival of SMEs: a study of biotechnology firms in south korea. Sustainability 9, 108.

Singer, D.S., Jacks, T., Jaffee, E., 2016. A U.S. "Cancer Moonshot" to accelerate cancer research. Science 353, 1105–1106.

Skardal, A., Shupe, T., Atala, A., 2016. Organoid-on-a-chip and body-on-a-chip systems for drug screening and disease modeling. Drug Discov. Today 21, 1399–1411.

Smaldino, P.E., McElreath, R., 2016. The natural selection of bad science. R. Soc. Open Sci. 3, 160384.

Smith, M., 2014. Treating HCV—is the price right? Medpage Today. Available from; http://www.medpagetoday.com/Gastroenterology/Hepatitis/44357.

Smith, A.G., 2016. Price gouging and the dangerous new breed of pharma companies. Harvard Business Rev. Available from; https://hbr.org/2016/07/price-gouging-and-the-dangerous-new-breed-of-pharma-companies.

Sneader, W., 1985. Drug Discovery: The Evolution of Modern Medicines. Wiley, Chichester, UK.

Sneader, W., 2005. Drug Discovery: A History. Wiley, Chichester, UK, pp. 8–31.

Sox, H.C., Greenfield, S., 2009. Comparative effectiveness research: a report from the institute of medicine. Annal. Intern. Med. 151, 203–205.

Spector, R., 2010. The war on cancer: a progress report for skeptics. Skeptical Inquirer 34.1. Available from; http://www.csicop.org/si/show/war_on_cancer_a_progress_report_for_skeptics.

Suennen, L., 2016. Silicon Valley joins the drug and device discovery party. Venture Valkyrie. Available from; http://venturevalkyrie.com/silicon-valley-joins-the-drug-and-device-discovery-party/.

Sumner, P., Vivian-Griffiths, S., Boivin, J., Williams, Venetis, C.A., Davies, A., et al., 2014. The association between exaggeration in health related science news and academic press releases: retrospective observational study. BMJ 349, g7015.

Tahir, D., 2016. Rumors, expectations surround Apple expansion into health care. Politico. Available from: http://www.politico.com/story/2016/10/apple-expansion-health-care-229111.

Taylor, N.P., 2017. Apple is testing a noninvasive blood glucose monitor: CNBC. FierceBiotech. Available from: http://www.fiercebiotech.com/medical-devices/apple-testing-a-noninvasive-blood-glucose-monitor-cnbc.

Thayer, K.A., Wolfe, M.S., Rooney, A.A., Boyles, A.L., Bucher, J.R., Birnbaum, L.S., 2014. Intersection of systematic review methodology with the NIH Reproducibility initiative. Environ. Health Persp. 122, A176–A177.

Thiel, P., Masters, B., 2014. Zero to one: notes on startups, or how to build the future. Crown Business, New York.

Timmins, N., 2013. The wisdom of the crowd. 65 views of the NHS at 65. Nuffield Trust. Available from; http://www.nuffieldtrust.org.uk/sites/files/nuffield/publication/130704_wisdom_of_the_crowd.pdf.

Topol, E., 2012. The Creative Destruction of Medicine: How the Digital Revolution Will Create Better Health Care. Basic Books, New York.

Treuille, A., Das, R., 2014. Scientific rigor through videogames. Trends Biochem. Sci. 39, 507–509.

Tsai, W., Erickson, S., 2006. Early-stage biotech companies: strategies for suevival and growth. Biotech. Healthcare 3, 49–53.

Tucker, M.E., 2015. IDF Atlas: about 415 million adults worldwide have diabetes. Medscape. Available from; http://www.medscape.com/viewarticle/855296.

Wachter, R., 2015. The Digital Doctor: Hope, Hype, and Harm at the Dawn of Medicine's Computer Age. McGraw-Hill Education, New York.

Wadman, M., 2013. NIH mulls rules for validating key results. Nature 500, 14–16.

Waldrup, M.M., 2016. The chips are down for Moore's law. Nature 530, 144–147.

Walker, J., 2015. Patients struggle with high drug prices. Wall St. J. Available from: http://www.wsj.com/articles/patients-struggle-with-high-drug-prices-1451557981.

Walker, J., McGinty, T., 2016. Biotech labs birth new drugs—and new fortunes. Wall St. J. Available from: http://www.wsj.com/articles/biotech-labs-birth-new-drugsand-new-fortunes-1466798095.

Watson, J., 1968. The Double Helix. Atheneum, New York.

Watson, J., 2009. To fight cancer, know the enemy. New York Times, Available from; http://www.nytimes.com/2009/08/06/opinion/06watson.html?_r=0.

Weinberg, R.A., 2014. Coming full circle—from endless complexity to simplicity and back again. Cell 157, 267–271.

Westphal, N., 2013. The economic burden of chronic disorders, a case study: how far are HONDAs driving healthcare costs? Decision Resources Group. Available from; http://decisionresourcesgroup.com/report/?id=676.

Whitehouse, P.J., 2016. The diagnosis and treatment of alzheimer's: are we being (Ir) responsible? In: Boenink, M., van Lente, H., Moors, E. (Eds.), Emerging Technologies for Diagnosing Alzheimer's Disease. Palgrave Macmillan, Basingstoke, UK, pp. 21–39.

Williams, M., 2007. Enabling technologies in drug discovery: the technical and cultural integration of the new with the old. Comp. Med. Chem. II 2, 265–287.

Williams, M., 2011. Productivity shortfalls in drug discovery: contributions from the preclinical sciences? J. Pharmacol. Exp. Ther. 336, 3–8.

Wilson, W.A. 2016. Scientific regress. First Things. Available from; https://www.firstthings.com/article/2016/05/scientific-regress.

Winquist, R.J., Mullane, K., Williams, M., 2014. The fall and rise of pharmacology—(Re-)defining the discipline? Biochem. Pharmacol. 87, 4–24.

Wray, S., Fox, N.C., 2016. Stem cell therapy for Alzheimer's disease: hope or hype? Lancet Neurol. 15, 133–125.

Wyden, R., Grassley, C., 2015. Wyden-Grassley Sovaldi investigation finds revenue-driven pricing strategy behind $84,000 hepatitis drug. United States Senate Committee on Finance. Available from; http://www.finance.senate.gov/ranking-members-news/wyden-grassley-sovaldi-investigation-finds-revenue-driven-pricing-strategy-behind-84-000-hepatitis-drug.

Yong, E., 2016. The plan to avert our post-antibiotic apocalypse. Atlantic Magazine. Available from; http://www.theatlantic.com/science/archive/2016/05/the-ten-part-plan-to-avert-our-post-antibiotic-apocalypse/483360/.

Ziemann, M., Eren, El-Osta, A., 2016. Gene name errors are widespread in the scientific literature. Genome Biol. 17, 177.

Index

C

S

Printed in the United States
By Bookmasters